ALLE·ZEIT·WACH
1842

Perspectives in Turbulence Studies

dedicated to the
75th birthday of Dr. J. C. Rotta

International Symposium
DFVLR Research Center, Göttingen, May 11-12, 1987

Editors: H. U. Meier and P. Bradshaw

Springer-Verlag
Berlin Heidelberg NewYork
London Paris Tokyo

Priv. Doz. Dr.-Ing. habil. H. U. Meier
DFVLR, Insitut für Experimentelle Strömungsmechanik
Bunsenstraße 10
3400 Göttingen
FRG

Prof. P. Bradshaw
Dept. of Aeronautics
Imperial College
Prince Consort Road
London SW7 2BY
United Kingdom

ISBN-13:978-3-642-82996-3 e-ISBN-13:978-3-642-82994-9
DOI: 10.1007/978-3-642-82994-9

Softcover reprint of the hardcover 1st edition 1987

2161/3020-543210

Meeting Organizers

H.U. Meier

P. Bradshaw

Advisory Committee

B. van den Berg	B.E. Launder
G. Drougge	O. Leuchter
T.K. Fanneløp	H. Oertel
H.H. Fernholz	V.C. Patel
D. Geropp	W. Rodi
E.H. Hirschel	A. Walz
H. Hornung	M. Wolfshtein

Symposium Sponsors

Corovin GmbH, D-3150 Peine
Dornier GmbH, D-7990 Friedrichshafen
Deutsch Niederländischer Windkanal (DNW), NL-8300 Emmeloord
Lambrecht GmbH, D-3400 Göttingen
Messerschmidt-Bölkow-Blohm (MBB), D-8000 München
Motoren Turbinen Union (MTU), D-8000 München 50
RoGAL mbH, D-4052 Korschenbroich
A. Thieß GmbH, D-3400 Göttingen

Dr. Ing. E. h. Julius C. Rotta

Preface

The present volume entitled "Perspectives in Turbulence Studies" is dedicated to

Dr. Ing. E.h. Julius C. Rotta

in honour of his 75th birthday.

J.C. Rotta, born on January 1, 1912, started his outstanding career in an unusual way, namely in a drawing office (1928 - 1931). At the same time he - as a purely self taught person - took a correspondence course in airplane construction. From 1934 to 1945 he worked in the aircraft industry on different subjects in the fields of flight mechanics, structures, aircraft design, and aerodynamics. In 1945 he moved to Göttingen and worked from that time at the Aerodynamische Versuchsanstalt (AVA, now DFVLR) and the Max-Planck-Institut für Strömungsforschung (1947-1958), interrupted only by a stay in the U.S. at the Glenn L. Martin Company (1954 - 1955) and a visiting professorship at the Laval University in Quebec, Canada (1956). Already during his activities in industry, Dr. Rotta discovered his special liking for aerodynamics. In Göttingen, he was attracted by Ludwig Prandtl's discussions about problems associated with turbulence and in particular his new contribution to fully developed turbulence, published in 1945. At that time, W. Heisenberg and C.F. v. Weizäcker published their results on the energy spectra of isotropic turbulence at large wave numbers. Since that time his main research interest in reasearch has been in turbulence problems. As early as 1951, he published a new theory of non-homogeneous turbulence in "Zeitschrift für Physik": this was a very important step forward in the construction of a turbulence model on which a full prediction procedure could be based. At that time, the computational procedures for exploiting his ideas were absent; moreover, the state of knowledge and interest in these topics among the scientific community was not favourable to the furtherance of Rotta's ideas. His paper was rediscovered at the Stanford Meeting in 1968, when new computational procedures and bigger computers were available. In 1962 his survey article in

"Progress in Aeronautical Sciences" on "Incompressible Turbulent Boundary Layers" was published and soon became a standard reference, not only for scientists but also for engineers who had to solve practical problems. This contribution also summarized his studies in this field of research and gave Dr. J.C. Rotta worldwide recognition as an expert in turbulent boundary layer research. A further volume entitled "Turbulente Strömungen", published in 1972 is now a standard text book. As a means of national recognition of his outstanding ideas, the honorary degree of a Doctor of Engineering was given to J.C. Rotta by the Technical University of Berlin in 1971. Until his retirement in 1976, Julius C. Rotta published more than 70 papers which all have one thing in common: the ideas are original, and the observations deep and stimulating. After his retirement, he joined his colleagues in the boundary layer section to continue his work. He has always been willing to share his profound knowledge, and to give courteous help to colleagues who sought his advice.

The effort of his friends to give him their recognition by participating in the Symposium and contributing survey papers to the Proceedings Volume is certainly not only based on the outstanding, inventive scientific contributions of J.C. Rotta, but also on his extreme modesty and his personality. All those who had the pleasure and privilege to work with him, and all his friends and colleagues, wish that he may be able to continue his work for many years and may influence our work with his inventive ideas.

Göttingen, May 11, 1987

Hans Ulrich Meier

Peter Bradshaw

Contents

Instability, Three-dimensional Effects and Transition in Shear Flows

J.T. STUART

Imperial College,
London.

Summary

The topic of instability of laminar flows and its connection with their transition to turbulence has been expressly with us for over 100 years. During the last few decades it has become increasingly clear that three-dimensionality and nonlinearity are of immense importance for transition, and some crucial experiments have been quite central in our understanding. Many theoretical ideas have emerged also, some of which have had a significant influence on the development of the subject. The present lecture considers these experimental and theoretical contributions, and attempts to assess the present state of our understanding, especially in relation to the problem of transition to turbulence.

It is with great pleasure that I dedicate this lecture to Professor J.C. Rotta, who has made so many contributions to the subject of turbulence and to our understanding of it.

1. Introduction

Although experimental research work on the subject of turbulence, and on its evolution from a laminar shear flow, may be said to have commenced in 1883 with the seminal paper by Reynolds [1], the present account will treat mainly the period from the early 1940s, when Schubauer and Skramstad [2] reported their remarkable discoveries. From a theoretical viewpoint, many important investigations were made during the decades between 1880 and 1950 (see, for example, the surveys of Stuart [3], [4]), by many engineers, scientists and mathematicians. The present account, however, will utilize as its point of origin the pioneering paper of Squire [5], published in 1933, which produced an important, albeit negative, result on the rôle of three-dimensional perturbations in two-dimensional boundary layers and shear flows. Our discussions will necessarily bring in the treatment of nonlinear processes. Indeed, studies of three-dimensional and nonlinear effects have dominated the subject during the last three or four decades. Thus Sections 2, 3 and 4 below will discuss respectively "Basic equations and linearized theory", "Observations and experiment" and "Nonlinear three-dimensional theories".

Although reference has been made above to perturbations on two-dimensional basic laminar flows, progress has been made also on the corresponding problem, when the basic laminar flow is three-dimensional in character as on the swept wing of a modern aircraft. This aspect of the subject will be discussed in Section 5.

A discussion of the overall state of the subject, together with conclusions, is given in Section 6.

2. Basic equations and linearized theory

We consider a three-dimensional, incompressible, viscous flow, one of whose major velocity components is in the x direction, with the y coordinate normal to a solid surface and therefore in the main direction of shear, while the z coordinate is normal to x and y in a right-handed sense. If $\underset{\sim}{u}$ is the velocity $\underset{\sim}{\omega}$ is the vorticity, p is the kinematic pressure, t is the time and R is a Reynolds number based on the maximum speed in the x direction and on an appropriate length, the equations of motion are

$$\frac{\partial \underset{\sim}{u}}{\partial t} - \underset{\sim}{u} \wedge \underset{\sim}{\omega} = - \nabla (p + \tfrac{1}{2} \underset{\sim}{u}^2) - R^{-1} \operatorname{curl} \underset{\sim}{\omega}, \qquad (2.1)$$

$$\underset{\sim}{\omega} = \operatorname{curl} \underset{\sim}{u}, \qquad (2.2)$$

$$\operatorname{div} \underset{\sim}{u} = 0. \qquad (2.3)$$

Equations (2.1), (2.2) and (2.3), together with appropriate boundary conditions define laminar flow, transitional processes and, indeed, turbulence.

Boundary layers and shear layers are not strictly parallel flows because of slow growth or spread due to viscosity, but it has been found valuable theoretically to consider them to be parallel, following the approximation first introduced by Prandtl (see [3], [4]). Thus we suppose a basic laminar flow to be given; in order to discuss its instability, we replace this flow by a "parallel" version, independent of x and z, and with the velocity profile appropriate to a given x-wise and z-wise location (see Fig. 1(a),(b)). Thus the basic flow is given by the velocity field

$$\bar{\underset{\sim}{u}} \equiv [\bar{u}(y),\ 0,\ \bar{w}(y)]. \qquad (2.4)$$

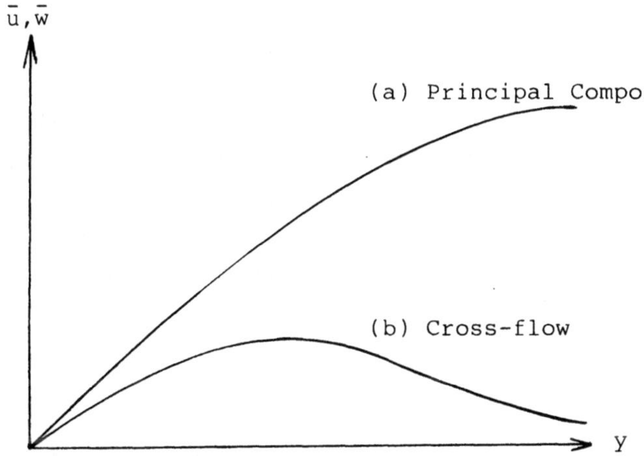

Fig. 1. (a) Velocity profile in principal direction;
(b) Cross-flow velocity profile

The velocity perturbation may be taken to have the form

$$\underset{\sim}{u} \equiv \varepsilon[u(y),\ v(y),\ w(y)]\ e^{i(\alpha x+\beta z-\alpha ct)}, \qquad (2.5)$$

and then the governing equations for u(y), v(y) and w(y), when linearized in the amplitude parameter ε, are

$$(\alpha\bar{u}-\alpha c+\beta\bar{w})(D^2-\gamma^2)v-vD(\alpha\bar{u}+\beta\bar{w})+iR^{-1}(D^2-\gamma^2)^2v = 0, \qquad (2.6)$$

$$(\alpha\bar{u}-\alpha c+\beta\bar{w})(\beta u-\alpha w)+iR^{-1}(D^2-\gamma^2)(\beta u-\alpha w)-ivD(\beta\bar{u}-\alpha\bar{w})=0, \quad (2.7)$$

$$\alpha u + \beta w - iDv = 0, \quad (2.8)$$

$$\gamma^2 \equiv \alpha^2 + \beta^2 \text{ and } D \equiv \frac{d}{dy}. \quad (2.9)$$

In the derivation of these equations, flow and surface curvatures have been ignored, as well as boundary-layer or shear-layer growth. Moreover, it should be noted that in (2.5) there would be no loss of generality in having the perturbation independent of z since, with the approximations stated, $\beta = 0$ in (2.5) could be achieved by appropriate rotation about the y axis [6, p. 551] so that x would be the coordinate of periodicity. For present purposes, however, it is convenient to retain $\beta \neq 0$ in (2.5) - (2.9).

Early work on this subject was concerned with the two-dimensional or plane case $\bar{w} \equiv 0$ [3,4], and the appropriate equations [6, p. 514] can be obtained by setting $\bar{w} \equiv 0$ in (2.6) - (2.9). Then (2.6) becomes essentially the classical Orr-Sommerfeld equation which, with appropriate boundary conditions, yields an eigenrelation of the form

$$F(\gamma^2, \alpha R, c) = 0. \quad (2.10)$$

Squire [5], however, made the seminal observation that (2.10) is merely the same eigenrelationship as for a corresponding plane case ($\beta \equiv 0$), namely

$$F(\gamma^2, \gamma\bar{R}, c) = 0, \quad (2.11)$$

provided we regard γ as an equivalent wave number for the plane case and set

$$\alpha R = \gamma\bar{R}; \quad (2.12)$$

thus $\bar{R}$ should be regarded as the notional Reynolds' number for the plane case.

Since, by (2.9), $\gamma > \alpha$, it follows that $\bar{R} < R$, so that for a three-dimensional perturbation to a flow $(\bar{u},0,0)$ of Reynolds number R, there is always a corresponding plane perturbation problem with a lower Reynolds number $\bar{R}$. Thus, in order to calculate a minimum Reynolds number for neutral stability ($c_i = 0$ in (2.5) with α and β real) it is necessary to consider plane perturbations only. This result is known as Squire's theorem, but applies only, it must be emphasized, to incompressible plane flows ($\bar{w} \equiv 0$).

In extension of Squire's work it is desirable to consider amplified modes ($c_i > 0$), as was recognised by Jungclaus [7], Watson [8], Michael [9], Magen and Patera [10] and Dhanak [11]. It is known that, if $c_i = 0$, there is a neutral curve for the plane case (2.10 with $\beta \equiv 0$, $\gamma \equiv \alpha$) and this is shown in the plane of α^2 against αR in Fig. 2(a) at the section $\beta = 0$. Then (2.10)

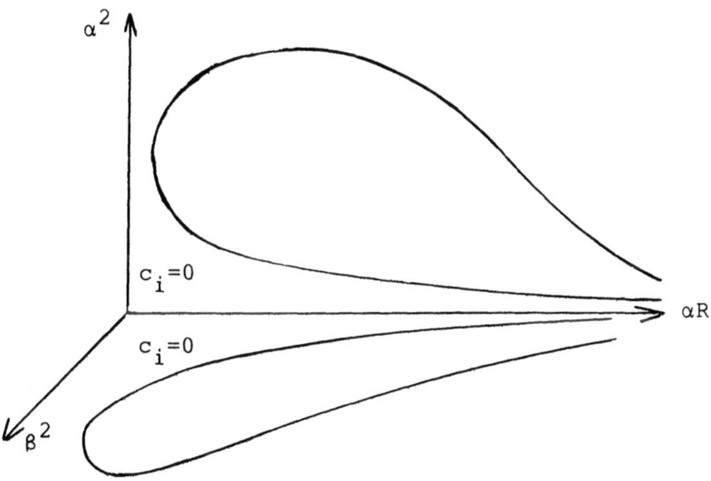

Fig. 2(a). Neutral curve $(\alpha^2, \alpha R, \beta=0)$ and neutral surface $(\alpha^2, \alpha R, \beta^2)$.

shows that the neutral surface for the general case $\beta \neq 0$ can be obtained by rotation about the αR axis. Within this curiously shaped surface of revolution, $c_i > 0$ (if α and β are real), so that there is exponential growth with time.

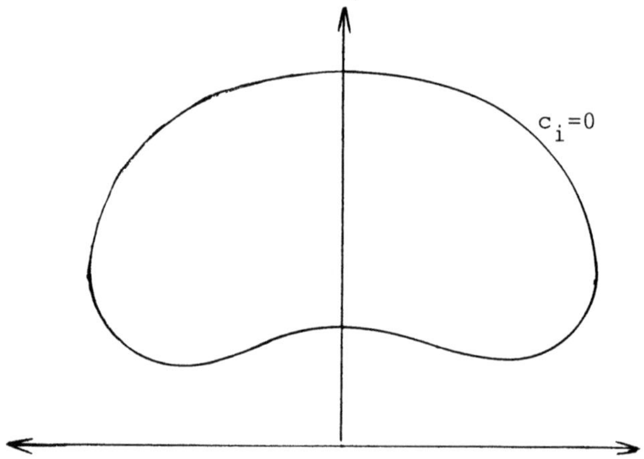

Fig. 2(b). Neutral curve (α,β; R fixed)

Of further interest is the section of the neutral surface for a given value of R, as indicated in Fig. 2(b). Provided $R > R_Q > R_c$, where R_c is the minimum critical Reynolds number (for $c_i = 0$), the neutral curve in the α-β plane has a "kidney" shape, so that the point of maximum amplification occurs at a location with $\beta \neq 0$. This contrasts with Squire's theorem, which is concerned essentially with the critical Reynolds number, and Fig. 2(b) indicates that three-dimensional disturbances do have great significance at sufficiently large Reynolds numbers.

In the above remarks the concept of instability is associated principally with α and β real but with c complex. An alternative scheme is to have β and αc (= ω) as real quantities, with α complex; then $\alpha_i < 0$ implies exponential instability in the x coordinate, and is rather more akin to the facts of experimental observation. The neutral surface of Fig. 2(a) and the neutral curve of Fig. 2(b) then correspond to $\alpha_i = 0$. For small growth rates, in the sense of both αc_i and α_i being small, Gaster [12] has shown that the relationship, between these two forms of growth rate, is linear in the group velocity (c_g), namely $\alpha c_i = - c_g \alpha_i$ for the situation when α_r (for the case of

α complex) equals α (real case).

At this point it is desirable also to remark on Gaster's [20] calculation of an evolving wave packet by superposition of linear modes calculated from (2.6) to (2.9). This achievement will be discussed in greater detail in the next section, in association with experimental observations.

The extension of linearized theory to take account of the nonlinear terms, which have been neglected in (2.6) to (2.9), will be treated later, but we emphasize here the crucial importance of such discussions. A particular case is that of axi-symmetric Hagen-Poiseuille flow which is widely believed to be stable against infinitesimal disturbances ($\alpha c_i < 0$); we shall return to this problem also.

3. Observations and experiment

Reynolds' [1] observations on the distortions of a filament of dye in Hagen-Poiseuille flow in a pipe of circular cross-section may be said to have stimulated scientific studies of turbulence and of its occurrence. Later Davies and White [13] studied experimentally the occurrence of turbulence in a pipe or channel of rectangular cross-section, finding a critical Reynolds number for the occurrence of turbulence. For the case of a rather wide channel, approximating therefore to plane Poiseuille flow, their observations were somewhat striking, showing a critical Reynolds number for turbulence of about 1000, much lower than any calculation by linearized theory would give in later years.

Confirmation, however, of the instability theories discussed briefly in the previous section came with the work of Schubauer and Skramstad in the 1940s [2] on boundary-layer instability. In these experiments a "vibrating ribbon" was mounted on the wall of a flat plate so as to stimulate the development of a perturbation to the plane boundary-layer velocity field by means of vortex shedding from the ribbon. The use of hot-wire anemometry then enabled observations to be made of the evolution of the wave in the main-flow direction. Thus the relevance of the diagram in Figure 2(a) for $\beta = 0$ was substantially verified. In these experiments one object was to keep the perturbation field and the basic flow field as reasonably two-dimensional as possible, in order to make comparisons with the corresponding theory of Tollmien and Schlichting [see 3,4]. Later, however, Klebanoff and Tidstrom [14] found a natural tendency for the wave motions, often called Tollmien-Schlichting waves, to develop a three-dimensional structure, perhaps by means of some sort of stimulus from the main stream outside the boundary layer.

The above-mentioned experiments led to the concept of natural transition being essentially a three-dimensional phenomenon, and suggested to Klebanoff, Tidstrom and Sargent that a "controlled" three-dimensional experiment should be performed [15]. This they did by rendering three-dimensional the wave produced by a vibrating ribbon, by means of strips of "Scotch" tape or "Sellotape" stuck to the surface below the ribbon. As a result it was found that the flow field downstream of the ribbon developed a system of three-dimensional, longitudinal vortices spaced periodically across the span. Such a structure, though less regular in its character, had been found earlier in [14] under "natural" conditions. In Fig. 3 the view shown is that in the

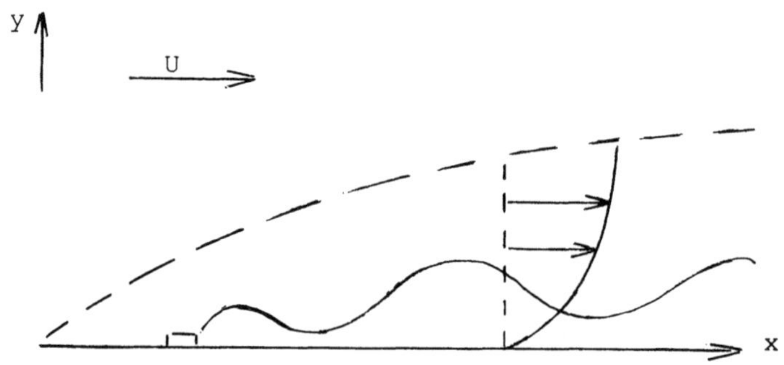

Fig. 3. Oscillations in a boundary layer

plane of the main-flow and vertical directions, x and y, with an oscillation emanating from the ribbon. The longitudinal-vortex structure, which is time-dependent and evolves in the x direction, is illustrated in Fig. 4, showing the closed pattern of streamlines periodic in the spanwise (z) direction. Similar phenomena were found by Kovasznay, Komoda and Vasudeva [16], together with a rather notable "vertical" shear layer as well as those of Klebanoff form.

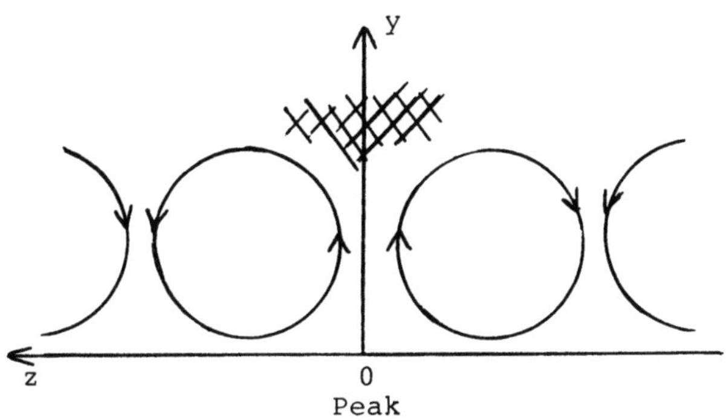

Fig. 4. Streamlines of secondary flow

Of especial significance are the spanwise locations where the secondary flow, which may have a velocity magnitude as little as 2% of the free-stream velocity, is directed outwards normal to the wall. These are the so-called "peaks", whereas the spanwise locations of secondary flow towards the wall are known as "valleys". At the peaks the vorticity is intensified in the outer part of the boundary layer and local shear layers are formed at the "peak" locations shown hatched in Fig. 4. From the local shear layers, patches of high-frequency velocity fluctuations emerge, which are convected downstream. These are embryonic turbulent spots; they occur at many locations in space and continually in time, and agglomerate as they are convected downstream.

Such turbulent spot development and evolution appears to be one crucial mechanism of breakdown of laminar flow to turbulence. Much earlier the concept of turbulent spots had been introduced and discussed by Emmons [17] both observationally and theoretically, including the suggestion that an intermittency factor was necessary to denote the fraction of time during which a given location of the transition region really is turbulent. The idea of intermittency was pursued by several groups of workers, including Schubauer and Klebanoff [18] who additionally elucidated many of the properties of turbulent spots and their development. Some results of their observational work are shown in Fig. 5.

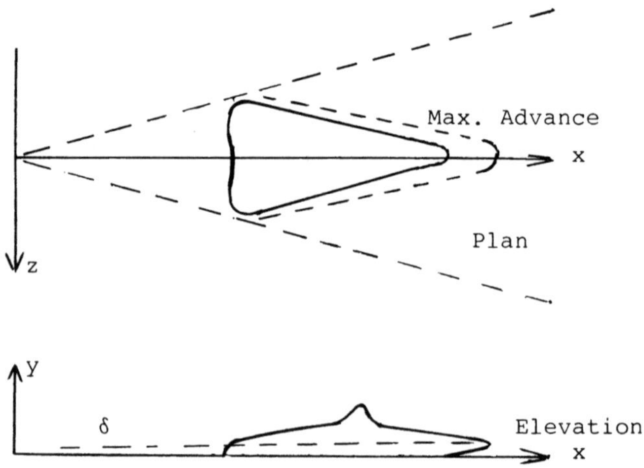

Fig. 5. Details of a turbulent spot

With a view to comparison with theory, Gaster and Grant [19] were led to stimulate a wave packet or "spot" of small amplitude in a flat-plate boundary layer and to follow its evolution down the plate, as shown in Fig. 6. As we shall see later, they were able to obtain remarkably good agreement with a calculation [20] based on a superposition of linearized modes, though nonlinear effects caused some deviation.

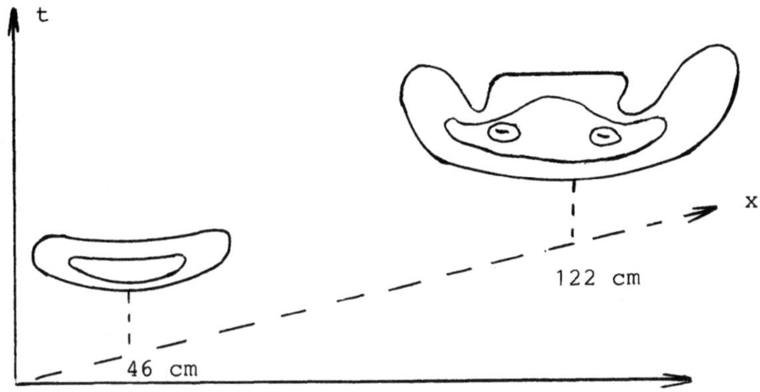

Fig. 6. Wave packet contours: experiment

In plane Poiseuille flow, Nishioka, Iida and Kanbayashi [21,22] also have drawn attention to the significant influence which three-dimensionality may have on the finite-amplitude wave development and transition for that type of flow. As suggested earlier by experimental observations of Davies and White [13], and confirmed by seminal experiments of Nishioka and his colleagues [see 3,4], a subcritical or finite-amplitude instability phenomenon occurs in plane Poiseuille flow. Three-dimensionality modifies the details of this.

Finally, a most significant development, which followed earlier theoretical work started by Raetz [see 4], has been the identification by Saric and Thomas [23] and by Kachanov and Levchenko [24] of the presence of resonances between wave components in boundary-layer instability, leading to what is sometimes called the staggered vortex pattern. These resonances can be associated with subharmonics of the fundamental wave, and the Klebanoff and subharmonic resonance mechanisms prior to transition may be associated with different fluctuating amplitude levels.

4. Nonlinear three-dimensional theories

The principal general concept, for developing the theory of instability beyond the linearized form of Section 2, is the weakly nonlinear theory. Significant advances in our understanding have arisen from both use of such theories, for both plane and fully three-dimensional perturbations, and the reader is referred to [3] and [4] for a partial discussion. This will perhaps form the main emphasis of the present section. Prior to that discussion, however, it is worthwhile to comment on two types of calculation, which are in part linear and which illustrate some of the experimental phenomena described in Section 3.

One central problem is the mechanism by which the local shear layers are formed in the outer part of the boundary layer between the longitudinal vortices, as shown in Figure 4. It was suggested by Kovasznay [see 16] that the appropriate process required the vortex lines, which initially are dominantly aligned in the spanwise (z) direction, to be convected outwards by the secondary flow at the peaks and then stretched by the divergent velocity field in the outer part of the boundary layer. The present writer [25] then followed up this suggestion by developing a linear model as follows.

We suppose that, as observed in the experiments of [15] and [16], the shear layers develop so quickly that viscosity has insufficient time to be effective. Thus the flow field is supposed to be inviscid and independent of x, the latter approximation leading to the (v,w) velocity field of the (y,z) coordinates uncoupling from the u component associated with the x coordinate. Thus, if v(y,z,t) and w(y,z,t) are supposed to be given, the u component is given by

$$\frac{\partial u}{\partial t} + v\,\frac{\partial u}{\partial y} + w\,\frac{\partial u}{\partial z} = 0, \qquad (4.1)$$

with the initial condition

$$t = 0,\; u = U(y), \qquad (4.2)$$

where it has been assumed that there is no pressure gradient in the x direction.

The initial state, $u = U(y)$, may be assumed to be Blasius boundary layer flow while, in accordance with experiment, v and w are periodic in z, with profiles in the y coordinate appropriate to the closed streamlines of Fig. 4. The exact calculations of [25] from (4.1) and (4.2) showed that an inflexional velocity profile would evolve for $t > 0$ at a peak, so as to produce a shear layer in the outer part of the boundary layer (Fig. 7).

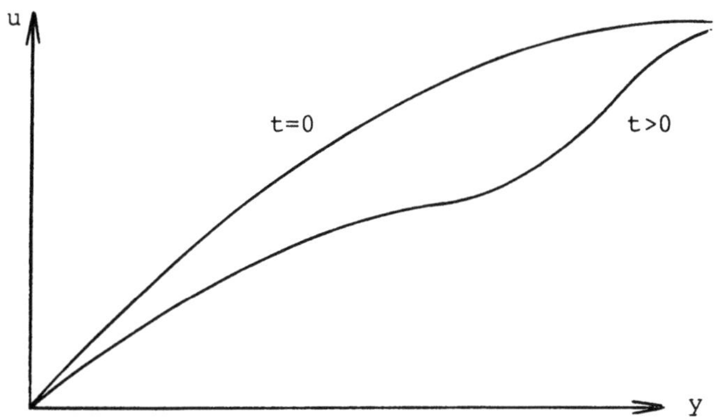

Fig. 7. Development of strong local shear layer

This feature is qualitatively similar to the observations sketched in Fig. 4. Moreover, if the secondary flow is assumed to have a magnitude of 2% of the free-stream velocity, in accordance with experimental observations, the local shear layer of Fig. 7 develops in the very short time scale of those experiments. Thus there seems little doubt of the truth and validity of Kovasznay's suggestion. More recently Wray and Hussaini have solved the initial-value problem for the nonlinear equations with viscosity included, but subject to some assumptions about the form of the initial velocity profile [26]. Their work shows rather substantial agreement with Kovasznay's experiments, with especial reference to the local vertical shear layer.

Another aspect of the inviscid problem discussed above concerns the fact that the local shear layer develops into a singularity, although this happens only after an infinite time. There is considerable interest in the study of the Euler inviscid equations, with especial reference to the possible development of a singularity at a finite time. The present writer [see 27,28] has solved a problem which is a nonlinear version of (4.1), (4.2), with dependence on x and a pressure gradient included. If $z = 0$ gives one peak location of the secondary field in Fig. 4, and if the initial spanwise field is $zw_0(x,y)$, then it can be shown that the velocity field $zw(x,y,t)$ for $t \geqslant 0$ is given by

$$w(x,y,t) = \frac{w_0(\varphi,\psi)}{1 + btw_0(\varphi,\psi)}, \qquad (4.3)$$

where b is a rate of strain and φ,ψ are characteristics of the convective derivative.

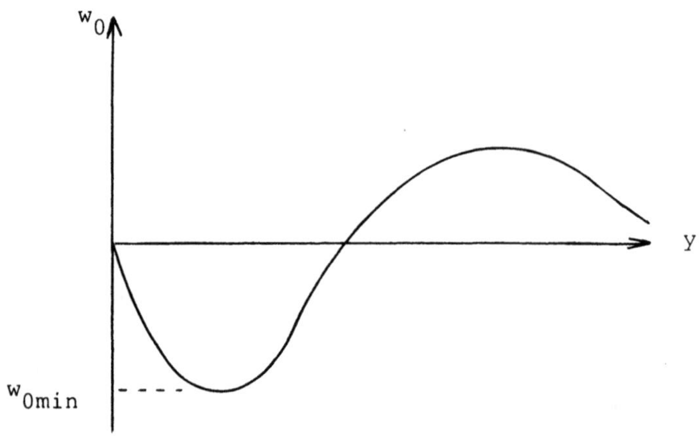

Fig. 8. Initial spanwise velocity profile

Typical data appropriate to the work of Klebanoff [15] and Kovasznay [16] has a shape for $w_0(x,y)$ as shown in Fig. 8; thus, provided the characteristic functions, φ and ψ, permit, w becomes singular when $t = -b^{-1}(w_{0min})^{-1}$, and a particular solution has been given [28] to show this explicitly. The relevance of this concept for boundary-layer transition and, perhaps, for sublayer eruptions is not known.

A second central problem is that of the development of a wave packet or spot of velocity fluctuations as it passes downstream. In Section 3 and Fig. 6 reference has been made to the experiments of Gaster and Grant [19], and also to Gaster's accompanying theory [20]. He synthesized a wave packet from linearized three-dimensional modes of the kind discussed in Section 2; contours of fluctuation amplitude are shown in Fig. 9 for two values of x, by sections in the z-t plane, where z is the spanwise coordinate. To a certain extent, boundary-layer growth

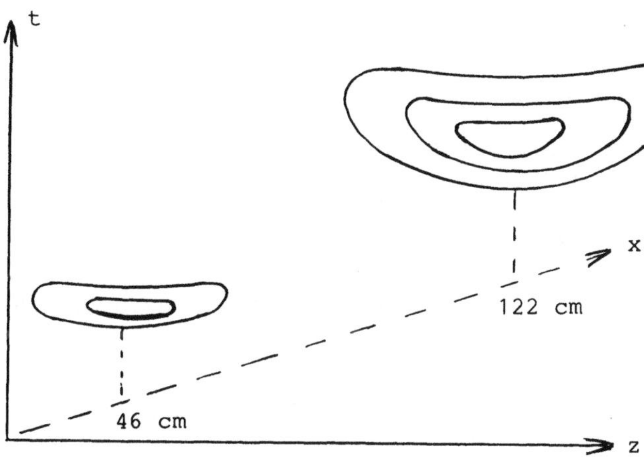

Fig. 9. Wave packet contours: theory

was incorporated in the calculation. It is instructive to compare the experimental and theoretical contours of Figs. 6 and 9, and to note the close quantitative agreement, except for the development of a two-peak phenomenon at 122 cm, a feature not present at earlier values of x. Almost certainly this is a feature of nonlinear disturbances of finite amplitude.

We turn now to the matter of weakly nonlinear theories. Our first concern should be with basic principles, and it is best therefore to turn to Fig. 2(a), or rather its equivalent in the α-R plane especially near to the critical Reynolds number, R_c. We consider a perturbation velocity field near to α_c and R_c, recognising that the work of Squire [5] and later workers has shown that the dominant mode is the plane case $\alpha = \alpha_c$, $\beta = 0$. Following earlier papers on weakly nonlinear theories [see 3,4], Davey, Hocking and Stewartson [29] showed that if the fundamental oscillation is modulated to have the form

$$A(\xi,\eta,\tau) \exp [i\alpha_c(x - c_{cr}t)], \tag{4.4}$$

while a mean pressure gradient perturbation, $\partial P_{01}/\partial\xi$, is proportional to

$$Q(\xi,\eta,\tau) - |A|^2, \tag{4.5}$$

then

$$\frac{\partial A}{\partial \tau} - a\frac{\partial^2 A}{\partial \xi^2} - b\frac{\partial^2 A}{\partial \eta^2} = A - \beta A|A|^2 - \gamma AQ, \tag{4.6}$$

$$\frac{\partial^2 Q}{\partial \xi^2} + \frac{\partial^2 Q}{\partial \eta^2} = \frac{\partial^2}{\partial \eta^2}|A|^2, \tag{4.7}$$

where a, b, β and γ are complex numbers. Here ξ, η and τ are defined by

$$\xi = \varepsilon^{1/2}(x - c_g t), \quad \eta = \varepsilon^{1/2}y, \quad \tau = \varepsilon t, \tag{4.8}$$

where ε is the growth rate αc_i, assumed to be positive, and c_g is the group velocity, $c + \alpha dc/d\alpha$. We comment here on the appearance of an effect on the amplitude A of a "lubrication" type of effect through the function Q [30].

It is known that the system of equations (4.6) and (4.7), subject to suitable initial conditions, does permit the development of a singularity when the coefficients have appropriate signs: in the plane case, $Q \equiv 0$, $\partial/\partial\eta \equiv 0$, for example, possible conditions are $a_r > 0$, $\beta_r < 0$. For plane Poiseuille flow, for which the above equations were developed, this is the situation and singular solutions [see 3,4] have been calculated which may have relevance.

On the other hand, if physical conditions are such that a plane-wave equilibrium solution of (4.6) and (4.7) is possible, which may require $\beta_r > 0$ in some circumstances, then it is possible to discuss the possible instability and evolution of such solutions by means of centre manifold theory (see, for example, Holmes [31,32]). Chaos, which may have connections with transition to turbulence, is a distinct possibility.

Returning to the problem posed by Fig. 4, we note that early attempts were made by Benney and Lin [33] and by Benney [34] to explain the occurrence and rôle of longitudinal vortices in the Klebanoff and Kovasznay experiments, although this work ran into criticism [35,36] on the grounds that inconsistent assumptions had been made about the eigenvalues of the interacting two- and three-dimensional modes considered. In my view, these criticisms have never been refuted, and the analysis of [35] and [36], which is in the spirit of the weakly nonlinear theories described by (4.4) - (4.7), gives the basis perhaps for a more acceptable explanation.(I hasten to add, however, that the theory based on (4.1) and (4.2) does nothing to confirm or refute the theory of [33,34], since it is concerned with evolutions in the presence of a longitudinal vortex system, which is, for some reason, extant.) Further work on the problem is to be found described in the work of Smith and Stewart [37] and in references described there. They have made substantial progress in relating experiment to theoretical work, which is based essentially on weakly nonlinear theory; thus there is now greater rational understanding of the phenomena described in [15] and [16].

The other major concept, however, is that of flow resonance, which was raised by Raetz [see 4] as early as 1959. If we consider a flow perturbation of the form

$$\sum_{n=1}^{3} u_n(y) \exp [i(\alpha_n x + \beta_n z - \alpha_n c_n t)], \tag{4.9}$$

where

$$\alpha_1 + \alpha_2 = \alpha_3, \quad \beta_1 + \beta_2 = \beta_3, \quad \alpha_1 c_1 + \alpha_2 c_2 = \alpha_3 c_3 \tag{4.10}$$

then there is resonance between the three constituents of (4.9), any pair of which can stimulate the third when complex conjugates are included. This idea was pursued by Craik in the 1970s [38,39] and has now come still more to the fore, especially because of the experimental verification given in [23] and [24]. More recently Smith and Stewart [37] have studied resonant problems, with especial reference to Craik's case of

$$\beta_1 = 0, \ \alpha_2 = \alpha_3 = \frac{1}{2}\alpha_1, \ \beta_3 = -\beta_2,$$

$$\alpha_2 c_2 = \alpha_3 c_3 = \frac{1}{2}\alpha_1 c_1. \tag{4.11}$$

and it is clear that substantial progress is being made for that mechanism of transition which involves resonances.

There is, moreover, another theoretical result which needs to be mentioned here. In plane Poiseuille flow, two-dimensional "threshold" modes are known to exist [see 3,4], and Orszag and Patera [40] have shown numerically that there is a strong three-dimensional instability of these "threshold" modes. In a seminal paper, Herbert [41] has demonstrated that a three-dimensional resonance of the type (4.9) - (4.11) is responsible for the Orszag instability. Thus the concept of resonant interactions is relevant for transition in plane Poiseuille flow, in providing a mechanism for departure of the flow from the subcritical threshold equilibrium states, which are possible at Reynolds numbers (of order 2500) much lower than the critical value given by linearized theory.

5. Three-dimensional basic flows

Equations (2.6) - (2.9) are still relevant for this situation, except that we may take $\beta \equiv 0$ by appropriate rotation of the coordinate frame about an axis normal to the solid surface. It needs to be emphasized, however, that it has been demonstrated by Faller [see 42 and papers cited there] and by Lilly [43] for the problem of instability of the Ekman boundary-layer flow that other known effects, such as Coriolis forces, which are ignored in (2.6) - (2.9), can be important in some ranges of wave number and Reynolds number. Thus a coupling is provided between (2.6) and (2.7) to yield a united sixth-order system. It is clear that the same point is true for other three-dimensional situations and this has been recognised by Malik, Wilkinson and Orszag [44] and by Kobayashi, Kohama and Takamadate [45] for the case of flow induced by a rotating disk [46].

In spite of the above reservations, however, it is relevant to note that, as H.B. Squire perhaps first recognised [see 6, p. 551], equations (2.6) - (2.9) indicate that the instability process governed by (2.6) depends only on the velocity profile ($\bar{u}(y)$ when $\beta \equiv 0$) in the direction of periodicity (e.g. Fig. 1(a) or 1(b)). This result leads to considerable simplifications, as a glance at Fig. IX.21 of reference [6, p. 552] indicates. In particular, the profile with an inflection point at the point of zero velocity plays an important rôle at large Reynolds number, in relation to disturbances stationary relative to the surface.

It is known from observations [6,46] that such modes of disturbance, which are stationary relative to the surface, are possible in a three-dimensional boundary layer. But observations make it clear also that perturbation modes are possible, which are travelling relative to the surface. To some extent different modes may be derived from different velocity profiles, each of which may be a linear combination of the curves shown in Figs. 1(a),(b), but subject to a coupling by curvature and Coriolis effects as mentioned earlier. Further experiments have been done by Wilkinson and Malik [47], and comparisons have been made between the observations of these authors and the theoretical work of [44], of Malik [48] and of Mack [49]. An admirable account of the linearized aspects of this subject is given by Mack [50], where it is argued that the well known streaks in the flow on a rotating disk (Fig. 10) are the representation of constant phase lines spreading from surface perturbations, in accordance with [47] and [49].

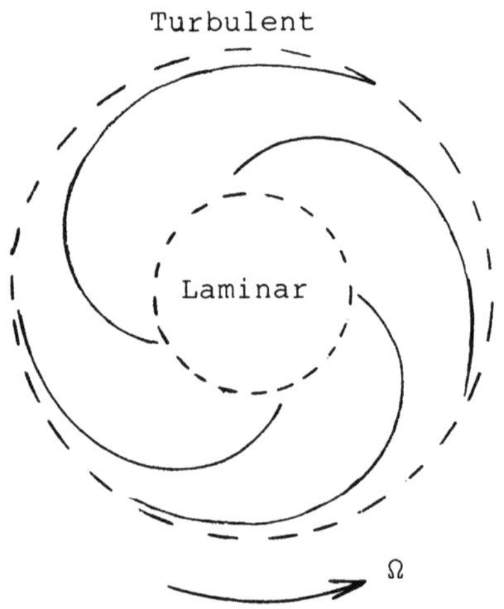

Fig. 10. Instability and transition on a rotating disk

Analytical theory of the instability processes for stationary modes on a rotating disk has been produced by Hall [51] and Mackerrell [52], for linear and weakly nonlinear cases, but clearly more work is needed on the nonlinear aspects of the problem before we can claim to have a real understanding of transition in general three-dimensional boundary-layer flows.

6. Discussion and conclusions

One problem which has been discussed only fleetingly in this account is that of the instability and transition in Hagen-Poiseuille flow or, indeed, plane Couette flow. It is widely believed, though not proved rigorously in my view, that these two flows are stable against two-dimensional disturbances. For one example of the substantial experimental evidence, the reader is referred to Wygnanski and Champagne [53]. For Hagen-

Poiseuille flow, however, Smith and Bodonyi [54] have developed a mechanism with a neutrally-stable mode of non-axisymmetric form of unit wave number in the azimuth, based upon the Benney-Bergeron nonlinear critical layer [55]. But a true instability neighbouring this mode has not yet been calculated or shown to be possible.

It is clear that linear and nonlinear theories have played a dominant rôle in theoretical work on boundary-layer transition up to the present, and more strongly nonlinear developments are needed, in spite of the successes so far made. Moreover, we know little or nothing of the relation of the Navier-Stokes equations and their solutions to the type of chaos known to occur in maps and in classes of ordinary differential equations, for example, the Lorenz system. On the other hand, we do have, through weakly nonlinear theory, a connection between nonlinear partial differential equations and sets of ordinary differential equations; and we do have a connection of the latter with maps through the Poincaré section. But what then is the connection with the Klebanoff-Kovasznay mechanism of transition on the one hand, and the resonance mechanism on the other? What is the mechanism of the development and agglomeration of wave packets, or spots of velocity fluctuations, or turbulent spots? The answers to such questions still lie in the future, but not, I hope, at too distant a future.

References

1. Reynolds, O.: An experimental investigation of the circumstances which determine whether the motion of water shall be direct or sinuous, and of the law of resistance in parallel channels. Phil. Trans. Roy. Soc. London A 174 (1883), 935-982.

2. Schubauer, G.B.; Skramstad, H.K.: Laminar Boundary Layer Oscillations and Transition on a Flat Plate. N.A.C.A. Rept. No. 909, Washington 1947.

3. Stuart, J.T.: Stewartson Memorial Lecture, Hydrodynamic Stability and Turbulent Transition. In: T. Cebeci (Ed.): Proc. 3rd Symp. on Numerical and Physical Aspects of Aerodynamic Flows, Long Beach, California, 1985, 1-8. To be published by Springer-Verlag, New York (1986).

4. Stuart, J.T.: Instability of flows and their transition to turbulence. 29th Ludwig Prandtl Memorial Lecture, 1 April 1986, Dortmund. Z. Flugwiss. to appear.

5. Squire, H.B.: On the stability for three-dimensional disturbances of viscous fluid flow between parallel walls. Proc. Roy. Soc. London A 142 (1933), 621-628.

6. Stuart, J.T.: Hydrodynamic stability. In: Laminar Boundary Layers, ed. L. Rosenhead, Clarendon Press, Oxford 1963.

7. Jungclaus, G.: On the stability of laminar flow with three-dimensional disturbances. Tech. Note Inst. Fluid Dyn. Appl. Math. Univ. Maryland, No. BN110, 1957.

8. Watson, J.: Three-dimensional disturbances of flow between parallel planes, Proc. Roy. Soc. London A 254 (1960), 562-569.

9. Michael, D.H.: Note on the stability of plane parallel flows. J. Fluid Mech. 10 (1961), 525-528.

10. Magen, M.; Patera, A.T.: Three-dimensional linear instability of parallel shear flows. Phys. Fluids 29 (1986), 364-367.

11. Dhanak, M.R.: On certain aspects of three-dimensional instability of parallel flows. Proc. Roy. Soc. London A 385 (1983), 53-84.

12. Gaster, M.: A note on the relation between temporally-increasing and spatially-increasing disturbances in hydrodynamic stability. J. Fluid Mech. 14 (1962), 222-224.

13. Davies, S.J.; White, C.M.: An experimental study of the flow of water in pipes of rectangular section. Proc. Roy. Soc. A 119 (1928), 92-107.

14. Klebanoff, P.S.; Tidstrom, K.D.: The evolution of amplified waves leading to transition in a boundary layer with zero pressure gradient. N.A.S.A. Tech. Note D-195, Washington, 1959.

15. Klebanoff, P.S.; Tidstrom, K.D.; Sargent, L.M.: The three-dimensional nature of boundary-layer instability. J. Fluid Mech. 12 (1962), 1-34.

16. Kovasznay, L.S.G.; Komoda, H.; Vasudeva, B.R.: Detailed flow field in transition. Proc. 1962 Heat Trans. Fluid Mech. Inst., 1-26, Stanford 1962.

17. Emmons, H.W.: The laminar-turbulent transition in a boundary layer, Part I, J. Aero. Sci. 18 (1951), 490-498.

18. Schubauer, G.B.; Klebanoff, P.S.: Contributions on the mechanics of boundary-layer transition. N.A.C.A. Rep. 1289, Washington, 1955.

19. Gaster, M.; Grant, I.P.: An experimental investigation of the formation and development of a wave packet in a laminar boundary layer. Proc. Roy. Soc. London A 347 (1975), 253-269.

20. Gaster, M.: A theoretical model of a wave packet in a boundary layer. Proc. Roy. Soc. A 347 (1975), 271-289.

21. Nishioka, M.; Iida, S.; Kanbayashi, S.: Proc. 10th Symp. Turb., Inst. Space Aero. Sci., Tokyo, 55 (1978).

22. Nishioka, M.; Asai, M.: Some observations of subcritical transition in plane Poiseuille flow. J. Fluid Mech. 150 (1985), 441-450.

23. Saric, W.S., Thomas, A.S.W.: Experiments on the subharmonic route to turbulence in boundary layers. In: T.Tatsumi (Ed.): Turbulence and Chaotic Phenomena in Fluids (I.U.T.A.M. Conference, Kyoto, 1983), 117-122, Elsevier 1984.

24. Kachanov, Yu.S.; Levchenko, V.Ya.: The resonant interaction of disturbances at laminar-turbulent transition in a boundary layer. J. Fluid Mech. 138 (1984), 209-247.

25. Stuart, J.T.: The production of intense shear layers by vortex stretching and convection: AGARD Report 514, 1965.

26. Wray, A.; Hussaini, M.Y.H.: Numerical experiments in boundary-layer stability. Proc. Roy. Soc. London A 392 (1984), 373-389.

27. Stuart, J.T.: Instability of laminar flows, nonlinear growth of fluctuations and transition to turbulence. In: T. Tatsumi (Ed.): Turbulence and Chaotic Phenomena in Fluids (I.U.T.A.M. Conference, Kyoto 1983), 17-26, Elsevier 1984.

28. Stuart, J.T.: Three-dimensional inviscid developments in the transition process. Whitehead Lecture, London Mathematical Society (1984), to be published.

29. Davey, A.; Hocking, L.M.; Stewartson, K.: On the nonlinear evolution of three-dimensional disturbances in plane Poiseuille flow. J. Fluid Mech. 63 (1974), 529-536.

30. Stuart, J.T.: Keith Stewartson: His life and work. Ann. Rev. Fluid Mech. 18 (1986), 1-14.

31. Holmes, C.A.: Bounded solutions of the nonlinear parabolic amplitude equation for plane Poiseuille flow. Ph.D. Thesis, University of London 1984.

32. Holmes, C.A.: Bounded solutions of the nonlinear parabolic amplitude equation for plane Poiseuille flow. Proc. Roy. Soc. London A 402 (1985), 299-322.

33. Benney, D.J.; Lin C.C.: On the secondary motion induced by oscillations in a shear flow. Phys. Fluids 3 (1960), 656-657.

34. Benney, D.J.: A nonlinear theory for oscillations in a parallel flow. J. Fluid Mech. 10 (1961), 209-236.

35. Stuart, J.T.: On the nonlinear mechanics of wave disturbances in stable and unstable parallel flows. Part I. The basic behaviour in plane Poiseuille flow. J. Fluid Mech. 9 (1960), 353-370.

36. Stuart, J.T.: On three-dimensional nonlinear effects in the stability of parallel flows. Adv. Aero. Sci. 3 (1962), 121-142.

37. Smith, F.T.; Stewart, P.A.: The resonant-triad nonlinear interaction in boundary-layer transition. J. Fluid Mech. (in the press).

38. Craik, A.D.D.: Nonlinear resonant instability in boundary layers. J. Fluid Mech. 50 (1971), 393-413.

39. Craik, A.D.D.: Nonlinear evolution and breakdown in unstable boundary layers. J. Fluid Mech. 99 (1980), 247-262.

40. Orszag, S.A.; Patera, A.T.: Secondary instability of wall-bounded shear flows. J. Fluid Mech. 128 (1983), 167-186.

41. Herbert, T.: Modes of secondary instability in plane Poiseuille flow. In: T. Tatsumi (Ed.): Turbulence and Chaotic Phenomena in Fluids (Kyoto 1983), 53-58, Elsevier 1984.

42. Faller, A.J.; Kaylor, R.E.: A numerical study of the instability of the Ekman boundary layer. J. Atmos. Sci. 23 (1966),466-480.

43. Lilly, D.K.: On the instability of the Ekman boundary flow. J. Atmos. Sci. 23 (1966),481-494.

44. Malik, M.R.; Wilkinson, S.P.; Orszag, S.A.: Instability and transition in rotating disk flow. A.I.A.A. J. 19 (1981), 1131-1138.

45. Kobayashi, R.; Kohama, Y.; Takamadate, C.: Spiral vortices in boundary layer transition regime on a rotating disk. Acta Mech. 35 (1980), 71-82.

46. Gregory, N.; Stuart, J.T.; Walker, W.S.: On the stability of three-dimensional boundary layers with application to the flow due to a rotating disk. Phil. Trans. A 248 (1955), 155-199.

47. Wilkinson, S.P.; Malik, M.R.: Stability experiments in rotating disk flow. A.I.A.A.J. 23 (1985), 588-595 (issued earlier as A.I.A.A. Paper 83-1760 (1983)).

48. Malik, M.R.: The neutral curve for stationary disturbances in rotating-disk flow. J. Fluid Mech. 164 (1986), 275-287.

49. Mack, L.M.: The wave pattern produced by point source on a rotating disk. A.I.A.A. Paper 85-0490 (1985).

50. Mack, L.M.: Boundary-layer linear stability theory. In: Special Course on Stability and Transition of Laminar Flow, AGARD Rep. 709 (1984).

51. Hall, P.: An asymptotic investigation of the stationary modes of instability of the boundary-layer on a rotating disc. Proc. Roy. Soc. A 406 (1986), 93-106.

52. Mackerrell, S.O.: A nonlinear asymptotic investigation of the stationary modes of instability of the three-dimensional boundary layer on a rotating disc. Proc. Roy. Soc. A, submitted (1987).

53. Wygnanski, I.J.; Champagne, F.H.: On transition in a pipe. Part I. The origin of puffs and slugs and the flow in a turbulent slug. J. Fluid Mech. 59 (1973), 281-335.

54. Smith, F.T.; Bodonyi, R.J. Amplitude-dependent neutral modes in the Hagen-Poiseuille flow through a circular pipe. Proc. Roy. Soc. A 384 (1982), 463-489.

55. Benney, D.J.; Bergeron, R.F.: A new class of nonlinear waves in parallel flows. Stud. Appl. Math. 48 (1969), 181-204.

Acknowledgement

This work was done partly at the Division of Engineering, Brown University, U.S.A., with support of the D.A.R.P.A. Program and of a N.A.T.O. Research Grant for travel, and with the strong interest of Professor J.T.C. Liu.

The Influence of Wind Tunnel Turbulence on the Boundary Layer Transition

H.U. Meier*, U. Michel**, H.-P. Kreplin*
* DFVLR, SM-ES, Bunsenstr. 10, 3400 Göttingen, FR Germany
**DFVLR, FS-ES, Müller-Breslau-Str. 8, 1000 Berlin 12, FR Germany

Abstract

The flow quality of three large, industrially used wind tunnels is investigated. The non-realizability of a complete evaluation concerning the influences of the wind tunnel turbulence on the laminar to turbulent boundary layer transition and/or the development of the turbulent boundary layer on models is discussed. As a new approach to the problem transition measurements on the DFVLR 1:6 prolate spheroid were carried out for axisymmetric flow conditions in three different wind tunnels at similar free-stream Reynolds numbers. Supplementary measurements of turbulence quantities by means of hot-wire anemometers and microphones lead to detailed information about the velocity and pressure fluctuation levels and enabled us to interpret the deviations in the transition locations measured in different facilities on the prolate spheroid.

1. Introduction

The turbulence of the air stream in wind tunnels is generally recognized as a variable of considerable importance as far as its influence on the measured results is concerned. It is a well-known phenomenon that wind-tunnel turbulence may affect the boundary-layer stability and consequently the location of the laminar to turbulent boundary-layer transition. Experimental evidence for this effect was established 1943 by Schubauer and Skramstad [1] who determined the transition Reynolds numbers for flat plate boundary layers at different free-stream turbulence intensities. According to their results, applied to the flow around airfoils, the transition region should move towards the leading edge with increasing free-stream turbulence at constant velocities. This effect is sometimes applied in wind-tunnel testing intentionally to move the transition. Such an imposed transition location may be in agreement with results observed in free flight tests at much higher Reynolds numbers. However, due to the higher free stream turbulence, the thicknesses of the turbulent boundary layers may increase considerably - depending on the local pressure gradient. This means that an increase of the wind-tunnel turbulence simulates a higher free-stream Reynolds number as far as the transition location is concerned. At the same time the increase in turbulence leads to thicker boundary layers which must be related to lower free stream Reynolds numbers as regards the outer flow. Several authors (Charney [2], Rotta [3], Meier & Kreplin [4], Hancock & Bradshaw [5] and Blair [6]) found within this context that the thickening of the boundary layer with increasing free stream turbulence is combined with a change of the boundary layer characteristics such as an increase of the local wall shear stress. This result would lead to erroneous drag measurements.

About 50 years ago, Taylor [7] and Dryden et al. [8] found that the effects of turbulence on the wind tunnel measurements cannot be satisfactorily correlated with the single property intensity. Thus, Dryden applied scales of correlations between the velocity fluctuations at neighbouring positions, as proposed by G.I. Taylor, and investigated experimentally the effects of this so-called "length-scale" of turbulence on the "critical" Reynolds number of spheres. In his comprehensive study he found that this Reynolds number varies not only with the turbulence intensity but also with the scale of turbulence. A similar qualitative result was obtained by Meier and Kreplin [4], namely that in addition to the intensity the structure of the turbulence can influence the local wall shear stress.

The above considerations were based on grid generated turbulence, which is a special type of turbulence. Kovasznay [9] has shown that three independent quantities contribute to the free-stream turbulence in wind tunnels, namely vorticity, pressure, and entropy fluctuations. Each of those quantities can influence the boundary-layer transition, however, grid generated turbulence is dominated by vorticity. Schubauer and Skramstad [1] and Ahuja et al. [10] have shown that boundary-layer transition can also be induced by pressure fluctuations. Unfortunately, vorticity and entropy cannot be measured directly, but measurements of velocity, pressure and temperature fluctuations are possible. It was shown by Michel and Froebel [11] how velocity fluctuations depend on vorticity and pressure fluctuations and that certain pressure fields contribute considerably to the velocity fluctuations.

In summarizing we can conclude that one has to distinguish between two different influences caused by wind tunnel turbulence, namely on

- the laminar to turbulent boundary-layer transition and on
- the turbulent boundary-layer characteristics.

At the same time these facts describe the complexity of the problem of predicting these influences on the boundary layer development in order to correct or interpret wind tunnel measurements. Even if we had a detailed information of the wind tunnel turbulence in the form of intensities, length scales and frequency spectra of vorticity, pressure and temperature fluctuations we still would not know how these different parameters change the transition process, the turbulent boundary layer characteristics and/or how they interfere with each other. On the other hand the quality of the analysis and interpretation of wind-tunnel measurements - in particular if laminar wing designs are considered - strongly depend on the understanding of the phenomena discussed.

As an approximate solution to attacking this problem we tried an unconventional way of testing, namely transition measurements in the axisymmetric flow over a prolate spheroid and free-stream turbulence measurements at well defined locations for specific flow conditions. While the transition measurements on the prolate spheroid allow a direct comparison of the flow quality in different large subsonic wind tunnels used by the aircraft industry, corresponding turbulence measurements may be expected to improve the analysis and interpretation of these transition measurements. The aim of these supplementing measurements is to get a new and deeper insight into the influence of the flow quality on the wind tunnel measurements. This combination will enable us to evaluate and to compare e.g. aircraft measurements carried out in different facilities in a more reliable way.

2. Methods for Evaluating the Effect of Free-Stream Turbulence in Low-Speed Wind-Tunnels

2.1 Critical Reynolds Numbers on Spheres

The drag coefficient of a sphere is defined by:

$$C_D = \frac{F}{\frac{\pi}{4}D^2\frac{1}{2}\rho U_\infty^2}$$

where F is the total drag, D the diameter of the sphere, ρ the density and U_∞ the free-stream velocity of the air. If the drag coefficient C_D is plotted as a function of the Reynolds number $R = U_\infty D/\nu$, where ν is the kinematic viscosity of the air, a sudden reduction of C_D with increasing Reynolds number from $C_D \simeq 0.5$ to $C_D \simeq 0.1$ can be observed. This sudden decrease of C_D occurs at the so-called critical Reynolds number Re_{crit}, within a range of values depending on the turbulence of the air stream. Prandtl suggested already in 1914 [12] that observations of such resistance curves for spheres give a means of comparing the air streams in different laboratories with respect to their turbulence. A summary of drag measurements on spheres from different wind tunnels, free flight tests and towing tests in air at rest is given by Hoerner [13]. Analyzing these results it is not possible to correlate the data points with respect to the intensity of turbulence as the only parameter. Taylor [7] found from a theoretical consideration that the critical Reynolds number should be a function of the quantity

$$Re_{crit} = f\left[Tu_1\left(\frac{D}{L_1}\right)^{\frac{1}{5}}\right]$$

where Tu_1 is the intensity and L_1 the scale of turbulence both derived from the velocity fluctuations u′ in the mean flow direction. Dryden et al. [8] measured the scale L_1 in addition to the intensity of the wind tunnel turbulence. Their experimental data obtained with two different spheres at wind tunnel turbulence conditions manipulated by means of screens are shown in Fig. 1. Except for the measurements made at a distance of less than 300 mm behind the screen, the observations for both spheres lie remarkably well on a single curve. However, it is not possible to apply this result for configurations like an airfoil or fuselage. In the case of an airfoil the ratio of the chord length to the scale of turbulence may be important. This parameter could influence the boundary layer development and could lead to different drag or pressure measurements for identical Reynolds numbers for airfoils of different size. A second effect of turbulence could be the influence of the length scale on the boundary layer transition process, a phenomenon which has not yet been investigated in a systematic experiment. From these considerations we may conclude that drag measurements on spheres alone will not suffice to understand the influence of the wind tunnel turbulence on the flow stability and turbulent boundary layer characteristics.

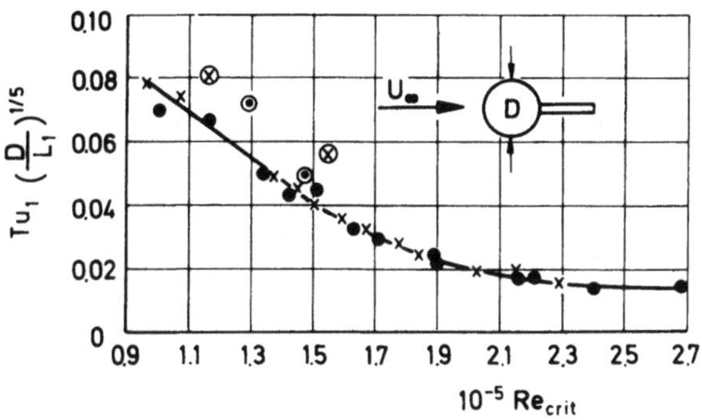

Fig. 1: Critical Reynolds numbers $Re_{crit} = U_\infty D/\nu$ of spheres as a function of $Tu_1(D/L_1)^{1/5}$ Ref. [8]

• D = 127 mm ⊙ distance from the screen
x D = 217 mm ⊗ $\leq$ 300 mm

2.2 Transition Measurements on the DFVLR Prolate Spheroid

In the DFVLR a long term research project on "Three-dimensional Viscous Effects" was carried out with a 2.4 m long 1:6 prolate spheroid which was investigated in different large European wind tunnels. For wall shear stress measurements the prolate spheroid is equipped with 12 flush-mounted surface hot film probes, see Fig. 2. The geometry of the sensors, their mounting on the surface and the principles of the corresponding measuring technique are described in detail in Ref. [14] and [15]. By means of 24 separate Constant Temperature Anemometers developed and built by the DFVLR the temperature difference of the films with respect to the tunnel flow temperature was adjusted to about 120 K. Under these conditions changes in the local wall shear stress (magnitude and/or direction) result in changes of the local hot film heat transfer which is indicated by changes of the anemometer voltage E. A universal calibration curve for all probes could not be established so that an individual calibration of each sensor was performed on the model. For this purpose measurements were carried out at different tunnel Reynolds numbers for axisymmetric flow conditions (zero angle of incidence). The corresponding wall shear stress values for each cross-section were obtained from boundary layer calculations for identical flow conditions applying the integral calculation method of Rotta [16]. The computations were based on the measured pressure distribution and the experimentally determined boundary layer transition location.

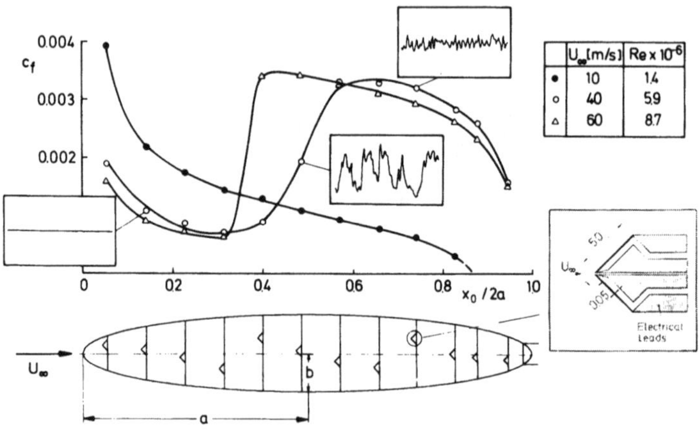

Fig. 2: Skin friction coefficient distributions and time signals for laminar, transitional and fully turbulent boundary layer flow. Measurements on the DFVLR 1:6 prolate spheroid carried out by means of surface hot films (cf. sketch) for axisymmetric flow conditions at different Reynolds numbers (a = 1.2 m, b = 0.2 m).

The transition is clearly indicated by a drastic increase of the anemometer mean output voltages and by fluctuations in the hot film signals even without any calibration of the probes. A representative result is given in Fig. 2, where wall shear stress distributions are plotted for three different Reynolds numbers, $Re = U_\infty 2a/\nu$.

In the framework of the research project "Three-dimensional Viscous Effects" numerous tests in different large European wind tunnels were carried out. Due to the described calibration procedure of the surface hot films we obtained a lot of data at axisymmetric flow conditions. The different locations of the boundary layer transition obtained at identical free-stream Reynolds numbers in different wind tunnels provide a valuable basis for a comparison of the flow quality. However, no information about the turbulence parameters involved can be derived from these results.

2.3 Measurements of Turbulence Quantities in the Free Stream of Wind Tunnels

As mentioned in the introduction, the turbulence in the empty test section of a wind tunnel can be described in terms of three independent contributions, vorticity, pressure, and entropy fluctuations, (Refs. [9], [11]). All other flow variables, like velocity, density, and temperature fluctuations depend on at least two of these independent quantities. For example, the velocity fluctuations are created by contributions from the pressure and vorticity fluctuations. It can be shown from the governing differential equations that the vorticity and entropy fluctuations in the test section of a wind tunnel are caused by the convection of frozen vorticity and entropy fields, (Ref. [11]). The propagation speed of these fields is the tunnel speed and their origin is upstream of the test section. The pressure fluctuations are governed by a convective wave equation which permits solutions which propagate in the flow direction with a velocity different from the tunnel velocity. Well known pressure waves in wind tunnels are plane sound waves that propagate with the speed of sound relative to the fluid and acoustical standing waves. Less well known are the pressure waves that are created in free jet facilities by the motion of large vortices in the shear layer. These pressure waves propagate with about 0.6 times the tunnel speed and are very effective generators for velocity fluctuations. This contribution may also be considerable in slotted wall test sections. The corresponding pressure field that is induced by the turbulent boundary layers on the walls of closed test sections is the limiting contribution for low turbulence wind tunnels.

Turbulence measurements were carried out in different tunnels with hot-wire anemometers and with microphones to determine the velocity and pressure fluctuation levels. The velocity fluctuations were determined with hot-wire anemometers. An x-wire probe was used to measure the fluctuating velocity components u' parallel to the mean velocity U_∞ and v' or w' perpendicular to the mean flow direction. The horizontal component v' was measured with the plane of the x-wire probe oriented horizontally and the vertical component w' with the plane oriented vertically. The two sensor wires were connected to two hot-wire anemometers DFVLR HDA III. The anemometer incorporates an analog polynominal linearizer circuit which yields an output voltage linear in the velocity. The outputs of two anemometers were fed into a turbulence level meter DFVLR TGM III for analog processing of the mean and fluctuating portions of the longitudinal and one transverse velocity component. Cut-off frequencies for the high-pass filter of 0.15 Hz and for the low-pass filter of 10 kHz were used. The fluctuating portions of the two velocity components were always available at output sockets of the turbulence level meter for real-time analysis in spectrum analyzers. The frequency spectra were determined in at least two different frequency bands in order to ensure sufficient resolution for small frequencies and a large total frequency bandwidth. The pressure fluctuations were measured by means of 1/4″ condensor microphones (Bruel & Kjaer, B&K 4135) fitted with a nose cone. The microphone and its preamplifier B&K 2633 were mounted in the place of the hot wire probe. The microphone amplifier (B&K 2133) was directly connected to the frequency analyzers.

3. Wind Tunnels Investigated

First investigations on the DFVLR prolate spheroid concerning three-dimensional viscous effects were carried out in the DFVLR 3m x 3m Low Speed Wind Tunnel (Niedergeschwindigkeits-Windkanal Göttingen, NWG) and the ONERA F1 Pressurized Low Speed Wind Tunnel. The analysis of measured transition data for axisymmetric flow conditions at different free-stream Reynolds numbers permit a comparison of the flow quality in these two facilities. A correlation of the transition with free-stream turbulence data was not possible, because relevant measurements were only available for the 3m x 3m Wind Tunnel in Göttingen. Detailed turbulence measurements, however, were carried out in two additional large subsonic wind tunnels, namely in the Deutsch-Niederländischer Windkanal (DNW) and the DFVLR Niedergeschwindigkeits-Windkanal at Braunschweig (NWB). Supplementary transition measurements on the prolate spheroid for axisymmetric flow conditions were performed in these two facilities. A brief description of these three facilities is given below. The differences in the principal design are shown in Fig. 3 and the main dimensions and tunnel data are given in Table 1.

The 3m x 3m Low Speed Wind Tunnel (NWG) is a typical Göttingen-type Wind Tunnel with an open test section, as originally designed by L. Prandtl. The fact that this wind tunnel had to fit into an existing building resulted in an unusual design of the diffusors with extremely sharp inflection corners. This may lead to flow separations inspite of the inserted vanes. A second source of flow separation may be the nozzle with its two-stage contraction.

The equivalent subsonic wind tunnel in Braunschweig (NWB) has approximately the same size of test section of 3.25m x 2.8m, but an open, closed, or slotted wall version can be chosen. This wind tunnel was remodelled recently and an improved flow quality is expected in particular in comparison with the wind tunnel in Göttingen (NWG).

The DNW is an atmospheric wind tunnel of the closed circuit type with three interchangeable, closed test-section configurations, [17]. The additional open-jet configuration was not investigated in our tests. The contraction ratio of the largest DNW test section (9.5m x 9.5m) is of the same order as those of the NWG and the NWB, cf. Table 1.

Quite extensive calibration tests have been carried out in the DNW. The mean flow data have been measured mainly in the 8m x 6m test section. At the usual model location the normalized deviations of the total head pressure lie within the bandwidth of $-0.001 < \Delta c_{pt} < 0.001$ where $\Delta c_{pt} = 2(p_t - p_{t\infty})/(\rho U_\infty^2)$ and the local total pressure is defined as $p_t = p + \rho U_\infty^2/2$ The corresponding static pressure distributions have been measured with a "flying" static probe. Over a length of about 12 meters the deviations from the static pressures measured at the test section center to vertical as well as to horizontal planes are within $\Delta c_p = \pm 0.002$. The flow angularities lie within the bandwidth of $\pm 0.2°$ and the uniformity of temperature is better than ± 0.2 K. Such detailed measurements of the mean flow data in the latest versions of the NWG and NWB are not available, while in all three facilities comprehensive investigations of the free-stream turbulence were performed.

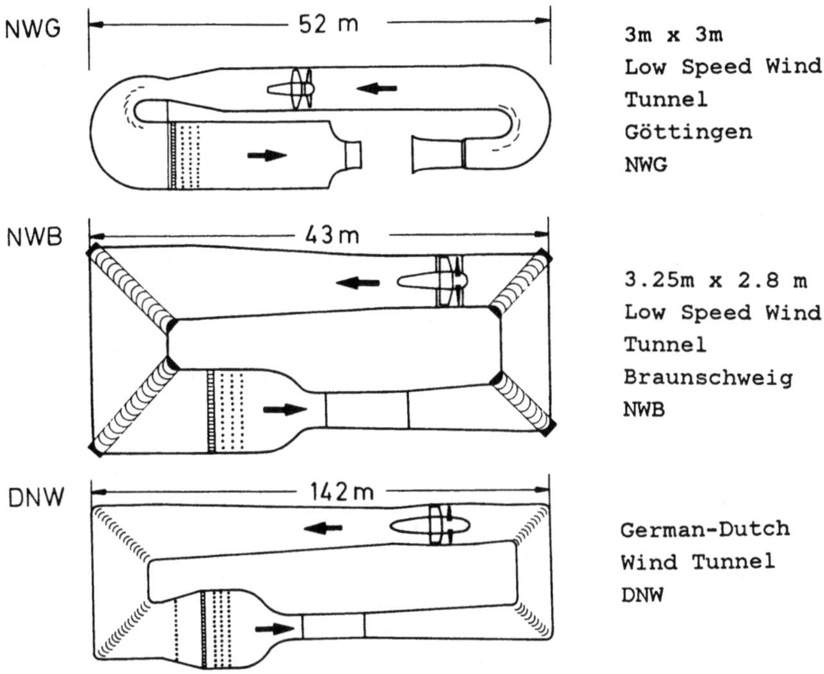

Fig. 3: Wind tunnels investigated

	DFVLR-GÖ NWG	DFVLR-BS NWB	GERMAN - DUTCH - WIND TUNNEL D N W		
Cross-Sectional Shape & Size (m x m)	3.0 x 3.0	3.25 x 2.8	6.0 x 6.0	8.0 x 6.0	9.5 x 9.5
Type	open	open closed slotted	closed	closed	closed
Length	6 m	6.2 m	9 m	16 m	15 m
Max. Air Speed	65 m/s	90 m/s	153 m/s	117 m/s	62 m/s
Contraction Ratio	5.4	5.6	12.0	9.0	4.8

Table 1: Main dimensions and flow specifications of the wind tunnels investigated

4. Results and Discussion

4.1 Transition Measurements on the Prolate Spheroid

The results of the wall shear stress measurements as a function of the free stream Reynolds number, based on the model length 2a = 2.4 m obtained on the prolate spheroid at zero incidence are shown in Figs. 4 - 7. In all wind tunnels the free-stream velocity was varied in the fully available range. The maximum velocity in the DNW was dependent on the specific test section chosen, while in the Braunschweig facility the maximum velocity in the closed test section was - due to better pressure recovery - slightly higher compared to the open one. The skin friction coefficient c_f reported in the figures is obtained here by normalizing the wall shear stress τ_w with the free stream dynamic pressure q_∞ rather than the local dynamic pressure q which is often applied in boundary layer calculations. In our plots only surface hot film data are presented which exclude the rear part of the body. In this regime a proper calibration of the sensors is not possible, because of flow separation.

In all results presented the region of the laminar to turbulent boundary layer transition is clearly indicated by the strong increase of the skin friction coefficients. The highest Reynolds number investigated in the Göttingen Wind Tunnel NWG was about $Re \simeq 9 \cdot 10^6$, Fig. 4a. For this flow condition the boundary layer remains laminar until the downstream location of $x_0/2a = 0.315$ as can be concluded from the wall shear stress measurements with surface hot film No. 4. At one measuring station further downstream, No. 5, the transition has already taken place. It is worth noting that the transition onset* at that location ($x_0/2a \simeq 0.4$) starts already at $Re \simeq 5 \cdot 10^6$. The Reynolds number range of $\Delta Re \simeq 4 \cdot 10^6$ in which the transition process occurs is equivalent to an increase of the free-stream velocity from $U_\infty = 35 m/s$ to $U_\infty = 60 m/s$. These remarkably large transition regimes can be observed at all measuring stations downstream of $x_0/2a = 0.4$, although this effect is diminished due to the increasing adverse pressure gradient, cf. Fig. 4b.

It is interesting that the transition regimes are obviously much smaller in the open test section of the Braunschweig Wind Tunnel NWB, Ref. [18]. This is mainly indicated by a much steeper increase of the wall shear stress in the transition regime, see Fig. 5. Without knowing any details about the free-stream turbulence, we can already state that the unsteady flow quality of the NWB Wind Tunnel is better than that of the equivalent facility in Göttingen (NWG). With this result the first supposition, reported in [19], had to be dropped, namely that the turbulence in the shear layer at the free jet boundaries causes large boundary layer transition regimes. An explanation will be given in section 4.2.

---

*As the transition onset we defined the location of the first increase of the wall shear stress and the occurrence of the first turbulent spots in the corresponding oscilloscope traces.

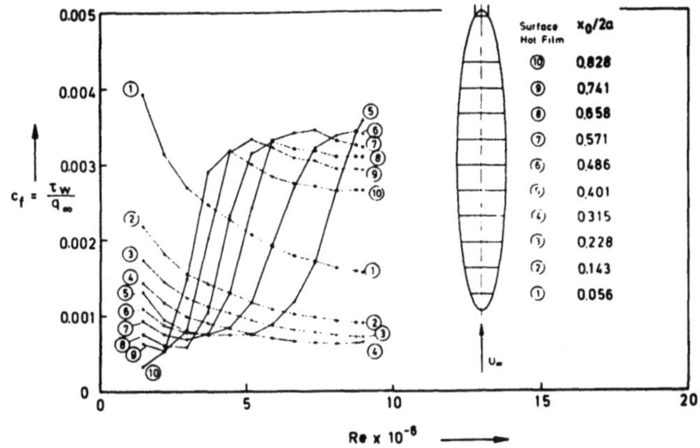

Fig. 4a: Skin friction coefficient distributions for different Reynolds numbers and positions on the prolate spheroid measured in the NWG

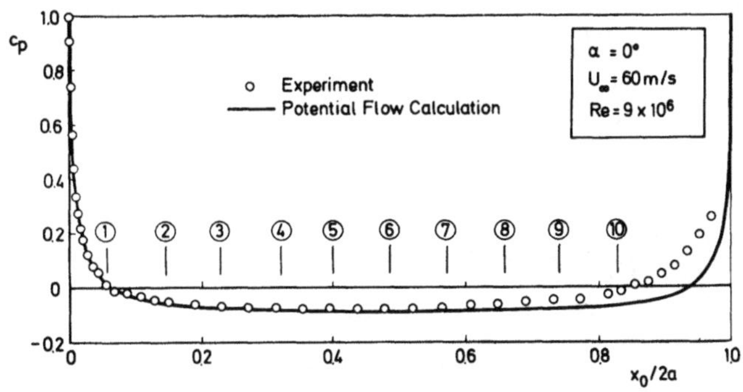

Fig. 4b: Measured and calculated wall pressure distributions, $c_p = \dfrac{2(p_w - p_\infty)}{(\rho U_\infty^2)}$

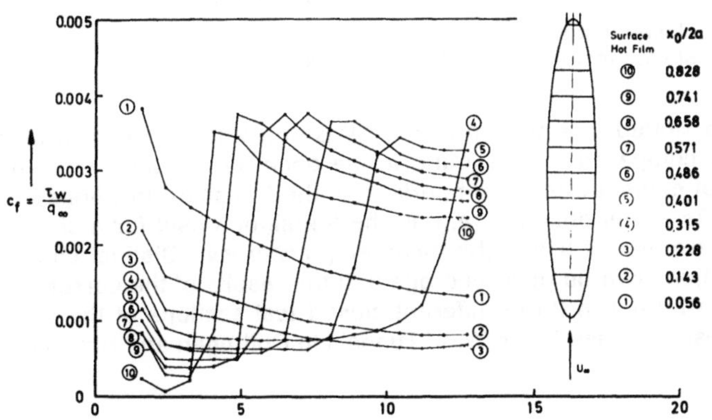

Fig. 5: Skin friction coefficient distributions for different Reynolds numbers and positions on the prolate spheroid measured in the NWB, open test section

a)

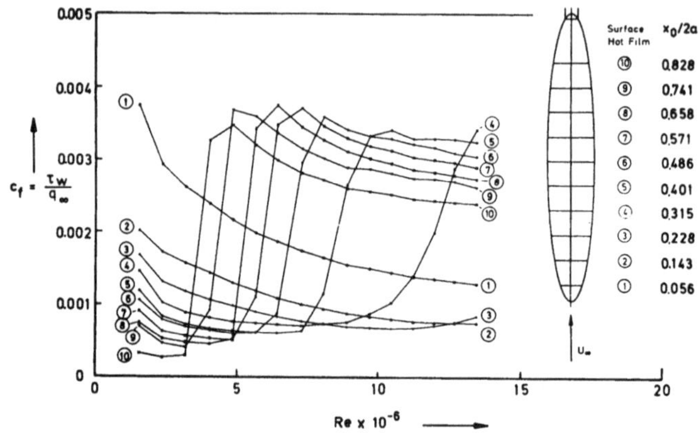

b)

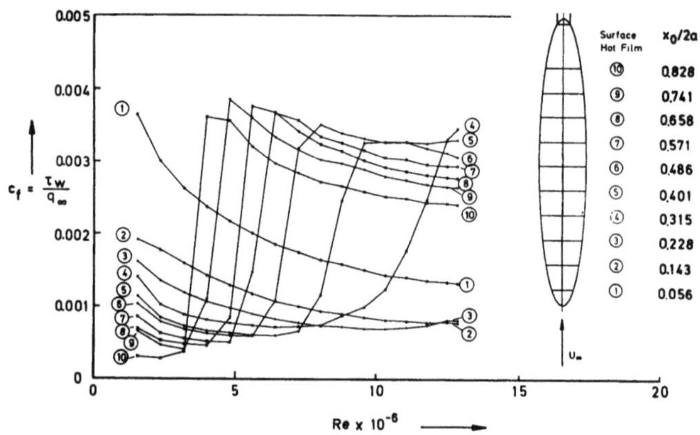

Fig. 6: Skin friction coefficient distributions for different Reynolds numbers and positions on the prolate spheroid measured in the NWB

a) closed test section
b) slotted test section (12% open area)

With a closed and slotted test section in the NWB one would expect to obtain a better flow quality compared to that found in the open test section. However, an improvement is not evident from a comparison of the transition locations in Figs. 5-6. The slight differences obtained, e.g. in the transition onset for a closed and open test section are almost within the measuring accuracy. Distinct changes in the transition locations can be observed again, if the results of the corresponding measurements carried out in three different closed test sections of the DNW are compared with those obtained in the Wind Tunnels in Göttingen and Braunschweig (NWG and NWB).

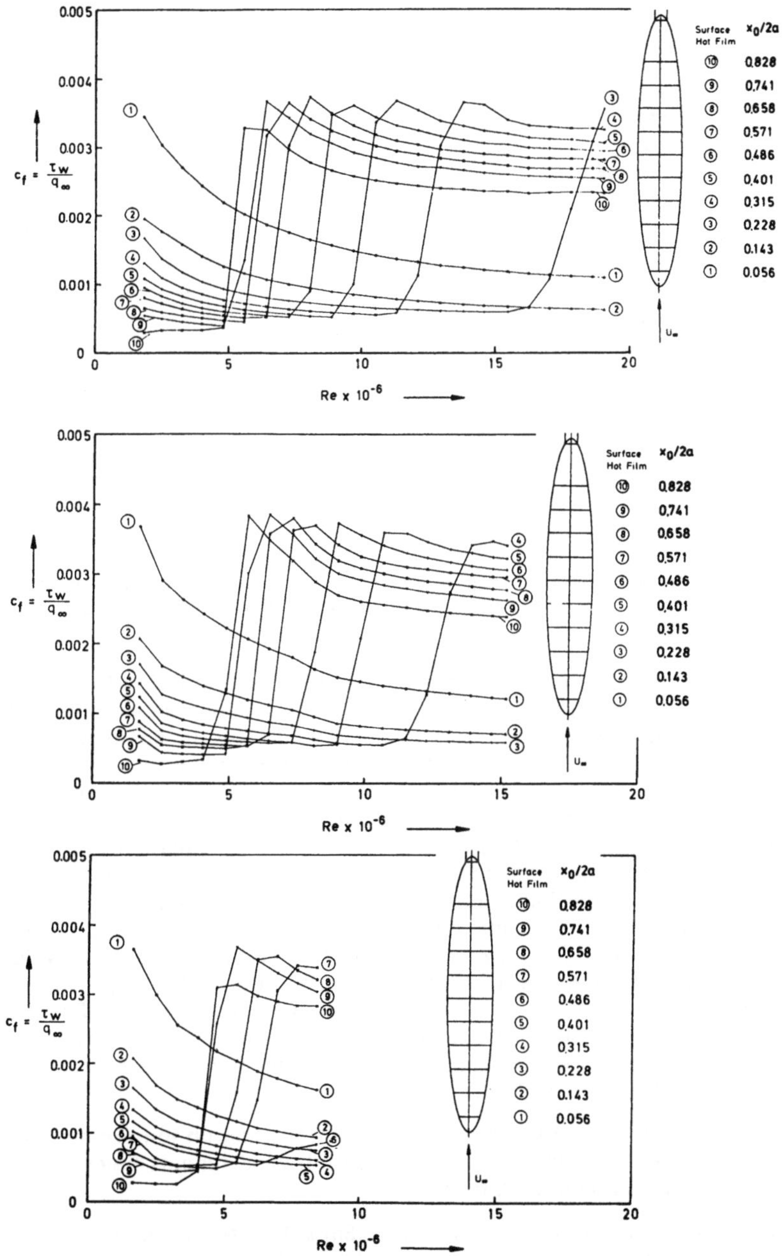

Fig. 7: **Skin friction coefficient distributions for different Reynolds numbers and positions on the prolate spheroid measured in the DNW**

a) 6m x 6m test section
b) 8m x 6m test section
c) 9.5m x 9.5m test section

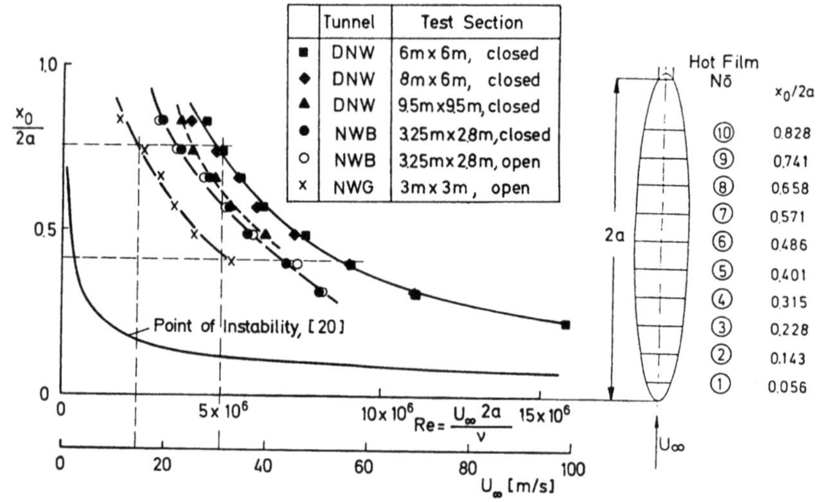

Fig. 8: Comparison of the transition onset measured in three different wind tunnels with open and closed test sections as a function of the free-stream Reynolds number

A detailed description of the DNW investigations is given in Ref. [19]. The analysis of the DNW data clearly indicates a shift of the transition location towards the nose region if the contraction ratio is decreased. This is an expected result because the turbulence intensity of the free stream increases with decreasing contraction ratio. In Fig. 8 the locations of the measured transition onsets obtained in different wind tunnels and test sections are plotted versus the Reynolds number Re and free-stream velocity U_∞. For comparison the points of instability based on laminar boundary layer calculations by Scholkemeier [20] are given. Obviously, the first instabilities in the laminar boundary layer occur much earlier than the measured transition onset. The experimental results can be interpreted as follows. For $x_0/2a = 0.75$ the free-stream velocity at which the transition onset is detected varies from $U_\infty \simeq 15 m/s$ in the NWG to $U_\infty \simeq 32 m/s$ in the DNW. Correspondingly, the location of transition onset moves from $x_0/2a = 0.42$ to $x_0/2a = 0.75$ at $U_\infty = 32 m/s$. In regions of favourable pressure gradients this phenomenon is even more pronounced. A comparison of the different test sections and different wind tunnels yields:

- The NWG yields an earlier boundary-layer transition on the prolate spheroid than the open test section of the NWB. Both wind tunnels have similar test section areas and contraction ratios.

- The transition location obtained in the 9.5m x 9.5m test section of the DNW is slightly shifted to higher Reynolds numbers relative to those obtained in the closed and open test sections of the NWB. The contraction ratios of the tunnel configurations considered are almost identical.

- An increase in the contraction ratio from 4.8 in the 9.5m x 9.5m DNW test section to 12 in the 6m x 6m test section results in a considerable downstream shift of the transition location.

4.2 Free-Stream Turbulence Measurements

The results of the free-stream turbulence measurements are presented in terms of frequency spectra. The spectra are plotted in a normalized form. The Strouhal number $St = f\,2a/U_\infty$ is plotted logarithmically on the horizontal axis, where f is the frequency, 2a is the length of the prolate spheroid, and U_∞ is the tunnel velocity. The power-spectral density $(\Delta\tilde{u}^2/\Delta f)$ times (f/U_∞^2) is plotted logarithmically on the vertical axis. The power-spectral density is the mean square $\Delta\tilde{u}^2$ of the velocity fluctuations within the frequency bandwidth Δf.

The spectra of the longitudinal and horizontal transverse velocity fluctuations u′ and v′ measured in the NWG for $U_\infty = 40m/s$ are plotted in Fig. 9. Each spectrum is generated from two single spectra measured in two different frequency bands. They are joined at St ≃ 4, which explains the sudden change in the scatter at this Strouhal number. The larger scatter at small Strouhal numbers had to be accepted in order to limit the integration time to about one minute.

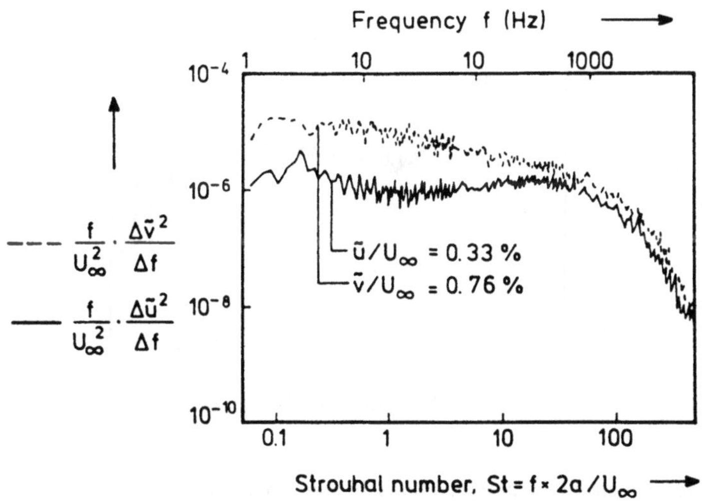

Fig. 9: Power spectra of u′- and v′ fluctuations in the NWG for $U_\infty = 40m/s$ $(x = 2m, y = z = 0)$

The spectra are dominated by vorticity fluctuations in the flow. For Strouhal numbers St < 10, the transverse v′-fluctuations are larger than the longitudinal u′-fluctuations as is to be expected behind the nozzle due to the influence of the flow contraction. It was found that the normalized u′-spectra are almost independent of tunnel speed U_∞. This independence from U_∞ was also found for the v′- and w′-spectra. In addition, the axial position x on the tunnel centerline has no influence on the spectra. These findings are typical for a tunnel that is dominated by vorticity fluctuations.

The situation is different for the open test section of the NWB. It can be seen from Fig. 10 that the spectral levels measured at $U_\infty = 50 m/s$ are much smaller than in the NWG for Strouhal numbers St > 1. A broad hump with several narrow peaks can be observed for 0.1 < St < 1. The hump is induced by the large vortices in the free shear layer. The narrow peaks are standing waves that are typical for free jet wind tunnels. It is a combination of organ pipe resonances in the tunnel circuit with an acoustic feedback between the collector and the nozzle. The influences of the free shear layer and of the standing waves are also present in the NWG, but they are swamped by the broad band spectrum resulting from the vorticity fluctuations. The narrow peaks for St > 100 are caused by sensor-wire vibrations. A corrected spectrum - indicated by dotted lines - was used to determine the turbulence intensity.

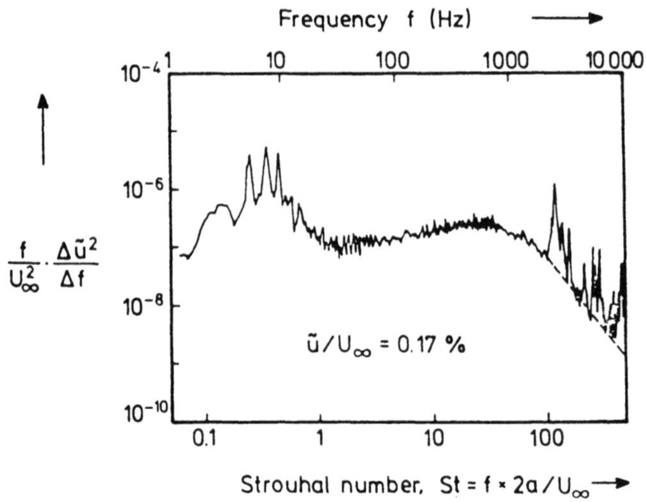

Fig. 10: Power spectrum of u′-fluctuations in the NWB with open test section for $U_\infty = 50 m/s$. Note shear layer deduced hump for St $\simeq$ 0.5 and peaks due to collector nozzle feedback. Peaks above St = 100 are caused by wire vibrations (x = 2.8 m, y=z=0)

The spectrum of the v′-fluctuations (not shown here) does not exhibit the low frequency hump and peaks, because standing waves and the shear layer vortices contribute predominantly to the fluctuations in the longitudinal direction.

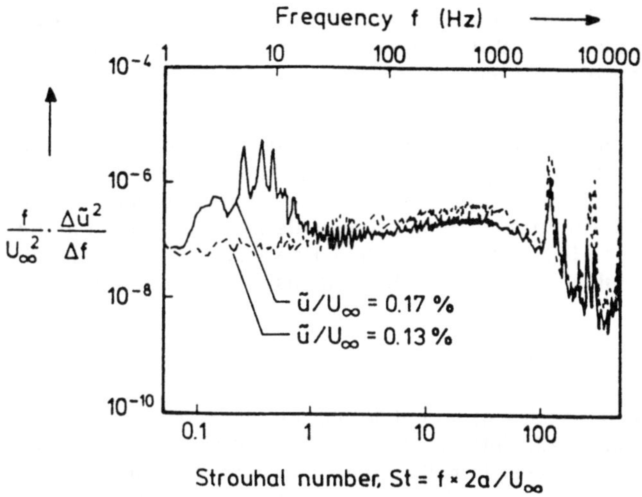

Fig. 11: Power spectra of u′-fluctuations in the NWB with open (—) and closed (---) test sections for $U_\infty = 50 m/s$ (x = 2.8 m, y=z=0)

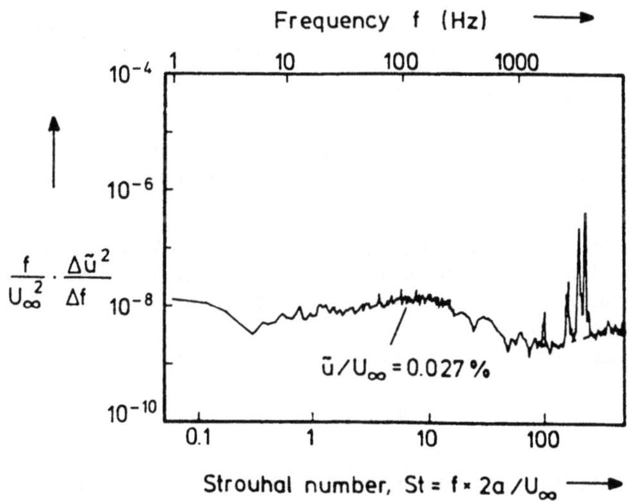

Fig. 12: Power spectrum of u′-fluctuations in the DNW 8m x 6m test section for $U_\infty = 40 m/s$ (x = 6.5 m, y=z=0)

In Fig. 11 the u′-spectra measured in the closed and open test sections of the NWB are compared with each other. Obviously, the closed test section configuration eliminates the effects of the free shear layer. There are almost no differences in the spectra for higher Strouhal numbers (St > 1). The rms turbulence intensity $Tu_1 = \tilde{u}/U_\infty$ is reduced from 0.17 % in the free jet test section to 0.12 % in the closed test section. In contrast to this the lateral turbulence levels $\tilde{v}/U_\infty$ of the two test sections are almost identical.

The corresponding u′-spectrum measured in the center of the 8m x 6m test section of the DNW is plotted in Fig. 12. The integrated turbulence intensity was calculated to $Tu_1 = 0.03$, only about one quarter of the equivalent value for the closed test section of NWB. No turbulence measurements were carried out in the smaller 6m x 6m and larger 9.5m x 9.5m test section. However, the free-stream turbulence in the DNW is dominated by vorticity fluctuations, as shown in Ref. [21].

4.3 Correlation of Transition with Turbulence Measurements

In Fig. 13 the u′-spectra of four different test sections measured in three wind tunnels are compared. This result can be correlated in some respect with the measured boundary layer transition data summarized in Fig. 11. For convenience we present the measured turbulence intensities Tu_1 together with the Reynolds numbers based on the transition onset at the free-stream velocity $U_\infty = 40 m/s$ in Table 2. In addition, the spectral levels $W\tilde{u}$ are given for St $\simeq$ 60 which corresponds to a frequency representative for Tollmien-Schlichting waves.

- The early transition in the NWG compared to the open test section in the NWB can be explained by the extremely high broad band turbulence in the NWG.
- The open and closed test sections of the NWB lead to almost identical transition locations on the prolate spheroid. The corresponding turbulence intensities differ clearly, cf. Table 2. However, the differences in the spectral levels of both test sections for St $\simeq$ 60 correlate with measurable changes of the transition onset.
- The extremely low turbulence intensity of the DNW is accompanied by a remarkable increase in the transition onset Reynolds number. If the free-stream turbulence is mainly caused by vorticity, we can apply the relation for the influence of the nozzle contraction ratio on the turbulence intensity proposed by Batchelor and applied by Beckwith and Rotta [22]

$$\frac{Tu_1}{Tu_S} = \frac{1}{c^2}\left[\frac{3}{4}(\ln 4c^3 - 1)\right]^{\frac{1}{2}}$$

which is valid for incompressible flow and c^3 much larger than 1. In this relation

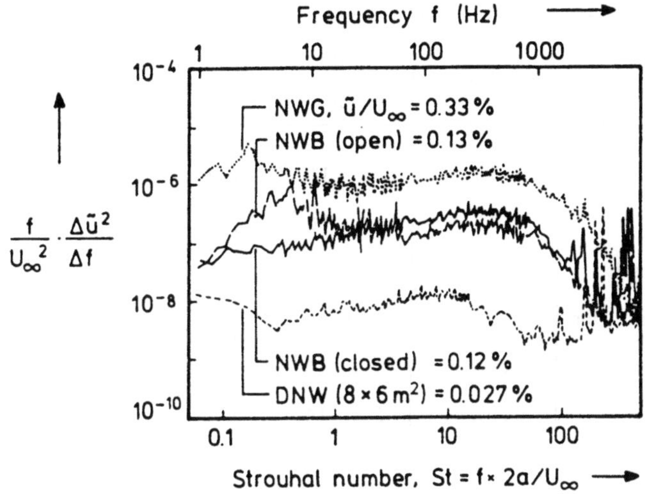

Fig. 13: Comparison of u′-spectra in four different test sections at $U_\infty = 40 m/s$. Note extremely low fluctuation level in DNW

$$c = \frac{A_T}{A_S} = \frac{\textit{Test section area}}{\textit{Settling chamber area}}$$

and Tu_S is the longitudinal turbulence intensity in the settling chamber. Calculated values of Tu_1 and $w\tilde{u} = (f/U_\infty^2)(\Delta\tilde{u}^2/\Delta f)$ for the 6m x 6m and 9.5m x 9.5m test sections are given in Table 2. The transition data of the DNW can obviously be correlated with both parameters, turbulence intensity Tu_1 or the spectral level $(f/U_\infty^2)(\Delta\tilde{u}^2/\Delta f)$. The transition data of all wind tunnels can only be correlated with the spectral level $w\tilde{u}$ at Strouhal numbers that are important for Tollmien-Schlichting waves.

- The measured pressure coefficients $c_{\tilde{p}}$ obtained by normalizing the rms pressure fluctuations $\tilde{p}$ with the free-stream dynamic pressure q_∞ are also given in Table 2. The pressure level in the NWG is larger than in the other tunnels investigated. The pressure coefficients in the NWB and DNW are almost equal and, consequently, are not the cause of the differences in the observed transition locations in the tunnels investigated which have a relatively low noise level.

Tunnel, Test Section	$Tu_1 = \tilde{u}/U_\infty$	$W_{\tilde{u}}$ for $St \simeq 60$ f	$c_{\tilde{p}}$	$x_0/2a$ transition onset
NWG, open	0.33 %	10^{-6}	0.35 %	(0.33)
NWB, open	0.17 %	10^{-7}	0.23 %*	0.46
NWB, closed	0.12 %	2×10^{-7}	0.24 %*	0.44
DNW, 9.5mx9.5m	0.09 %	9×10^{-9}**		0.48
DNW, 8mx8m	0.03 %	3×10^{-9}	0.20 %	0.56
DNW, 6mx6m	0.02 %	2×10^{-9}**		0.57

Table 2: Position of $x_0/2a$ of transition onset for $U_\infty = 40 m/s$ in comparison with turbulence intensity $Tu_1 = \tilde{u}/U_\infty$ and pressure coefficient $c_{\tilde{p}} = \tilde{p}/q_\infty$ and spectral level $W_{\tilde{u}} = (f/U_\infty^2)(\Delta\tilde{u}^2/\Delta f)$ for $St \simeq 60$
(* at $U_\infty = 50 m/s$, ** calculated [22])

6. Conclusions

It was found that the normalized spectra of the longitudinal velocity fluctuations are sufficient to explain the measured transition results. The Reynolds numbers at which the transition onset was detected can be correlated perfectly with the spectral level in the frequency range representative for the Tollmien-Schlichting waves. It is demonstrated that transition cannot be correlated with the parameter turbulence intensity measured in the total frequency band 0.1Hz < f < 10kHz. This result can be explained by the existence of large intensity levels at low frequency in some wind tunnels, which are e.g. typical for open jet facilities. These low frequency fluctuations obviously do not influence the transition process.

6. References

[1] Schubauer, G.B.; Skramstad, H.K.: Laminar boundary layer oscillations and stability of laminar flow; National Bureau of Standards, Paper 1772, and J. Aero. Sci. 14, pp. 69-78, 1947.

[2] Charnay, G.; Comte-Bellot, G; Mathieu, J.: Development of a turbulent boundary layer on a flat plate in an external turbulent flow; AGARD, CP 93, Paper No. 27, 1971.

[3] Rotta, J.C.: Ein theoretischer Beitrag zum Einfluß der äußeren Turbulenz auf die turbulente Grenzschicht; DFVLR IB 222-81 A 22, 1981.

[4] Meier, H.U.; Kreplin, H.-P.: Influence of free-stream turbulence on the boundary layer development; AIAA J., Vol. 18, No. 1, pp. 11-15, 1980.

[5] Hancock, P.E.; Bradshaw, P.: The effect of free stream turbulence level in turbulent boundary layers; J. Fluids Eng., Vol. 105, pp. 284-289, 1983.

[6] Blair, M.F.: Influence of free-stream turbulence on turbulent boundary layer heat transfer and mean profile development; Part I - Experimental data; J. of Heat Transfer, Vol. 105, pp. 33-47, 1983.

[7] Taylor, G.I.: The statistical theory of isotropic turbulence; J. Aero. Sci. 4, No. 8, pp. 311-315, 1937.

[8] Dryden, H.L.; Schubauer, G.B.; Mock, W.C. jr.; Skramstad, H.K.: Measurements of intensity and scale of wind-tunnel turbulence and their relation to the critical Reynolds number of spheres; NACA Rep. 581, 1937.

[9] Kovasznay, L.S.G.: Turbulence in supersonic flow; J. Aero. Sci. 20, pp. 657-674 and 682, 1953.

[10] Ahuja, K.K.; Whipkey, R.R.; Jones, G.S.: Control of turbulent boundary layer flows by sound; AIAA Paper No. 83-0726, 1983.

[11] Michel, U.; Froebel, E.: Definition, sources, and lowest possible levels of wind-tunnel turbulence; AGARD CP-348, pp. 11.1-11.12, 1984.

[12] Prandtl, L.: Der Luftwiderstand von Kugeln; Nachr. d. Kgl. Ges. der Wissensch., Göttingen, Math.-phys. Kl. pp. 177, 1914.

[13] Hoerner, S.F.: Fluid dynamic drag; 1st edition published by the author, New York 1958, 2nd edition Washington, DC, 1965.

[14] Meier, H.U.; Kreplin, H.-P.: Experimental investigations of the boundary layer transition and separation on a body of revolution; Z. Flugwiss. Weltraumforsch., Vol. 4, pp. 65-71, 1980.

[15] Kreplin, H.-P.; Vollmers, H.; Meier, H.U.: Measurements of the wall shear stress on an inclined prolate spheroid; Z. Flugwiss. Weltraumforsch., Vol. 6, pp. 248-252, 1982.

[16] Rotta, J.C.: FORTRAN-IV Rechenprogramm für Grenzschichten bei kompressiblen ebenen und achsensymmetrischen Strömungen; DLR-FB 71-51, 1971.

[17] Jaarsma, F.; Seidel, M.: The German Dutch wind tunnel DNW - design aspects and status of construction; 11th ICAS Congress, Lisbon, Sept. 1978, Paper No. B3-07, 1978.

[18] Kreplin, H.-P.; Meier, H.U.; Baumgarten, D.: Wall shear stress measurements on a prolate spheroid at zero incidence in the DFVLR Low Speed Wind Tunnels NWG and NWB; DFVLR, IB 222-87 A01, 1986.

[19] Kreplin, H.-P.; Meier, H.U.; Mercker, E.; Landhäußer, A.: Wall shear stress measurements on a prolate spheroid at zero incidence in the DNW Wind Tunnel; DFVLR-Mitt. 86-06, 1986.

[20] Scholkemeier, F.-W.: Die laminare Reibungsschicht an rotationssymmetrischen Körpern; Diss. Braunschweig 1943, shortened Version in Arch. d. Math. 1, pp. 270-277, 1949.

[21] Michel, U.; Froebel, E.: Investigations of the sources of velocity fluctuations in the German Dutch Wind Tunnel DNW; DFVLR, IB 22214-86 B4, 1986.

[22] Beckwith, I.E., Rotta, J.C.: Effect of contraction on turbulence in the working sections; In: AGARD Advisory Rep. No. 83, 1975.

Stability Investigation in Nominally Two-dimensional Laminar Boundary Layers by Means of Heat Pulsing

Zhou Ming de, Liu Tian Shu
Nanjing Aeronautical Institute (NAI)
Nanjing, P.R. China

Abstract

An experimental study on the boundary layer instability was carried out in flat-plate boundary layers by means of controllable heat pulsing, introduced into the critical layer. The experimental results indicate that the two-dimensional linear theory which predicts the stability boundaries can be applied successfully in the initial period of disturbance amplification even in the presence of considerable three-dimensional components of disturbances. Compared to the "Klebanoff" (K-type) transition mechanism, the subharmonic routes started from lower fluctuation levels of initial disturbances or from lower Reynolds numbers. The maximum amplification of the disturbances occurred at particular oblique-wave angles, which agreed well with the prediction of Craik's mechanism of a resonant triad. The potential and limitation of the heat-pulsing technique are also discussed in this paper.

---

*)This work was carried out in the framework of the DFVLR (Deutsche Forschungs- und Versuchsanstalt für Luft- und Raumfahrt) - CAE (Chinese Aeronautical Establishment) cooperation. The paper summarizes the work documented in Refs. [5], [6], [7]..

1. Introduction

An understanding of the evolution of disturbances in laminar boundary layers is of prime importance not only for the explanation of flow phenomena observed in the experiments and the development of new techniques for numerical testing but also for practical purposes like laminar flow control. Laminar-turbulent transition is usually regarded as a consequence of the following steps:

1. The appearance and amplification of two-dimensional Tollmien-Schlichting (TS) waves. The linear theory based on the two-dimensional disturbance assumption can be applied successfully in this flow regime.

2. The nonlinear amplification of the waves and the formation of longitudinal vortices. The nonlinear or weakly nonlinear theories have reached great progress in calculating such wave interaction problems.

3. The occurrence of turbulent spots and their spreading to a fully developed turbulent boundary layer. Theoretically, the direct numerical simulation based on the Navier Stokes equations is starting to play an important part in some important physical points.

Although excellent new experimental contributions have been made towards the understanding of instability phenomena, many open questions remain. Some of these are closely associated with the technique used for producing artificial disturbances. An artificial roughness, for example, may cause a momentum loss in the boundary layer flow or a vibrating ribbon introduce waves with a small randomness in the spanwise amplitude distribution. The main purpose of this paper is to introduce a new technique for producing controllable disturbances, namely the heat pulse technique applied in the critical layer and to investigate the flow response to these disturbances. Due to the decisive importance of the first and second steps in the laminar-turbulent transition process, the discussions in this paper will concentrate on these two steps only.

2. A description of techniques for the generation of controllable disturbances

Already in 1948, Schubauer [1] applied the vibrating ribbon technique in boundary layer stability research. This is a generally accepted technique to produce controllable disturbances in laminar boundary layers. However, the mechanical device interferes strongly with the flow even when it is at rest and the disturbance produced includes a slight randomness in the spanwise amplitude distribution, which may affect the steadiness of three-dimensional flow patterns.

Liepmann et al. [2], [3] proposed the application of heat pulse devices flush mounted on the model surface for laminar flow control. The experiments were carried out in water. In wind tunnels, the effectiveness of heat-pulse devices decreases considerably because the heat transfer from the heated element to the air flow is much lower and the temperature dependence of the viscosity is much smaller than in water. Furthermore, the pulsing frequencies have to be at least one order higher than those in water. Maestrello [4] reported a successful transition fixing in wind tunnels by means of surface heaters. However, the flow response described in [4] is obviously independent of the input frequency and the level of input power is of the same order as that needed in artificial transition by means of constant heating, cf. [5].

In our investigation, we applied a wire of diameter d = 0.03 mm as a heat-pulsing unit which was positioned closely to the wall in the critical layer. The disturbances produced are based on the following theoretical consideration:

From the two-dimensional incompressible boundary layer equation for a flat plate with zero pressure gradient, we have:

$$\frac{\partial u}{\partial x} + \frac{\partial v}{\partial y} = 0 \qquad (1)$$

$$\rho \frac{du}{dt} = \frac{\partial}{\partial y} (\mu \frac{\partial u}{\partial y}) \qquad (2)$$

where the viscosity μ is temperature dependent. Without heating, equation (2) becomes

$$\rho \frac{du}{dt} = \mu \frac{\partial^2 u}{\partial y^2} \tag{3}$$

When heat is introduced at point P in the boundary layer as shown in Fig. 1, the viscosity μ is no longer constant and equation (2) becomes:

$$\rho \frac{du}{dt} = \mu \frac{\partial^2 u}{\partial y^2} + \frac{\partial u}{\partial y} \frac{d\mu}{dT} \frac{\partial T}{\partial y} \tag{4}$$

where T is temperature.

Equation (4) differs from equation (3) by an additional term on the right hand side. For air $d\mu/dT > 0$, and $\partial u/\partial y > 0$ for flat plate boundary layers. Thus, the sign of this additional term depends on $\partial T/\partial y$ only.

In Fig. 1 the possible deformation of the boundary layer profile due to the local heat transfer is sketched by the dashed line. This implies that $\partial T/\partial y < 0$ for $y>y_p$ and $\partial T/\partial y>0$ for $y<y_p$. Therefore, in the vicinity of y_p the flow accelerates at the side near to the wall and decelerates at the other side. Consequently, a local vorticity disturbance is introduced into the flow.

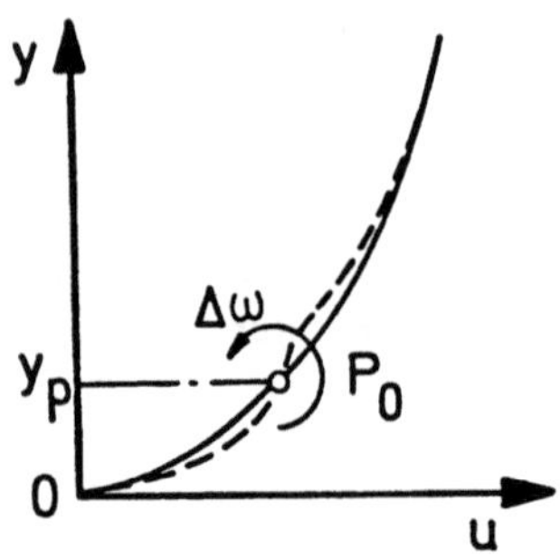

Fig. 1: Possible deformation of the velocity profile caused by a local heat source

3. Test Set-Up and Data Analysis

Two kinds of heating devices were used in our experiments in order to introduce a local heat transfer into the boundary layer:

1. A single wire of 15 mm length was arranged at the center line of a plate, Fig. 2, 180 mm downstreamn of the leading edge,

2. A two-row array of wires, specially designed to investigate the influence of wave interactions.

The wires were connected to a pulse generator in order to control the frequencies and intensity. It should be mentioned that three-dimensional components of disturbances existed not only in the case of the array as designed but also in the case of a single wire.

The experiments were carried out in the 1.5m x 0.3m Low Turbulence Wind Tunnel of the DFVLR-AVA, Göttingen (Fig. 3). Its maximum free-stream velocity is 45 m/s and the longitudinal turbulence intensity is about 0.05 %. The test surface consists of a 800 mm long, 300 mm wide and 32 mm thick wooden flat plate with a 6:1 elliptical nose (Fig. 2). The artificial disturbances were introduced 180 mm (for single wire) or 124 mm (for the array) downstream of the leading edge, symmetrically to the center line of the plate. The flow responses were measured by a set of surface hot films further downstream. A supplementary flow visualization study is now being made in the 1m x 1m x 10m three-dimensional Smoke Tunnel by means of the laser light sheet technique in Nanjing Aeronautical Institute, P.R. China.

Before results obtained with heat pulsing at both rows of wires can be presented, some phase relations of the different pulse signals and heat waves are to be illustrated. The local phase shift $\Delta\phi$, introduced in [6], [7] is used to characterize the phase relation between the two waves. In this context "local" stands for the streamwise position of pulse wire row (2). Thus, the local phase shift is the phase shift between the heat wave which has been convected from row (1) to row (2) and the heat pulse created at row (2).

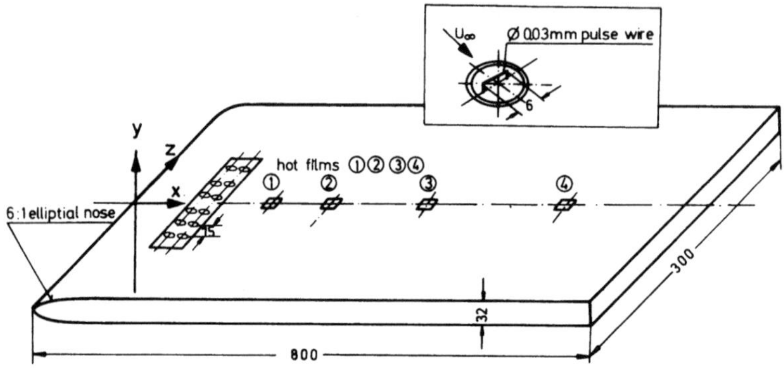

Fig. 2: Flat plate model equipped with heat pulse wires and surface hot films (dimensions in mm)

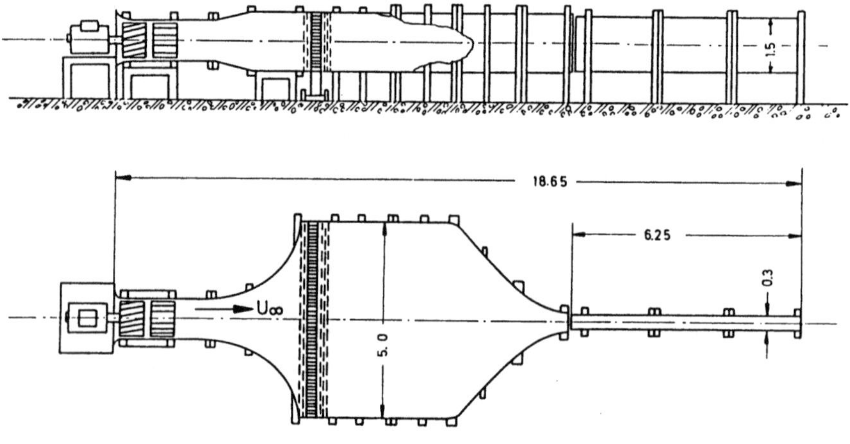

Fig. 3: Low Turbulence Wind Tunnel of DFVLR-AVA, Göttingen (all dimensions in m)

The convection of waves in the boundary layer is illustrated in Fig. 4. The convection velocity c is the velocity in the critical layer

$$c = 0.35\ U_\infty . \tag{5}$$

This was confirmed experimentally in [6], [7]. The travelling time Δt_1 from row (1) to row (2) is given by

$$\Delta t_1 = d/c, \tag{6}$$

where d is the distance between the first and second row. The corresponding phase shift due to convection is

$$\Delta\phi_1 = 2\pi(\Delta t_1/T) = 2\pi f\ \Delta t_1, \tag{7}$$

where T and f are the period and frequency of the waves. The fixed distance between the two rows, d = 16 mm, and a free-stream velocity of U_∞ = 16 m/s resulted in a time shift Δt_1 = 2.8 ms. Together with the phase shift $\Delta\phi_2$ which was generated electronically, we obtain the local phase shift as

$$\Delta\phi = \Delta\phi_1 + \Delta\phi_2 = 2\pi f\ (\Delta t_1 + \Delta t_2) \tag{8}$$

with the adjustable time shift Δt_2 generated by the phase shift unit. For a local phase shift of $\Delta\phi$ = 0, 2π, 4π ...a pulse emitted at row (1) just reaches row (2) when a second pulse is started there. This causes an increase of the amplitude of the heat pulse at a constant pulsing frequency (Fig. 5a). Accordingly, a doubling of the pulsing frequency is obtained for a local phase shift of $\Delta\phi = \pi$, 3π 5π..., (Fig. 5b).

4. Results and discussions

4.1 The flow response to nominally two-dimensional disturbances

Calculations based on linear stability theory under the condition of two-dimensional disturbances have been carried out by different researchers. As a matter of fact in reality these disturbances will never be entirely two-dimensional. For this reason it is meaningful to check the validity of the linear theory by experimental investigations in which the disturbances include considerable three-dimensional components from the very beginning.

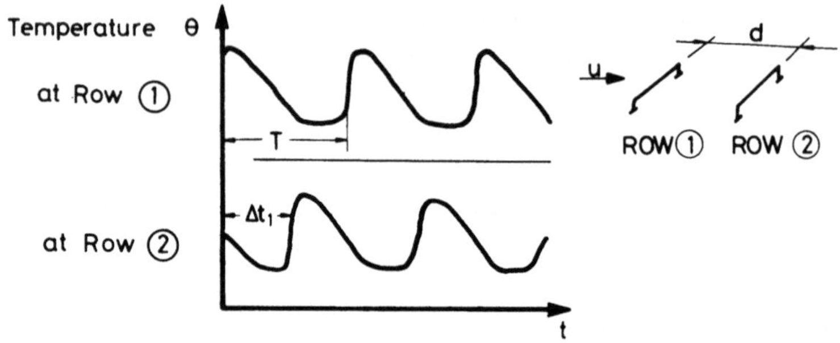

$c \sim 0.35 \cdot U_\infty$ Velocity in the critical layer

$\Delta t_1 = \frac{d}{c}$ Running time of the convective heat wave from row ① to row ②

$$\Delta \Phi_1 = 2\pi \frac{\Delta t_1}{T} = 2\pi f \Delta t_1$$

Fig. 4: Phase shift due to convective transport of heat waves in the critical layer

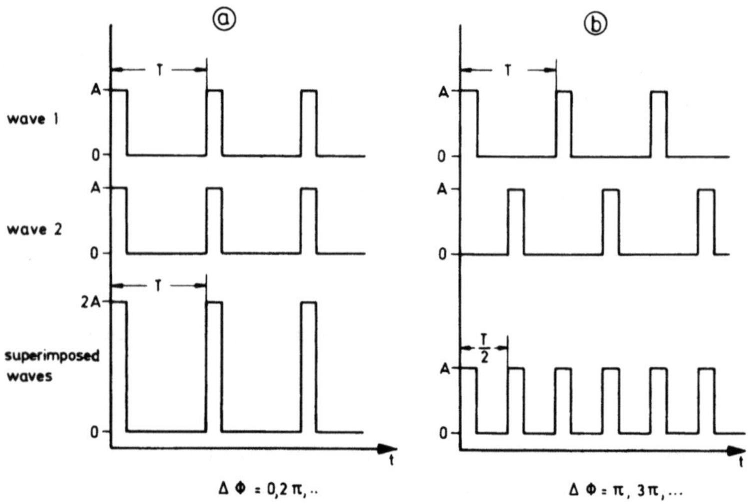

Fig. 5: Local phase shift adjusted to increased heat wave amplitude (a) and double pulsing frequency (b)

At first, a single wire was positioned at various distances from the wall. For constant input power and frequency the flow response measured with a surface hot film downstream is shown

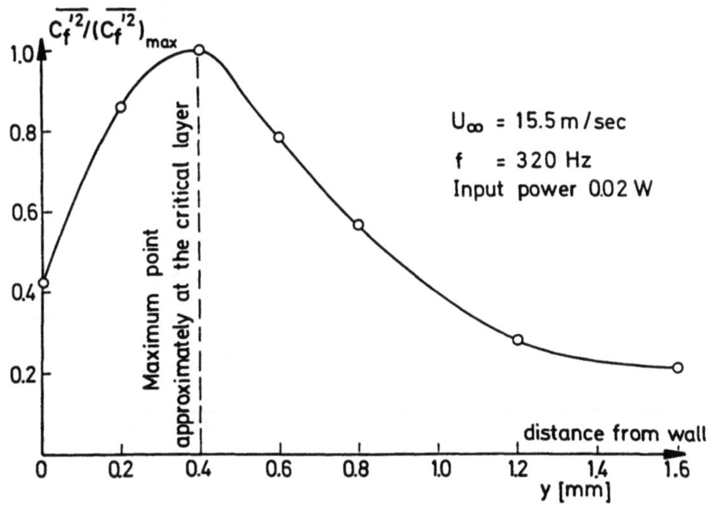

Fig. 6: The influence of wire distance from the wall on the flow response

in Fig. 6. The maximum point of the local mean square values of the skin friction coefficient $\overline{c_f'^2}/(\overline{c_f'^2})_{max} = 1$ indicates the most sensitive region of the boundary layer as far as the efficiency of the disturbances is concerned. It was found to be in the critical layer of the boundary layer. In order to optimize the effectiveness of the pulsing device, all pulse wires were placed in the critical boundary layer region throughout all the other experiments reported in this paper.

Firstly, a single wire and then an array with only one row working were used as pulsing devices. Experiments were carried out for various frequencies at given Reynolds numbers.

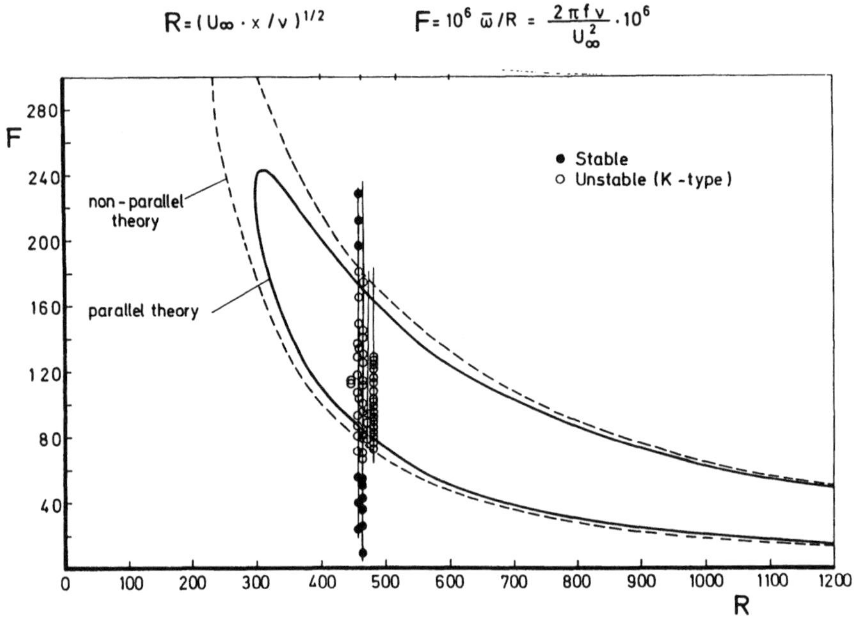

Fig. 7: Experimental data relative to the neutral curve

The rms-values obtained from the surface hot film just downstream of the pulse wires were used as a criterion for flow stability. The results are shown in Fig. 7, where the nondimensional frequencies F are plotted versus R, the square root of the Reynolds number. Qualitatively, the results are in good agreement with the neutral curve except that the instability region shown by experiments is a little larger. This is considered to be caused by the finite disturbances in the experiments instead of the infinitesimal ones in the linear theory.

In addition, the minimum input power to cause a full transition at a given point downstream was used as an indication of the growth rate of disturbances. The results showed that the maximum growth rate occurred approximately in the region of most strongly amplified disturbances in the neutral curve.

4.2 The phase velocity

The propagation speed of disturbances was measured when the first row of the array worked as a pulsing device and one wire of the second row was applied as a hot wire sensor. The measured phase shifts $\Delta\phi_1$ for various pulsing frequencies are shown in Fig. 8. The constant $a = c/U_\infty = 0.34$ was obtained by a least squares fit to the experimental results. The value of $a = 0.34$ is in agreement with the estimate from linear theory.

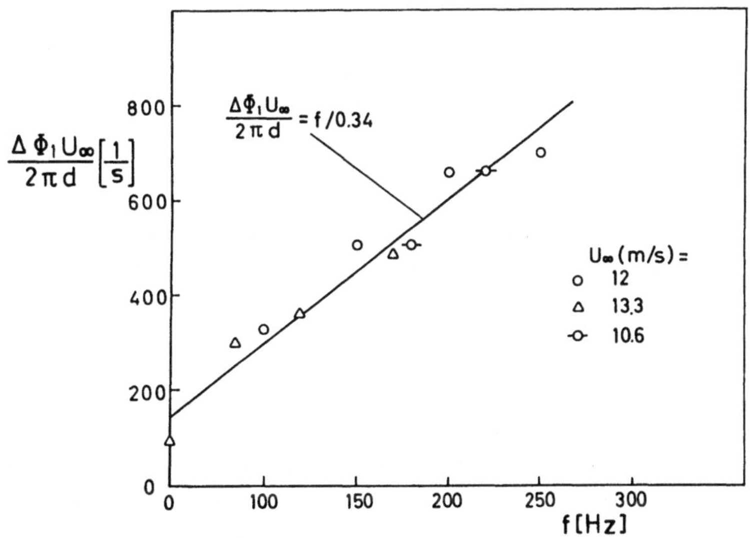

Fig. 8: Measured phase shift $\Delta\phi_1$ for various pulsing frequencies

The above-mentioned results add up to an important conclusion, namely that the two-dimensional linear theory which predicts the stability boundary is valid in the initial period of disturbance amplifications even if there exist considerable three-dimensional components of disturbances from the very beginning.

4.3 Typical signals of different amplification mechanisms

Experimental evidence has shown that different paths can lead disturbances from initially two-dimensional Tollmien-Schlichting waves to transition in wall bounded shear flows. These paths are distinguished by the nature of secondary three-dimensional disturbances that result in different characteristic patterns of λ-shaped vortex loops as sketched in Fig. 9.

The main results obtained by Klebanoff et al. [9] were spanwise alternating "peak-valley" patterns. The λ vortices were aligned along the peaks and repeated with the wave length λ_x of the Tollmien-Schlichting waves (Fig. 9a). This is known as a "K"-type pattern. This pattern provides a strong growth of disturbances, however, it assumes a maximum of streamwise initial rms fluctuation u' exceeding a threshold value of about 1 % of the basic reference velocity.

Craik [10] investigated one type of subharmonic transition mechanism, the "C"-type pattern, which is associated with the resonant triads of Tollmien-Schlichting waves in an unstable boundary layer. The triads considered are those comprising a two-dimensional Tollmien-Schlichting wave and two oblique waves propagating at equal and opposite angles to the flow direction, and such that all three waves have the same phase velocity in the downstream direction. This mechanism is characterized by a staggered λ-shaped pattern (Fig. 9b). For such resonant triads, remarkably powerful wave interactions take place which cause a continuous and rapid transfer of energy from the primary shear flow to the disturbances.

Herbert [11], [12], [13] analysed a different type of subharmonic transition mechanism, i.e. "H"-type pattern (Fig. 9c), which is associated with a secondary instability of the streamwise periodic flow to three-dimensional disturbances of normal-to-the-wall vorticity type. The λ-shaped pattern is also of staggered type, however, this mechanism can be distinguished from Craik's model by a broad band of wave lengths. This can be amplified if the Tollmien-Schlichting amplitude exceeds a relatively low threshold value.

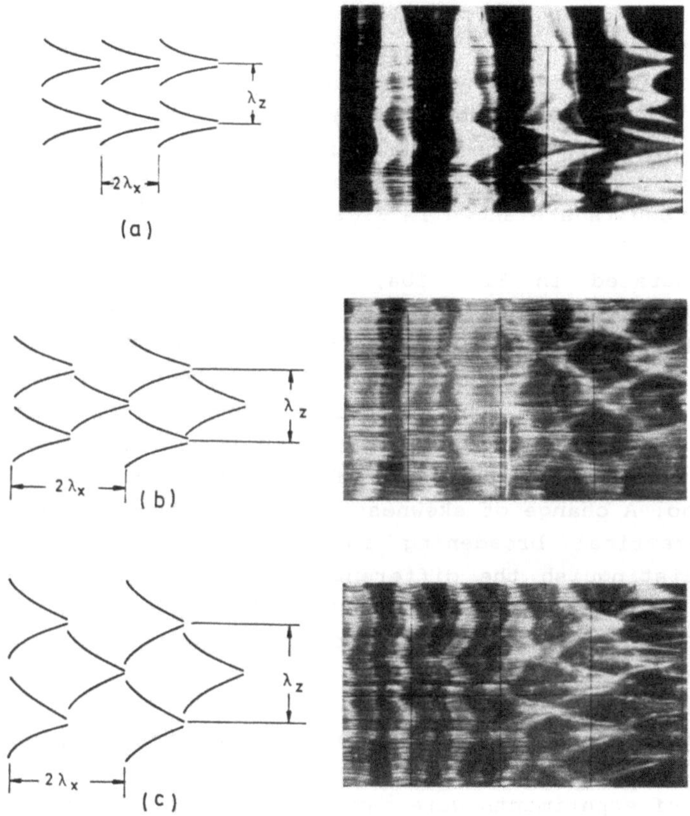

Fig. 9: Patterns of Λ vortices (flow visualization by Saric et al. [14])

a) aligned peak-valley pattern with wavelength λ_x - "K" type (maximum rms $\frac{u'}{U_\infty} \approx 1$ %).

b) staggered pattern with wavelength $2\lambda_x$ - "C" type (maximum rms $\frac{u'}{U_\infty} \approx 3$ %; $\lambda_z/\lambda_x = 0.7$)

c) staggered pattern with wavelength $2\lambda_x$ - "H" type (maximum rms $\frac{u'}{U_\infty} \approx 0.4$ %; $\lambda_z/\lambda_x = 1$)

Saric et al. [14], [15] carried out experimental studies on these distinct transition types. Their results showed that the subharmonic routes started at much lower fluctuation levels than the "K"-type breakdown at the same Reynolds numbers. In order to investigate the transition mechanisms discussed, the analysis of surface hot film signals was applied. The corresponding results are shown in Figs. 10a, b, c.

As demonstrated in Fig. 10a, the frequencies of the flow response can be distinguished in the initial stage of disturbance amplification. The intermittency of the "K"-type pattern is characterized by the formation of turbulent spots, whereas no turbulent spot could be found in the presence of subharmonic disturbances. The corresponding probability density functions shown in Fig. 10b indicate a difference between the two patterns, too. A change of skewness in the "K"-type and a continuous symmetrical broadening in the subharmonic mechanism clearly distinguish the different routes. For more evidence, the corresponding spectra are shown in Fig. 10c. The subharmonic frequency is half of the input frequency.

4.4 The flow response to the three-dimensional disturbance components

Two sets of experiments were carried out with a single wire as a pulsing device. In the first set of experiments, the free-stream velocity and the pulsing frequency were kept constant. Hot film signals were recorded and analysed with a gradual increase of input power. A selection of the time signals is presented in Fig. 11. The Tollmien-Schlichting waves appeared at relatively low input power. With an increase of input power, subharmonic signals can be observed, corresponding to the "H"-type transition modes. A further increase of input power causes a higher level of initial fluctuation and the subharmonics disappear. Instead of that, a fundamental frequency with higher harmonics can be observed, which characterizes a "K"-type breakdown. With a further increase of input power, a fully developed turbulence occurs.

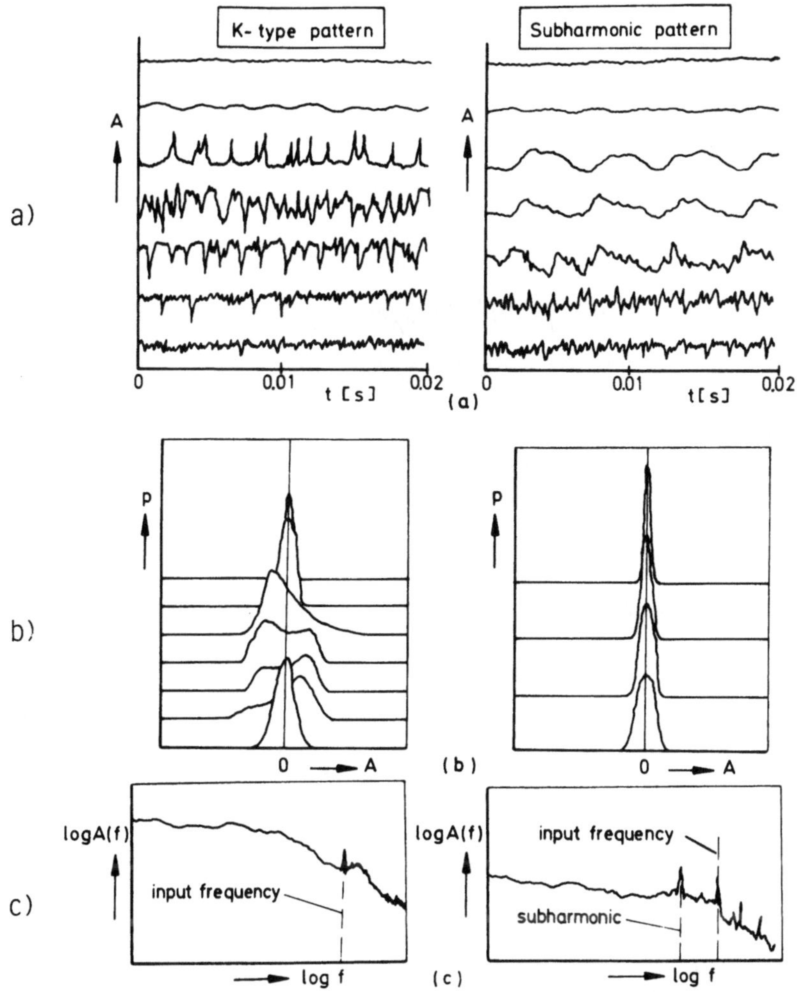

Fig. 10: Difference of the hot film signals between K-type breakdown and the subharmonic route of transition

a) time signals A(t)
b) probability functions p(A)
c) frequency spectra

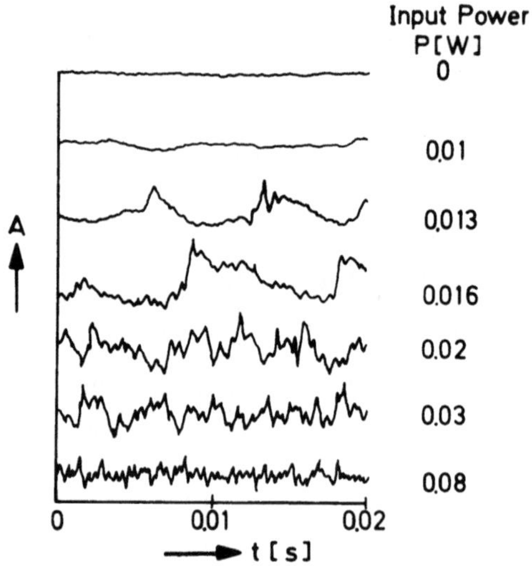

Fig. 11: Influence of input power on the time signals; U_∞ = 16.1m/s, f = 320 Hz

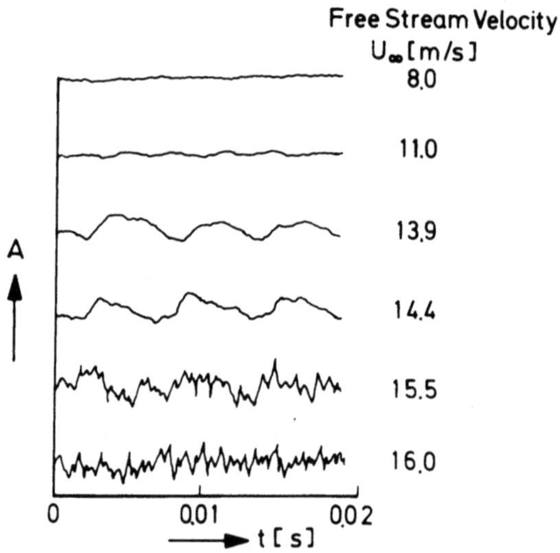

Fig. 12: Influence of free-stream velocity on the time signals, P = 0.018 W, f = 320 Hz

In the second series of experiments (Fig. 12), the input power and the frequency were kept constant. Increasing the free-stream velocity had a similar effect on the whole development of the flow stability to an increase of the input power.

The above-mentioned development was consistent with experiments made by Saric et al. [14], [15] in the sense that the staggered pattern occurred at much lower fluctuation levels than the peak-valley splitting at the same Reynolds number. Moreover, our experiments showed that the subharmonic mechanism occurred at a lower Reynolds number than the "K"-type breakdown under the condition of equal initial fluctuation level. Thus, it is extremely important to pay enough attention to the subharmonic mechanism, both in transition research and in laminar flow control.

The two-row array of pulse wires (Fig. 2) was especially designed to investigate the influence of the wave interaction. A collection of results is shown in Fig. 13. Compared to the neutral curve predicted by the linear theory, the experimentally detected unstable region is strongly enlarged. The non-linear interaction between the two rows seems to play an important role in disturbance amplifications.

With a given pulsing frequency, the least input power to cause a full transition at a certain point downstream was found by adjusting the local phase shift. Here, the least input power indicated the maximum amplification, which occurred at an optimum local phase shift. Assuming that the heated wires are point heat sources, the given geometry of the sensor arrangement leads to a relation between $\Delta\phi$ and the oblique wave angle Θ:

$$tg\Theta = 0.34\ \Delta\phi\ U_\infty/(360^\circ\ S\ f). \tag{9}$$

Thus, this optimum local phase shift corresponds to an optimum oblique wave angle which, combined with the two-dimensional wave component, might set up the resonant triads.

Systematical experiments were carried out for various frequencies. The optimum local phase shift $\Delta\phi$ which was found to cause transition is plotted versus the pulsing frequency f in Fig. 14. Craik's model predicted a wave angle for resonant triads

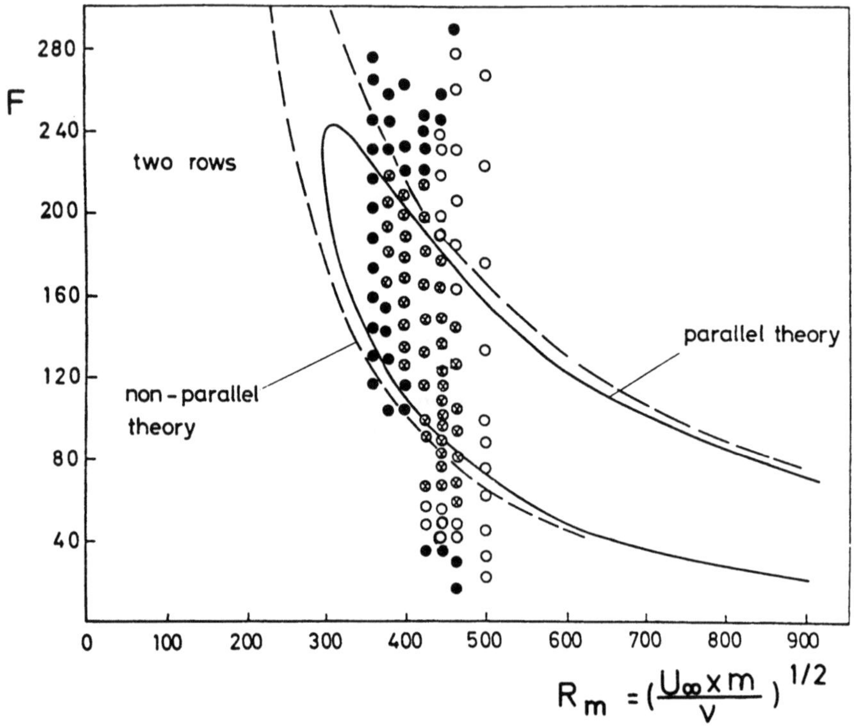

Fig. 13: Excitations relative to the neutral curve for $\gamma = \frac{1}{5}$ and $P \leq 2$ Watt (two rows);
• excitations which cause no transition at $x_3 = 420$ mm (P = 2 Watt);
o excitations which cause the K-type transition at $x_3 = 420$ mm ($P \leq 2$ Watt);
⊗ excitations which cause the subharmonic route to transition at $x_3 = 420mm$ ($P \leq 2$ Watt)

that slightly decreased from $\theta = 50.5°$ at branch 1 to $\theta = 47.6°$ at branch 2 of the neutral curve [13]. For comparison, a straight line corresponding to $\theta = 50.5°$ is shown in the figure with the relation:

$$\Delta\phi = 360° \, Sf \, tg\theta / 0.34 \, U_\infty \tag{10}$$

The Reynolds numbers covered in our experiments were of the same order as those used in the theoretical prediction. So, the maximum amplification found in the experiments may be interpreted as a result of resonant triads.

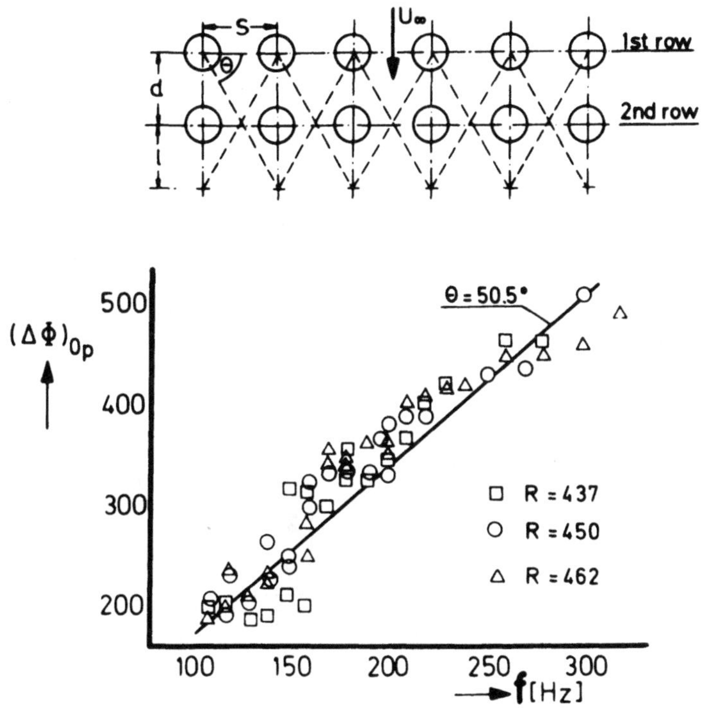

Fig. 14:

a) Formation of possible oblique waves
b) The measuring of oblique wave angle to cause resonant triads

5. Potential and limitation of the heat pulsing technique

The following points list some of the advantages of the technique itself and in particular its possible application in flow stability research.

- As demonstrated in Figs. 5a, b the amplitude and the frequency of the local disturbances can be increased by a superposition technique. In Fig. 15 examples of the superposition of waves and comparison of the flow response between two cases are given. The results obtained clearly indicate that the limitation in frequency and amplitude of the pulsing device can be increased to much higher levels by a multi-wire technique.

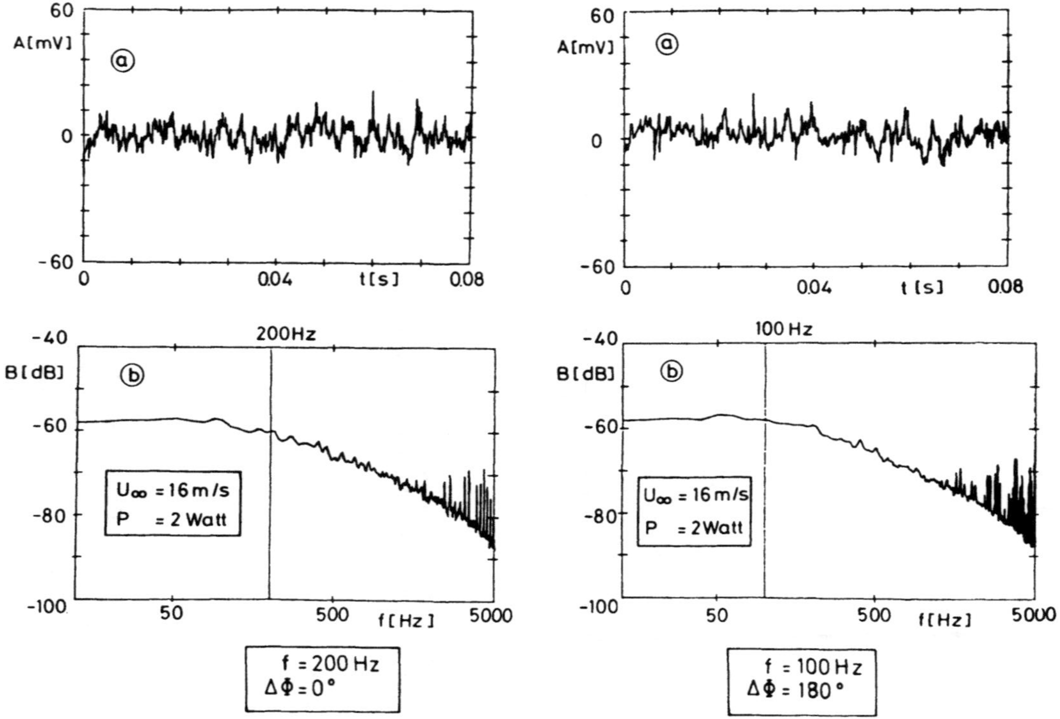

Fig. 15: The effective double frequency produced by wave superposition

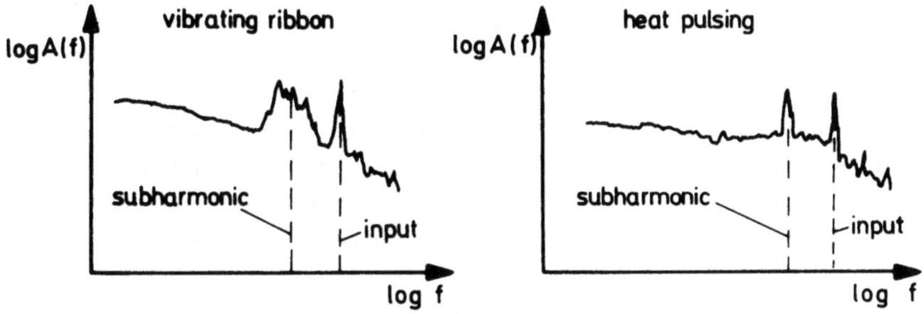

Fig. 16: Comparison of the spectrum downstream of a vibrating ribbon with a heat pulsing device

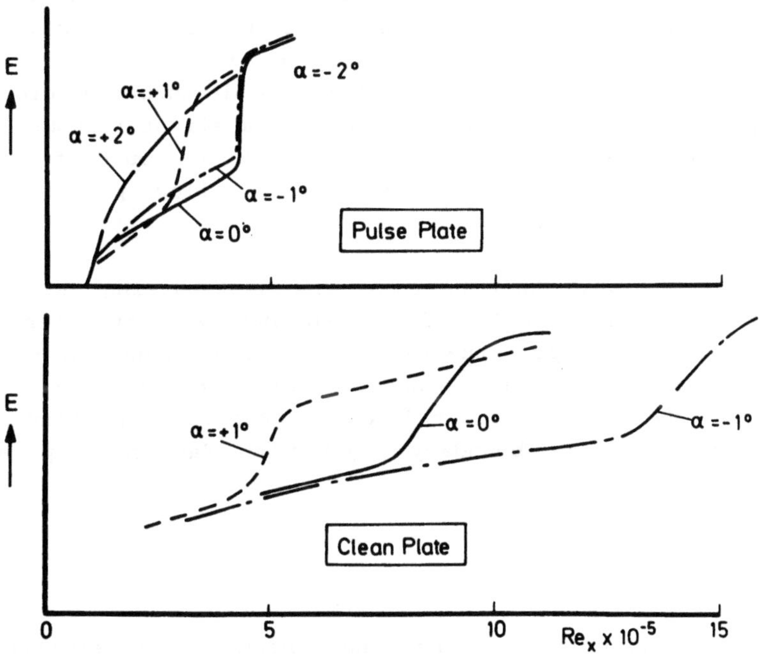

Fig. 17: The interference of multi-wire arrangement

- The momentum loss caused by the heat pulse itself is extremely small. Mean velocity profiles with and without heat pulsing were measured by means of the LDA technique. Results obtained just downstream of the pulse wire showed no measurable difference between the corresponding mean velocity profiles. This result implies that the heat pulse itself causes only a negligible momentum loss.

- A heat-pulsing device can produce more steady three-dimensional flow patterns than a vibrating ribbon. A comparison of our spectrum with those obtained by Kachanov et al. [16] is shown in Fig. 16. When the subharmonic occurs, the spectrum of the flow response downstream of the vibrating ribbon is always very wide. However, the bandwidth of subharmonics in the case of the heat-pulsing device is almost one order narrower than the former. The reason for this is that the spanwise amplitude distribution of disturbance produced by heat pulsing is fixed by the geometric design of the device. On the other hand the disturbances produced by a vibrating ribbon will include a slight randomness in the spanwise amplitude distribution, which would affect the steadiness of the three-dimensional flow patterns and broaden the spectrum.

However, like any other devices, the heat-pulsing device has its limitations. The multi-wire arrangement will interfere with the flow and even spoil the results in some cases. An example is shown in Fig. 17. The mechanical interference caused by the pins and the wires without heating promotes the transition considerably in the boundary layer with favourable pressure gradients.

6. Conclusions

- The experimental evidence has shown the feasibility of creating controllable disturbances in flow stability research by means of applying heat-pulsing in the critical layer of the laminar boundary layer.

- The validity of the two dimensional linear theory in the initial period of disturbance amplification was confirmed even if there exist considerable three-dimensional disturbance components from the very beginning.

- The subharmonic routes to transition start from a lower fluctuation level of initial disturbance or lower Reynolds number than that known from the so-called "K"-type mechanism.

- The maximum amplification of disturbances occurs at particular oblique wave angles which agree well with the prediction of Craik's mechanism of the resonant triads.

7. References

[1] G.B. Schubauer, H.K. Skramstad: Laminar boundary layer oscillations on a flat plate; NACA Rep. No. 909, 1948.

[2] H.W. Liepmann, G.L. Brown, D.M. Nosenchuck: Control of laminar instability waves using a new technique; J. Fluid Mech. 118, pp. 187, 1982.

[3] H.W. Liepmann, D.M. Nosenchuck: Active control of laminar-turbulent transition; J. Fluid Mech. 118, pp. 201, 1982.

[4] L. Maestrello: Active transition fixing and control of the boundary layer in air, AIAA-85-0564, 1985.

[5] Zhou Ming de, H.U. Meier, A. Maier: A preliminary study on artificial transition by heat pulsing; DFVLR IB 222-84 A 36, 1984.

[6] Liu Tian shu, H.U. Meier: Artificial boundary layer transition generated by heat pulsing, DFVLR IB 222-86 A 21, 1986.

[7] Wang Kai li, H.-P. Kreplin, H.U. Meier: Further experiments on artificial boundary layer transition forced by means of heat pulsing, DFVLR IB 222-86 A 36, 1986.

[8] C.C. Lin: On the stability of laminar flow and its transition to turbulence, Boundary Layer Research Symp. Freiburg, Ed.: H. Görtler, Springer Verlag Berlin, pp. 144, 1957.

[9] P.S. Klebanoff, K.D. Tidstrom, L.M. Sargent: The three-dimensional nature of boundary layer instability; J. Fluid Mech. 12, pp. 1, 1962.

[10] A.D.D. Craik: Non-linear resonant instability in boundary layers; J. Fluid Mech. 50, Part 2, pp. 393, 1971.

[11] T. Herbert: Secondary instability of plane channel flow to subharmonic three-dimensional disturbances; Phys. Fluids, 26 (4), pp. 871, 1983.

[12] T. Herbert: Subharmonic three-dimensional disturbances in unstable plane shear flows; AIAA-83-1759, 1983.

[13] T. Herbert: Analysis of the subharmonic route to transition in boundary layers; AIAA-84-0009, 1984.

[14] W.S. Saric, A.S.W. Thomas: Experiments on the subharmonic route to turbulence in boundary layers; Proc. IUTAM Symp. on "Turbulence and Chaotic Phenomena in Fluids", Kyoto, Japan, 1983.

[15] W.S. Saric, V.V. Kozlov, V.Y. Levchenko: Forced and unforced subharmonic resonance in boundary layer transition; AIAA-84-0007, 1984.

[16] Y.S. Kachanov, V.Y. Levchenko: The resonant interaction of disturbances at laminar-turbulent transition in a boundary layer, J. Fluid Mech. 138, pp. 209, 1984.

Three-dimensional Stability of Boundary Layers

WILLIAM S. SARIC and HELEN L. REED

Mechanical and Aerospace Engineering
Arizona State University
Tempe, AZ 85287 USA

Summary

The most recent efforts on the stability and transition of three-dimensional flows are reviewed. These include flows over swept wings, rotating disks, rotating cones, yawed bodies, corners, and attachment lines. The generic similarities of their stability behavior is discussed. It is shown that the breakdown process is very complex, often leading to contradictory results. Particular attention is paid to opposing observations of stationary and traveling wave disturbances.

1. Introduction

Besides the intrinsic mathematical and experimental challenges that characterize stability work, the main driving force for the study of boundary-layer stability is the understanding, prediction, and control of transition to turbulence. The origins of turbulent flow and transition to turbulence remain the most important unsolved problems in fluid mechanics and aerodynamics. The process of the breakdown of the laminar flow is three-dimensional and may be described by the following simplified discussion.

1.1 Basic Ideas

Disturbances in the freestream, such as sound or vorticity, enter the boundary layer as steady and/or unsteady fluctuations of the basic state. This part of the process is called *receptivity* (Morkovin [1]) and although it is still not well understood, it provides the vital initial conditions of amplitude, frequency, and phase for the breakdown of laminar flow. Initially these disturbances may be too small to measure and they are observed only after the onset of an instability. The type of instability that occurs depends on Reynolds number, wall curvature, sweep, roughness, and initial conditions. The initial growth of these disturbances is described by *linear* stability theory. This growth is weak, occurs over a viscous length scale, and can be modulated by pressure gradients, mass flow, temperature gradients, etc. As the amplitude grows, three-dimensional and nonlinear interactions occur in the form of *secondary* instabilities. Disturbance growth is very rapid in this case (now over a convective length scale) and breakdown to turbulence occurs.

The essential idea is that the understanding of the transition process will only come from the consideration of *three-dimensional* effects in the stability process even though the basic state may be one-dimensional and the primary instability is two-dimensional. Moreover, the nature of transition is critically tied to the upstream initial conditions. The important features of this problem for one- and two-dimensional basic states have been recently reviewed by Tani [2], Herbert [3], Singer et al. [4], and Saric [5].

With an emphasis on the design of energy efficient airfoils, the past five years have seen a renewed interest in problems of stability and transition in swept-wing flows. These flows are three-dimensional in nature and are, of course, subject to three-dimensional instabilities. As recent symposia and conferences have indicated, there is far more current interest in these problems than ever before. Therefore, the present paper concentrates on work in three-dimensional boundary layers and will attempt to bring up to date the status of the new work in this area.

1.2 Three-Dimensional Boundary-Layer Stability

When a boundary-layer flow is fully three dimensional it exhibits behavior quite different to that of the corresponding two-dimensional flow. Of particular interest are the stability characteristics of these three-dimensional flows where inviscid criteria may produce a stronger instability than the usual Tollmien-Schlichting waves. Examples of 3-D flows of practical interest include swept wings, rotating cones, corners, inlets, and the rotating disk. It appears that these flows exhibit a rich variety of stability behavior that is generic to 3-D boundary layers. A consistent characteristic of the instabilities is the presence of streamwise vorticity within the shear layer. This streamwise structure produces a strong spanwise modulation of the basic state that gives rise to secondary instabilities. The reader should also see Kohama [6], who makes many of the same points in his review of the subject.

The thrusts of this paper are to: (1) illustrate the generic nature of the 3-D flow stability, (2) demonstrate that the breakdown process is very complicated, (3) show that secondary instabilities can produce unexpected results, and (4) justify the use of multiple (independent) techniques for observing stability behavior.

The flow over a swept wing is a typical example of a three-dimensional boundary layer. This type of 3-D flow is susceptible to four types of instabilities that lead to transition. They are leading-edge contamination, streamwise instability, centrifugal instability, and crossflow instability. Leading-edge contamination occurs along the attachment line and is caused by disturbances that propagate along the wing edge (Poll [7-9], Hall et al. [10], Hall and Malik [11]). Streamwise instability is associated with the chordwise component of flow and is quite similar to processes in two-dimensional flows, where Tollmien-Schlichting (T-S) waves generally develop (Mack [12]). This usually occurs in zero or positive pressure-gradient regions on a wing. Centrifugal instabilities occur in the shear flow over a concave surface and appear in the form of Görtler vortices (Floryan and Saric [13], Hall [14]). However, Hall [15] has conclusively shown that the Görtler vortex instability is unimportant in the concave region of swept wings when the angle of sweep is large compared to $Re^{-1/2}$. This situation is easily realized in wings of moderate sweep and thus one would expect crossflow or T-S breakdown of the laminar flow.

Much of the earlier work on stability and transition is contained in the review papers of Mack [12], Arnal [16,17], and Poll [8]. Mack [12] is a monograph on stability theory and is the primary source of basic information on the subject. Arnal [16] concentrates on transition prediction in 2-D flows while Poll [8] covers transition work in 3-D flows. The recent work of Arnal [17] is an extensive review of transition in 3-D flows and as such is complementary to the material of this paper. Because of these excellent reviews, only the most recent papers are quoted here.

In this paper, attention is first focused, in Section 2, on the problems of flow over a swept-wing that exhibit *crossflow instability*. Problems of rotating disks and cones are discussed in Sections 3 and 4. Attachment-line contamination and corner flow problems are important for wing-body problems and are reviewed in Sections 5 and 6. Otherwise, T-S and Görtler instabilities are not discussed as primary instabilities (the literature is vast) but only in terms of secondary instabilities.

2. Swept-Wing Flows

The focus of this section is on the crossflow instability which occurs in strong negative pressure gradient regions on swept wings. In the leading-edge region both the surface and flow streamlines are highly curved. The combination of pressure gradient and wing sweep deflects the inviscid-flow streamlines inboard (as shown in Gregory et al. [18]; Poll [8]; Arnal [17]). This mechanism re-occurs in the positive pressure gradient region near the trailing edge. Because of viscous effects, this deflection is made larger in the boundary layer, and causes *crossflow*, i.e. the development of a velocity component inside the boundary layer that is perpendicular to the inviscid-flow velocity vector. This profile is characteristic of many different three-dimensional boundary-layer flows. The crossflow profile has a maximum velocity somewhere in the middle of the boundary layer, going to zero on the body surface and at the boundary-layer edge. This profile exhibits an inflection point (a condition which is known to be dynamically unstable) causing so-called crossflow vortex structures to form with their axes in the streamwise direction. These crossflow vortices all rotate in the same direction and take on the form of the familiar "cat's eye" structure when viewed in the stream direction. Descriptions of this instability are given in the classic paper by Gregory, Stuart and Walker [18] and in the reports by Mack [12] and Poll [8].

2.1 Stability and Transition Prediction

In the past ten years considerable progress has been achieved in calculating the stability characteristics of 3-D flows. The state-of-the-art transition prediction method still involves linear stability theory coupled with an e^N transition prediction scheme (Mack [12]; Poll [8]). N is the integration of the linear growth rate from the first neutral stability point to a location on the body. Thus e^N is the ratio of the amplitudes at the two points and the method correlates the transition Reynolds number with N. Malik and Poll [19] extend the stability analysis of three-dimensional flows, analyzing the flow over a yawed cylinder, to include curvature of the surface and streamlines. They show that curvature has a very stabilizing effect on the disturbances in the flow. This is compared with the experimental results of Poll [9] which show good agreement with the transition prediction scheme. They, as does Reed [20], also find that the most highly amplified disturbances are traveling waves and not stationary waves. Here again Malik and Poll [19] obtain good agreement with Poll's [9] recent experimental work where Poll identifies a highly amplified traveling wave around one kHz near transition. Malik and Poll obtain N factors between 11 and 12 for the fixed-frequency disturbances which agreed with the work of Malik, Wilkinson and Orszag [21] on the rotating disk. In both cases (the disk and cylinder), when the extra terms involving curvature and Coriolis effects are omitted in the stability analysis, the N factors are much larger which illustrates the need to do the realistic stability calculations.

Michel, Arnal and Coustols [22] develop transition criteria for incompressible two- and three-dimensional flows and in particular for the case of a swept wing with infinite span. They correlate transition onset on the swept wing using three parameters: a Reynolds number based on the displacement thickness in the most unstable direction of flow, the streamwise shape parameter, and the external turbulence level. They simplify the problem by not including curvature effects and assuming locally parallel flow and even with these simplifications, the comparison with experiment shows good agreement.

Arnal, Coustols, and Jelliti [23] suggest a method for calculating the beginning of transition as well as the transition region itself. Their theory includes the influence of sweep angle on transition and discusses laminarization by wall shaping and suction. Michel, Coustols, and Arnal [24] present a transition criteria based on stability theory for streamwise instability, crossflow, and leading edge contamination. They use an intermittency method condition and a mixing length scheme to calculate the transition region which has the advantage over a sudden transition calculation in predicting boundary-layer thickness.

Poll [25] extends the ideas of Poll [26] and generates approximate criteria for transition through a process of physical arguments based on existing knowledge, a number of approximate empirical correlations to describe the process of instability onset and transition onset in a 3-D flow. For the onset of instability, data for correlation purposes is generated by applying linear stability theory to a two-parameter family of crossflow profiles using the height of the inflection point and the second derivative of velocity near the wall. It is found that the minimum critical Reynolds number is strongly dependent on the inflection point location. Computations are performed for typical swept-wing profiles. For transition onset, a simple correlation can be made using only three parameters.

2.2 Transition Experiments

The current experimental work of Poll [9] focuses on the crossflow instability where he shows that increasing yaw has a very destabilizing effect on the flow over a swept cylinder. He characterizes the instability in two ways. The first is by *fixed* disturbances visualized by either surface evaporation or oil-flow techniques. These disturbances are characterized by regularly spaced streaks aligned approximately in the inviscid-flow direction, leading to a "saw-tooth" pattern at the transition location. The second way is with unsteady disturbances in the form of a large-amplitude high-frequency harmonic wave at frequencies near one kHz. At transition near the wall surface, he obtains disturbance amplitudes greater than 20% of the local mean velocity. Initially he tries to use two parameters to predict transition. They are the crossflow Reynolds number and a shape factor based on the streamwise profile. However, based on the results of his research, he found that two parameters alone are not enough to predict transition, and that one needs at least three parameters to accurately describe the crossflow instability.

Michel, Arnal, Coustols and Juillen [27] present some very good experimental results on the crossflow instability, conducted on a swept airfoil model (the complete details of the experimental setup are given in Coustols [28]). By surface visualization techniques they show regularly spaced streaks that are aligned practically in the inviscid-flow direction, with a "saw-tooth" pattern near the transition area. They perform hot-wire measurements on the stationary waves. Their results show a spanwise variation of the boundary layer before transition that becomes chaotic in the transition region. The variations are damped in the turbulent region. They also find a small peak in the spectra around one kHz (like Poll [9]), which is due to a streamwise instability. In addition to this they provide some theoretical work on the secondary velocities, and show counter rotating vortices in the streamwise direction. However, when these components are added to the mean velocities the vortices are no longer clearly visible.

Kohama [29] studies the transition process on a concave-convex wall typical of the lower surface of a Laminar Flow Control supercritical airfoil. Using hot-wire and smoke-wire techniques in the concave region, he observes the Taylor-Görtler instability in the concave region when there is no sweep and the crossflow instability for a sweep angle of 47 degrees. No Görtler vortices were observed in the swept case corresponding to the prediction of Hall [15]. Intermediate angles are not tested.

The design of modern Laminar Flow Control (LFC) transports depends on the prediction of the growth of the various characteristic disturbances using the e^N method. These designs are being carried out using advanced computer codes. Typically, the upper surface of the two-dimensional airfoil is characterized by an extensive supersonic flat-pressure region preceded by a leading-edge negative pressure peak and followed by a gradual shock-free recompression to subsonic flow with a subsequent rear pressure rise of the Stratford type. Consequently, the crossflow instability will dominate fore and aft, while the T-S instability affects the mid-chord region. Pfenninger [30] comments that the classification of the stability problem into independent parts is physically acceptable as long as different strongly amplified disturbances do not occur simultaneously. At this time there is no suitable criterion for establishing transition when both crossflow and T-S waves are present. For the lack of anything better, for engineering design, a linear relationship is assumed between the N-factors for steady crossflow transition (N_{cf}) and T-S transition (N_{TS}). Usually one assumes

$$N_{TS} = 12 - (1.2)\,N_{cf}$$

with some error bands, as a transition criterion. Understanding what occurs when both crossflow and T-S waves coexist is necessary.

2.3 Role of Spanwise Variations

A major unanswered question concerning swept-wing flows (besides the aforementioned steady/unsteady discrepancy between theory and experiment) is the interaction of crossflow vortices with T-S waves. If the vortex structure continues aft into the mid-chord region where T-S waves are amplified, some type of interaction could cause premature transition. In fact, the unsteadiness at transition observed by Poll [9] and Michel et al. [27] could be due to this phenomenon. Indeed early LFC work of Bacon et al. [31] shows a somewhat anomalous behavior of transition when sound is introduced in the presence of crossflow vortices. Klebanoff et al. [32] show that the onset of three-dimensionality is quickly followed by breakdown of the laminar flows, and various instabilities have been found to interact. These interactions have been reviewed for plane channels and boundary layers by Herbert [33].

It is well known that streamwise vortices in a boundary layer strongly influence the behavior of other disturbances. Pfenninger [30] suggests that amplified streamwise vortices produce spanwise periodicity (three-dimensionality) in the boundary layer that causes resonance-like growth of other secondary disturbances.

A series of analyses in 1980 addressed this issue of the destabilizing nature of spanwise periodicity. Nayfeh [34] shows that Görtler vortices produce a double-exponential growth of T-S waves. Herbert and Morkovin (35) show that the presence of T-S waves produces a double-exponential growth of Görtler vortices, while Floryan and Saric (36) show a similar behavior for streamwise vortices interacting with Görtler vortices. Malik [37] gives a good review of these efforts and lays the groundwork for the more general attack of these problems using complete Navier-Stokes solutions to account for secondary distortion of the basic state. For example, Malik [37] in his computational simulation is unable to find the interaction predicted by Nayfeh [34]. Srivastava [38] in a calculation similar to Nayfeh, also shows growth rates lower by an order of magnitude than those predicted by Nayfeh [34]. However, in a recent paper, Nayfeh [39] has clarified the parameter range over which this interaction may occur. The T-S/Görtler problem has been examined in curved channels by Hall and Bennett [40] and Daudpota et al. [41]. This latter paper is a weakly nonlinear analysis and a Navier-Stokes simulation in a region where both amplified Görtler vortices and T-S waves exist. They show that four types of interactions are possible with four stationary states for each interaction. Relative amplitude conditions are given for each type of interaction and final state configuration. The underlying theme of all of these papers is the strongly destabilizing nature of spanwise variations on weakly growing 2-D waves.

In addressing the problem of 3-D boundary layers, Reed [42] analyzes the crossflow/T-S interaction in the leading-edge region by using a parametric-resonance model. Reed shows that the interaction of the crossflow vortices with T-S waves produces a double exponential growth of the T-S waves. [The wavelengths predicted for the given conditions chosen are larger than realistic however, again indicating the need for more investigation.] Reed [20] shows a crossflow/crossflow interaction that is responsible for observations in the experiments of Saric and Yeates [43]. The results of Bacon et al. [31] and Reed [20, 42] clearly show the need to experimentally study problems of this kind. These papers are discussed later in the context of the results from Saric and Yeates [43], Bippes and Nitschke-Kowsky [44,45,46], and Arnal and Jullien [47].

2.4 Stability Experiments

Saric and Yeates [43] established a three-dimensional boundary layer on a flat plate that is typical of infinite swept-wing flows. This is done by having a swept leading edge and contoured walls to produce the pressure gradients. The experimentally measured C_p

distribution is used along with the 3-D boundary-layer code of Kaups and Cebeci [48] to establish the crossflow experiment and to compare with the theory. Some of the results of Saric and Yeates [43] are discussed below because they illustrate that not everything is as it should be in three-dimensional boundary layers.

Boundary-layer profiles are taken at different locations along the plate with both the slant-wire and straight-wire probes. Reduction of both the straight-wire and slant-wire data at one location produces a crossflow profile which provides comparison with the theory. The velocity component perpendicular to the inviscid-flow velocity vector is called the crossflow velocity. By definition, since the crossflow profile is perpendicular to the edge velocity, the crossflow velocity is zero in the inviscid flow.

Disturbance measurements of the mean flow are conducted (Saric and Yeates [43]) within the boundary layer by making a spanwise traverse (parallel to the leading edge) of the hot wire at a constant y location with respect to the plate. These measurements are carried out at many different x and y locations using two different mean velocities. The results show a *steady* vortex structure with a dominant spanwise wavelength of approximately 0.5 cm. The corresponding spectrum for this disturbance measurement shows a sharp peak at a wavelength of about 0.5 cm, but it also shows a broad peak at a larger wavelength of 1.0 cm, generally at a lower amplitude. The cause of this broad peak at the larger wavelength is explained by the linear-theory predictions (Dagenhart [49]) for crossflow vortices. This 0.5 cm wavelength does not agree with the flow-visualization results nor with the theoretical calculations of the MARIA code (Dagenhart [49]).

The naphthalene flow-visualization technique shows that there exists a steady crossflow vortex structure on the swept flat plate. The pattern of disturbance vortices is nearly equally spaced and aligned approximately in the inviscid-flow direction. The wavelength of the vortices is on the scale of 1 cm and this spacing agrees quite well with the calculated wavelength from the MARIA code.

While the flow visualization clearly indicates a spanwise wavelength of 1 cm on the surface, the spectra of the hot-wire measurements show a dominant sharp peak at 0.5 cm and a smaller broad-band peak at 1 cm. This apparent incongruity can be explained with the wave interaction theory of Reed [20], who uses the actual test conditions of this experiment. Reed shows that it is possible for a parametric resonance to occur between a previously amplified 0.5 cm vortex and a presently amplified 1 cm vortex and that measurements taken near the maximum of the crossflow velocity would show a strong periodicity of 0.5 cm. Moreover, Reed's wall-shear calculations and streamline calculations show the 0.5 cm periodicity dying out near the wall and the 1 cm periodicity dominating. These phenomena are not observed by Michel et al. [27] who measure phenomena not measured by Poll nor Saric and Yeates.

Two important points need to be emphasized. First, one must, whenever possible, use multiple independent measurements. This was the only way the 0.5 cm and the 1.0 cm vortex structure could be reconciled. Second, the steadiness of the vortex structure in the wind tunnel experiments in contrast to the unsteady predictions of the theory, indicates that some characteristic of the wind tunnel is fixing the vortex structure. This is directly analogous to the biasing of the K-type secondary instability in channel flow (Singer et al. [4]).

Nitschke-Kowsky [44] finds traveling waves in the initial state of the boundary layer on a swept flat-plate model with an imposed negative pressure gradient. The measured frequencies are accurately predicted by linear theory as the most amplified (Bieler and Dallmann [50]). In further experiments, Bippes [45] and Bippes and Nitschke-Kowsky [46] find that the traveling waves propagate in a unique direction which is different from that of the mean flow and the coexisting stationary waves. The traveling waves are found to originate in the same chordwise location as the first appearance of crossflow instabilities. Initially only the frequency range can be established, whereas further downstream the direction of propagation is determined. Under certain conditions, two frequency ranges are amplified and propagate in different directions.

Arnal and Juillen [47] describe the recent transition studies at ONERA/CERT, in particular, the hot-wire and hot-film measurements on two different swept-wing configurations. Crossflow vortices are visualized with characteristics in agreement with linear stability theory. Traveling waves are observed in the nonlinear range prior to transition. In further work, interactions between crossflow and streamwise instabilities are studied. Small amplitude, high frequency oscillations are found superimposed on the crossflow. Other work reported includes the development of transition criteria including the effects of freestream turbulence.

What we see from independent work at three different facilities is that there is no rule regarding the appearance of steady or unsteady crossflow vortices. The wave-doubling of Saric and Yeates [43] was not observed elsewhere. Perhaps some very weak freestream vorticity or roughness is providing the fix for the crossflow vortex structure. Certainly the unsteadiness that is observed by everyone just prior to transition is due to a secondary instability. Perhaps some guidance in this area can come from extensions of the work of Reed [20,42], Malik [37], and Fischer and Dallmann [51]. All of this serves notice that stability and transition phenomena are *extremely* dependent on initial conditions.

2.5 Interaction Theory

To analytically model the unsteady crossflow instability and interactions, a three-dimensional analysis based on small-disturbance theory and Floquet theory is used. The undisturbed state consists of the leading-edge boundary-layer flow over a swept wing with wall mass and heat transfer, the solution of which is provided by the code of Kaups and Cebeci [48].

The linear disturbance equations which govern the shape and variation of the crossflow vortices are solved with the normal-mode assumption. In particular, the eigenvalue problem provides the dispersion relation relating frequency to streamwise and spanwise wavenumber. For crossflow vortices the wave angle $P = \tan^{-1}(b_{V_r}/a_{V_r})$ is a few degrees less than ninety degrees with respect to the local freestream direction. [Here, $\{a_{V_r}, b_{V_r}\}$ is the real part of the {chordwise, spanwise} component of the crossflow wavenumber vector (a_V, b_V).] The condition of group velocity being real is also satisfied. A collocation method employing Chebyshev polynomials is used to solve the eigenvalue problem. In our analysis, unsteady crossflow vortices with frequency f_V are found to be most unstable.

The growth of secondary instabilities in the presence of finite-amplitude A_V, unsteady crossflow vortices is considered in a study of parametric resonance. Onto the basic state (that is, the undisturbed three-dimensional boundary-layer flow with a superposed flow corresponding to unsteady streamwise vortices) two infinitesimal-amplitude A_T, oblique, traveling, harmonic waves are superposed. For finite-amplitude crossflow disturbances, it is assumed that $O(A_T) < O(A_V) < O(1)$, nonlinear distortion of the vortices is neglected, weak variation of crossflow wave amplitude on a viscous time scale is ignored, and Floquet theory is applied. The almost periodic form of the basic state allows normal-mode solutions for the secondary instability. The quantities a_i, b_i, f_i, $i=1,2$ are the chordwise wavenumber, spanwise wavenumber, and frequency of the two waves, respectively. For resonance then $a_{1r}=a_{V_r}+a_{2r}$, $b_{1r}=b_{V_r}+b_{2r}$, and $f_{1r}=f_{V_r}+f_{2r}$.

Results in the leading-edge region of a swept wing and for the experimental conditions above indicate a crossflow/crossflow interaction; that is, the primary unsteady crossflow vortices in the basic state interact with secondary disturbances of half the primary wavelength. Moreover, the wave angle associated with the amplified secondary disturbances is also in the crossflow direction ([Reed [20]). These findings along with calculations of wall shear stress (Reed [20]) explain the anomalies in the experiments of Saric and Yeates [43]. The theory is also supported by the Navier-Stokes calculations of Malik [37] who shows a wave doubling in the rotating disk problem.

Fischer and Dallmann [51] apply a theory of secondary instability by Floquet methods to flow over a swept wing. Considering a primary zero-frequency disturbance of sufficient amplitude, secondary-disturbance oblique waves become dominant and possibly play an important role in the transition process. They find a significant influence of higher harmonics on the fundamental secondary waves.

2.6 Summary

Detailed crossflow stability experiments are few in number, the most recent being the work reviewed in section 2.4. Basic research is still required as the nature of crossflow vortices is not completely understood at this point. Pertinent unanswered questions and puzzles include: For a given flow configuration (that is, Reynolds number, Mach number, sweep angle, disturbance amplitude level,...), what range of three-dimensional disturbances is most unstable and what kind of interactions between disturbances are possible? What are the mechanisms involved in the interactions? For some flow conditions (for example leading edge flows), are theory-predicted-unsteady disturbances possible in an experimental environment or is there a selection mechanism peculiar to each facility that forces the disturbances to be steady? Is three-dimensional theory well-posed; are growth rates and wavelengths accurately predicted? When interactions are involved, how do three-dimensional disturbances propagate (group velocity ratio, initial conditions?); how does one compute amplification factors? What are the effects of compressibility, curvature, and nonparallelism on three-dimensional disturbances? How are transition phenomena in three dimensions related to the formation of three dimensional structures in two-dimensional boundary-layer transition? In three-dimensional experiments how does one introduce controlled three-dimensional disturbances? All of these questions are important because the design of modern LFC airfoils depends on the accurate prediction of disturbance growth. For instance, natural-laminar-flow airfoils are particularly susceptible to the crossflow instability because of their dependence on pressure-gradient tailoring for transition delay. Furthermore, passive controls such as suction (e.g. Saric [52]), are known to be more effective in controlling the familiar Tollmien-Schlichting instability than crossflow due to the different natures of the instabilities (viscous versus inflectional). Consequently, accurate control-system power requirements can be determined and optimized only after the nature of the instability is fully understood.

3. Rotating Disk

A model problem exhibiting the same rich variety of instabilities as the swept wing is the rotating disk. As the disk spins, flow moves in axially, a three-dimensional boundary layer builds up on the surface, and the fluid is cast off the edge like a centrifugal pump. The boundary layer is of constant thickness allowing simpler applications of theory, experiment, and computations. Much of our knowledge of crossflow has and will continue to develop from study of the disk.

The crossflow instability exists on the disk and co-rotating vortices form, spiraling outward with their axes along logarithmic spirals of angle 90°+e with respect to the radius of the disk. Typically e is from 11 to 14 deg. Gregory et al. [18] visualized the vortex spirals by the wet-china-clay technique and Federov et al. [53] by naphthalene sublimation. The latter group, however, found e to be 20 deg, contrary to all other reports. Wilkinson and Malik [54] and Kobayashi et al. [55] did detailed hot-wire traces and found the vortices to be steady with respect to the disk.

Wilkinson and Malik [54] fixed the vortex position close to the disk center by a single roughness element and showed that steady wave patterns emanate from point sources on the disk. It is pointed out by Morkovin [56] that china clay, naphthalene, and roughness would tend to "favor fixed, steady patterns over regular and irregular moving patterns". (See the discussion of Mack [57] below.)

For the rotating disk, Malik et al. [21] calculate temporal eigenvalues, which they convert to spatial using a group-velocity transformation (Gaster [58]). They then calculate N-factors using the real part of the group velocity. They included curvature and showed that streamline curvature and Coriolis forces have a very stabilizing effect on disturbances. They obtain an N-factor of approximately 11 in the transition correlation scheme which seems quite reasonable and matches the N-factors of two-dimensional flows for transition. When the extra terms, involving curvature and Coriolis effects, were omitted from the stability analysis, the N-factors were much larger (on the order of 17) and initially placed doubt on the validity of the transition prediction method. They also show that for the rotating disk, fixed disturbances produce the highest amplification rates in agreement with the experiments. According to them, for rotating disks there is no discrepancy between theory and experiment regarding steadiness or unsteadiness being characteristic of the most amplified disturbance. However, Mack [57] finds traveling waves to have higher amplification rates than stationary waves.

Defining a Reynolds number by $r(W/v)^{1/2}$, where r is the radius, W is the angular speed, and v is the kinematic viscosity, the experimental critical value below which all small disturbances dampen has been reported anywhere between 280 and 530 for vortices observed with e=14 deg. Federov et al. [53] reported a Reynolds number range of 182 to 242 for his e=20 deg vortices. Differences in surface roughness and detection technique (hot wire, flow visualization, acoustic detection) among experiments are believed responsible for this variation in the reports. Transition occurs between 500 and 560.

The hot-wire measurements of Wilkinson and Malik [54] show the critical Reynolds number to be around 280. Early linear stability analyses do not account for curvature effects and consequently are not able to predict the experimental results. When these effects are included in a spatial analysis (Malik et al. [21]), the predicted Reynolds number is 287, in good agreement with experiment. In an independent temporal analysis including curvature terms, Kobayashi et al. [55] find a value of 261.

Malik [59] calculates the neutral curve more accurately for stationary disturbances including the effects of streamline curvature and Coriolis forces. He finds the minimum critical Reynolds number of 285.36 in agreement with the results of Malik et al. [21] and a vortex angle of 11.4 deg at the critical point. He also notes a second minimum on the lower branch. The associated vortex angle is 19.45 deg, similar to that of Federov et al. [53]. The upper branch can be associated with the asymptotic solution of Stuart (Gregory et al. [18]), the lower branch is associated with the wave angle corresponding to the direction of zero mean wall shear.

Hall [60] investigates stationary instabilities asymptotically. He finds, in addition to the "inviscid mode found by Gregory et al.", a "stationary short-wavelength mode" whose structure is "fixed by a balance between viscous and Coriolis forces and cannot be described by an inviscid theory". His procedure takes nonparallel effects into account. He finds good agreement with Malik [59] in the high-Reynolds-number limit and concludes that his theory could be a useful tool in finding structures in general three-dimensional boundary layers.

A major contribution to the understanding of the rotating-disk flow is that of Mack [57] who, following Gaster [58], studied the stability characteristics of these vortices theoretically by assuming a white-spectrum, zero-frequency source distribution over Wilkinson and Malik's [54] area of roughness. He then let the disturbance differential equations filter and amplify the spectrum into a wave-interference pattern that turns out to be very similar to the pattern observed by Wilkinson and Malik [54]. The critical Reynolds number predicted by his theory again depended on whether or not curvature was included. With curvature he found a value of 273, and the detailed characteristics of his results are in excellent agreement with experiment. The significant conclusion from this work is that surface roughness (even the minutest of particles) appears to be a strong forcing agent for streamwise vorticity fueling the crossflow, whereas other effects such as freestream vorticity do not seem to be as crucial (Morkovin [56]). Various wave patterns merge together that form the pattern seen in flow visualization. The pattern is "the result of the superposition of the entire spectrum of normal modes of zero frequency, both amplified and damped" (Mack [57]).

In an experiment using hot wires and companion flow visualization, Kohama [61] finds disagreement with the statement of Malik et al. [21] that the number of vortices (n) is a function of rotation Reynolds number that increases linearly. He reports only a very slight increase in n that does not agree with the formula provided by Malik et al. With a trip wire inserted, he also finds that the flow is fully turbulent outboard of the wire.

Concerning the study of nonlinear stability and interactions of waves on the rotating disk, Malik [37] used a Fourier-Chebyshev spectral method in a Navier-Stokes simulation. He was able to find the crossflow/crossflow (second harmonic) interactions similar to those predicted by Reed [20] and found experimentally by Saric and Yeates [43] in swept-wing flows. Itoh [62] predicts the same for the disk in an independent study using weakly nonlinear theory. In an experimental investigation, Kohama [61,63] finds ring-like vortices on the surface of each spiral vortex.

Some puzzles remain to be solved regarding rotating-disk stability. Experimenters have found different spiral angles, critical Reynolds numbers, and number of vortices around the disk. The theory of Itoh [62] suggests a rather wide range of wavenumbers with positive amplification rates. The rotating-disk experiments of Federov et al. [53] show vortex spirals at angles, spacings, and Reynolds numbers not given by any linear theory. Moreover, traveling waves (not observed experimentally) have higher amplification rates than stationary waves according to theory, roughness appears to play some kind of major role, and the importance of secondary instabilities is unclear. More basic research is obviously required for complete understanding.

4. Rotating Cone

Of related interest is the flow over a spinning axisymmetric body such as a cone. Again one observes streamwise vorticity, spiral streaks in the transition region of the boundary layer (Mueller et al. [64], Kobayashi et al. [65], Kobayashi and Kohama [66], and Kohama [67]). These same spirals were observed by Kohama and Kobayashi [68] for rotating spheres. For a spinning ogive nose cone in a uniform flow, Mueller et al. speculate that this is the result of crossflow instability but when illuminated with a slit strobe light, Kohama finds the vortices to be counter-rotating, as do Kobayashi et al. by hot-wire measurements. Kobayashi and Izumi [69] find counter-rotating vortices on the rotating cone in a still fluid by companion computations and flow visualization.

When the cone is rotating in a quiescent fluid, Kohama and Kobayashi [70] observe counter-rotating vortices (dominant centrifugal instability) for total included angles less than 60 deg down to 0 deg (rotating cylinder) and co-rotating vortices (dominant crossflow instability) for angles greater than 60 deg up to 180 deg (rotating disk). In a uniform flow this changeover angle is observed to be smaller. A similar result is found by Kobayashi and Izumi [69]. When the boundary-layer velocity profile is measured on a 30 deg cone, no inflectional point can be found [71].

Kobayashi [72] pioneered the theoretical studies of rotating cones. For rotating cones in axial flow, Kobayashi et al. [65] compare theory with their flow visualizations and hot-wire measurements and report qualitative consistency between the two for critical Reynolds number, spiral-vortex-axis direction, and number of vortices (theoretical values are smaller than experiment). For increasing rotational speed, critical Reynolds number decreases. For rotating cones in still flow, Kobayashi and Izumi [69] carry out a linear stability analysis and find critical Reynolds number, spiral-vortex direction, and number to increase and approach rotating-disk values as the included angle is increased.

Considering secondary instabilities, downstream of initially straight vortices, Kohama [67, 73] observes horseshoe patterns along the axes of the spiral vortices on an ogive nose cone, as does Kohama [71] on a cone. The horseshoes originate on the surface where large velocity gradients are expected (associated with inflection points created by the vortices) and travel along the spirals with some phase speed relative to the wall. Eventually they break down to turbulence. These patterns are attributed to a T-S secondary instability. (In their photographs, Kobayashi et al. [65] and Kobayashi and Izumi [69] show these patterns but offer no comments.)

5. Other Geometries

Besides swept wings and rotating bodies, other complex geometries with three-dimensional boundary layers have been investigated both experimentally and theoretically. Axisymmetric bodies at incidence provide a particularly interesting group of such flows. Eichelbrenner and Michel [74] studied transition experimentally on an ellipsoid at various angles-of-attack. At angle-of-attack, the transition front was modified by three-dimensional effects. Meier et al. [75] studied a prolate spheroid at incidence experimentally and Lekoudis and Kinard [74] studied the conditions of Meier et al. [76] using linear stability theory. Here, at low angle-of-attack, the instability is basically dominated by a two-dimensional instability, while at higher incidence, the crossflow becomes stronger and dominates the transition process. Transition prediction schemes have been considered by Gleyzes et al. [77], Jelliti [78], Cebeci [79], and Arnal et al. [80].

Another relevant problem in three-dimensional flow stability is that involved with the intersection of surfaces, e.g. wing/body junctures, corners, and inlets. In these situations, boundary layers from different surfaces merge to produce a three-dimensional flow and it is expected that many of the same characteristics as prescribed above for other three-dimensional boundary-layer flows occur here also. The most recent analysis of this problem is contained in Lakin and Hussaini [81]. Their linear analysis is a fundamental attempt to tackle the corner-flow problem which is made extremely formidable by difficulties in solving the basic state. Moreover, there is essentially no body of experimental knowledge of stability in corners and junctures. It appears that good prototypes for wing/body junction, inlet, and wingtip flows must be found if progress is to be made in this area.

6. Attachment-Line Stability

For swept wings, disturbances produced in corners may propagate along the leading edge and affect stability elsewhere, giving rise to so-called leading-edge contamination. This problem was first investigated experimentally by Gaster [82] and Pfenninger and Bacon [83]. Recent experimental work of Poll [7,8] focuses strictly on the leading-edge contamination problem. Poll [7] studies the effect of placing trip wires normal to the attachment line. He defines a length scale h given by

$$h = [\, v \,/\, (dU_e/dx)_{x=0} \,]^{1/2}$$

where U_e is the edge velocity in the streamwise direction, v is the kinematic viscosity and x is in the chord direction. He uses the parameter d/h where d is the diameter of the wire. From this parameter he determines a maximum ratio of 1.55 below which the wire feeds disturbances along the attachment line until turbulent bursts occur. Above this value he discovers that the wire introduces turbulent bursts directly at the trip wire. There is a value of d/h equal to 0.8 or less where the wire has no effect on the transition process. Poll's [8] recent work defines a Reynolds number based on the edge velocity V_e parallel to the leading edge and the length scale mentioned above. Based on this definition he obtains a critical Reynolds number of 250, below which propagation of disturbances along the attachment line does not occur.

Recently, Hall, Malik, and Poll [10] and Hall and Malik [11] have attacked this problem with a linear stability solution and both a weakly nonlinear solution and a complete Navier-Stokes simulation. Although they considered only a 2-D instability, they were able to predict and explain the experimental data of Pfenninger and Bacon [83]. Of particular interest is that they were able to explain the absence of upper branch neutral stability modes as being due to a subcritical bifurcation along most of the upper branch. They also hint at the idea that the stabilizing effects of suction predicted from linear theory may not hold here because the suction may lead to a larger band of nonlinear unstable modes. This work is important because it is another illustration of the power of combined analysis and computation dedicated to experimental results. It lays the foundation for more detailed experimental work and the extension of the theory to three dimensional disturbances.

7. Conclusions

Three-dimensional flows all exhibit similar characteristics (streamwise vorticity) and all appear to depend heavily on initial conditions. Here we have reviewed the current knowledge for some basic flows and have seen that many questions remain to be solved. The encouraging aspect of this is the increased research effort in this area that is perhaps prompted by energy efficient aircraft.

We observed that it is possible that "disturbance sources" such as roughness could favor the stationary mode, the end conditions could inhibit traveling disturbances in the experiments, or the theory could be inadequate. There is room for more Navier-Stokes simulations that could be used closely with carefully controlled experiments and perhaps offer explanations such as those provided by Singer et al. [4] in the two-dimensional case.

In general, it is uncertain how to define transition in three dimensions. Clearly, the attempts at transition correlation are just necessary stop-gap measures to permit the designer to carry out his work. A firm understanding of transition will not come from these efforts. The details of the transition process in three dimensions are still missing. For example, it is important to know whether in three dimensions the formation of three-dimensional structures (K-type and H-type) and characteristic stages (e.g. 1-spike stage) will occur as in two-dimensional flow transition. For controlled experiments it is even unclear how to introduce controlled disturbances into the three-dimensional boundary layer and whether these can be made representative "modes" for "natural transition." The possibilities of and mechanisms for secondary instabilities remain to be determined, and we have not even begun to discuss the roles of unsteadiness, strong pressure gradients, incipient separation, and laminar/turbulent separation bubbles.

Acknowledgments

This work is supported under NASA Grants NAG-1-280, NAG-1-402, and NAG-1-731 from NASA-Langley Research Center.

References

1. Morkovin, M.V.: On the many faces of transition. *Viscous Drag Reduction* ed: C.S. Wells, Plenum, 1969.

2. Tani, I.: Three-dimensional aspects of boundary-layer transition. *Proc. Indian Acad. Sci., vol. 4,* (1981) 219.

3. Herbert, T.: Three-dimensional phenomena in the transitional flat-plate boundary layer. *AIAA Paper No. 85-0489*, 1985.

4. Singer, B.A.; Reed, H.L.; Ferziger, J.H.: Effect of streamwise vortices on transition in plane-channel flow. *AIAA Paper No. 87-0048*, 1987.

5. Saric, W.S.: Boundary-layer transition to turbulence: the last five years. *Proc. Tenth Symposium on Turbulence,* Univ. Missouri-Rolla, September, 1986.

6. Kohama, Y.: Some similarities in the breakdown process of the primary instability between 2-D and 3-D boundary layers. submitted to: *Phys. Chem. Hydrodynamics,* 1987.

7. Poll, D.I.A.: Transition in the infinite swept attachment line boundary layer. *Aeronautical Quart., Vol. XXX,* (1979) 607.

8. Poll, D.I.A.: Transition description and prediction in three-dimensional flows. *AGARD Report No. 709* (Special course on stability and transition of laminar flows) VKI, Brussels, 1984.

9. Poll, D.I.A.: Some observations of the transition process on the windward face of a long yawed cylinder. *J. Fluid Mech., Vol. 150,* (1985) 329.

10. Hall, P.; Malik, M.R.; Poll, D.I.A.: On the stability of an infinite swept attachment-line boundary layer. *Proc. R. Soc. Lond., A395,* (1984) 229.

11. Hall, P.; Malik, M.R.: On the instability of a three-dimensional attachment-line boundary layer: weakly nonlinear theory and a numerical approach. *J. Fluid Mech., Vol. 163,* (1986) 257.

12. Mack, L.M.: Boundary-layer linear stability theory. *AGARD Report No. 709* (Special course on stability and transition of laminar flows) VKI, Brussels, 1984.

13. Floryan, J.M.; Saric, W.S.: Stability of Görtler vortices in boundary layers. *AIAA J., Vol. 20,* (1982) 316.

14. Hall, P.: The linear development of Görtler vortices in growing boundary layers. *J. Fluid Mech., Vol. 130,* (1983) 41.

15. Hall, P.: The Görtler vortex instability mechanism in three-dimensional boundary layers. *Proc. R. Soc. Lond., A399,* (1985) 135.

16. Arnal, D.: Description and prediction of transition in two-dimensional incompressible flow. *AGARD Report No. 709* (Special course on stability and transition of laminar flows) VKI, Belgium, 1984.

17. Arnal, D.: Three-dimensional boundary layers: laminar-turbulent transition. *AGARD Report* (Special course on Calculation of three-dimensional boundary layers with separation) VKI, Belgium, 1986.

18. Gregory, N.; Stuart, J.T.;and Walker, W.S.: On the stability of three-dimensional boundary layers with applications to the flow due to a rotating disk. *Phil. Trans. R. Soc. Lond., A248,* (1955) 155.

19. Malik, M.R.; Poll, D.I.A.: Effect of curvature on three-dimensional boundary layer stability. *AIAA Paper No. 84-1672*, 1984.

20. Reed, H.L.: Disturbance-wave interactions in flows with crossflow. *AIAA Paper No. 85-0494*, 1985.

21. Malik, M.R.; Wilkinson, S.P.; Orszag, S.A.: Instability and transition in rotating disk flow. *AIAA J., Vol.19,* (1981) 1131.

22. Michel, R.; Arnal, D.; Coustols, E.: Stability calculations and transition criteria in two- or three-dimensional flows. *Laminar-Turbulent Transition,* ed. V.V. Kozlov, Springer, 1985.

23. Arnal, D.; Coustols, E.; Jelliti, M.: Transition en tridimensionnel et laminarisation de la couche limite sur une aile en fleche. *22eme Colloque d'Aerodynamique Appliquee,* 1985.

24. Michel, R.; Coustols, E.; Arnal, D.: Calculs de transition dans les ecoulements tridimensionnels. *Symposium on Numerical and Physical Aspects of Aerodynamic Flows, TP 1985-7,* 1985.

25. Poll, D.I.A.: Approximate criteria for instability and transition in crossflow dominated boundary layers. *AIAA Paper No. 87-1339,* June 1987.

26. Poll, D.I.A.: On the effects of boundary-layer transition on a cylindrical afterbody at incidence in low-speed flow. *Aero. J. Roy. Aero. Soc.,* October 1985.

27. Michel, R.; Arnal, D.; Coustols, E.; Juillen, J.C.: Experimental and theoretical studies of boundary-layer transition on a swept infinite wing. *Laminar-Turbulent Transition,* ed. V.V. Kozlov, Springer, 1985.

28. Coustols, E.: Stabilite et transition en encoulement tridimensionnel: cas des ailes en fleche. *Doctorate thesis,* L'Ecole Nationale Superieure de L'Aeronautique et de l'Espace, 1983.

29. Kohama, Y.: Three-dimensional boundary-layer transition on a concave-convex curved wall.*Proc. IUTAM Symposium on Turbulence Management and Relaminarization,* Bangalore, 19-23 Jan, 1987.

30. Pfenninger, W.: Laminar flow control - Laminarization. *AGARD Report No. 654,* (Special Course on Drag Reduction), 1977.

31. Bacon, J.W. Jr.; Pfenninger, W.; Moore, C.R.: Influence of acoustical disturbances on the behavior of a swept laminar suction wing. Northrup Report NOR-62-124 (1962).

32. Klebanoff, P.S.; Tidstrom, K.D.; Sargent, L.M.: The three-dimensional nature of boundary-layer instability. *J. Fluid Mech., Vol. 12,* (1962) 1.

33. Herbert, T.: Analysis of secondary instabilities in boundary layers. *Proc. Tenth U.S. National Congress of Applied Mechanics,* Univ. Texas, June 1986.

34. Nayfeh, A.H.: Effect of streamwise vortices on Tollmien-Schlichting waves. *J. Fluid Mech., Vol. 107,* (1981) 441.

35. Herbert, T.; Morkovin, M.V.: Dialogue on bridging some gaps in stability and transition research. *Laminar-Turbulent Transition,* ed: R. Eppler and H. Fasel, Springer, 1980.

36. Floryan, J.M.; Saric, W.S.: Wavelength selection and growth of Görtler vortices. *AIAA J., Vol. 22,* (1984) 1529.

37. Malik, M.R.: Wave interactions in three-dimensional boundary layers. *AIAA Paper No. 86-1129*, 1986.

38. Srivastava, K.M.: Effect of streamwise vortices on Tollmien-Schlichting waves in growing boundary layers. *DFVLR Report IB 221-85-A-07*, March 1985.

39. Nayfeh, A.H.: Influence of Görtler vortices on Tollmien-Schlichting waves. *AIAA Paper No. 87-1206*, June, 1987.

40. Hall, P.; Bennett, J.: Taylor-Gortler instabilities of Tollmien-Schlichting waves and other flows governed by the interactive boundary-layer equations. *J. Fluid Mech., Vol.171*, (1986) 441.

41. Daudpota, Q.I.; Zang, T.A.; Hall, P.: Interaction of Görtler vortices and Tollmien-Schlichting waves in a curved channel flow. *AIAA Paper No. 87-1205*, June, 1987.

42. Reed, H.L.: Wave interactions in swept-wing flows. *AIAA paper no. 84-1678*, 1984.

43. Saric, W.S.; Yeates, L.G.: Experiments on the stability of crossflow vortices in swept-wing flows. *AIAA Paper No. 85-0493*, 1985.

44. Nitschke-Kowsky, P.: Experimentelle untersuchungen zu stabilitat und umschlag dreidimensionaler grenzschichten. *DFVLR-FB 86-24*, 1986.

45. Bippes, H.: Hot-wire measurements in an unstable three-dimensional boundary layer. *DFVLR IB 222-86 A 31*, 1986.

46. Bippes, H.; Nitschke-Kowsky, P.: Experimental study of instability modes in a three-dimensional boundary layer. *AIAA Paper No. 87-1336*, June 1987.

47. Arnal, D.; Juillen, J.C.: Three-dimensional transition studies at ONERA/CERT. *AIAA Paper No. 87-1335*, June 1987.

48. Kaups, K.; Cebeci, T.: Compressible laminar boundary layers with suction on swept and tapered wings. *J. Aircraft, Vol. 14*, (1977) 661.

49. Dagenhart, J.R.: Amplified crossflow disturbances in the laminar boundary layer on swept wings with suction. *NASA TP-1902*, 1981.

50. Bieler, H.; Dallmann, U.: Prediction and analysis of primary instability of a three-dimensional swept plate boundary layer. *AIAA Paper No. 87-1337*, June 1987.

51. Fischer, T.M.; Dallmann, U.: Theoretical investigation of secondary instability of three-dimensional boundary layers flows. *AIAA Paper No. 87-1338*, June 1987.

52. Saric, W.S.: Laminar Flow Control with suction: theory and experiment. *AGARD Report No. 723* (AGARD Special Course on Aircraft Drag Prediction and Reduction), VKI, Belgium, 1985.

53. Federov, B.I.; Plavnik, G.Z.; Prokhorov, I.V.; Zhukhovitskii, L.G.: Transitional flow conditions on a rotating disk. *J. Eng. Phys., Vol.31*, (1976) 1448.

54. Wilkinson, S.P.; Malik, M.R.: Stability experiments in rotating-disk flow. *AIAA Paper No. 83-1760*, 1983.

55. Kobayashi, R.; Kohama, Y.; Takamadate, Ch.: Spiral vortices in boundary-layer transition regime on a rotating disk. *Acta Mechanica, Vol. 35,* (1980) 71.

56. Morkovin, M.V.: *Guide to Experiments on Instability and Laminar-Turbulent Transition in Shear Layers,* in press, 1987.

57. Mack, L.M.: The wave pattern produced by point source on a rotating disk. *AIAA Paper No. 85-0490,* 1985.

58. Gaster, M.: A theoretical model of a wave packet in the boundary layer on a flat plate. *Proc. R. Soc. Lond., A347,* (1975) 271.

59. Malik, M.R.: The neutral curve for stationary disturbances in rotating-disk flow. *J. Fluid Mech., Vol. 164,* (1986) 275.

60. Hall, P.: An asymptotic investigation of the stationary modes of instability of the boundary layer on a rotating disc. *Proc. R. Soc. Lond., A406,* (1986) 93.

61. Kohama, Y.: Crossflow instability in rotating disk boundary layer. *AIAA Paper No. 87-1340,* June 1987.

62. Itoh, N.: Stability calculations of the three-dimensional boundary-layer flow on a rotating disk. *Laminar-Turbulent Transition,* V.V.Kozlov (ed.),Springer, 1985.

63. Kohama, Y.: Study on boundary-layer transition of a rotating disk. *Acta Mechanica, Vol. 50,* (1984) 193.

64. Mueller, T.J.; Nelson, R.C.; Kegelman, J.T.; Morkovin, M.V.: Smoke visualization of boundary-layer transition on a spinning axisymmetric body. *AIAA Paper No. 81-4331,* 1981.

65. Kobayashi; R., Kohama, Y.; Kurosawa, M.: Boundary-layer transition on a rotating cone in axial flow. *J. Fluid Mech., Vol. 127,* (1983) 341.

66. Kobayashi, R.; Kohama, Y.: Vortices in boundary-layer transition on a rotating cone. *Laminar-Turbulent Transition,* ed. V.V. Kozlov, Springer, 1985.

67. Kohama, Y.: Flow structures formed by axisymmetric spinning bodies. *AIAA J., Vol. 23,* (1985) 1445.

68. Kohama, Y.; Kobayashi, R.: Boundary-layer transition and the behavior of spiral vortices on rotating spheres. *J. Fluid Mech., Vol. 137,* (1983) 153.

69. Kobayashi, R.; Izumi, H.: Boundary-layer transition on a rotating cone in still fluid. *J. Fluid Mech., Vol. 127,* (1983) 353.

70. Kohama, Y.; Kobayashi, R.: Behavior of spiral vortices on rotating axisymmetric bodies. Report of Institute of High Speed Mechanics, Tohoku University, Japan, *Vol. 47,* (1983) 27.

71. Kohama, Y.: Behavior of spiral vortices on a rotating cone in axial flow. *Acta Mechanica, Vol. 51,* (1984) 105.

72. Kobayashi, R.: Linear stability theory of boundary layer along a cone rotating in axial flow. *Bull. Japan Soc. Mech. Eng., 24,* (1981) 934.

73. Kohama, Y.: Turbulent transition process of the spiral vortices appearing in the laminar boundary layer of a rotating cone. *Phys. Chem. Hydrodynamics, Vol. 6, No. 5*, (1985) 659.

74. Eichelbrenner, E.; Michel, R.: Observations sur la transition laminaire-turbulent en trois dimensions. *La Recherche Aeronautique, No.65*, 3-10, 1958.

75. Meier, H.U.; Kreplin, H.P.; Vollmers, H.: Development of boundary layer and separation patterns on a body of revolution at incidence. *2nd Symposium on Numerical and Physical Aspects of Aerodynamic Flows*, Long Beach, CA, 1983.

76. Lekoudis, S.G.; Kinard, T.: The stability of the boundary layer on an ellipsoid at angle of attack. *AIAA Paper No. 86-1045*, 1986.

77. Gleyzes, C.; Cousteix, J.; Aupoix, B.: Couches limites tridimensionnelles sur des corps de type fuselage. *Rapport Technique OA No. 4/5025* AYD, ONERA,1985.

78. Jelliti, M.: Thesis in preparation.

79. Cebeci, T.: Problems and opportunities with three-dimensional boundary layers. in: *Three-dimensional boundary layers AGARD Report No. 719*, 1984.

80. Arnal, D.; Coustols, E.; Juillen, C.: Experimental and theoretical study of transition phenomena on an infinite swept wing. *Rech. Aerosp. 1984-4* (1984).

81. Lakin, W.D.; Hussaini, M.Y.: Stability of the laminar boundary layer in a streamwise corner. *Proc. R. Soc. Lond., A393*, (1984) 101.

82. Gaster, M.: On the flow along swept leading edges. *Aeronaut. Q., Vol. XVIII,* (1967) 165.

83. Pfenninger, W.; Bacon, J.W.: Amplified laminar boundary layer oscillations and transition at the front attachment line of a 45° flat-nosed wing with and without boundary layer suction. *Viscous Drag Reduction*, ed. C.S. Wells, Plenum, 1969.

Coherent Structures in Turbulent Boundary Layers

DONALD COLES

Graduate Aeronautical Laboratories
California Institute of Technology
Pasadena, CA 91125

Summary

Research on coherent structure in boundary layers has identified characteristic signatures of several dynamic processes associated with two scales, an overall scale and a sublayer scale. The most useful variables are vorticity and shearing stress. After a brief review of this research, a preliminary report is given on work at GALCIT on a flow having a controlled pattern for the large eddies.

Introduction

If the concept of coherent structure is ever to lead to any substantial increase in understanding of turbulent flow, this concept must be consistent with the presently accepted description of such flows in terms of the Reynolds-averaged equations and the associated similarity laws. These similarity laws for the boundary layer have developed historically as a description of the mean velocity profile, presumably because it was relatively easy a few decades ago to measure velocity but relatively difficult to measure shearing stress in the interior of a flow. The absence of similarity laws for the shearing stress is a manifest obstacle to orderly development of the phenomenology of the subject. Moreover, it is this stress whose description will be most affected by identification of mechanisms connected with coherent structure.

Several definitions of coherent structure have been proposed by different writers. Among them are

Coles [1] "Coherent structure can be any flow pattern which survives the operation of ensemble averaging over realizations having a common phase reference ... in some suitable moving frame." (Emphasis in practice on celerity and topology.)

Blackwelder [2] "A coherent eddy structure consists of a parcel of vortical fluid occupying a confined spatial region such that a distinct phase relationship is maintained between the flow variables ... as the structure evolves in space and time." (Emphasis in practice on vorticity and detection.)

Hussain [3] "A coherent structure is a connected turbulent fluid mass with instantaneously phase-correlated vorticity over its spatial extent." (Emphasis in practice on vorticity and interaction with incoherent motions.)

In the case of the turbulent boundary layer, the similarity laws recognize two distinct scales for the flow; the overall thickness and the sublayer thickness. The search for structure has followed two lines of development, corresponding to these two characteristic scales. One line originated in work on large-scale phenomena in the outer flow reported by Kovasznay and co-workers [4][5]. The other line originated in work on small-scale phenomena near the wall reported by Kline and co-workers [6][7]. Various investigations still tend to concentrate on one scale or the other, with most of the weight on the side of the sublayer scale. Almost all experiments in the turbulent boundary layer have so far been carried out at relatively low Reynolds numbers, for the sake of a thicker sublayer, but this practice is not necessarily unsound as long as the two scales are well separated. There is considerable experimental evidence that intermediate scales may also need to be considered in the search for structure, but the need is not apparent from the form of the similarity laws and will not be discussed at any length in the present narrative.

The essential problem from the beginning has been to detect some repeatable and recognizable signature, or syndrome of signatures, that defines the coherent motions in time and space. Four papers can be cited to illustrate the various strategies adopted by different investigators. Blackwelder and Kaplan [8] focused on the signature of the smoothed streamwise component of velocity in or near the sublayer, emphasizing a sequence |slow decrease/rapid increase with overshoot/slow decrease| or its inverse. Willmarth and Lu [9] focused on the occurrence of rapid excursions of the product $-\overline{u'v'}$ to values much greater than the mean. Brown and Thomas [10] focused on large-scale maxima in the smoothed shearing stress at the wall. Fukunishi [11] focused on negative extrema in the smoothed streamwise component of velocity in the outer part of the boundary layer.

Flow in the Sublayer

By far the larger effort so far has been invested in study of the sublayer and of a poorly defined phenomenon called "bursting". The sublayer is the region adjacent to the wall and extending about 50 wall units away. A dimensional quantity is made dimensionless in wall units by using u_τ for reference velocity, ν/u_τ for length, and ν/u_τ^2 for time, where $u_\tau = (\bar{\tau}_w/\rho)^{1/2}$ and $\bar{\tau}_w$ is a spatial, temporal, or statistical mean value. One generally accepted notation is superscript +; e.g., $u^+ = u/u_\tau$.

The property of the sublayer associated with the term coherent structure is the presence of organized streamwise vorticity. Viewed along the streamwise direction, this vorticity alternates in sign with a mean period λ^+ (two counter-rotating cells) of about 90 to 100, as shown in the upper sketch in Fig. 1 (after Blackwelder [2]). Each cell can be visualized as roughly square in cross section. The distribution of cell sizes in a given flow, however, is very broad. Numerous measurements, summarized, for example, by Kim et al. [7] and by Smith and Metzler [12], have established that the average value of λ^+ for a smooth wall is sensibly independent of Reynolds number, pressure gradient, and outer boundary conditions. On the other hand, the scale λ^+, like the rest of the sublayer structure, is strongly affected by the presence of drag-reducing polymers and by surface roughness.

Although there is no consensus so far on the origin of the sublayer vortices, there is a consensus on their ubiquity. Kline et al. [6] and Smith and Metzler [12], using hydrogen-bubble flow visualization in a plan view, did not have to pick and choose among frames to find the pattern; any frame would do. The same is true for the hot-wire rake measurements by Gupta, Laufer, and Kaplan [13] and for the surface-element measurements by Lee, Eckelman, and Hanratty [14]. Probably the best evidence is the photograph in Fig. 2 (Cantwell, private communication) showing a view of the sublayer structure as seen from below. A dense suspension of aluminum flakes is used for visualization. The value of λ^+ in Fig. 2, as determined by an optical correlation method, is about 86. The Reynolds number based on momentum thickness is about 1150. The overall thickness δ relative to λ can be estimated at the top of the photograph, where a mirror at 45° is used to display the intersection of the boundary layer with the free surface. The streamwise scale of the largest eddies is perhaps

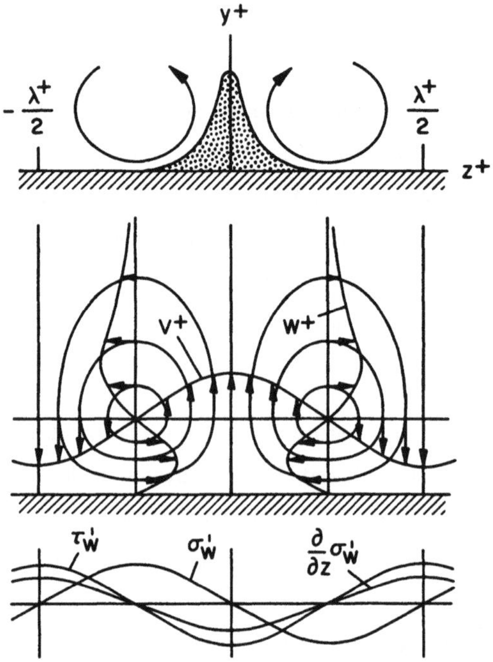

Fig. 1. Schematic view of sublayer vortices, looking in the stream direction. Top: accumulation of passive scalar contaminant in low-speed streak (after Blackwelder [2]). Center: suggested pattern of secondary flow. Bottom: relative phase of fluctuations in streamwise and spanwise stress at the wall.

2δ to 3δ, a value that eventually has to be reconciled with a larger value appearing in a later section of this paper.

In the definitions quoted in the introduction, coherent structure is associated explicitly or implicitly with vorticity of the appropriate scale. It is convenient to think first in terms of a deterministic description of the sublayer vortices, as shown in the center sketch in Fig. 1. Little is known about the magnitude of the streamwise vorticity $\xi^+ = \partial w^+/\partial y^+ - \partial v^+/\partial z^+$ which represents a non-trivial secondary flow superposed on the mean spanwise vorticity at the wall, $\zeta^+ = -\partial u^+/\partial y^+ = -1$. According to the sketch, the streamwise vorticity changes sign at values of y corresponding to the two extrema in w. This property is one of several that legislate in favor of defining a vortex in terms of closed streamlines rather than vorticity magnitude.

Fig. 2. Visualization of flow in sublayer using aluminum flakes (Cantwell, private communication). Reynolds number R_θ is about 1150; sublayer scale λ^+ is about 86. View at top shows intersection of boundary layer with free surface.

The deterministic model in Fig. 1 has several implications in any attempt to understand the effect of a wall on turbulent fluctuations. In particular, there are important consequences for the mean-velocity profile and the transport mechanism near the wall. With the aid of some tedious algebra, the first few terms of power-series expansions can be obtained for the turbulent shearing stress and other mean quantities near a smooth wall. For a two-dimensional steady mean flow,

$$-\overline{u'v'} = \left[\frac{3}{2\mu^2} \frac{d}{dx} \overline{(\tau_w')^2} + \frac{3}{\mu^2} \overline{\tau_w' \frac{\partial}{\partial z} \sigma_w'} \right] \frac{y^3}{6} + \ldots \tag{1}$$

where τ_w' is the fluctuation in $\mu\, \partial u/\partial y$ and σ_w' is the fluctuation in $\mu\, \partial w/\partial y$. The first term in parentheses in Eq. (1) is zero for channel flow, and for a boundary layer at constant pressure can be shown to be so small as to be unmeasurable. It is the second term that is at issue, as discussed in a somewhat tentative way by Hinze [15, p 621]. For a featureless flow, it might seem plausible to argue that this second term vanishes by symmetry, much as the Reynolds stresses $-\rho\,\overline{u'w'}$ and $-\rho\,\overline{v'w'}$ vanish for a flow that is two-dimensional in the mean. Given a value for one of the fluctuating quantities, the value for the other is equally likely to be positive or negative, and the correlation is therefore zero. That this argument is not correct in the case of Eq. (1) has been shown in a qualitative way by Chapman and Kuhn [16]. However, the real point is that the flow is not featureless, as pointed out by Lee et al. [14] and others. For the deterministic model in Fig. 1, the fluctuations τ_w' and $\partial\sigma_w'/\partial z$ are in fact strictly in phase, as shown at the bottom of the figure.

It follows from this evidence that the first term in an expansion for $-\rho\,\overline{u'v'}$ is definitely a term in y^3, and the corresponding term in an expansion for $\bar{u}$ is definitely a term in y^4, just as for the Blasius boundary layer. In fact, some of the best evidence for the value of the coefficient of this term comes not from experiment but from recent numerical solutions of the Navier-Stokes equations. In wall units, with $\tau_w'^{+} = \tau_w'/\rho u_\tau^2$ and $\sigma_w'^{+} = \sigma_w'/\rho u_\tau^2$,

$$\frac{-\rho\,\overline{u'v'}}{\tau_w} = \frac{1}{2}\left[\overline{\tau_w'^{+}\frac{\partial}{\partial z^{+}}\sigma_w'^{+}}\right](y^+)^3 + \ldots = A\,(y^+)^3 + \ldots \tag{2}$$

Spalart and Leonard ([17] and private communication) find $A = 1.0 \times 10^{-3}$. Chapman and Kuhn [16] find $A = 0.7 \times 10^{-3}$ for their model 1. The profile formula of Spalding [18] yields $A = \kappa^4 e^{-\kappa c}/6$; if $\kappa = 0.41$ and $c = 5.0$, the corresponding numerical value is $A = 0.6 \times 10^{-3}$. The deterministic sublayer model proposed by Coles [19] implies $A = 1.0 \times 10^{-3}$, although it should be noted that this model requires a value of unity for the correlation $-\overline{u'v'}/u'v'$ very near the wall, whereas some of the numerical solutions just cited, as well as the experiments by Eckelmann [20], yield a value of about 0.4. In any event, the general implication is that the term in $(y^+)^3$ in Eq. (2) is of order unity when y^+ is about 10 to 12. The radius of convergence of the series is certainly finite, and probably lies within the sublayer, perhaps near $y^+ = 15$.

The paper by Coles [19] includes an extensive survey of fluctuation measurements within the sublayer in pipes, channels and boundary layers. The data for u'^{+} show no definite effect of Reynolds number over a range of about two decades. This finding suggests that an expansion of $\bar{u}^+$ in powers of y^+ is free of parameters at least as far as the term in $(y^+)^4$, thus supporting Prandtl's idea that the mean velocity profile can be expressed quite generally in the form $\bar{u}^+ = f(y^+)$ from the wall through the sublayer and beyond to the logarithmic region.

Finally, and most important, the sublayer model suggests that the sublayer vortices account for essentially all of the non-molecular transport in a substantial part of the sublayer. In short, there are three transport mechanisms at work near a wall. First is laminar or molecular transport, which gives way to transport by coherent, almost deterministic streamwise vortices, which gives way to transport by eddies in the outer flow. My conjecture is that the sublayer vortices are present because they are the mechanism provided by nature to allow the fluctuations of larger scale in the outer flow to satisfy the no-slip condition at the wall.

Bursting and Ejection

The deterministic model just described has the same relationship to reality as the hard spherical molecule in the kinetic theory of gases; i.e., it is a powerful conceptual model and perhaps a useful first approximation. Reality, according to Fig. 2, is a confused and barely recognizable version of the model. There is a large dispersion in the local size and strength of the vortices in response to the irregular outer motion. One common kinematic feature is accumulation of a tracer such as dye (added very near the wall) in the shaded region of the top sketch in Fig. 1. This region is called a low-speed streak, and its ubiquity suggests that the sublayer vortices are everywhere strong enough to participate in the transport process. Under certain conditions, a strong local shear layer, or more properly a streamwise shear ribbon, can develop at the top of the streak. This layer may be subject to instability of the Kelvin-Helmholtz type, and may develop oscillations detectable in a photograph or in a time trace, say at $y^+ = 15$. Such oscillations are of the essence of bursting. This mechanism, together with transport by the vortices themselves, accounts for the observed strong turbulence production in the sublayer, as required by the local peak in the product $-\rho\, \overline{u'v'}\, \partial\bar{u}/\partial y$.

Of the two main schemes for signature detection one seems to detect a slightly inclined shear layer near the top of the shaded region in Fig. 1. This is the VITA technique of Blackwelder and Kaplan. This shear layer is most intense when the velocity outside the sublayer is higher than the mean. The other scheme looks for large values of the product $u'v'$. This is the quadrant method of Willmarth and Lu. The two schemes are now known to be equivalent. Both recognize that the process of instability and bursting is extremely local, requiring a probe whose dimension in wall units is small compared with λ^+. The bursting event is an elusive target.

The two detection schemes have been compared, with particular attention to the effect of threshold values on statistics, by Bogard and Tiederman [21], Subramanian et al. [22], Alfredsson and Johansson [23], and especially Guezennec [24]. It is important to notice that energetic bursting is quite rare. In fact, it is something of an anomaly that sublayer streaks are common but bursting is not. Various measurements, summarized, for example, by Bogard and Tiederman, show an average period of several times δ/u_∞. These authors suggest that the pdf of the period is bimodal, with rapid sequences separated by longer intervals. On the other hand, numerous measurements, beginning with the work by Kline et al. [6] and by Grass [25], show that the bursting event is often accompanied by ejection of turbulent fluid whose trajectory can be followed throughout most of the boundary layer. This property has led to two hypotheses, both unproven. One is that the bursting process is self-driven and self-contained within the sublayer, perhaps through some local instability, and that the large-scale outer eddies are formed by coalescence of debris from bursting events. The other, to which I subscribe, is that the sublayer activity is the signature of the passage of an outer eddy.

The Outer Flow

The large coherent structure in a boundary layer, if it exists, is concealed in a tremendous clutter of noise. The mechanisms that might produce such structures are unknown, although they presumably involve an instability, perhaps like the one in the late stages of transition in a laminar boundary layer. If so, the main difference is that the turbulent figure is more nearly a climax or saturated state than the laminar figure.

Almost the only attacks so far on the large-structure problem are those by Brown and Thomas [10], by Fukunishi [11], and by Guezennec [24]. Fukunishi's approach is the most transparent. If the structure being sought is a large V-shaped vortex leaning downstream, it should pump fluid upward between the legs of the V, and the velocity of this fluid should be less than the average. A scheme that triggers on a low streamwise velocity is destined to find the structure, with the built-in advantage that the technique is largely self-centering in the spanwise direction. Guezennec has successfully linked the bursting signature at the wall with the presence of large counter-rotating streamwise eddy pairs that occupy much of the boundary-layer thickness. The direction of rotation of the eddies and the direction of the transport normal to the wall between them can have either sign, corresponding to the events called ejection and sweep.

For the purposes of this paper, it is sufficient that certain impressions emerge from the discussion, particularly with respect to scale. Any characteristic time can be measured in wall units ν/u_τ^2 or in outer units δ/u_∞. At the low Reynolds numbers of typical experiments, the ratio of the two kinds of units is about 50:1. The signature discovered by Blackwelder and Kaplan is observable in a boundary layer at dimensionless intervals of about 150 to 250 in wall units, or 3 to 5 in outer units. The typical duration of the signature is about 50 in wall units, or 1 in outer units. The rapid acceleration at the center of the event occurs in a time of about 5 in wall units, or 0.1 in outer units. These data suggest that large co-rotating eddies whose scale is of order δ or more are frequently in contact at internal interfaces, where relatively thin shear layers of either sign are generated. It is not clear whether these estimates are characteristic of the largest eddies in the flow, or whether, as suggested by experience with the turbulent spot, they are characteristic of intermediate scales that tend to become more important at larger Reynolds numbers.

The Synthetic Turbulent Boundary Layer

The term "synthetic turbulent boundary layer" was proposed by Coles and Barker [26] to describe the flow produced by generating a regular array of turbulent spots in the laminar boundary layer near the leading edge of a flat-plate model. This array flows downstream with a known phase reference, so that ensemble averages at constant phase can be obtained without prior knowledge of any characteristic signature of the large-scale motions. The original synthetic flow was constructed in water, using momentary jets from the surface to generate the spot pattern. The present version was constructed in air, using a two-lobed camshaft to displace small pins momentarily into the laminar boundary layer. In both cases the pattern of spots is hexagonal.

Our work on coherent structure in the synthetic flow has been carried out in two stages, the first being concerned with the label "coherent" and the second being concerned with the label "structure". In the first stage, Savas [27] systematically varied the streamwise and spanwise scales of the hexagonal pattern and used a rake of 24 hot wires to observe the modulation of the mean intermittency at constant phase in the outer flow at various stations along the plate. A decrease of the modulation below a reasonable threshold was taken to indicate effective loss of coherence. Savas concluded that the pattern is most coherent when the streamwise and spanwise scales have a ratio (in his notation) of about $u_\infty \tau/\zeta = 3.2{:}1$. This is essentially the ratio chosen for the detailed structural measurements reported here. The measurements have been carried out by J. Arakeri, and full details will appear in his thesis. I am pleased to have this opportunity to call attention to Arakeri's research, which I consider to be a major contribution to the subject of coherent structure.

Experimental Protocol

Most of the measurements were made in two primary traverses. One was along the flow direction in the central plane of symmetry, directly downstream from one of the pins, with boundary-layer profiles obtained as a function of phase at 11 stations: $z = 0$, $x = 117.8\ (4.0)\ 157.8$ cm. The other traverse was in the spanwise direction, with profiles obtained at 29 stations: $x = 117.8$, $z = -7.0\ (0.5)\ 7.0$ cm. The spanwise traverse was carried out twice, once with an X-wire oriented in the uv-plane (normal to the surface) and once with the same X-wire oriented in the uw-plane (parallel to the surface). At each station, data from two hot wires and one heated surface element were recorded for 1000 camshaft revolutions, or 2000 cycles of the periodic pattern, at various standard values of y. Altogether, the data accumulated on magnetic tape during these measurements amount to between 2 and 3 Gbytes. Sampling was controlled by a 200-line optical encoder on the camshaft, so that 100 samples were recorded within each cycle of the basic pattern. The pulse train from the encoder also served as one input to a phase-locked-loop drive for the camshaft. The second

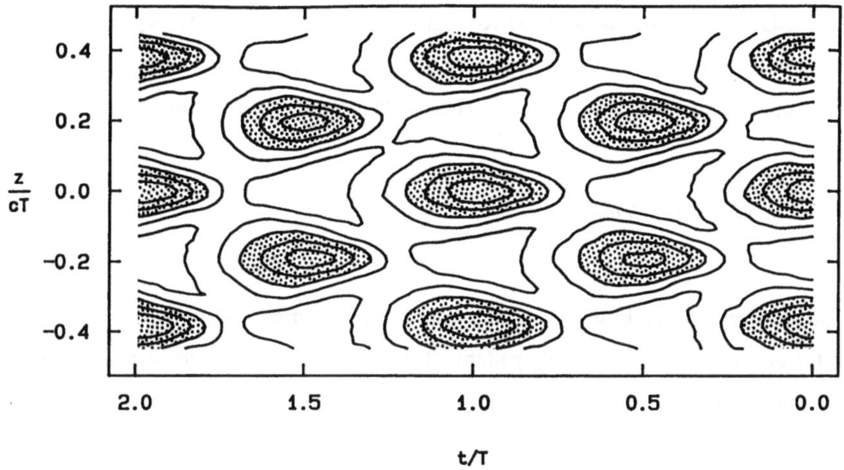

Fig. 3. Contour plot of $<\gamma>$ in plan view at $x = 117.8$ cm, $y = 1.93$ cm for one camshaft revolution. Scales for z and $x = ct$ are matched. Contour levels are 0.1 (0.1) 0.5. Shaded area corresponds to $<\gamma> \geq 0.3$.

input to the phase-locked loop was a crystal-controlled pulse train at a frequency of 2,000,000/640 = 3,125 Hz; this was also the sampling frequency on each channel. The period T of the periodic pattern was therefore 32 msec. The free-stream velocity was 1200 cm/sec.

Supporting measurements included a spanwise traverse at $x = 117.8$, $y = 1.93$, $z = -14.5$ (0.5) 14.5 cm to verify the periodicity of the eddy pattern by observing the mean intermittency at a fixed distance from the surface over the full spanwise range of the traverse. The results are shown as contours of constant intermittency in Fig. 3. Regions where $<\gamma>$ is greater than 0.3 are shaded. To ensure that the eddy pattern is represented accurately, identical dimensionless scales are used for z and for $x = ct$. The abscissa corresponds to one full camshaft revolution or two periods of the hexagonal pattern. The methods used to determine the celerity c are described below.

Another spanwise traverse was made at $x = 117.8$, $y = 0.04$, $z = -7.0$ (0.5) 7.0 cm with a single wire placed at a value of y corresponding very nearly to 15 wall units. The sampling frequency for this latter traverse was increased to 25 kHz in order to implement the VITA technique of Blackwelder and Kaplan [8] for signature detection in the sublayer. The phase reference in this case was generated by using the index pulse from the encoder as input to a J-K flip-flop and recording the output bit. These data have not yet been processed.

The Plane of Symmetry; Celerity.

The streamwise traverse in the plane of symmetry was carried out primarily to establish a value for the celerity c of the eddy pattern. The 100 samples in a cycle were averaged five samples at a time, to reduce the number of phase intervals in a cycle to 20, and the two cycles of each camshaft revolution (corresponding to the two lobes on the camshaft) were superposed. The population of the ensemble of data samples per phase at each point in space was thus increased from 1000 to 10,000. This population was large enough to insure negligible scatter for the mean velocity and acceptable scatter for the Reynolds stresses.

At each station, for each of the 20 phases of the periodic signal, the mean-velocity profile at constant phase was fitted to the standard wall-wake formula of Coles [28], and the parameters $\langle\delta\rangle$, $\langle u_\tau\rangle$, and $\langle\Pi\rangle$ were determined. The same operation was carried out for the time average over one period to obtain $\bar{\delta}$, $\bar{u}_\tau$, and $\bar{\Pi}$. These values cannot be distinguished, for any practical purpose, from the mean of the 20 averages at constant phase; i.e., $\overline{\langle\delta\rangle}$, etc. It was not at all obvious at the outset that this fitting operation would suit the properties of the present highly structured flow. Nevertheless, the results are quite satisfactory; the fit is excellent, as shown for $x = 117.8$ cm in Fig. 4. To demonstrate the overall quality of the data, the variation of $\langle\Pi\rangle$ with phase and downstream distance is displayed in Fig. 5. The lowest curve corresponds to Fig. 4. The other curves were measured at increments of 4 cm in x, ending at $x = 157.8$ cm. The solid lines are five-term Fourier series fitted to the data.

To establish the celerity, the phase corresponding to the maximum and minimum values of $\langle\Pi\rangle$ in Fig. 5 and to the two crossings of the mean value is plotted in Fig. 6. The straight lines all have the same slope but are otherwise a least-squares fit. Similar plots were prepared for $\langle R_\delta\rangle = u_\infty\langle\delta\rangle/\nu$, $\langle C_f\rangle = 2\langle u\rangle_\tau^2/u_\infty^2$, and $\langle R_Y)$, where Y is the value of y for which the measured intermittency was equal to 0.5. Sixteen values were thus obtained for

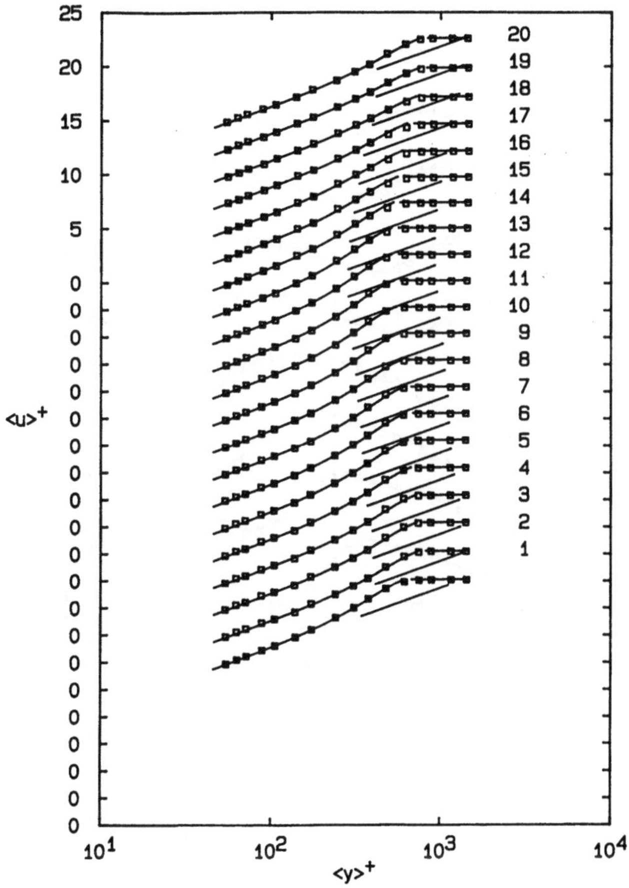

Fig. 4. Ensemble-mean velocity profiles in wall coordinates for phases 1-20 of the cycle. Probe is at $x = 117.8$ cm in plane of symmetry. Lowest curve with filled symbols is time-mean profile. Solid lines are fitted curves from wall-wake formula. Labels identify phase.

the dimensionless celerity or phase velocity defined by the formula $d(x/u_\infty T)/d(t/T) = c$ or $dx/dt = c\, u_\infty$. The mean was $c = 0.842$, with a root-mean-square deviation of 0.020. The measured celerity is lower than the value 0.88 estimated by Savas [27], presumably because the present data describe a more limited and more upstream region of the synthetic flow. The physical celerity, $c\, u_\infty = 1010$ cm/sec, and the period, $T = 0.032$ sec, imply the streamwise wave length $\lambda_x = c\, T = 32.3$ cm. The spanwise wave length ζ or λ_z is 12.2 cm for these measurements, so that $\lambda_x/\lambda_z = 2.65$ and $u_\infty T/\lambda_z = 3.15$.

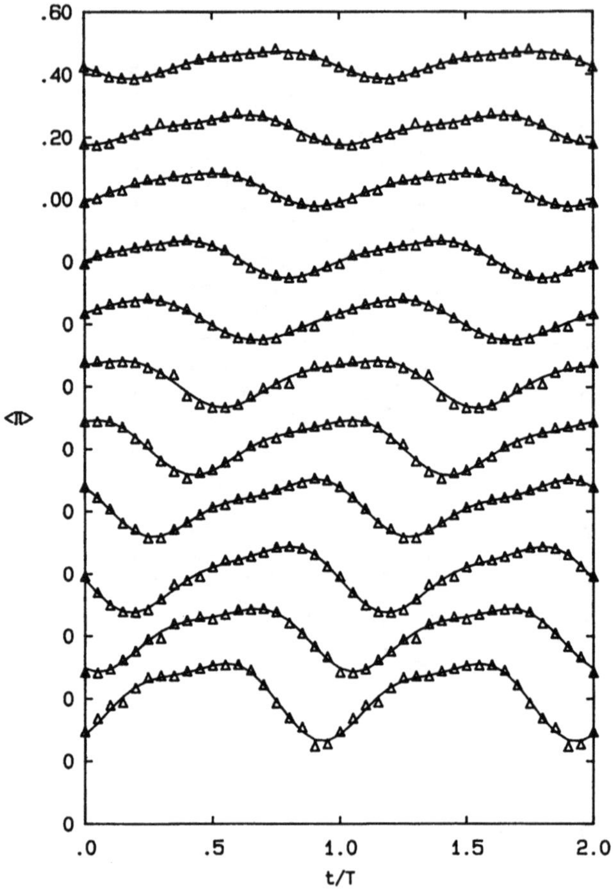

Fig. 5. Variation of the profile parameter <Π> with phase and downstream distance in the plane of symmetry. Lowest curve is for $x = 117.8$ cm; increment in x is 4 cm. Abscissa t/T is relative time, with T the period of the cycle. Solid lines are fitted five-term Fourier series.

Loss of Coherence

It is apparent in Fig. 5 that the coherence of the periodic pattern, as measured by the amplitude of the observed modulation, decreases as the pattern flows downstream. However, the shape of the curves does not change noticeably; the fitted lines in Fig. 6 remain non-uniformly spaced and parallel. A semi-logarithmic plot of normalized amplitude suggests that the modulation of all three profile parameters decreases by about a factor of two for a

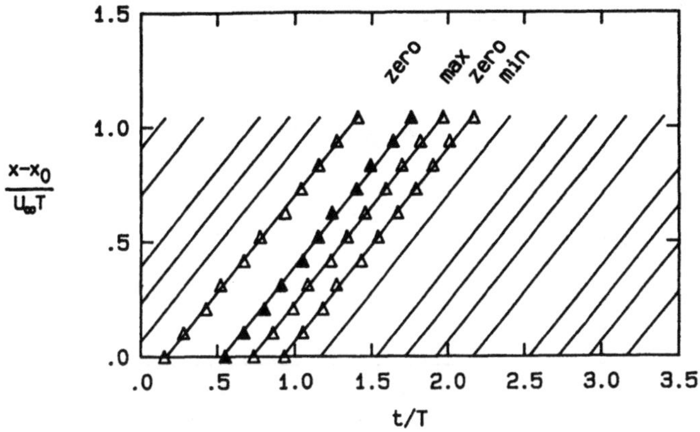

Fig. 6. Celerity plot for the profile parameter <Π>, showing locus of maximum value (filled for identification), minimum value, and two crossings of mean value (see Fig. 5). Slope of straight lines is 0.842.

downstream displacement of one spatial period (32.3 cm). This decrease does not necessarily mean that the real modulation is decreasing, only that dispersion is smoothing the averages at constant phase.

The main premise of the present research is that the turbulent spot and the large eddy in a natural turbulent boundary layer have closely related structures. This assumption can be tested indirectly by comparing suitably averaged properties of the natural and synthetic flows. Let the natural flow be represented by the formulas recommended by Coles [28], with the addition of an empirical formula for Π; namely,

$$\Pi = 0.62 - 1.21\, e^{-\delta^+/290} \tag{3}$$

The agreement for the time-mean quantities is excellent (3 percent) for the friction coefficient, fair (8 percent) for the layer thickness, and less than fair (15 percent) for the magnitude of the wake component, which accounts for about 10 percent of the velocity profile. The mean values here are formed only for the plane of symmetry, not for the full doubly-periodic pattern.

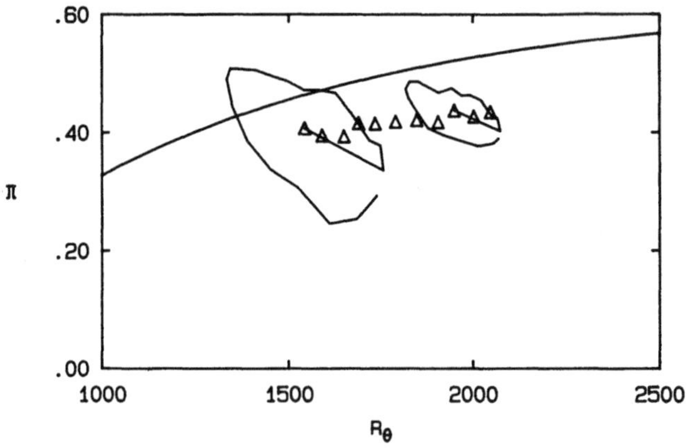

Fig. 7. Trajectory of profile parameter $<\Pi>$ at stations $x = 117.8$ cm and $x = 149.8$ cm. Radial line connects data point for phase 1 to corresponding mean. Trajectory is counterclockwise.

For the variable Π, the counterclockwise trajectory of $<\Pi>$ against $<R_\theta>$ is shown in Fig. 7 for two stations 32 cm apart. The data point for phase 1 of 20 is connected to the corresponding mean value for identification. The trajectory moves across the grain of the mean curve and shows substantial departures from equilibrium.

Structure in the Plane of Symmetry

Partly to perfect various computer programs, and partly to gain some sense of structure, a number of quantities have been derived from the data at the intersection of the two major traverses; i.e., at the station $x = 117.8$, $z = 0$ cm. These measurements belong to the stream-wise traverse, and the hot-wire data include u and v only. The boundary layer is depicted as changing in time rather than in space, because there is no useful alternative at present without a further massive increase in the quantity of data. One variable of interest is the instantaneous streamline pattern as viewed by an observer moving at the celerity c. It was found that the measurements of the mean normal component $<v>$ or $\bar{v}$ were not usable, to the required accuracy of a small fraction of a degree for flow angle, and the streamlines were therefore calculated from the continuity equation by way of the stream function

$$<\psi> = \int_0^y <u> dy \tag{4}$$

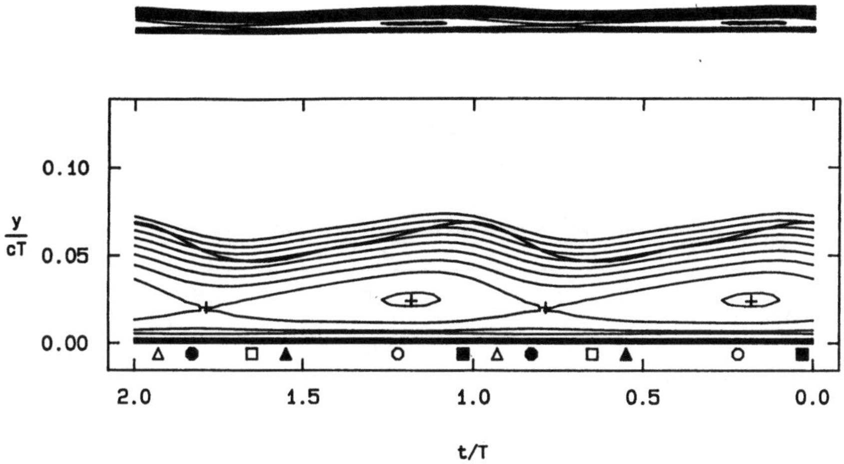

Fig. 8. Mean streamlines in plane of symmetry as viewed by observer moving at celerity c. Upper figure is to proper scale; lower figure is expanded vertically by 5:1. Spanwise velocity $<w>$ is not accounted for. Layer is depicted with downstream distance replaced by time; net entrainment is zero. Symbols at bottom show location of maximum (filled) and minimum (open) values for: △ $<\Pi>$; □ $<R_\delta>$; 0 $<C_f>$.

This preliminary formulation ignores the derivative $\partial<w>/\partial z$, which is of the same order as $\partial<u>/\partial x$ and $\partial<v>/\partial y$ but which has not yet been determined from the data of the spanwise traverse. With this proviso, the streamlines appear as shown in Fig. 8. The figure is shown twice; at the top with identical scales for $x=ct$ and y, to emphasize that the flow is always and everywhere a boundary layer in the traditional sense (i.e., $<v> << <u>$ and $\partial/\partial x << \partial/\partial y$), and at the bottom with the vertical scale expanded by 5:1 to reveal some internal details. The three presently accessible components of the Reynolds-stress tensor, $<u'u'>$, $<u'v'>$, and $<v'v'>$ in the plane of symmetry, are shown as contour plots in Fig. 9. The top contour in each of Figs. 9a-9c, incidentally, is not a level curve for the stress, but is the boundary-layer thickness $<\delta>$ measured quite independently and plotted in the same coordinates.

Structure off the Plane of Symmetry

Finally, the profile-fitting operation has been completed for all of the data of the spanwise profile traverse, which covered a distance of 14 cm. Contour plots of $<\Pi>$, $<R_\delta>$, and $<C_f>$

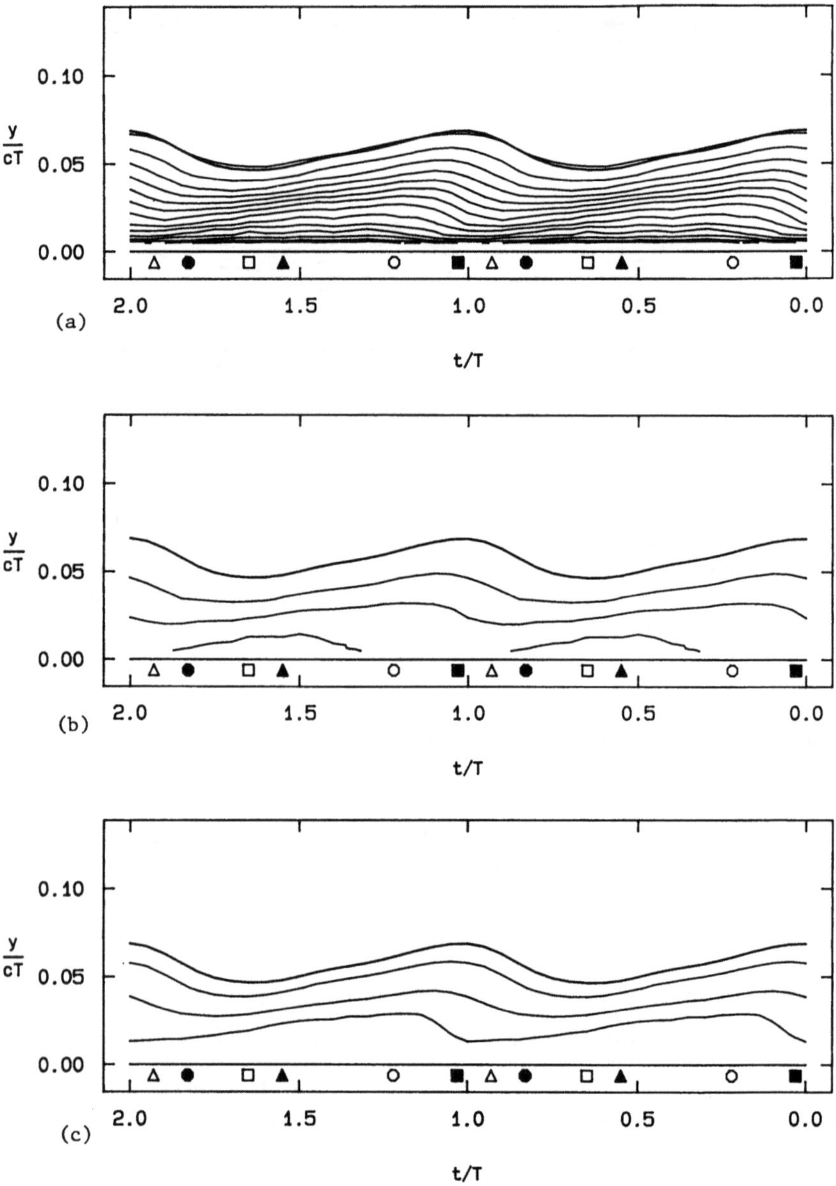

Fig. 9. Three Reynolds stresses in plane of symmetry as a function of phase. (a) $\langle u'u' \rangle$; (b) $-\langle u'v' \rangle$; (c) $\langle v'v' \rangle$. Contour levels are 0.0005 (0.0005) 0.0055. Upper curve (nominally contour zero) is layer thickness $\langle \delta \rangle$. Symbols at bottom as in Fig. 8.

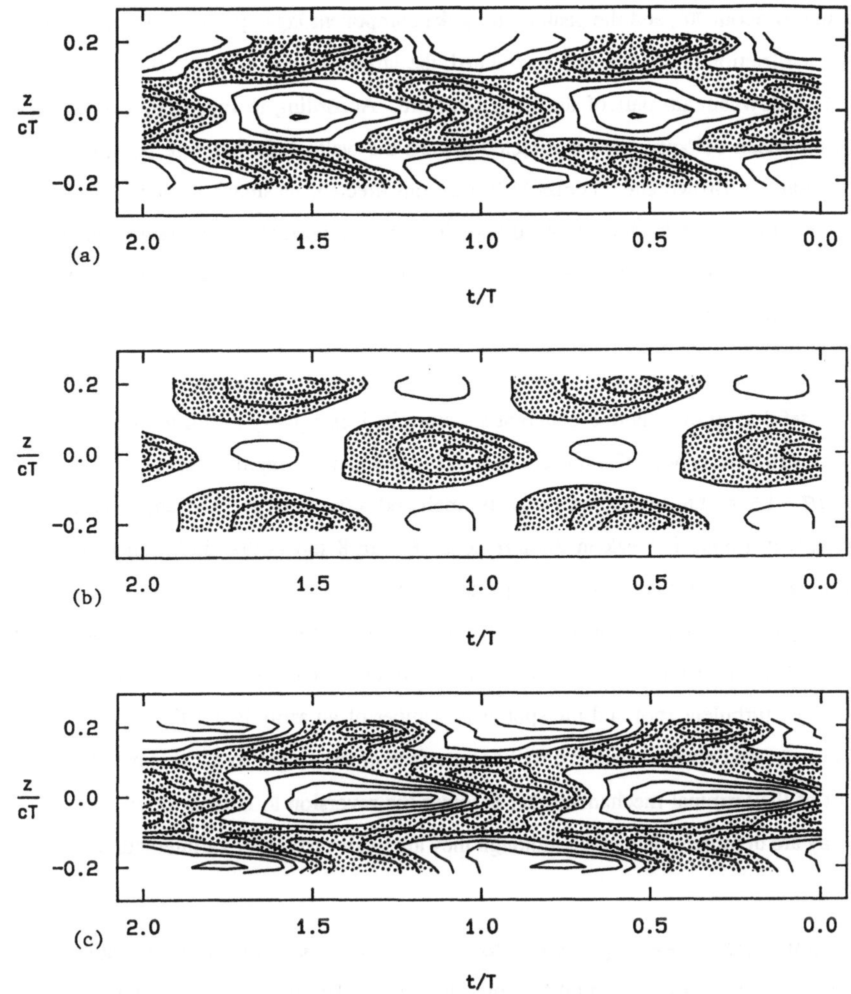

Fig. 10. Contour plots of three profile parameters in plan view at $x = 117.8$ cm. (a) $\langle \Pi \rangle$; contour levels are 0.7 (0.1) 1.3; shaded if value is less than mean. (b) $\langle R_\delta \rangle$; contour levels are 0.9 (0.1) 1.2; shaded if value is greater than mean. (c) $\langle C_f \rangle$; contour levels are 0.97 (0.01) 1.03; shaded if value is greater than mean.

are shown in Fig. 10. The quantities $\langle \Pi \rangle$ and $\langle R_\delta \rangle$ are evidently much more strongly modulated than the quantity $\langle C_f \rangle$. In the case of Π, regions where the variable is less than its mean value are shaded; in the case of R_δ and C_f, regions where the variable is greater than its mean value are shaded. The most important finding so far is the phase relationship between surface friction and layer thickness. The peak in friction leads the peak in layer thickness by about 70°, and the peak in the wake component lags by about 190°. Granted that the surface friction is not strongly modulated, the largest values lie near $t/T = 1.0$, under the leading or downstream part of the large structure. According to Fig. 8, this is the region where the mean velocity $<v>$ is directed toward the wall. It is therefore possible that the observations by Brown and Thomas [10] were misdirected, inasmuch as their detection criterion of a maximum in the smoothed wall friction was used to study features of the flow downstream from the detection point rather than upstream, where the associated large structure now seems to lie.

If it is useful to know where a coherent structure is, it is equally useful to know where the coherent structure is not. The present data in Fig. 10 identify regions between the eddies, at about $t/T = 1.5$ or 1.6, say, where $<R_\delta>$ is small and $<\Pi>$ is large. The large value of $<\Pi>$ suggests that mixing is weak in these regions. Figure 8 shows that the mean flow is away from the wall, and a comparison of Figs. 9 and 10 shows that the Reynolds shearing stress must have quite steep gradients close to the wall. The dominant turbulence-production term $-\rho <u'v'> \partial<u>/\partial y$ has not yet been worked out. In any event, one clear difference between the isolated turbulent spot and the spot as a member of an array is the absence of the maximum in C_f that appears near the trailing interface of the single spot. This maximum must be part of the terminal mechanism that enforces the reversion to laminar flow. The spots in the array are definitely in contact, although they are in an early stage of interaction.

Considerable data processing has still to be done. The spanwise mean velocities will be determined first. These are small but important, especially in any effort to find patterns of large-scale vorticity. To determine the location of energetic local events in or near the sublayer, the VITA technique and the uv-quadrant technique will be applied to various measurements made close to the wall. The point of view in these operations is that the largest available scale is the most logical point of departure for any exploration of coherent structure outside the sublayer. Whether or not this point of view is productive remains to be seen.

Acknowledgment

The research described in this paper was supported by the National Science Foundation under Grant MEA-8315042.

References

1. Coles, D.: Prospects for useful research on coherent structure in turbulent shear flow. Proc. Indian Acad. Sci. (Eng. Sci.) 4 (1981) 111-127.
2. Blackwelder, R.F.: The bursting phenomenon in bounded shear flow. In Turbulent Shear Flows. Von Karman Inst. for Fluid Dyn. Lecture Series 1983-03 (1983).
3. Hussain, A.K.M.F.: Coherent structures and turbulence. J. Fluid Mech. (in press).
4. Kovasznay, L.S.G.; Kibens, V.; Blackwelder, R.F.: Large-scale motion in the intermittent region of a turbulent boundary layer. J. Fluid Mech. 41 (1970) 283-325.
5. Blackwelder, R.F.; Kovasznay, L.S.G.: Time scales and correlations in a turbulent boundary layer. Phys. Fluids 15 (1972) 1545-1554.
6. Kline, S.J.; Reynolds, W.C.; Schraub, F.A.; Runstadler, P.W.: The structure of turbulent boundary layers. J. Fluid Mech. 30 (1967) 741-773.
7. Kim, H.T.; Kline, S.J.; Reynolds, W.C.: The production of turbulence near a smooth wall in a turbulent boundary layer. J. Fluid Mech. 50 (1971) 133-160.
8. Blackwelder, R.F.; Kaplan, R.E.: On the wall structure of the turbulent boundary layer. J. Fluid Mech. 76 (1976) 89-112.
9. Willmarth, W.W.; Lu, S.S.: Structure of the Reynolds stress near the wall. J. Fluid Mech. 55 (1972) 65-92.
10. Brown, G.L.; Thomas, A.S.W.: Large structure in a turbulent boundary layer. Phys. Fluids 20 (suppl.) (1977) S243-S252.
11. Fukunishi, Y.: Influence of ordered motions on the structure of outer region of the turbulent boundary layer. In Tatsumi, T. (ed.) Turbulence and Chaotic Phenomena in Fluids. Elsevier (1984) 371-376.
12. Smith, C.R; Metzler, S.P.: The characteristics of low-speed streaks in the near-wall region of a turbulent boundary layer. J. Fluid Mech. 129 (1983) 27-54.
13. Gupta, A.K.; Laufer, J.; Kaplan, R.E.: Spatial structure in the viscous sublayer. J. Fluid Mech. 50 (1971) 493-512.
14. Lee, M.K.; Eckelman, L.D.; Hanratty, T.J.: Identification of turbulent wall eddies through the phase relation of the components of the fluctuating velocity gradient. J. Fluid Mech. 66 (1974) 17-33.
15. Hinze, J.O.: Turbulence (2nd ed.). McGraw-Hill 1975.

16. Chapman, D.R.; Kuhn, G.D.: The limiting behaviour of turbulence near a wall. J. Fluid Mech. 170 (1986) 265-292.

17. Spalart, P.R.; Leonard, A.: Direct numerical simulation of equilibrium turbulent boundary layers. In Turbulent Shear Flows 5. Springer (in press).

18. Spalding, D.B.: A single formula for the "law of the wall". J. Appl. Mech. 28 (1961) 455-457.

19. Coles, D.: A model for flow in the viscous sublayer. In Smith, C.R.; Abbott, D.E. (eds.) Coherent Structure of Turbulent Boundary Layers. Lehigh Univ. (1978) 462-475.

20. Eckelmann, H.: Experimentelle Untersuchungen in einer turbulenten Kanalstromung mit starken viskosen Wandschichten. Mitt. M-PI und AVA Gottingen Nr. 48 (1970).

21. Bogard, D.G.; Tiederman, W.G.: Burst detection with single-point velocity measurements. J. Fluid Mech. 162 (1986) 389-413.

22. Subramanian, C.S.; Rajagopalan, S.; Antonia, R.A.; Chambers, A.J.: Comparison of conditional sampling and averaging techniques in a turbulent boundary layer. J. Fluid Mech. 123 (1982) 335-362.

23. Alfredsson, P.H.; Johansson, A.V.: On the detection of turbulence-generating events. J. Fluid Mech. 139 (1984) 325-345.

24. Guezennec, Y.G.: Documentation of large coherent structures associated with wall events in turbulent boundary layers. Ph. D. thesis. Illinois Institute of Technology (1985).

25. Grass, A.J.: Structural features of turbulent flow over smooth and rough boundaries. J. Fluid Mech. 50 (1971) 233-255.

26. Coles, D.; Barker, S.J.: Some remarks on a synthetic turbulent boundary layer. In Murthy, S.N.B. (ed.) Turbulent Mixing in Nonreactive and Reactive Flows. Plenum (1975) 285-292.

27. Savas, O.; Coles, D.: Coherence measurements in synthetic turbulent boundary layers. J. Fluid Mech. 160 (1985) 421-446.

28. Coles, D.: The young person's guide to the data. In Coles, D.; Hirst, E.A. (eds.) Computation of Turbulent Boundary Layers. Vol.II. Stanford (1969) 1-45.

Turbulence Modelling Using Coherent Structures in Wakes, Plane Mixing Layers and Wall Turbulence

A.E. PERRY

Department of Mechanical Engineering

University of Melbourne

SUMMARY

Coherent structures in turbulent flows are reviewed mainly from work carried out at the University of Melbourne and possible uses of the concept for predicting the evolution of turbulence are discussed. Emphasis is placed on the so-called "coherent substructures" in plane mixing layers and wall turbulence. The use of turbulence spectra for making inferences concerning the properties of the coherent substructures is outlined.

INTRODUCTION

There appears to be some controversy as to what coherent structures are and how they should be defined. This problem of definition is discussed at length by Hussain [1] in his article "Coherent Structures - Reality and Myth". The title alone suggests some uneasiness and confusion about the concept. Workers in the field of turbulence have become accustomed to working with ill defined concepts. From discussions with many colleagues the concensus of opinion is that even turbulence has no precise definition but it has the following properties: it is unsteady, three dimensional and random. There should be included a further property and this is that it consists of a range of scales of eddying motions and this range increases with Reynolds number.

Coherent structures are usually associated with the large scale motions of turbulence and much of the early work was in free shear flows. A Kármán vortex street behind a cylinder at high Reynolds number is an array of coherent structures. If smoke were introduced into the flow, these structures would show up as bulges. The smoke from a chimney in cross-flow reveals a streamwise array of large bulges which are thought to be coherent structures. The experiment of Brown and Roshko [2] in plane mixing layers showed that the shear layer rolls up into an array of large eddies whose length scale increases in the downstream direction.

There is also a feeling that there is an orderliness about coherent structures. The Kármán vortex street is an example where the structures appear to recur in an almost periodic manner. There is a belief that coherent structures contain a significant fraction of the turbulent kinetic energy and that they are responsible for a large part of the transport of scalar quantities through the flow. Indeed it is felt by many, that their existence should be recognised and incorporated in any modelling of turbulent flow.

However, as Lumley [3] has pointed out, there are two divergent schools of thought regarding coherent structures. Firstly there is the Brown & Roshko [4][2], Roshko [5], Cantwell [6] Browand & Troutt [7] school which supports the view that these structures are more characteristic of all turbulence than we previously thought and that we had overlooked their existence and importance in the past and made measurements using ordinary statistical methods which tended to smear out and conceal the organisation truly present in the flow. It is felt that conditional sampling should replace the standard methods so that these structures could be properly educed. To quote Lumley [3]: "the opposing school (for example Bradshaw [8] and Hussain [9]) feels that there is substantial evidence that the degree of organization in these flows decreases as the flows age; that one reason we were not previously aware of the existence of these well organised energetic structures is that we never measured in the early part of the turbulent flows saying that the flows were not yet fully developed; that, in fact, the well-organised structures observed may be attributable in part to the care with which these flows have been set up - that is, if extreme care is taken to remove adventitious disturbances from the oncoming flow (as is now more usual than formerly), the instability of the initial flow will be of a single type ...... There is some feeling that the structures present in the early flow are probably more characteristic of the initial instability while those present in the fully developed turbulence are probably characteristic in some sense of the fully developed flow...."

Work recently carried out at Melbourne using phase averaging techniques behind bodies of different types and various devices (such as propellors and flapping flags) supports the view that the structures which appear for a considerable distance downstream depend on what you "put in" even though the flow is turbulent (see Perry & Steiner [10] and Steiner & Perry [11]). These structures cannot be regarded as being characteristic of turbulence but their existence cannot be denied. Although they are probably far more energetic than the genuine naturally occurring large scale structures which might ultimately evolve further downstream, their study is not without interest, because of their technological importance. The structures behind devices such as propellors or wind mills can be very orderly for a considerable distance but these are not the only patterns which are considered coherent. Lumley suggests (private communication) that a coherent structure is a pattern which is recognizable and recurs throughout the flow. Lumley [3] also suggests that there need not be any orderliness about these structures and that their position and scale within a flow is quite random. Also their eduction by conditional sampling methods is fraught with difficulty. Commonly recurring velocity signatures need not necessarily have any real significance but might be caused by chance juxtapositions of these coherent structures. Hence there is considerable uncertainty about the interpretation of conditionally sampled data. The author believes that the hair-pin or horseshoe structures so often attributed to wall turbulence fit in very closely with the description of Lumley [3]. Their scales range from the very smallest (the Kline scaling, Kline et al. [12]) to the very largest which is of the order of the shear layer thickness. They are all in a sense attached to the boundary but their streamwise and spanwise position is quite random. These structures have a recognizable shape but this shape is rather variable. Because these structures are not necessarily large, Hussain [1] calls these coherent substructures.

In this paper an attempt is made to incorporate coherent structures in the modelling of turbulence. This has been done mainly for gaining an understanding and a description of the phenomena. However, models of a "predictive" nature might eventually evolve and this is outlined. The motivation for this latter activity is that modern day computers are not sufficiently large or fast to produce economically, full direct simulations of the Navier-Stokes equations at practical ranges of Reynolds numbers. However, this situation is slowly changing (e.g. see the full direct simulations of Kim & Moin [13] and Kim [14]). Even with such computations we still have the very difficult task of knowing how to present the results in the most compact and meaningful manner. We also have to decide whether the results are valid and genuinely reflect a phenomenon governed by the Navier-Stokes equations. This is where an understanding is important.

PHASE AVERAGED STRUCTURES

The most convincing eduction of coherent structures occurs in flow situations where a periodic vortex shedding process is present and the data is sampled on the basis of the phase of the vortex shedding. The stronger the periodicity, the less is the washout of data and the further downstream these structures can be educed. Figure 1(a) shows phase-averaged structures behind a bluff plate which were recently measured at Melbourne. A flying hot-wire was used which scanned the flow and the phase of vortex shedding was detected by a stationary wire which was placed close behind the body but outside the turbulent wake region. The vortical structures beyond the cavity region become apparent as shown in figure 1(b) when the velocity of the observer is the same as that of the structures. The signal from the phase detecting wire is band-pass filtered and then passed through a Schmitt trigger. This gives pulses which indicate the start of a new vortex shedding cycle and the time interval between successive pulses is divided up into 16 phase boxes as was done by Cantwell & Coles [16].

(a)

(b)

Fig.1. Phase averaged velocity vector field streamlines behind a bluff plate at a Reynolds number of 20,000. (a) As seen by an observer at rest relative to body. (b) As seen by an observer moving with the shed vortices. After Perry & Stenier [10].

The data from the flying hot-wire is sorted into various phase boxes. What is convincing about the method is that the power spectral density of the phase detector signal may show a number of peaks. This is particularly true if the body is three dimensional. However, a phase averaged pattern emerges for the structures downstream only if the correct frequency peak is band-pass filtered. If this is not carried out, then a completely featureless streamline pattern will result downstream. In other methods of educing structures which do not rely on an approximate periodicity, the detection system can have its parameters (threshholds, hold times, etc.) adjusted over wide ranges and a pattern of some sort will always emerge.

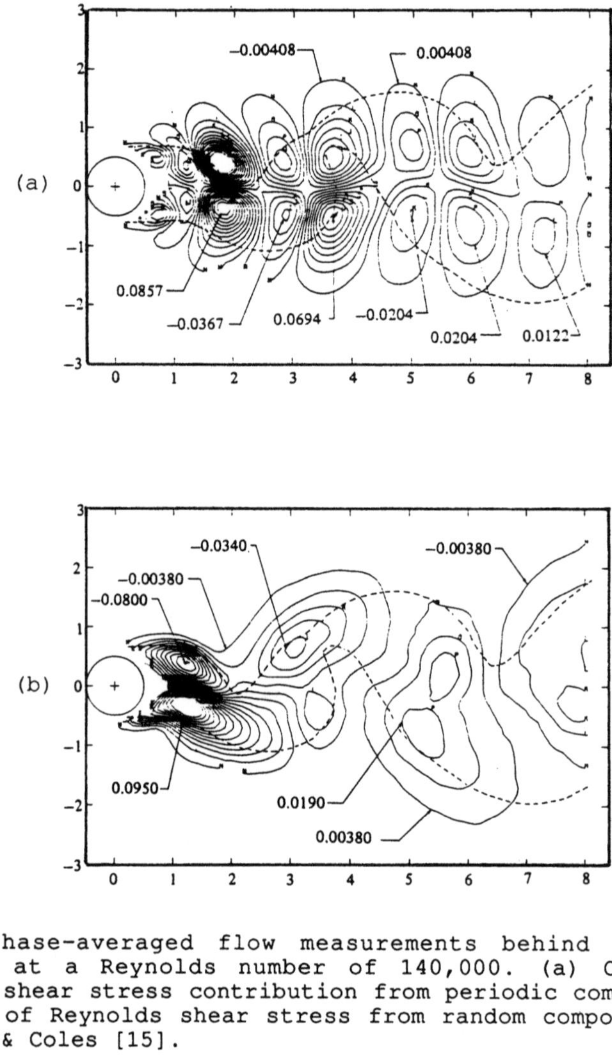

Fig.2. Phase-averaged flow measurements behind a circular cylinder at a Reynolds number of 140,000. (a) Contours of Reynolds shear stress contribution from periodic component. (b) Contours of Reynolds shear stress from random component. After Cantwell & Coles [15].

Some form of optimization is needed and the meaning of the educed pattern or ensemble-averaged signature is unclear although recently Kim [14] appears to have successfully applied the quadrant analysis of Lu & Willmarth [15] to the full direct simulation calculations of Kim & Moin [13] for channel flow and educed a horseshoe vortex. However this was at a relatively low Reynolds number where the range of eddy scales is limited.

With the phase detection method various salient features of the velocity vector field such as saddles, centres (or foci) can be related to other phase averaged quantities. Figure 2 shows some phase averaged results of Cantwell & Coles [16] for flow behind a circular cylinder. The dashed lines are the .5 contour of intermittency factor. The shape bounded by these dashed lines resemble smoke patterns observed in flow visualisation experiments. In figure 2 contours of the Reynolds shear stress from the random fine scale motions and from the periodic motions are mapped out in space. The contribution to the Reynolds shear stress from the periodic motions was approximately 40% of the global stresses (i.e. the total time averaged values). This would diminish further downsteam where phase jitter causes washout of data. Phase averaging shows how various turbulence processes are distributed in space and how they are related to the large scale structures. Any convincing computational method would either need to have these features incorporated or they should emerge naturally in the computed results particularly for those flows which are close to the source of the initial disturbance.

COHERENT SUBSTRUCTURES

Let us now discuss the types of coherent structures which are very difficult to educe by conditional sampling but are believed to be present from flow visualization studies. These are the horseshoe vortices in turbulent boundary layers observed by Head & Badyopadhyay [17], the hairpin structures in turbulent spots observed by Perry et al. [18] and the longitudinal vortices which ride on the back of the large scale roll-ups in a plane mixing layer as observed by Breidenthal [19]. These structures have a recognizable shape, they recur throughout the flow and they have a characteristic orientation. The author believes that in most flow situations these structures are responsible for the Reynolds shear stress and for most of the energy containing motions. The effect that these structures have on the power spectral density, turbulence intensities and mean vorticity will be discussed.

VELOCITY SIGNATURES AND SPECTRA

Power spectral density analysis shows how the energy is distributed among the scales of the motions and this leads to useful scaling laws.

Let us consider for a moment that turbulence consists (in part) of straight parallel vortex rods aligned in some characteristic direction and the velocity signatures are to be generated by the use of the Biot-Savart law. Figure 3 shows such a vortex rod in isolation and imagine it being swept in the x direction past a stationary observer. The signature observed at different spanwise positions (i.e. different values of y) are shown. Typical velocity signatures are given. The u_1 signature is always an even shaped bell function (no matter how the vortex rod is orientated) whereas the u_2 and u_3 signatures are a combination of an even- and odd-shaped function. For cuts further away from the rod, the amplitude of the signature decreases but its length scale increases. The u_2 and u_3 signatures reach a peak amplitude if we cut through the centre of the rod but the u_1 signature is zero on such a cut. The u_1 signature reaches a peak amplitude as we graze the side of the rod if solid body rotation is assumed within the rod. The flat region on the u_1 signature for a cut inside the rod is due to solid body rotation. This becomes rounded if a Gaussian distribution of vorticity is assumed as given by the Burgers [20] solution for a stretched vortex filament or tube. The interesting point about the signatures for cuts outside the rod (and this probably accounts for most of the flow), is that the square of the modulus of the Fourier transforms for the transient signatures for all three components is an exponential-like function. Figure 4(a) shows this distribution for the bell shaped u_1 velocity signature. Figure 4(b) shows the same for a burst of bell-shaped signatures in a periodic array. Here the square of the modulus consists of a series of spikes with the same exponential-like curve in figure 4(a) acting as an envelope.

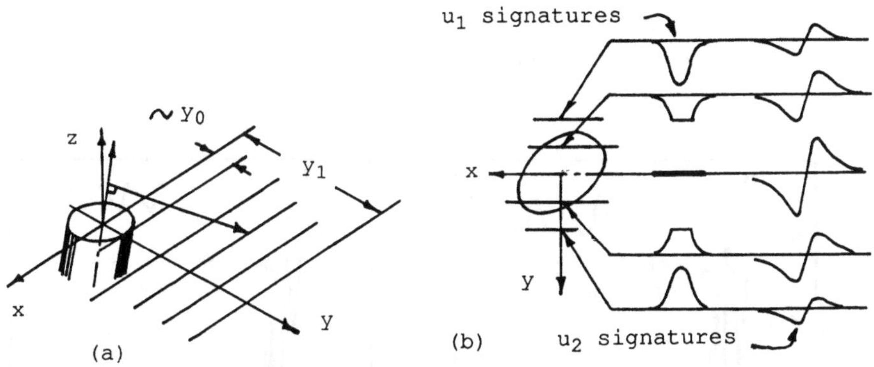

Fig.3. Velocity signatures generated by an isolated vortex filament. (a) Definition of coordinates. (b) Signatures of u_1, and u_2 produced along lines of constant y. u_2 signatures have similar properties to u_3 signatures.

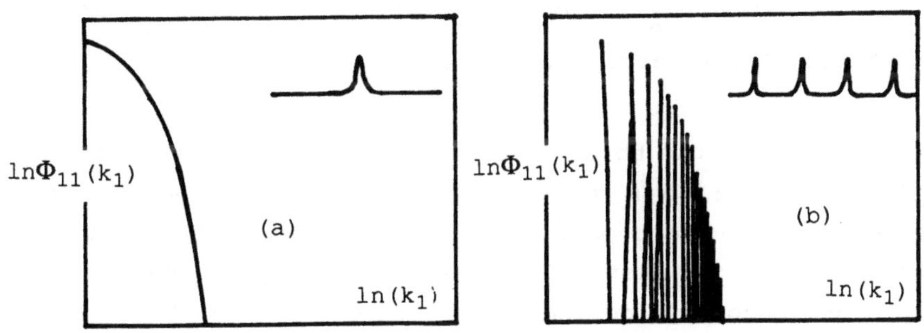

Fig.4. Power spectral density $\Phi_{11}(k_1)$ produced by a fast Fourier transform of (a) an isolated u_1 pulse in the sampling period (b) by periodic train of pulses.

It can be shown that a randomly-spaced array of such bell signatures has a smoothed power spectral density which is proportional to the exponential-like curve given in figure 4(a). We could use multiple point smoothing in wavenumber space or we could achieve this smoothing by ensemble averaging a large number of realizations. Although so far we have assumed that the vortex filaments are straight and infinite in extent, the signatures given by a periodic array of vortex loops with the probe close to the vortex cores give spectra with exponential-like envelopes. This was done for a negatively buoyant coflowing wake by Perry & Tan [21] at a Reynolds number of order 1000 and subjected to periodic disturbances (see figure 5).

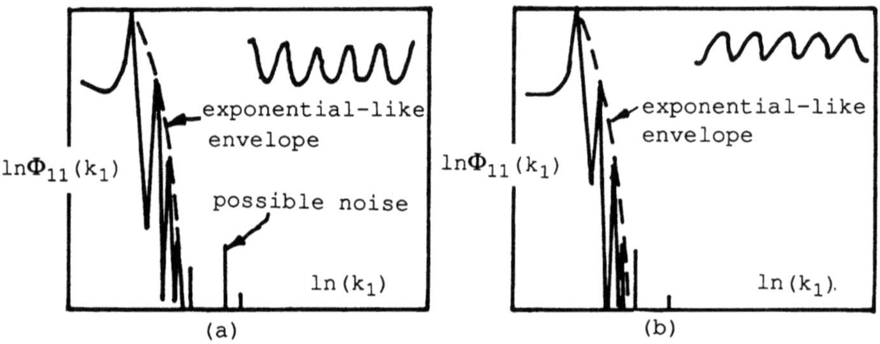

Fig.5. Power spectral density of velocity signatures generated by a periodic array of vortex loops in a coflowing wake with probe close to the vortex cores. (a) u_1 signatures (b) u_2 signatures. After Perry & Tan [21].

When the probe cuts through the core, individual vortex sheets would give rise to a saw-tooth signature. This occurs if the vortex is young and viscous diffusion has not had time to produce a Gaussian distribution. Approximating the vortex scroll with a series of cylindrical vortex sheets, a signature as shown in figure 6(a) is produced. A useful formula in spectral analysis is the Bracewell [22] rule. If n is the number of times a signature needs to be differentiated to produce a delta function then the smoothed power spectral density for an array of such signatures will have a slope of - 2n on a log-log plot. A saw tooth signature needs to be differentiated once and this gives the -2 power law as shown in figure 6(b) from an experiment. This rapidly disappears as we move further downstream.

Perry & Chong [23] carried out a crude analysis based on the above potential flow spectral results and assumed that vortex rods occured in pairs as shown in figure 7, with opposite signs of circulation. This vortex pair will be regarded as an "eddy" and its far field effect is weak, giving a finite amount of kinetic energy in the induced velocity field in planes crossing the rods.

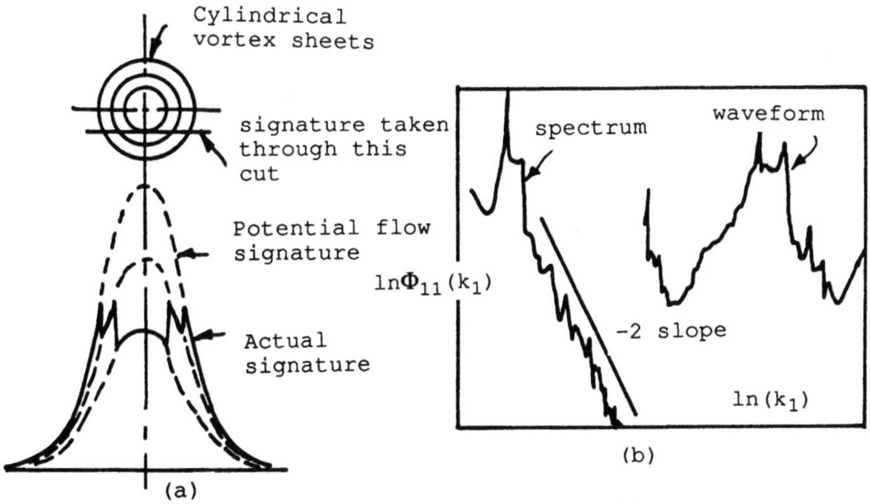

Fig.6. Saw-tooth signatures through a vortex core (a) theoretical signature generated by cylindrical vortex sheets. (b) experimental results. After Perry & Tan [21].

The analysis used signatures generated by an isolated line vortex with arbitrary limits set on certain integrals to simulate the effect of the presence of another line vortex of opposite circulation some distance away. For a random array of vortex pairs, each with a characteristic length scale y_0 and velocity scale V, the power spectral density $\Phi_{11}(k_1y_0)$ obtained by ensemble averaging the square of the modulus of the Fourier transforms of all the signatures is given by

$$\frac{\Phi_{11}(k_1y_0)}{V} \sim \frac{f}{k_1y_0} \cdot \{\exp(-2k_1y_0) - \exp(-2k_1y_0(\frac{y_1}{y_0}))\} \qquad (1)$$

The suffix 11 denotes a spectrum for the streamwise velocity fluctuations, k_1 is the streamwise wavenumber and $\Phi_{11}(k_1y_0)$ is the energy per unit nondimensional wavenumber k_1y_0. From now on the argument of a spectral function will denote the unit quantity over which the energy density is measured. Here y_0 is the core radius assuming solid body rotation, y_1 is the distance between two vortices in a vortex pair and f is a weighting factor related to the population density of vortex pairs. The symbol ~ means "to scale with" or "is proportional to". Also in this approximation y_1/y_0 is considered to be large.

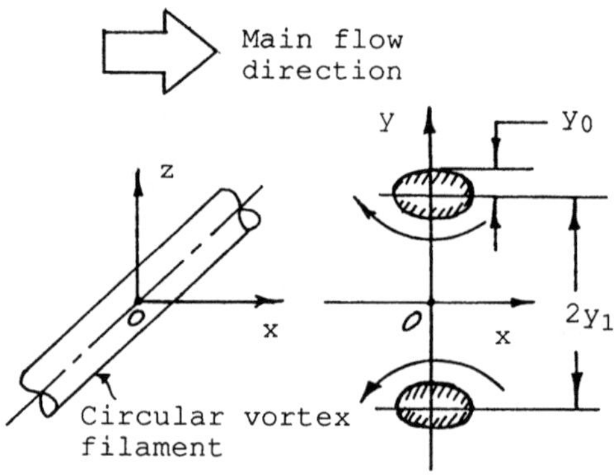

Fig.7. A vortex pair which constitutes an eddy.

Recently Perry et al. [24] carried out a more sophisticated analysis where the vortex loop is given a more realistic shape (like a horseshoe or Λ-shaped vortex) and have included the effects of a boundary in the case of boundary layer flows by the use of image vortices. Furthermore, a Gaussian distribution of vorticity was used in the vortex cores and the cores need not necessarily be small. The results of the analysis are qualitatively the same as the simple analysis (at least for the streamwise velocity fluctuations) and so for the discussion here the simple analysis will be used. Equation (1) is shown plotted in figure 8(a) with the usual log-log coordinates and it can be seen that at low wavenumbers it is flat and at high wavenumbers it has an exponential like fall off. In figure 8(b) (1) is shown in a premultiplied form on a log-linear plot i.e. $(k_1y_0)\Phi_{11}(k_1y_0)$ versus $\log(k_1y_0)$. This plot has the desirable feature that the area under the curve is the energy contribution given by this random array of eddies. If u_1 is the streamwise velocity fluctuation, Φ_{11} is assumed to be normalized such that

$$\overline{u^2} = \int_0^\infty \Phi_{11}(k_1y_0)\,d(k_1y_0) = \int_{-\infty}^\infty (k_1y_0)\Phi_{11}(k_1y_0)\,d\ln(k_1y_0) \qquad (2)$$

An overscore denotes a time average.

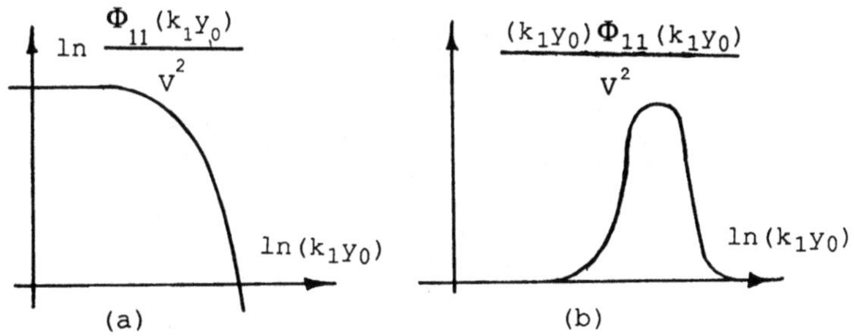

Fig.8. Power spectral density produced by a random array of vortex pairs. (a) Log-log plot (b) premultiplied spectra on log-linear plot. (Sketches only).

APPLICATION TO PLANE MIXING LAYERS

Perry et al. [25] made use of the above concept of a coherent structure in the development of a predictive scheme for the evolution of spectra in a plane mixing layer and this will be critically discussed. To the author's knowledge this is the first "predictive" scheme which incorporates coherent structures. Based on the flow visualization results of

Breidenthal [19] and Jimenez, et al. [26] a plane mixing layer is considered to be made up of vortex loops as shown in figure 9. Much of the vorticity is longitudinally orientated but it is the spanwise components which give the familiar roll-ups oriented across the span as observed by Brown & Roshko [2]. As the layer evolves these spanwise roll-ups are believed to pair and this results in a linear growth rate of the shear layer as illustrated in figure 10. Actually in the layer studied by Perry et al. [25] these spanwise roll-ups were very jumbled and three dimensional even close to the splitter plate. Although the spanwise roll-ups undeniably pair (at least in the Brown & Roshko [2] experiments), a crucial question to ask is whether or not the three-dimensional vortex loops pair.

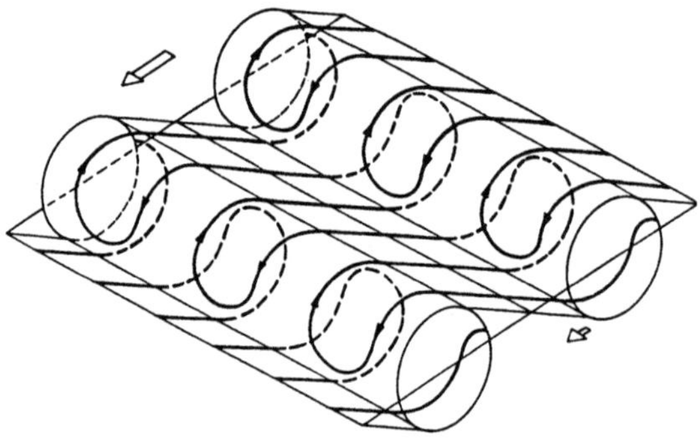

Fig.9. Model for streamwise vorticity distribution in a plane mixing layer. After Jimenez et al. [26].

Fig.10. Plane mixing layer vortex roll-up. Vortices in the act of pairing are not shown.

Perry & Tan [21] have found that in negatively buoyant coflowing wakes which are stimulated with a fixed frequency f of disturbance with a subharmonic f/2 superimposed, three-dimensional vortex loops do pair in a similar fashion as in the carefully controlled two-dimensional pairing experiments of Ho & Huang [27].

In the model of Perry et al. [25] it is assumed that, from the spectral viewpoint, most of the turbulent kinetic energy comes from the longitudinal eddies. This is probably true only after sufficient streamwise development. The smallest eddies form at the trailing edge and have a length scale y_0 which is dependent on viscosity and a characteristic velocity scale U_0 which is the velocity jump across the layer. The spanwise and streamwise spacing of these eddies is also assumed to scale initially with y_0. These are the only eddies present right at the splitter plate and are said to belong to the 1st hierarchy of eddy scales. This would give rise to a power sectral density as given by equation (1). However, as we move downstream some of these eddies pair, giving eddies of the same characteristic velocity scale U_0 but with their length scale and circulation doubled. Such eddies are said to belong to the second hierarchy and as the population of the second hierarchy grows the population of the first hierarchy is depleted as we move downstream. This process of "eddy migration in hierarchy space" continues to yet higher hierarchies giving a range of eddy scales. A simple "migration policy" was devised based on dimensional analysis and this satisfied all the self-preserving flow constraints i.e. it gave a linear increase in shear layer thickness with distance x downstream, an invariant maximum turbulence intensity $\overline{u_1^2}/U_0^2$ with x, a production of energy which scaled with the dissipation and although the scale of the first hierarchy was viscosity dependent, the evolution of the layer was invariant with Reynolds number for the energy containing motions. Even the correct formula for the spreading angle could be derived from the analysis.

It turns out that the contribution each hierarchy makes to the turbulent kinetic energy is proportional to the population density of the first hierarchy eddies which make up the given hierarchy. If the eddies are imagined to line up in neat spanwise rows (this is simply a device for keeping count of the eddies), then this density is the number of rows per unit length x (or number of "basic eddy rows" per unit length as referred to by Perry, et al. [25]). The spanwise spacing between the cores of a vortex pair doubles in the pairing process and so also does the spanwise distance between vortex pairs in a given row. The migration policy is based on the idea that the total number of 1st hierarchy basic eddy rows per unit length must remain constant irrespective of how they are distributed among the hierarchies so as to satisfy the self-preserving flow constraint of a maximum $\overline{u_1^2}/U_0^2$ being invariant with x. The rate of accumulation of basic eddy rows in a given hierarchy is given by the difference of basic eddy rows flowing in and those flowing out of the hierarchy. The greater the population density in a given hierarchy the more likely pairing will occur causing basic eddy rows to flow out into the next hierarchy. By solving a group of coupled differential equations the migration can be calculated and the evolution of the spectra and shear layer length scale can be computed.

Figure 11(a) shows how the population of basic eddy rows migrate in time (or streamwise distance) and this influences the weighting factor f in (1). In fact the power spectral density is given by

$$\frac{\Phi_{11}(k_1 y_0)}{U_0^2} \sim \sum_{n=1}^{N} \frac{f(n)}{k_1 y_0}\{\exp(-2k_1 y_0 2^{n-1}) - \exp(-2k_1 y_0 (\frac{y_1}{y_0}) 2^{n-1})\} \qquad (3)$$

Here n is the the hierarchy number and N is the total number of hierarchies. The factor 2^{n-1} is used since the length scales are assumed to double in the pairing process. Since geometrical similarity for all vortex pairs is assumed, then y_1/y_0 is regarded as a universal constant. From (3) it can be shown that the resulting power spectral density is obtained by using the curve shown in figure 8(b) to give a series of curves one octave apart as in figure 11(b) and multiplying them by the appropriate weighting factor f(n) shown in figure 11(a) and then summing up the distributions to give figure 11(c). The asymptotic distribution gives a -2 power law, and the range of wavenumbers over which this occurs increases with streamwise distance.

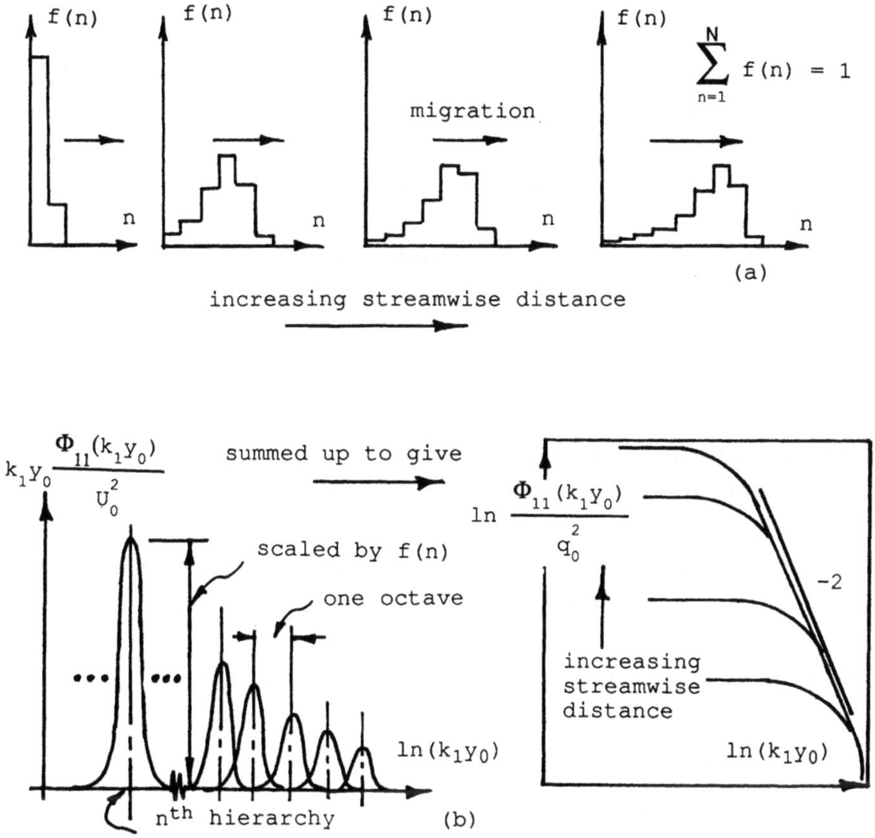

Fig.11. Method of constructing spectra given the evolution of f(n). (a) f(n) evolution by migration process (b) spectral construction.

The spectra satisfy certain similarity laws. For instance we have spectral collapse at low wavenumbers according to

$$\frac{\Phi_{11}(k_1 l_s)}{U_1^2} = f_1(k_1 l_s) \tag{4}$$

whereas, at high wavenumbers

$$\frac{\Phi_{11}(k_1y_0)}{q_0^2} = f_2(k_1y_0) \tag{5}$$

where $q_0 = U_0(f(1))^{1/2}$ and l_s is the average length scale of the hierarchies weighted according to their basic eddy row population density and this is assumed to scale with the shear layer thickness. Making the usual assumption of how the dissipation ε is related to the finest scale motions and also assuming that production scales with the dissipation gives

$$\varepsilon \sim \nu \frac{q_0^2}{y_0^2} \sim \frac{U_0^3}{l_s} \tag{6}$$

If there exists a region of overlap between equations (4) and (5) then (4), (5) and (6) lead to -2 power spectral region which is consistant with the asymptotic prediction from the migration analysis. In fact, the non-asymptotic calculations corresponding to the experimental range observed by Perry et al. [25] give a slope close to the classical Kolmogoroff result of a -5/3 law and the computed spectra scaled with the classical Kolmogoroff velocity and length scales reasonably well and agreement with experiment was good.

These spectra serve to illustrate some important general properties of power spectral density plots. Over most of the range of wavenumbers we have a -2 power law and this arises completely from the distribution of eddy scales. Unlike the -2 power law given earlier for a rolled-up vortex sheet, this distribution is unaffected by signature shape. However at the very highest wavenumbers, the spectra are influenced mainly by the high wavenumber end of the Fourier transforms of the velocity signatures of the smallest eddies. At very low wavenumbers the flat part of the spectrum arises from the property of the signatures of the larger eddies in the flow.

Although the model just outlined appears to give all the right answers the basic ideas are completely opposed to the classical assumptions. The model is speculative and it should not be taken too seriously. Nevertheless, it forces us to ask crucial questions and to think about things in a different light. The basic assumption made in the model is that the finest scale motions have a fixed Reynolds number $U_0 y_0/\nu$ which is universal and that the smallest scale y_0 remains constant throughout the flow development although the population of such eddies in the lowest hierarchy are continually being depleted. The Kolmogoroff theory states that the finest scale motions have a fixed universal Reynolds number $q_0 y_0/\nu$ and that $y_0 \sim \varepsilon^{-1/4}\nu^{3/4}$ and $q_0 \sim \nu^{1/4}\varepsilon^{1/4}$. This would give rise to the finest scale motions increasing their length scale with streamwise distance, in fact $y_0 \sim x^{1/4}$. In the model the formation of the large scale motions comes from a continual pairing of the finer scale motions. Also, all motions have been assumed to be coherent i.e. all eddies on average have a fixed orientation which is quite different from the classical Kolmogoroff assumption that motions, even in the inertial subrange, are statistically isotropic.

The Achilles heel of the model is that $U_0 y_0/\nu$ is universal but during the formation process near the trailing edge of the plate, other length scales such as boundary layer thicknesses are probably involved. Also the pairing process is also probably accompanied by a vortex cancellation process where longitudinal vortices of opposite sign are forced to pair and an eddy count shows that this is possible even though the population density of basic eddy rows remain fixed. This would give rise to a considerable amount of incoherent motions made up of "dead eddy debris". Of course, the spanwise component of vorticity can never cancel and the associated circulation and velocity jump across the shear layer must remain fixed. Also new spanwise vorticity cannot be created since, as Lighthill [28] pointed out, in an incompressible constant density fluid all vorticity is generated at solid boundaries.

A further death process must also be present for individual vortex pairs particularly in the first hierarchy. Figure 7 shows a vortex pair and it can be shown that vorticity diffusion must cause the dimension y_0 to grow i.e. $y \sim (\nu t)^{1/2}$ which for a fixed y_1, must lead to the death of the eddy. The characteristic life time T_L of the eddy must be $T_L \sim y_1^2/\nu$. However, if the vortex pair is being stretched by the large scale spanwise roll-ups, the growth of the vortex cores might be suppressed. For instance, the Burgers [20] solution for an isolated vortex filament subjected to a constant axisymmetrical strain rate along the filament axis causes y_0 to initially shrink and then to asymptote to a fixed radius. If the strain rate is constant, then a marked length of vortex filament must increase exponentially with time. Perry & Chong [23] have shown that for a vortex filament whose marked length increase linearly with time, the radius y_0 initially shrinks but will ultimately increase according to $y_0 \sim (\nu t)^{1/2}$. Our vortex pair in the shear layer is probably being subjected to a plane strain rate where the vortex cores are being squeezed normal to the line joining the two core centres. In any case the author asserts that in order to prevent the vorticity cancellation, it is necessary for the vortices to increase their length at least exponentially with time. Such a strain rate occurs at the saddle points between two successive roll-ups but unfortunately this strain rate is not sustained. It can be shown quite easily that the strength of the plane strain rate decreases as we move downstream since the characteristic velocity scale of the large scale roll-ups remain fixed but their length scales increases as we move downstream.

In the model, the finest scale motions must persist even though their population is being depleted. However, from the above discussion they must die if they have failed to pair. Similarly this must ultimately happen in the next hierarchy and so on. Perhaps the model conjectured for wall turbulence (see later) is applicable here. That is, if vortex pairing (of the longitudinal vortices) does occur, it is confined to the energy containing and low wavenumber range with a classical Kolmogoroff model for the inertial and dissipation range. The supply of eddies for the migration or pairing process may be coming from instabilities in the spanwise vorticity. Spectral measurements alone cannot clarify these issues.

One important question regarding the classical -5/3 region is whether this is due to the presence of a range of geometrically similar eddies or is it due to the characteristic signature of eddies. There are models, e.g. Lundgren [29], where the -5/3 law is thought to result from a Kaden like spiral roll-up of vortex sheets. The author currently supports the former view rather than the latter but perhaps both effects are present.

WALL TURBULENCE

As mentioned earlier, it has been postulated that wall turbulence consists of a forest of horseshoe vortices. These vortices are attached to the boundary and are imagined to "slip" relative to this boundary and retain their angle of inclination of approxmately 45^0 in the downstream direction as they are convected downstream. According to Perry & Chong [23] and Perry et al. [24] there exists a range of eddy scales i.e. hierarchies. In each hierarchy, the average lateral and streamwise spacing is proportional to the scale of the eddies and that in a zero pressure gradient boundary layer all hierarchies are geometrically similar and have the same characteristic velocity scale U_τ, the wall shear velocity. The length scale of the smallest hierarchy is proportional to the Kline scale ($\sim \nu/U_\tau$) if the wall is smooth.

In each hierarchy there will be a range of eddy shapes. From a very crude Biot-Savart law calculation Perry and Chong showed that an isolated Λ-vortex and its image undergoes a self-induced stretching process where the legs of the vortex are brought together and each leg appears to be undergoing a local axisymetrical stretching although in the large, the legs undergo a plane strain like motion as shown in figure 17(d). This is quite different from the stretching in the plane mixing layer. The length of the leg increases linearly with time and this combined with the observation of the legs coming together give rise to eddy death. Thus, the height of the highest eddy within a hierarchy is limited and scales with the scale of the hierarchy. For the moment let us assume that a hierarchy can be characterized by a single representative eddy shape.

These attached coherent eddies are thought to be responsible for the mean vorticity, Reynolds shear stress and for most of the energy containing motions. Perry et al. [24] postulate that they are not the only motions present but there are detached incoherent motions which do not contribute to the mean vorticity or a Reynolds shear stress and are locally isotropic. These eddies are responsible for the dissipation and will be discussed in more detail later. How the hierarchies form is a mystery but vortex pairing has been suggested by Perry and Chong with dead eddies being continually replaced with new vortices being generated at the wall. The pairing model is consistent with the p.d.f. of hierarchy length scales being a series of delta functions of equal area being distributed in a geometric progression (with a factor of 2) and having a constant velocity scale U_τ.

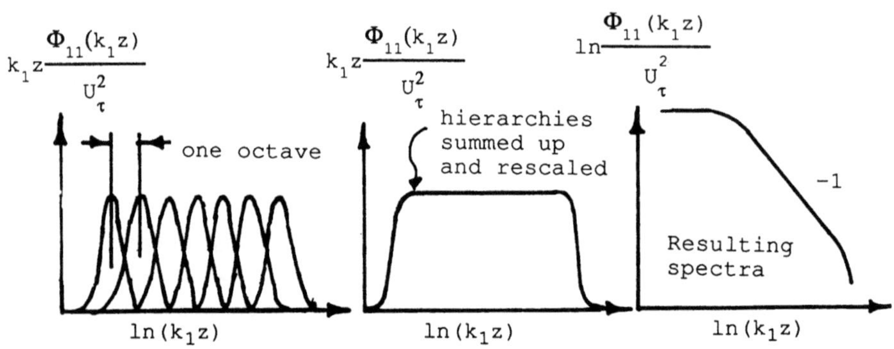

Fig.12. Construction of spectra for wall turbulence.

Any geometric progression will give the correct spectral behaviour and will also give (when smoothed) the logarithmic mean velocity profile in the wall region. In fact Townsend's [30] assumption of a turbulent boundary layer being made up of a superposition of geometrically similar eddies whose length scale p.d.f. is a continuous inverse power law is equivalent to this model. This p.d.f. law is the continuous version of the discretized geometrical progression mentioned earlier. This gives rise to a Reynolds shear stress distribution which varies only slightly with distance z from the wall in the logarithmic wall region. It turns out that with these assumptions the contribution each hierarchy makes to the streamwise component of turbulence intensity $\overline{u_1^2}/U_\tau^2$ is constant and so $f(n)=1$ in equation (3). In fact if we use (3) (which incorporates our discretized p.d.f.) we obtain an a -1 power law for the spectrum over most of the energy containing range as illustrated in figure 12. The resulting spectral behaviour for the streamwise and spanwise components of velocity fluctuations from the model agree with experiment, namely, in the logarithmic wall region there is an outer flow scaling,

$$\frac{\Phi_{11}(k_1\Delta_E)}{U_\tau^2} = f_0(k_1\Delta_E) \qquad (7)$$

which gives collapse of data at low wavenumbers. Here Δ_E is the boundary layer thickness. There is an inner flow scaling where data collapses at high wavenumbers i.e.

$$\frac{\Phi_{11}(k_1 z)}{U_\tau^2} = f_i(k_1 z) \qquad (8)$$

If a region of overlap exists between these two laws then the -1 power law spectrum can be derived analytically from these two laws. However, what the model lacks are the fine scale dissipative motions which are needed to conform with the experimental result that energy production and dissipation are approximately in balance in the logarithmic region. The attached eddies alone do not give this result. These fine scale motions are thought to be made up of the debris of dead eddy material being convected from the very near wall region throughout the flow by the more active attached coherent motions. Therefore assume a classical Kolmogoroff region where

$$\frac{\Phi_{11}(k_1\eta)}{\upsilon} = f_k(k_1\eta) \tag{9}$$

and υ and η are the usual Kolmogoroff velocity and length scales respectively. If production and dissipation of energy are in balance then a relationship between the Kolmogoroff scales and the usual wall variables z and U_τ can be derived. This, together with assuming a region of overlap between (8) and (9) gives the -5/3 law. Experimental data shows strong evidence for the existence of the -5/3 law for all three velocity components and the range of wavenumbers over which it occurs increases with z^+ (where $z^+ = zU_\tau/\nu$). This similarity scheme has been developed to a fairly complete form for the logarithmic wall region by Perry, et al. [24] and the Kolmogoroff scaling is applicable throughout the layer but in the outer flow, production and dissipation are not equal and so we have no simple formula for the dissipation.

One strong feature of the attached eddy model which can readily be tested is the comparison of the spectra for streamwise fluctuations and fluctuations normal to the wall. Figure 13 shows three scales of our representative attached eddy and the instantaneous streamlines induced by each eddy in isolation is illustrated. The boundary condition at the wall means that u_3, the component of velocity normal to the boundary is zero for z -> 0. Here we are using inviscid boundary conditions i.e. there is slip at the wall which is accounted for by the existence of a viscous sublayer. These boundary conditions were used by Townsend [30] in his attached eddy hypothesis. Consider the signatures registered by the hot-wire probe at distance z. Eddies whose scale is much less than z will not register a signature since it can be shown by a Biot-Savart law calculation that such an eddy with its image has a very small far-field effect. An eddy whose length scale is of order z will register a high u_1, u_2 and u_3. Here u_2 is the spanwise velocity fluctuation. However, an eddy whose length scale is much larger than z will register a high u_1 and u_2 but very small u_3 because of the wall boundary condition. Townsend [30] would refer to this as an "inactive" motion since it contributes little to the Reynolds shear stress. As far as u_1 and u_2 signatures are concerned, all eddies of height z and above are registered by the probe. If the largest eddy scales with the boundary layer thickness Δ_E, then the lower z/Δ_E is, the larger is the range of scales observed or using our discretized p.d.f., the more hierarchies are observed. This means that N in equation (3) represents the number of observed hierarchies and y_0 is a representative length scale for the smallest observed hierarchy and would scale with z, the distance our observation is from the wall. In the case of the plane mixing layer, it was assumed that all hierarchies were stretched across the flow and experimental results of Perry et al. [25] indicate this. However, this is not the case in wall turbulence. As the flow Reynolds number increases, the smaller will be the viscous zone and the scale of the smallest hierarchy ($\sim \nu/U_\tau$) compared to the layer thickness Δ_E. This means that we can make z/Δ_E smaller and still be in the fully turbulent wall region. Thus the number of observed hierarchies will increase giving a larger range of wavenumbers over which the -1 spectral law is observed. In fact the Kármán number of the flow, $\Delta_E U_\tau/\nu$ is actually proportional to the ratio of the largest hierarchy to the smallest hierarchy.

Contrary to all of the above, the number of observed hierarchies will always be approximately one for the u_3 fluctuations. This gives rise to all spectra for u_3 in the wall region to scale with an inner scaling only, i.e.

$$\frac{\Phi_{33}(k_1 z)}{U_\tau^2} = f_3(k_1 z) \tag{10}$$

without any -1 power law region but dimensional analysis shows that this can include a Kolmogoroff region with the usual viscosity dependent high wavenumber cut off.

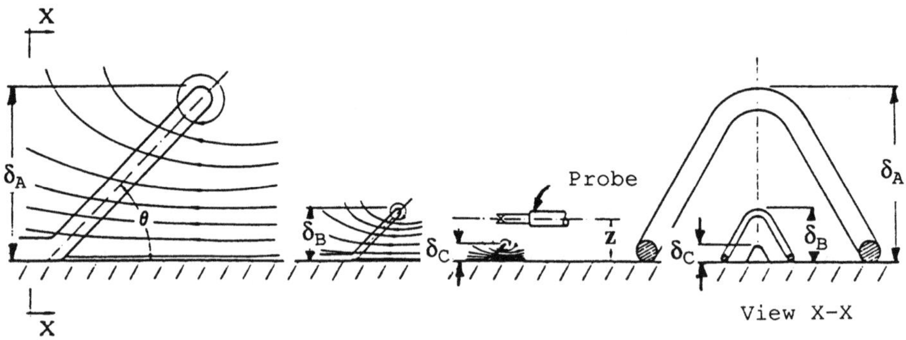

Fig.13. Representative eddies of wall turbulence. Three scales shown. After Perry et al. [24].

Recently, all of this analysis has been extended to the whole region of turbulent flow in the boundary layer. A more sophisticated Biot-Savart law calculation was carried out for the representative eddy shown in figure 14.

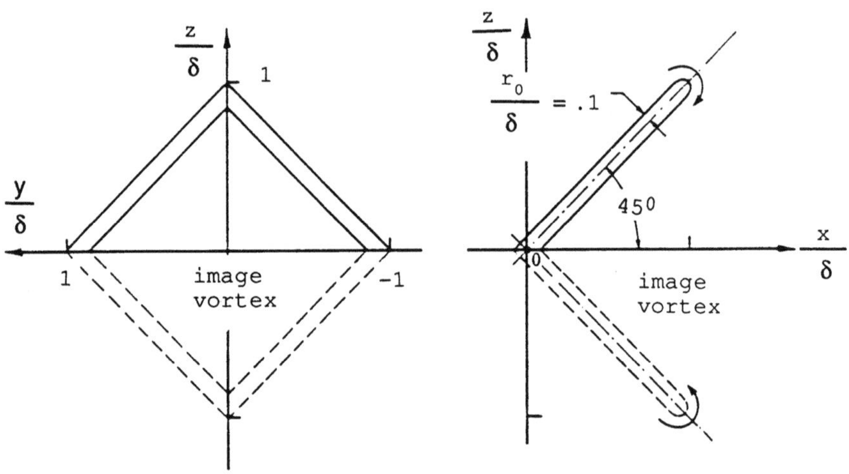

Fig.14. Representative eddy used in the Biot-Savart law calculations for spectra. r_0 is the standard deviation radius for the assumed Gassian distribution of vorticity.

This has a Gaussian distribution of vorticity in the vortex cores and an image vortex was included to give the correct boundary conditions. Let us now use the symbol δ, to denote the scale of the representative eddy (i.e. hierarchy scale). Velocity signatures were computed versus x/δ for a large range of z/δ and y/δ. The resulting computed spectra are shown in figure 15 for u_1 and u_3 i.e. Φ_{11} and Φ_{33} respectively. A continuous inverse power law p.d.f. was used rather than the discretized p.d.f. Also the weak far field effect of the eddy for $\delta << z$ came in automatically whereas in the crude analysis of Perry & Chong [23], the signatures were assumed to switch off once $\delta < z$. These results do not include the isotropic detached eddying motions i.e. there is no Kolmogoroff inertial subrange nor any Kolmogoroff viscosity dependent cut off. The high wavenumber cut-off is due entirely to the signatures of the smallest observed hierarchy of attached eddies.

In figure 16 are the spectra for flow in a pipe. Although the analysis is more appropriate for a boundary layer, at least the trends are qualitatively similar. At the time of writing the author and his colleagues were in the process of preparing some very encouraging results from boundary layer flow measurements. One interesting feature of the model calculations is that it can be seen that the -1 spectral power law is only an asymptotic law. One would need unrealistically high Reynolds numbers before the spectra truely asymptotes to such a law. This is also brought out in the experimental data. This law is a consequence of the distribution of energy among the scales. However the shape of the u_3 spectra depends on velocity signature shape as also does the u_1 spectra for large z/Δ_E where the number of hierarchies is small.

The experimental evidence for all of the above scaling laws is encouraging in spite of experimental difficulties with probe spatial resolution and the assumption of the Taylor hypothesis for transforming frequency to wavenumber.

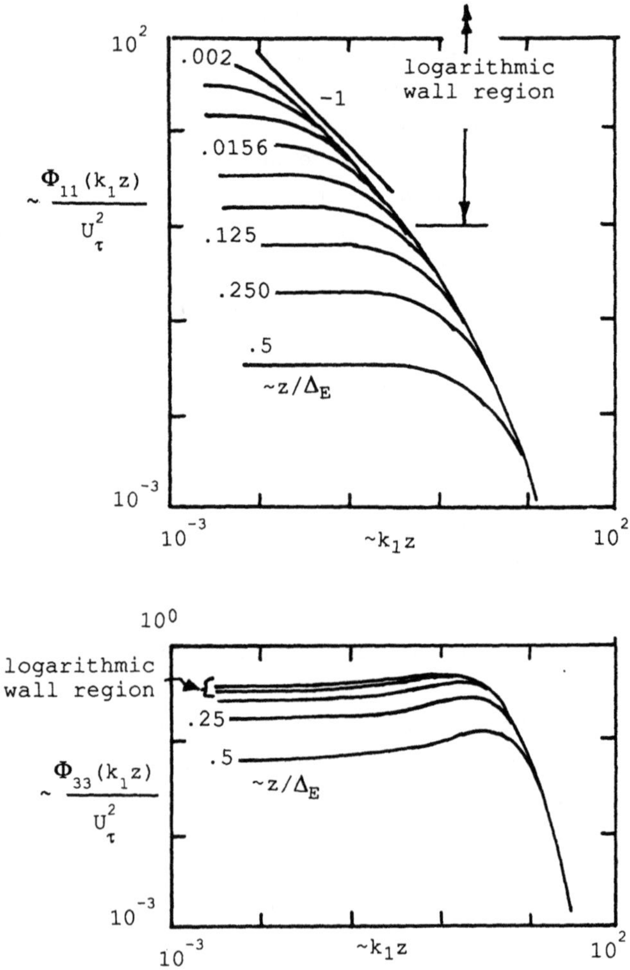

Fig.15. Computed spectra using the representative eddy shown in fig.14. No weighting function W was used (see later). Logarithmic wall region occurs for $z/\Delta_E \lesssim .1$ and $z^+ \gtrsim 100$.

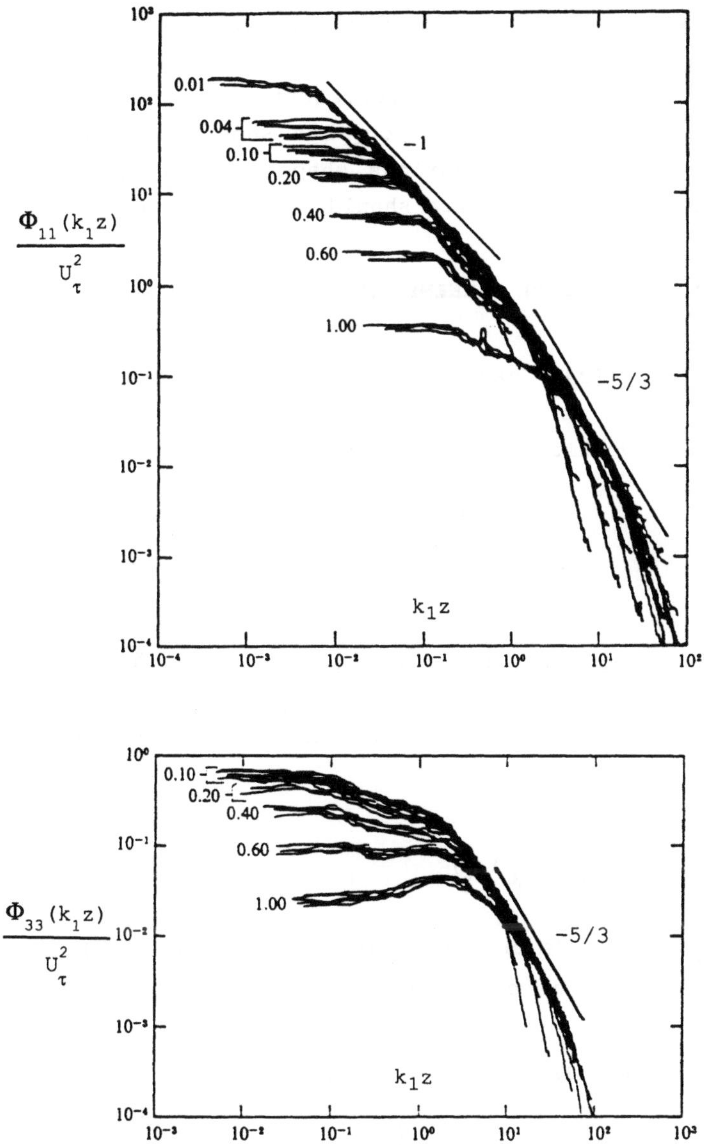

Fig.16. Experimental spectra obtained from a pipe. After Perry et al. [24].

In these calculations a fixed eddy shape was used. As mentioned earlier, the eddies in a given hierarchy are undergoing a stretching process and the representative eddy should not resemble a single realizable eddy but should be made up of an assemblage of eddies of various shapes. However, the general conclusions regarding spectra should be unaffected.

TOWARDS A PREDICTIVE SCHEME IN WALL TURBULENCE

Perry et al. [24] have recently been experimenting with different shapes for the representative eddies. It can be shown that the mean vorticity through the layer is given by

$$\frac{dU_1}{dz} = \int_{\delta_1}^{\Delta_E} \xi_H P_H(\delta)\, d\delta \tag{11}$$

where $P_H(\Delta)$ is the p.d.f. of hierarchy scales, δ_1 is the scale of the smallest hierarchy (proportional to the Kline scaling) and ξ_H is the vorticity intensity function for the hierarchy. In fact

$$\xi_H = \frac{U_\tau}{\delta} f\left(\frac{z}{\delta}\right) \tag{12}$$

where $f(z/\delta)$ depends on the representative eddy shape assuming that all vorticity is confined to the vortex loops.

The inverse power law p.d.f. is given by

$$P_H(\delta) = \frac{M}{\delta} \tag{13}$$

where M is a universal constant. This leads to the velocity defect being given by

$$U_D^* = \frac{U_{1E}-U_1}{U_\tau} = \int_0^{\lambda_E} \int_{\lambda_1}^{\lambda_E} M\, h(\lambda) e^{-\lambda} d\lambda d\lambda_E \qquad (14)$$

where U_{1E} is the free stream velocity, U_1 is the local mean velocity and $\lambda = \ln(\delta/z)$, $\lambda_1 = \ln(\delta_1/z)$, $\lambda_E = \ln(\Delta_E/z)$ and $h(\lambda) = f(z/\delta)$. Note that $\lambda_1 = \lambda_E - \ln \Delta_E/\delta_1$ and $\Delta_E/\delta_1 \sim \Delta_E U_\tau/\nu$, the Kármán number. Hence for a fixed Kármán number λ_E and λ_1 are related.

Figure 17 shows various eddy shapes as one would see them by looking upstream. The corresponding vorticity functions $f(z/\delta)$ and $h(\lambda)e^{-\lambda}$ are also shown. The resulting velocity-defect distributions are shown in figure 18 and are compared with the Hama [31] velocity defect law although the Coles [32] law of the wall and wake model would serve equally as well. It was found that the only eddy shape which gave the Hama defect law was the bow-legged eddy labelled (c). However, it will be seen from spectral considerations that the inverse power law p.d.f. distributions need modification as indicated in figure 19(a). That is, more weighting should be given to the highest hierarchies and this shows up as a low wavenumber bump in the spectrum measured in the logarithmic wall region. This bump can hardly be seen on a log-log plot given in figure 19(c) but the premultiplied spectra on a log-linear plots shows this quite clearly as shown in figure 19(b). This bump is not as obvious in pipe flow and it appears that it is controlled by the large scale geometry of the flow situation (i.e. its size depends on whether we are in a pipe, boundary layer or duct). If the p.d.f. distribution is modified as shown in figure 19(a), it should be possible to obtain the correct spectral behaviour and at the same time have the eddy shapes labelled (e) giving the Hama velocity defect law distribution. This is an assemblage of Π-eddies undergoing a stretching process and was chosen purely for simplicity. So far such spectral calculations were done using the crude Perry & Chong model.

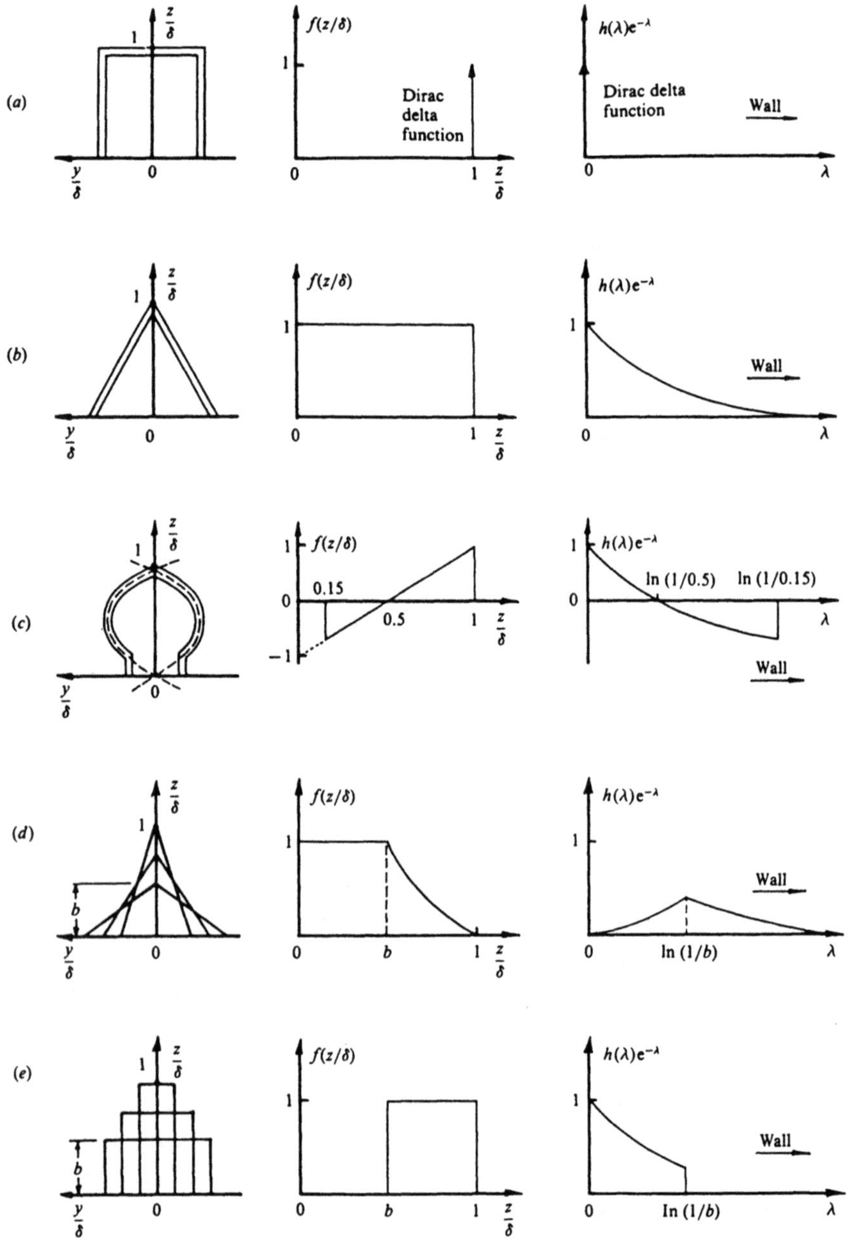

Fig.17. Projection in the y-z plane of various representative eddy geometries together with their $f(z/\delta)$ and $h(\lambda)e^{\lambda}$ distributions (see text). After Perry et al. [24].

Let now $P_H(\delta)$ be given by

$$P_H(\delta) = \frac{M}{\delta} W(\frac{\delta}{\Delta_E}) \tag{15}$$

and $W(\delta/\Delta_E) = 1$ as $\delta/\Delta_E \rightarrow 0$

W is a weighting function and if $w(\lambda - \lambda_E) = W(\delta/\Delta_E)$ then

$$U_D^* = \int_0^{\lambda_E} \int_{\lambda_1}^{\lambda_E} M\ h(\lambda) w(\lambda-\lambda_E) d\lambda d\lambda_E \tag{16}$$

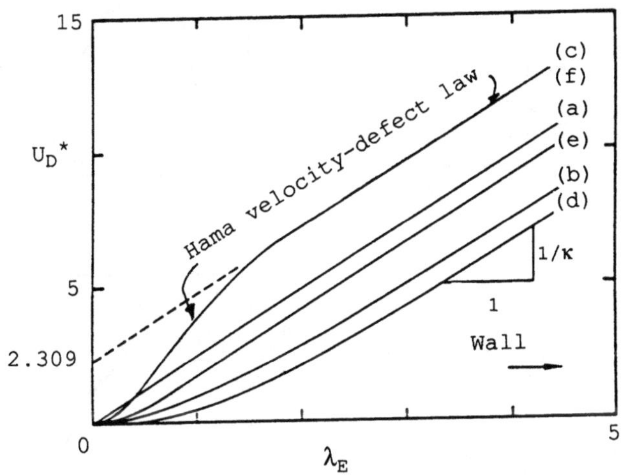

Fig.18. Velocity defect distributions using representative eddies shown in fig. 17. After Perry et al. [24].

The physical meaning for this weighting function could be that there exists a higher density population of representative eddies in the higher hierarchies, a higher characteristic velocity scale or simply that there is a loss of geometrical similarity. Of course, like in the plane mixing layer, we can drop the requirement that the hierarchies are geometrically similar in all aspects which includes the average eddy spacing. We only require that all representative eddies are geometrically similar and their population density or velocity scale can vary between hierarchies. This can be accounted for by the weighting factor w. It might give us a clue to a possible link between the Coles [32] wake function parameter and the low wavenumber motions in the wall region. It may therefore be possible to model the evolution of a turbulent boundary layer in an adverse pressure gradient in the following way.

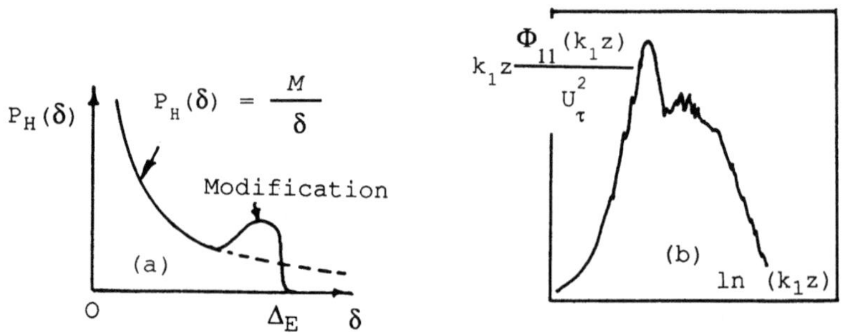

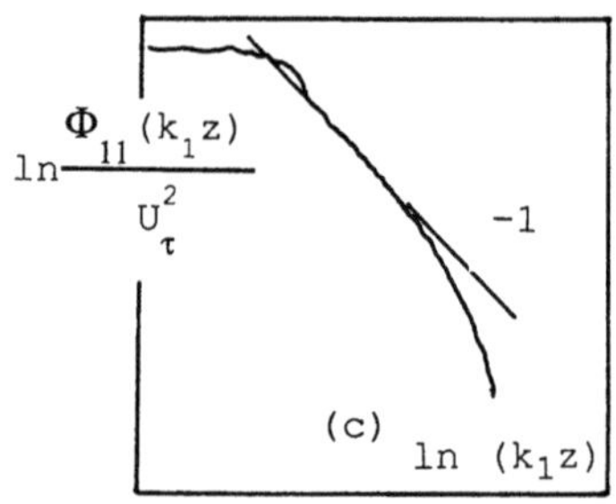

Fig.19. Possible relationship between hierarchy p.d.f. and spectra.

(a) Modification to p.d.f.
(b) premultiplied spectra in logarithmic wall region in a turbulent boundary layer.
(c) Log-log plot of same spectra. After Perry et al. [24]

Typical eddy intensity functions are shown in figure 20 and the behaviour at $z/\delta \rightarrow 0$ is a consequence of the Townsend boundary condition discussed earlier. These functions $I_{ij}(z/\delta)$ are defined such that

$$\frac{\overline{u_i u_j}}{U_\tau^2} = \int_{\delta_1}^{\Delta_E} I_{ij}\left(\frac{z}{\delta}\right) P_H(\delta)\, d\delta \tag{17}$$

and if our weighting function is included then

$$\frac{\overline{u_i u_j}}{U_\tau^2} = \int_{\lambda_1}^{\lambda_E} M\, I_{ij}(\lambda)\, w(\lambda - \lambda_E) d\lambda \tag{18}$$

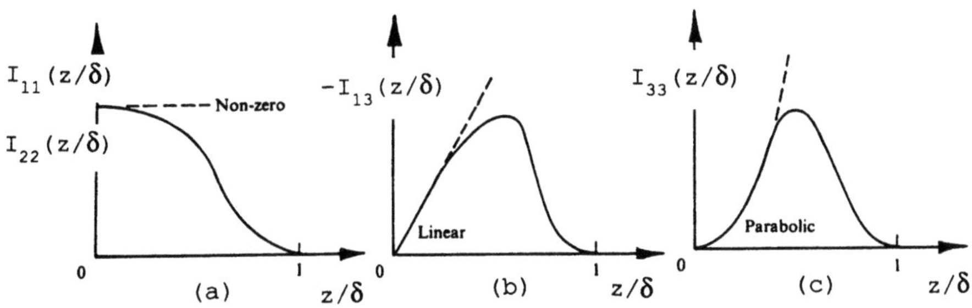

Fig.20. Typical eddy intensity functions (Perry et al. [24]).

Contributions from the Kolmogoroff inertial subrange cannot at this stage be included but this should not matter for the nondimensional Reynolds shear stress $-\overline{u_1 u_3}/U_\tau^2$. Thus, the equations needed to form a closure scheme have emerged. These equations are

$$U_D^* = \int_0^{\lambda_E} \int_{\lambda_1}^{\lambda_E} M\, h(\lambda) e^{-\lambda} w(\lambda - \lambda_E)\, d\lambda d\lambda_E$$

or

$$\frac{dU_D^*}{d\lambda_E} = \int_{\lambda_1}^{\lambda_E} M\, h(\lambda)\, e^{-\lambda} w(\lambda - \lambda_E)\, d\lambda \tag{19}$$

and
$$\frac{\overline{u_1 u_3}}{U_\tau^2} = \int_{\lambda_1}^{\lambda_E} M\, I_{13}(\lambda)\, w(\lambda - \lambda_E)\, d\lambda$$

Thus if all eddy shapes are assumed to be invariant with hierarchy scale even with an imposed pressure gradient we have a link between the velocity-defect distribution and the Reynolds shear stress distribution. For a fixed eddy shape there are no empirical constants except for M and this is found by setting

$$\frac{dU_D^*}{d\lambda_E} = \frac{1}{\kappa} \quad \text{for} \quad \lambda_E \gg 0 \ \text{ and } \ \lambda_1 \ll 0 \tag{20}$$

where κ is the Kármán constant. Any further adjustments made to the model would be equivalent to adjusting the shape of the eddies. Perhaps a differential or integral prediction scheme could be devised based on (19) in combination with the law of the wall, the Coles law of the wake and the Reynolds momentum equations. Normal stresses could also be included in the momentum balance if some means are found for the inclusion of the inertial subrange. However, the contribution from the inertial subrange may not be important.

The difficulty with the method is to find the appropriate representative eddy shape or, more correctly, the appropriate assemblage of eddy shapes in a given hierarchy if vortex stretching is to be included. Some simple eddy shapes could be tried out first. The hope is that these will be universal, i.e. invariant with hierarchy scale even with imposed pressure gradients. Otherwise all of the above analysis would need to be rederived to account for this variance. The complexity of the problem would then become considerable and we would need to know something about eddy shape dynamics. Perhaps then the method with fixed eddy shapes might be applicable to a limited class of flows, e.g. flows with weak pressure gradients. The formulation given by (19) is consistent with the ideas expressed by Townsend [33] that turbulence must be modelled with an account made for the fact that physical processes occurring at a point are related to processes remote from that point. In other words, the flow cannot be analysed on a point by point basis such as is implicit in those methods which use transport coefficients or other types of coefficients related to the local properties of the flow. The flow should really be analysed as an integrated whole with regard for the distribution of quantities remote from the point of interest. The integral equations given in (19) satisfy these requirements.

CONCLUSIONS

It can be seen that the use of the coherent structure concept shows considerable promise in the description of turbulence. However, the evolution of a range of scales for the coherent substructures is still a mystery as are the incoherent motions usually associated with the Kolmogoroff region. The use of coherent structures for prediction methods is still very much in its infancy and some crude tentative attempts have been outlined here.

ACKNOWLEDGEMENTS

The author wishes to acknowledge colleague Dr. M.S. Chong for many fruitful and stimulating discussions and the Australian Research Grants Scheme for financial assistance.

REFERENCES

[1] Hussain A.K.M.F. Coherent structures - Reality and Myth. Report FM-17. Department of Mechanical Engineering. University of Houston Central Campus. Houston,Texas. July 1982. Revised: April 1983.

[2] Brown. G. & Roshko A. On density effects and large structure in turbulent mixing layers. J. Fluid Mech. (1974) 64: 775-816.

[3] Lumley. J.L. Coherent structures in turbulence. In Transition and Turbulence (ed. R. Meyer) N.Y. Academic Press (1981); 215-242.

[4] Brown G. & Roshko A. Symposium on Turbulent Shear Flows. London; Imperial College (1971) p. 23.1

[5] Roshko A. The plane mixing layer; flow visualization results and three dimensional effects. In Proceedings, An international conference on the role of coherent structures in modelling turbulence and mixing. Madrid: University Politecnica 1981.

[6] Cantwell B.J. Organized motions in turbulent flow. Ann. Rev. Fluid Mech. (1981) 13. 457-515

[7] Browand. E.K & Troutt. T.R. A note on spanwise structure in the two-dimensional mixing layer. J. Fluid Mech. (1980) 917: 771-781.

[8] Bradshaw P. The effect of initial conditions on the development of a free shear layer. J. Fluid Mech. (1966) 26: 225-236.

[9] Hussain A.K.M.F. Coherent structures and studies of perturbed and unperturbed jets, in Proceedings, An international conference on the role of coherent structures in modelling turbulence and mixing Madrid; University Politechnica (1981)

[10] Perry. A.E. & Steiner T.R. Large-scale vortex structures in turbulent wakes behind bluff bodies. Part 1. Vortex formation process. J. Fluid mech. (1987). In the press.

[11] Steiner T.R. & Perry A.E. Large-scale vortex structures in turbulent wakes behind bluff bodies. Part 2. Far-wake structures. J. Fluid Mech. (1987) In the press.

[12] Kline S.J., Reynolds W.C., Schraub F.A., Rundstadler P.W. The structure of turbulent boundary layers. J. Fluid Mech. (1967) 30; 741-773.

[13] Kim J. & Moin J. "The structure of the vorticity field in turbulent channel flow" Part 2. Study of ensemble averaged fields. J. Fluid Mech. (1985), 162; 339-363.

[14] Kim J. Evolution of a vortical structure associated with the bursting event in a channel flow. Fifth Symposium on Turbulent Shear Flow. Cornell University. August 7-9 (1985) 9.23.

[15] Lu S.S. & Willmarth. W W. Measurements of the structure of the Reynolds stress in a turbulent boundary layer.J. Fluid Mech. (1973) 60; 481.

[16] Cantwell B. & Coles D. An experimental study of entrainment and transport in the turbulent near wake of a circular sylinder. J. Fluid Mech. (1983) 136; 321-374.

[17] Head M.R. & Bandyopadhyay P. New aspects of turbulent boundary layer structure. J. Fluid Mech. (1981) 107, 297-337.

[18] Perry. A.E. Lim. T.T. & Teh E.W. A visual study of turbulent spots J. Fluid Mech. (1981) 104; 387-405.

[19] Breidenthal R. Structure in turbulent mixing layers and wakes using chemical reaction. J. Fluid Mech. (1981) 109, 1-24.

[20] Burgers J. M. A mathematical model illustrating the theory of turbulence. Adv. Appl. Mech. (1948) 1; 197-199.

[21] Perry A.E. & Tan D.K.M. Simple three-dimensional vortex motions in coflowing jets and wakes. J. Fluid Mech. (1984) 141; 197-231.

[22] Bracewell R.N. The Fourier transform and its application. McGraw-Hill (1978).

[23] Perry A.E. & Chong M.S. On the mechanism of wall turbulence. J. Fluid Mech. (1982) 119; 173-217.

[24] Perry A.E., Henbest. S & Chong M.S. A theoretical and experimental study of wall turbulence. J. Fluid Mech. (1986) 165: 163-199.

[25] Perry A.E. Chong M.S. & Lim T.T. Vortices in turbulence. In Vortex Motion. Proceedings of a colloquium held at Goettingen on the occasion of the 75th Anniversary of the Aerodynamische Versuchsanstalt in November 1982. Eds. H.G.Hornung and E.A. Muller. Friedr. Vieweg & Sohn, Braunschweig/Wiesbaden.

[26] Jimenez J., Cogollos M. & Bernal L. A perspective view of the plane mixing layer. J. Fluid Mech. (1985) 152; 125-143.

[27] Ho C.M. & Huang L.S. Subharmonics and vortex merging in mixing layers. J. Fluid Mech. (1982) 119; 443.

[28]Lighthill M.J. Attachment and separation in three-dimensional flow. In Laminar Boundary Layers (ed. L. Rosenhead). Oxford Univ. Press, Oxford, U.K. (1963); 72-82

[29] Lundgren T.S. Strained spiral vortex model for turbulent fine structure. Phys. Fluids (1982) 25; 2193-2203.

[30] Townsend A.A. The Structure of Turbulent Shear Flows. (1986) Cambridge University Press.

[31]Hama F.R. Boundary layer characteristics for smooth and rough surfaces. Trans. Soc. Naval Arch. Mar. Engrs. (1954) 62; 333-358.

[32] Coles D. The law of the wake in the turbulent boundary layer. J. Fluid Mech. 1; 191-226.

[33] Townsend A.A. Equilibrium layers and wall turbulence. J. Fluid Mech. (1961) 11; 97-120.

Onset and Development of Coherent Structures in Turbulent Shear Flows

S.P. Bardakhanov, V.V. Kozlov

Institute of Theoretical and Applied Mechanics,
Sibirian Branch of USSR Academy of Sciences,
Novosibirsk, USSR

Introduction

During the last few decades the understanding of turbulent shear layer structures has undergone considerable changes. Earlier turbulence was understood to be in general a chaotic or irregular process. Such understanding comes from fundamental experiments of Osborne Reynolds who suggested to describe the flow field as consisting of a mean velocity field and temporal fluctuations. This method served as a basis for the statistical theory of turbulence developed in works [1-7]. In [8,9] it was proposed to build the theory of turbulence based on "first principles". At the same time the study of turbulence has acquired a new aspect, and it is accepted that in the light of contemporary data turbulent flows do not seem completely chaotic; this evolution of the outlook is connected with the realizing of the role of large-scale regular formations or coherent structures which appear at least in some, and possibly in all the types of the developed turbulent shear flows. What characterizes a new trend is the preservation of information about phases in order to detect and study structures. This is what distinguishes the new ideas from the averaging by Reynolds when such information is lost. It is possible that the most important aspects of the existence of deterministic structures in turbulent flows is the possibility of control over turbulence by means of direct interference with these large structures. Such a control could lead to considerable technical achievements [10].

The determination of coherent structures is by no means settled yet. Let us keep to the determination given in [11]. "Long-living regular large-scale formations having a small-scale turbulence in the background which possess a high degree of universality for the given flow type are called coherent structures".

At present little can be said about theoretical descriptions of turbulent flows based on coherent structures, though attempts have been made to build the models [11-13] which explain some experimental works. It is shown in [17], for example that theoretical studies of coherent structures in streamline flows are only in the initial stage at present. Kovasznay has noted [14] that there is no theory for coherent structures at present. It is clear that to create such a theory one must have experimental results concerning the existence of the conditions of formation of the structures, their interaction with each other and external fields.

1. Connection of the processes of the onset and development of coherent structures with hydrodynamic stability theory

The first works to stress the importance of coherent structures for turbulent flows were the experiments of Brown and Roshko investigating plane turbulent mixing layers [15]. Brown and Roshko observed in photos clear large-scale formations which looked like overlapping waves coils or eddies. The pictures of flows testifying to the existence of coherent structures in streams, wakes and other flows have been obtained long ago (Dr. Rotta has also made his contribution having in his time studied the prototypes of coherent structures in a turbulent flow in a tube [16] but the works of Brown and Roshko permit a different approach to the earlier obtained results which testified to the existence of coherent structures in turbulent flows and marked the beginning of a new stage in the study of turbulence).

A most detailed research on streamlines and mixing layers is given in [17,18]. It proved, for example, that the generation and suppression of turbulence discovered for the first time in [19] are explained by the interaction of an acoustic field with large-scale formations in turbulent streams. It is to be noted that up to now there exists a confusion of ideas: by coherent structures one understands regular formations in "fully" developed turbulent shear layers as well as in transitional and even laminar shear layers [18]. Here in our opinion, it is necessary to draw a demarcation line: under coherent structures one should understand formations in developed turbulent flows or, at least, in flows where laminar and transition regions are known to be located far upstream. Otherwise one will always doubt as in [20,21], whether these structures are not merely remains of formations which result from the instability of the laminar flow.

Having drawn such a demarcation line, one can speak about an instability of the turbulent flow. The close connection between stability and turbulence is discussed in [10,13,17]. It is directly observed in [18] that a coherent structure results from the instability of some sort. In fact, each kind of instability (the initial flow regime either being laminar or turbulent) potentially can generate coherent structures. For this reason although when talking about hydrodynamic stability we think of laminar flows, an analogy between laminar and turbulent flows lies in the fact that turbulent shear layers can also be unstable [13].

Thus, the approach based on the analogy with a hydrodynamic theory of laminar flow stability but where the initial state will be a developed turbulent flow seems to be most promising. (In experimental conditions such a "developed" turbulent state can be realized when using an adequate turbulizator placed rather far off upstream). The velocity pulsations of the original boundary layer in the uncontrolled case are thought to be a source of initial disturbances from which coherent structures result, for example, in the mixing layer due to the shear layer instability. Moreover, the onset of coherent structures in such a flow may also be attributed to disturbances called forth by the acoustic disturbances and vibrations of the flowed about body which are always to be found in real flows. If a certain part of the instability range which is characteristic of the flow overlaps in the initial disturbance range, the onset of coherent structures can be expected. It is natural to expect such a generation in the region of sharp change, i.e. of integral values of the flow. As an example of such inhomogeneities there can serve the region of the splitter plate edge in the flow of a mixing layer type or the region of convergence of the boundary layers leaving the trailing edge of a streamlined body. Due to the random distribution in frequencies of initial disturbances the corresponding flow disturbances will be considerably scattered which is observed in the mixing layer [15]; nevertheless, the initial disturbances from a certain frequency range are likely to generate coherent structures in all turbulent shear flows with inhomogeneities.

Consequently, it can be said that coherent structures result from some flow instability in respect to external disturbances. It is known that in laminar flows the onset of the eigenoscillations of the boundary layer or Tollmien-Schlichting waves and, consequently, the further development of the flow and transition to turbulence considerably depends on the external conditions. In [22,23] the problem of receptivity of the boundary layer to the external disturbances was formulated: what mechanisms and what different small external flows generate the eigenoscillations of the boundary layer? Probably, for turbulent flow one can also formulate a problem of the transformation of external disturbances to turbulent flow oscillations analogous to the receptivity problem for laminar flows. Some attempts at posing the problem have already been made [24].

The description of the development of the eigenoscillations that appear in unstable turbulent flow must take into account the unstationary dynamics [18] and here the problem of the turbulent flow stability fundamentally differs from that of laminar one. In fact, as opposed to the laminar flow with a stationary initial mean velocity profile, in turbulent flow the instantaneous profile considerably differs from the mean one. For this reason in case of turbulent flow it is hardly possible to speak directly about the instability of a local mean velocity profile. Nevertheless, such attempts are being made, for example in [13] dealing with the development of large-scale instabilities in a turbulent wake the mean velocity distribution is chosen initially. In this work the problem is finally reduced to the Orr-Sommerfeld equation with a turbulent velocity. As far as the correctness of the problem in such a setting is concerned there exists no unanimous accord at present.

The theory of the hydrodynamic stability is known to have received experimental confirmation after the experiments by Schubauer and Skramstad, who introduced artificial disturbances in a laminar boundary layer. The experiments with artificial disturbances in turbulent shear layers are likely to play the same role, making allowances for the instability of a turbulent flow. The generation of disturbances by a vibrating ribbon in a turbulent flow in a plane channel was conducted in [25]. In a series of works the disturbances were generated in jets and mixing layers by means of acoustic fields, vibrations and spark discharges (for example [19,20,26-29]). It is to be discussed how the formations appearing thereat are connected with the structures developing from the natural disturbances. Widespread is the opinion that the excitation plays the initiating role at the formation of structures in the beginning of the flow. In other words, if, for example, the excitation is provided by a periodic field, the latter singles out some structure, fixes its initial phase, which allows to study the development of the structure in space and time. In fact, it shows that structure properties appearing at the excitation under control and structures developing from the natural background of disturbances are identical. Evidently, at least for disturbances of different frequencies which develop independently from each other if the mean flow is not affected, this affirmation holds because since the structure evolution in space and time is determined by the given major flow, the evolution of the disturbed structure cannot differ from the evolution of the "natural" structure. Thus, the superposition of the external disturbances under control upon the flow allows to study in more detail regular vortex-like formations proper to the flow. From the practical point of view the choice of parameters of the corresponding external factors can make possible the control of the developed turbulent flow. Below, we present the attempts to study the conditions of the onset and development of the coherent structures when superposing the controlled eddies in some types of turbulent shear flows.

2. Onset and development of eddy disturbances in a turbulent wake

The experiments to study the flow structure in the wake were conducted in subsonic aerodynamic tubes ITPM SO AN SSSR. The scheme of the experiments is given in Fig. 1. A thin plate with 300 mm length with sharp edges and the maximal thickness of 3 mm, a symmetrical Joukovsky with 292 mm span an identical profile at an angle of attack $\alpha = 8.5°$, and a thick plate, with a span of 300 mm and a semisymmetrical nose and tail were considered as typical test bodies. To realize a developed turbulent boundary layer on the bodies a special turbulizer (2) was used. Acoustical disturbances were generated by means of the loudspeaker (4) from the sound generator. The pulsation characteristics and mean flow parameters were registrated by the thermoanemometer probe (5).

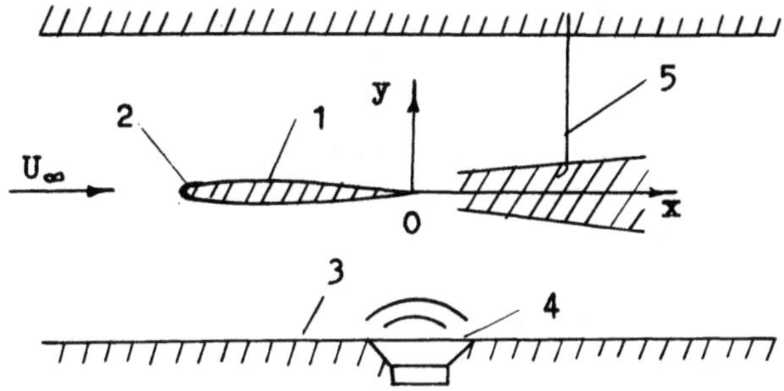

Fig. 1: Scheme of the experiment: 1 - typical test body, 2 - turbulizator, 3 - wall of the working part, 4 - loudspeaker, 5 - thermoanemometer probe

The signals were processed by the complex of the thermoanemometer equipment "DISA" and by frequency analyser FAT-1.

a) Wake behind a thin plate

In Fig. 2 there are given measured mean velocity profiles in the near wake behind the thin plate at velocity $U_\infty = 10$ m/s. It is obvious that the flow in this region is characterized by a strong spatial inhomogeneity and by flexion points in the averaged profile. These distributions are analogous to those obtained earlier in [30,31]. But then the coherent structures were not observed in a turbulent wake behind the thin plate. At the same time in Ref. [32] the supposition was made that the

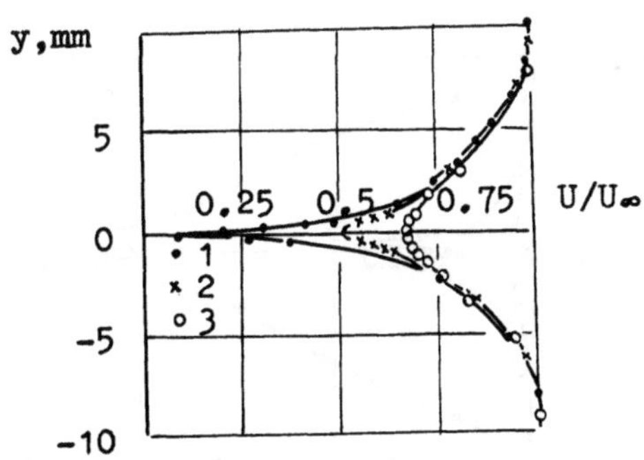

Fig. 2: Mean velocity profiles in the wake behind a thin plate: 1 - x=1 mm; 2 - x=10 mm; 3 - x=30 mm

transition of the near wake in the asymptotic far off wake is determined by the dynamics of the intermixing large-scale structures: at the same time in the work it was noted that the large-scale structures in a so-called intermediate region of the wake were not studied up to now but are doubtless to be studied in future.

When switching the acoustic field in the spectrum of the signal from the probe located in the wake, there appears a clear shaped peak at the frequency of the acoustic disturbance (Fig. 3). The conducted phase measurements testify to the eddy like nature of the disturbance which caused the appearance of a peak in the spectrum of the wake.

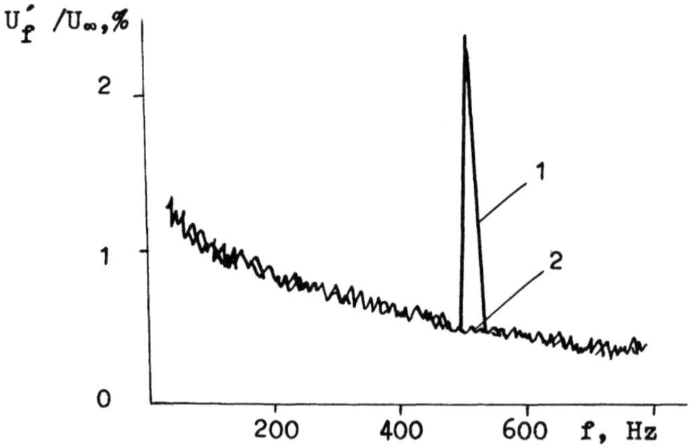

Fig. 3: The spectrum of disturbances in the wake behind the plate: x=10 mm, y=0.5 mm, f=518 Hz, A_S=101 dB; 1 - with acoustics; 2 - without acoustics, Δf=4 Hz

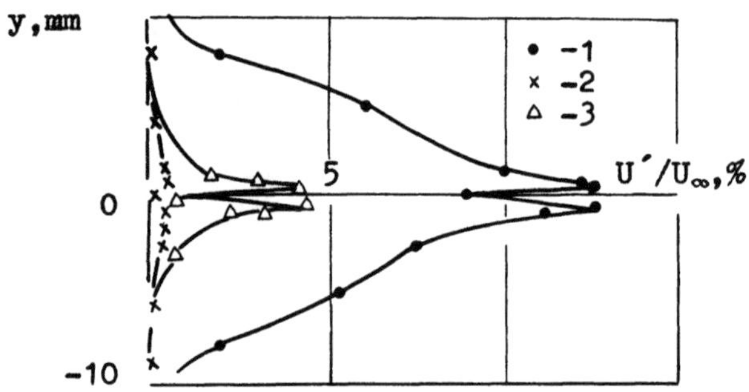

Fig. 4: Distribution of velocity pulsation intensity in the wake behind a thin plate. 1 - U'/U_∞ integral in spectrum; 2 - U_f'/U_∞ in 4 Hz band without acoustics; 3 - U_f'/U_∞ in 4 Hz band with acoustics (f=518 Hz)

In Fig. 4 the results of the measurements in the wake of the longitudinal component of the velocity pulsations U_f'/U_∞ with a bandwidth Δf = 4 Hz with acoustics and without are given. For the sake of comparison in the same figure is shown the distribution of intensity of pulsations of the longitudinal velocity component U'/U_∞ integral in spectrum. Here we note the qualitative coincidence of the distribution along transversal coordinate Y of the amplitude of integral intensity of transversal component of the velocity pulsation and that engendered by the sound field of the structure. Thereat the location of the maxima of these disturbances approximately coincide with inflexion points of the mean velocity (Fig. 2). Phase measurements under the acoustic effect showed that the disturbance's phase is random in time in the turbulent boundary plate layer, where in the spectrum the disturbance's amplitude at the given frequency practically does not exceed the turbulent pulsation; and only in the wake near x=0 does it acquire a definite value, which is connected with the growth of the determined component in the disturbances spectrum. The amplification law in this region is approximately linear and the processing of the results obtained shows that the phase velocity of these disturbances spreading strongly differs from the velocity of spreading of sound vibrations and constitutes ≈ 0.9 from the external field velocity and is characteristic of eddy disturbances.

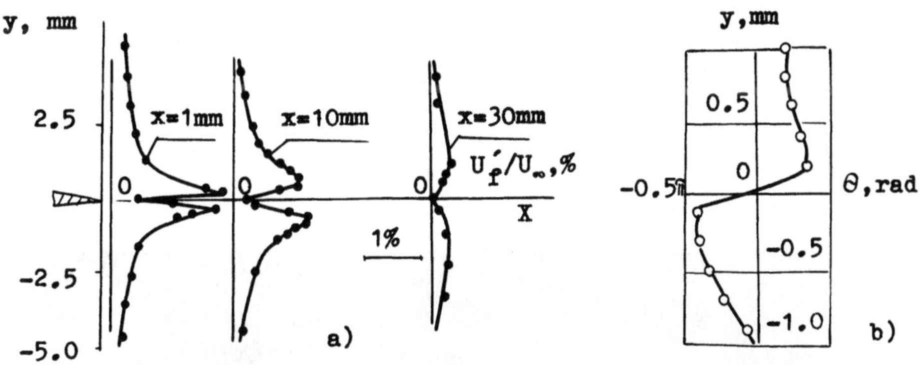

Fig. 5: a) Distribution of amplitude of disturbances of velocity f=518 Hz, A_s=101 dB; b) Distribution of disturbances phase of velocity (x=5 mm)

Fig. 5 gives the measurement results of the distribution of the amplitude of the observed disturbance without turbulent background and the phase distribution depending on the transversal coordinate Y.

It can be noted that in the region Y=0 the phase acquires a jump close to 180° which corresponds to the distribution of the disturbance's amplitude in this region (the minimum is in the region Y=0). It means that the engendered eddy disturbance for the longitudinal component of the velocity pulsation is practically antisymmetric.

In Fig. 6 the results of the visualization of the given structure are presented. When no acoustic excitation is present the turbulent wake represents a strongly irregular flow consisting of a set of "clots" superposed on each other. When the acoustic field is being superposed in the wake there appear regular wave like in the pictures because the deterministic component exceeds strongly the random one.

These pictures resemble those obtained in the work [15] which gave the impulse to coherent structures studies. It is to be noted that the distance between the waves crests or eddies or the lengths of disturbance waves coincide well with the lengths of the waves obtained by means of thermoanemometer measurements. Thus, the studies carried out showed that under the acoustic influence in the turbulent wake behind a thin plate there appear coherent structures [33,35,37]. The mere fact of the existence of such a transformation is interesting in itself but apart from this the use of an acoustic field and other controlled effects as sources of artificial disturbances makes possible the studies of the development of the wake in detail as well as the coherent structures connected with it, which are considered as regular disturbances existing in the turbulent wake.

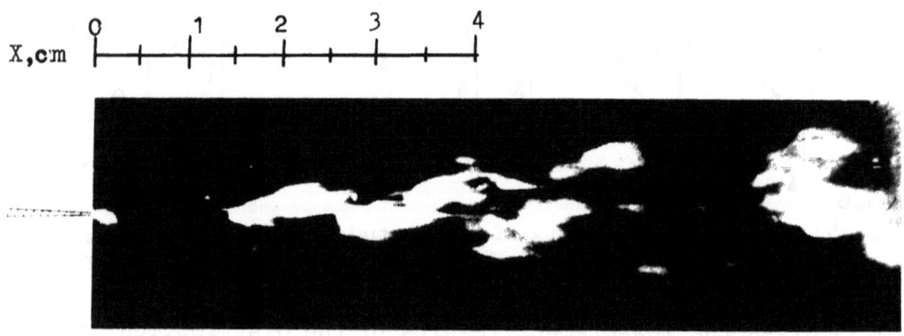

Fig. 6: Visual flow studies in the wake. U_∞=10 m/s; under acoustic effect, f=518 Hz

The disturbances of some frequencies can spread far in such a flow. In the work [35] a problem was posed: to study the development of the appearing coherent structures, their interaction with each other and with the mean flow as well as with the spectral modes of the turbulent flow, since it is these questions which arise when trying to build theoretical models describing the development of the like structures (see, for example [13,25]).

In the previous experimental work when considering the onset of the coherent structures the object was to attain the largest possible amplitudes of these formations, so as to maximally influence the turbulent flow structure (mean velocity, angle of jet amplification, mixing layer etc.). On the other hand, when theoretically describing we tried to maximally simplify the task and for this reason we neglected the influences undergone by the mean velocity and the turbulent spectrum. Thus, directions of theoretical researches and experiments were directly opposed to one another. In [35] the problem was set in the following way. First there was generated a coherent structure with a large amplitude, i.e. a structure which caused substantial changes in the turbulent wake structure, and then the amplitude threshold was lowered; limits were established when such effects disappeared. Then, in addition to one structure another one was excited in the flow, and their interaction studied.

The distributions of the mean velocity in the boundary layers when merging are typical of developed turbulent flow on a smooth plate without a pressure gradient and are close to the 1/7 power law, and the maximum of integral pulsations intensity on the boundary of a viscous sublayer is approximately equal to 10 %, which agrees with the measurements of other authors, [36]. Turbulent wake amplifies along the transversal coordinate and possesses the characteristic mean velocity profiles with flexion points in the vicinity of which the maximum of integral intensity of velocity pulsations are located.

To study the influence of the coherent structure on the mean flow the acoustic field with the sound frequency f=318 Hz was superposed on the turbulent wake. When the sound amplitude was equal to 116.5 dB, 111.7 dB, 101.5 dB in succession, the influence on the mean flow was measured at the distance x=15 mm from the plate in the point of the transversal coordinate where the amplitude of the given structure is at its maximum.

It is to be noted that at the amplitudes of sound equal to 116.5 dB and 111.7 dB and at coherent structures 6.6 % and 4.6 % (in a 4 Hz band) the mean flow is observed to be under influence. In the first case the velocity in this point decreases by 1.4 % and in the second by 1.1 %. Downstream this influence decreases but persists nevertheless.

When the sound amplitude decreases to 101.5 dB and the amplitudes of the coherent structure generated to 1.7 % the mean velocity does not change either for x=15 mm, or further downstream (did not exceed 0.6 %, which practically coincided in the present experiments with the error of measurement). At the same time the influence on the turbulent spectrum of the coherent structure amplitude level was studied. At the amplitudes of the structures 6.6 % the low frequency spectrum region is observed to be considerably affected (Fig. 7a). At the amplitude 1.7 % no effect is observed (Fig. 7b).

It is to be noted that the spectrum measurements in Fig. 7a-b are given for the one realization. Nevertheless, the spectra measurements for several realizations showed that there exists the overlapping of these spectra in a high-frequency domain, but in the low-frequency domain the upper boundary of the sum of realizations with the acoustic effect, which proves the existence of the influence mentioned earlier.

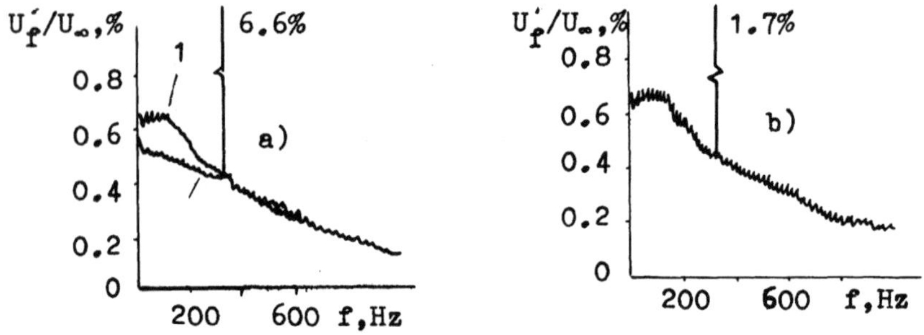

Fig. 7: Velocity pulsation spectra in the wake at different levels of acoustic field: a) A=116.5 dB; b) A=101.5 dB; 1 - without acoustic influence; 2 - with acoustic effect

The second coherent structure was obtained in the wake when superimposing the acoustic field containing sound at two frequencies. Thereat the total intensity of the sound was equal to 116.5 dB. Thus, now there existed in the wake two coherent structures with frequencies 89 Hz and 318 Hz. The location of the maxima of the two structures approximately corresponds to the position of flexion points in mean velocity profiles and of integral intensity maxima of velocity pulsations (Figs. 2, 4). Therefore universality of the given distributions for these frequencies is to be considered. (Such an universality is characteristic of the eigendisturbances of laminar shear flows, for example, for Tollmien-Schlichting waves of different frequencies). When the amplitudes of both coherent structures (in a 4 Hz band) are less than 2 % in all the domain of the wake under study, the structures do not influence the mean flow and the turbulent spectrum either when they appear together, or when separately. The example of such a spectrum is given in Fig. 8. The presence or absence of the second structure did not seem to affect the amplitude of the first. There is possibly generation of subharmonics, multiples and also sums, differences and frequencies divisibles by the last two of the harmonics of the major structures.

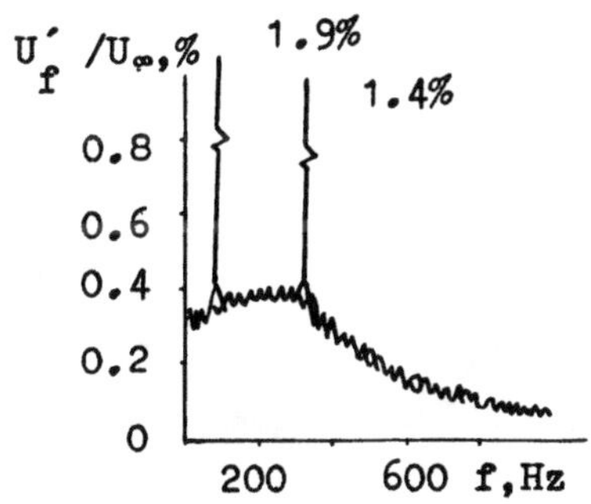

Fig. 8: The spectrum of the longitudinal component of the velocity pulsations in the wake (x=150 mm, y=5mm), the acoustic fields of two frequencies (f=318 Hz, f=89 Hz, A=116.5 dB) being superimposed at the same time

The spectra were measured at the position of the maximum integral intensity of the velocity pulsation. It can be said that in the vicinity of the maximum combined frequencies are not observed. It is also to be noted that the evolution of sound pulsations spectrum takes place in the longitudinal direction, i.e. natural disturbances moving off the trailing edge lower their low frequency still further, whereas those of high frequency gradually get extinguished - the spectrum loses its shape.

In Fig. 9 phase measurements for these two structures are given. Phase velocities determined according to wave lengths λ_1 and λ_2 and according to corresponding frequencies f_1 and f_2 were in both cases the same and equalled 0.9 U_∞. In the same figure it is seen that the influence of one structure on the phase velocity of the other (within the limits of the phase measurement error which constituted about 10 %) is not observed (for each of the frequencies the phase was measured first with and then without the other). In the same way a reciprocal influence of the structures on each other's curves of attenuation is shown (Fig. 10). From the results given in Fig. 10 one can deduce that the structures appearing under the acoustic influence have the same character of increase and attenuation as natural disturbances, i.e. those of lower frequencies attenuate more slowly than those of high frequency, which is confirmed by the spectrum deformation in an undisturbed flow.

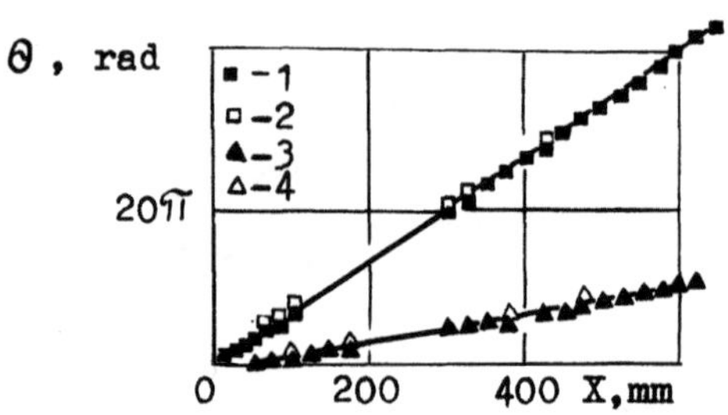

Fig. 9: Dependence of the phase of disturbance on the longitudinal coordinate. 1, 3 - at separate superimposition of teh acoustic field with frequencies 318 Hz and 89 Hz, respectively. 2, 4 - at simultaneous superimposition of the acoustic field with frequencies 318 Hz and 89 Hz

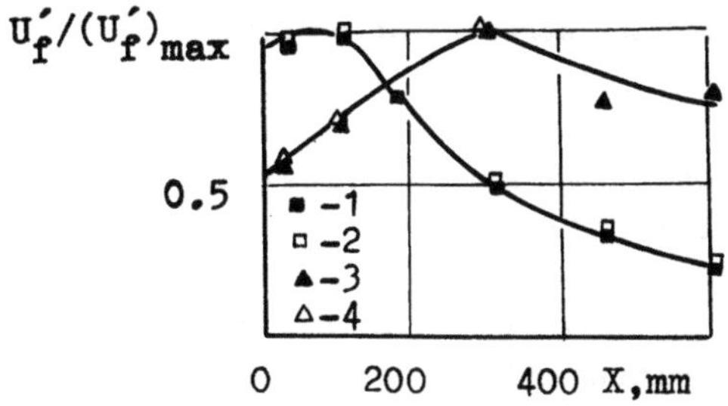

Fig. 10: Dependence of the amplitude of the coherent structure on the longitudinal coordinate. 1, 3 - at separate superimposing of the acoustic field with frequencies 318 Hz and 89 Hz, respectively. 2, 4 - at simultaneous superimposition of the acoustic field with frequencies 318 Hz and 89 Hz

Thus, the studies of the flow in a turbulent wake behind a thin plate showed that the acoustic disturbances can be transformed in eddy disturbances or coherent structures. When the amplitude of the coherent structure does not exceed 2 % of the undisturbed flow velocity, within the measurement accuracy the coherent structure is not observed to influence the mean flow or the turbulent spectrum. At the same amplitudes the superposition principle is observed - two or more structures can exist separately, i.e. their reciprocal influence is not observed. The distribution of the pulsation intensity in space as well as the character of attenuation and the phase velocities remain the same at their independent development. The generation of combined frequencies is not observed. One is to think that a turbulent flow is a medium where the majority of processes are conditioned by non-linear interactions. But experimental data show the linearity of the generation of coherent structures and their development in the turbulent wake, as in the case with laminar flows, though the level of the amplitudes of structures formally does not let us speak about the linear theory.

b) Turbulent wake behind the aerodynamic profile

The analogous results concerning the transformation of acoustic disturbances into eddies were also obtained for the turbulent wake behind the profile [37,38].

On the whole the picture of the distribution of the mean velocity and the longitudinal component of the velocity pulsations is qualitatively similar to the structure of the flow with and without acoustic influence for the above-described case of the turbulent flow about a thin plate. Changing the angle of attack of the given profile up to 8.5° led to a considerable non-symmetry of the mean flow on the profile itself as well as in the wake behind it. With the appearance of acoustic disturbances of the amplitude A_s=112 dB the mean flow was observed to be considerably influenced, which led to the deformation of mean velocity profiles in the wake. Thereat the mean flow in the boundary layer practically did not change.

At the same time at the given frequency of the eddy disturbance a considerable amplitude appeared (Fig. 11), which changed the integral pulsation intensity of the transverse velocity component. The determinate structure appearing under acoustic influence is strongly non-symmetrical which seems to be connected with the fact that the highest non-homongeneity of the mean flow in the wake is observed on the lower side of the profile.
The phase velocity of the given coherent structure growing at the angle of attack α=8.5° approximately equals 0.6 U_∞. All the measurements described above were conducted at turbulent degree ≈ 0.3 %. In case of a zero angle of attack in the wake the measurements were conducted for a smooth model, but the turbulence degree being ≈ 3.0 %. In this case and with no turbulizator on the model a turbulent boundary layer is observed beginning from the nose and the mean velocity distribution in the turbulent wake qualitatively coincides with the distribution in the wake behind the model with the turbulizator. With the superposition of the acoustic field one can observe the generation of the structure similar to that described above.

The studies conducted showed that in the turbulent wake behind the profile with a sharp rear edge the acoustic oscillations

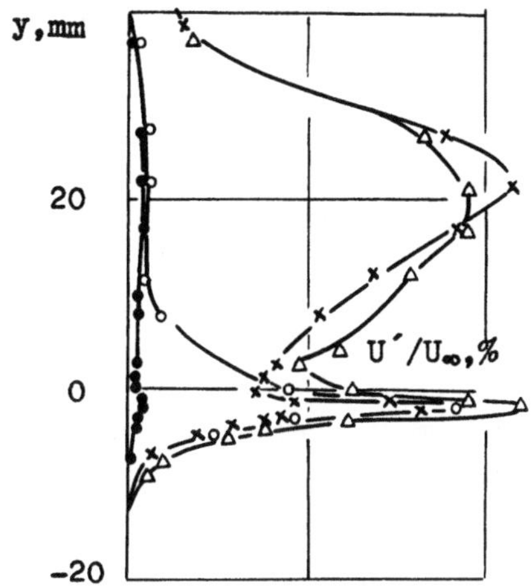

Fig. 11: Distribution of pulsation intensity in the wake behind the Joukowski profile under the angle of attack (α=8.5°). X=5mm. 1 - U'/U_{∞} - is the one integral in spectrum, 2 - U'/U_{∞} - is the one integral in spectrum with acoustics, 3 - U_f'/U_{∞} - in 4 Hz band without acoustics, 4 - U_f'/U_{∞} - in 4 Hz band with acoustics

are transformed into coherent disturbances which can affect the integral characteristics of the flow, if the intensity of the sound is rather high. The same happens also in natural conditions: when the wake behind the body in the airflow is likely to be influenced by acoustic sources located close by, the analogous phenomenon will take place.

That is why this influence is to be taken into account when considering the processes which take place in real flows.

c) Coherent structures in wakes behind bodies in separated flow

The above-described cases of flows in wakes are characterized by the following: when an ordinary spectral analysis from the thermoanemometer probe takes place no frequencies corresponding to coherent structures are observed without external influence.

But the structure in these flows exist and their presence can be discoverd, for example, under visual studies. The study of the acoustic field effect on onset and development of coherent structures in a developed turbulent wake behind a bluff body, where the coherent structures manifest themselves in the spectral analysis present a certain interest. The scheme of the experiments therein corresponded to that described above. The main measurements with the thermoanemometer are given for the velocity U_∞ = 30 m/s.

The distribution of mean velocity and integral pulsation intensity in the boundary layer is that characteristic of the developed boundary layer and the spectra of velocity pulsations do not contain any discrete frequencies. At the same time in the spectrum of the signal from the thermoanemometer probe in the wake contains a discrete characteristic frequency 255 Hz which we will call a major one (see Fig. 12).

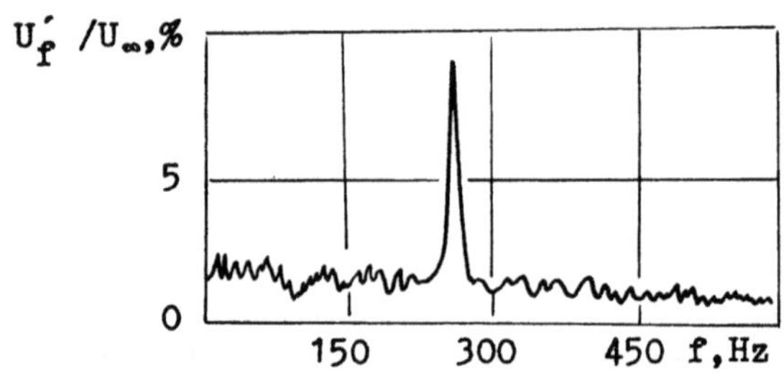

Fig. 12: Spectrum of velocity pulsations without acoustic effect. x=20 mm, y=15 mm

In Fig. 13a there are given the distributions of mean velocity in the vicinity of the model (in the distance x = -15; -10; -5 and 40 mm of the beginning of coordinates). These distributions gradually change to the distributions in the wake. Apart from this, the velocity profiles in the range x = -10 mm and x = -5 mm have a peculiar shape which indicates the separation boundary layer in these cross-sections. In Fig. 13b there is given the distribution of disturbance intensity at the major frequency (in the 4 Hz band) in the region of separation and in the wake. It is seen that the amplitude of disturbances gradually grows along the longitudinal coordinate. The superposition of the acoustic field with the frequency corresponding to the major one intensities the disturbances. Then, the disturbances corresponding to the main frequency pass over from the separation region into the wake; the maxima of the disturbance intensity along the transversal coordinate and in the wake approximately correspond to the location of flexion points in the mean velocity profiles.

The disturbance with the characteristic frequency 255 Hz is of eddy nature, i.e. is a coherent structure which is confirmed by phase measurements, and the phase begins to disperse only in the separation region. Phase measurements along the transversal coordinate Y conducted in the wake behind the body showed that the difference between the disturbances phases at differents sides of the wake is close to 180°.

Visual studies were conducted by means of a smoke wire, the velocity being $U_\infty = 17$ m/s. Thereat on the body and in the wake

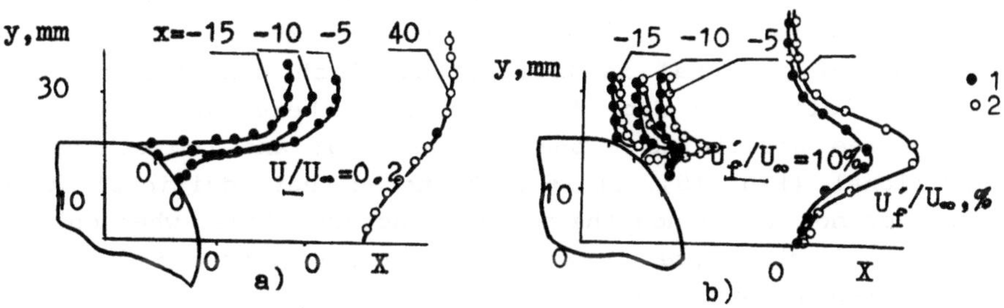

Fig. 13: Distributions of mean velocity and pulsation intensity; 1 - without sound; 2 - with sound

a developed turbulent flow is realized and distributions of mean and pulsation characteristics are analogous to those in Fig. 13. In the wake even without superposition of influences under control there exist regular formations with eigenfrequency approximately equal to 129 Hz (Fig. 14). Thereat the signal spectrum is analogous to that given in Fig. 12. The superposition of an acoustic field leads to synchronization, and the structures which came off the body at a frequency varying over a finite range are now spaced more regularly.

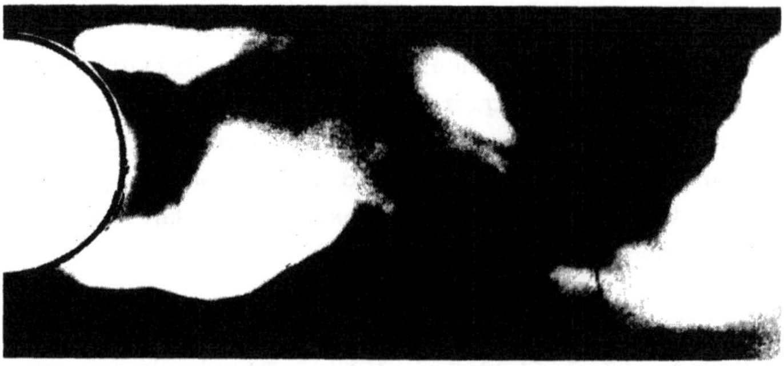

Fig. 14: Flow in the wake without acoustic influence

When the acoustic field of the frequency f = 175 Hz which differs from the major one is superposed on the flow (U_∞ = 30 m/s), an additional structure can appear in the flow which exists in the flow simultaneously with the major structure, as shown by phase measurements. Just as happens with the major structure, the phase of the additional structure starts changing only in the separation region. Moreover, when by mixing inputs to the loudspeaker a signal is transmitted which contains two frequencies differing from the major one, there can exist in the flow three structures - one is the major and two others are additional (Fig. 15). It is to be noted that additional structures do not influence the major structure either when only an additional structure is excited or when a major and additional structures are excited.

In Fig. 16 the results of measurements are given, from which it is seen that the flow under study has a range of frequencies beyond which no structures are excited.

Moreover, in the given flow it is discovered that the major structure is "captured" by the external acoustic field. From the results given in Fig. 17 it is seen that in some narrow range of frequencies the intensity of the major structure depends on the frequency of the external acoustic influence.

Thus, the studies [39] show that acoustic influence at the frequency of the major structure leads to its amplification.

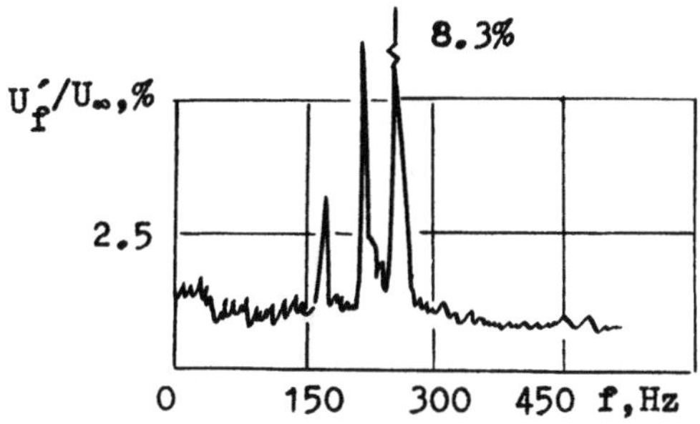

Fig. 15: Spectrum of velocity pulsations with acoustic influence at frequencies f=175 Hz and f=216 Hz. $A_{175}=A_{216}=120$ dB. x=20mm, y=15mm

The transformation of acoustic disturbances in eddy ones takes place near the separation point of the boundary layer in the stern part of the model. With the superposition of the acoustic field with another frequency apart from the major structure there can be excited other structures and these structures can develop without interacting with the major structure. For this flow there exists a frequency range of the flow receptivity towards acoustic disturbance. In the flow under study the phenomenon of "captivity" is observed, where in a narrow range of frequency, the external acoustic signal entrains the frequency of the main structure.

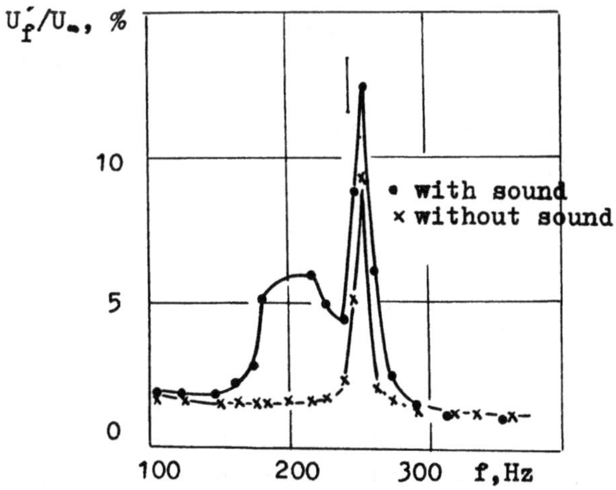

Fig. 16: The curve of receptivity to acoustic disturbances. x=20 mm, y=15 mm, A=120 dB

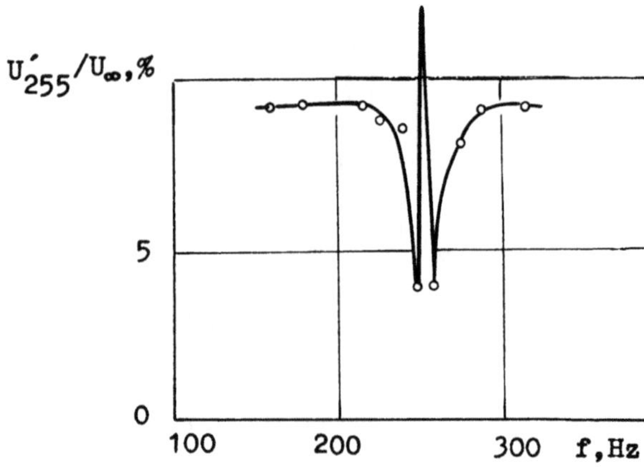

Fig. 17: Influence of additional structures on the amplitude of the major structure

d) The control of the coherent structure onset in the turbulent wake

As was already mentioned, in the turbulent wake behind a thin plate under acoustic irradiation there appear coherent structures and it shows that in turbulent flows there can also arise a problem about receptivity of the flow: the external disturbances can be transformed in eigenoscillations of the given flow, and that is why external disturbances of different nature can contribute differently in generation of coherent structures. Of all disturbing factors the most important for the given case seem to be the following: the turbulence of boundary layers coming off the plate (in the region of their convergence there forms a wake); acoustic disturbance, vibration of the surface. If the first factor in our experiment cannot practically be changed since on the plate we get velocity and disturbance distributions typical of a developed turbulent flow, then two other kinds of disturbances can play a significant role separately and jointly.

As a rule under acoustic influence there appears a definite level of surface vibration, and with the irradiation of a thin plate of a small mass, vibrations can have a significant value. There is known a work [26] where in turbulent mixing layer at very high amplitudes of thin flap oscillations at the trailing edge of the splitter plate (of the order of several hundreds microns) coherent structures were observed. To understand the role of vibration in the process of coherent structure forming a special study was necessary.

In the given part of the work the problem was settled to establish first of all, what disturbances play a decisive role (at primary acoustic influence), secondly, if under vibration influence there appear coherent structures, whether they do not possess some special qualities.

The disturbances were introduced by a special vibrator. To estimate the contribution of the vibrational disturbance the following experiment was carried out. At first there were measured all the distributions of the mean velocity and integral intensity of velocity pulsations in the section x = 3 mm, which qualitatively correspond to those obtained earlier in other sections. Then, for the frequency of acoustic disturbance f = 718 Hz and for the sound amplitude 116 dB there were measured amplitude vibration of the hind plate edge (the sound and the vibration were monochromatic) and the amplitude of the induced coherent structure at x = 3 mm. Then the sound was switched off, and by means of vibrator the same amplitude of amplitude vibration was achieved. The amplitude of vibrations thereat constituted 5.5 µm. In this case the signal of the coherent structure at the given frequency was not registered. This shows that in this case the main contribution at the transformation gives rise to acoustic disturbances.

At the same time with the increase of the vibration amplitude up to 22 µm one could obtain the coherent structure amplitude approximately of the same value as at the acoustic influence, thereat the sound level was equal to the background level (87 dB), and the vibrational influence was monochromatic. The distribution of this structure along the transversal coordinate Y proved to be analogous to the distribution of disturbances arising at the acoustic influence (Fig. 18).

For the sake of comparison in the same figure the intensity of the turbulent background in the given frequency is given.

Phase measurements along the transversal coordinate showed that this disturbance or structure is also asymmetric and the phase jump is equal approximately 180°. I.e. at vibrational as well as at acoustic influence in the flow, there appear antisymmetric disturbances or coherent structures which look like a singular path of eddies, which grow with the same phase velocity as confirmed by the phase measurements.

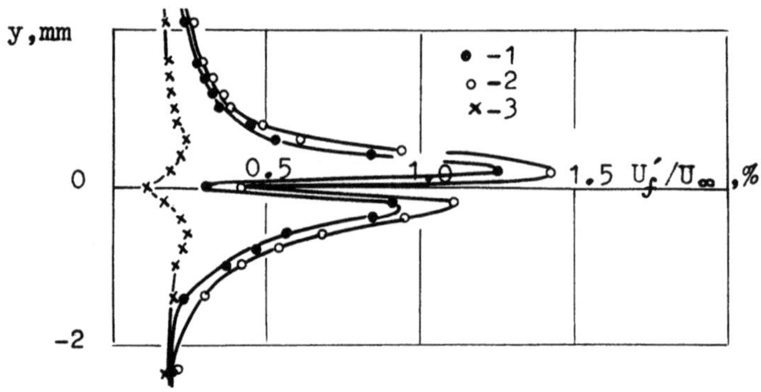

Fig. 18: Distribution of velocity pulsation intensity in 4 Hz band. f=718 Hz. 1 - acoustic influence; 2 - vibrational influence; 3 - turbulent phone

For the two amplitudes of vibrational effect which differ by two times there were studied the nature of attenuation, phase velocities as well as distribution of disturbances along the transversal coordinate. Thereat corresponding initial intensities of disturbances, characterizing coherent structures, constituted 1.7 % and 3.4 % in the section x = 3 mm. It proved that the curves of attenuation, distribution along the transversal coordinate, phase velocities of distribution along longitudinal and transversal coordinates do not depend on the amplitude of the coherent structure (coherent levels of disturbances in each section normed according to the initial significance). The same can be possibly said about a coherent structure appearing under acoustic and vibrational influences. It can be supposed that for the vibrational influence the principle of superposition of two or more frequencies of the same level can also be valid, as was obtained under acoustic influence.

All the previous results show that coherent structures can be formed linearly and it is possible that superposition of structures can be used for their suppression. To verify the possibility of amplification (or suppression) of the amplitude of the induced structure by means of a simultaneous application of the acoustic field and vibrational plate of one frequency the change of phase correlation between them was carried out. For this purpose the following treatment was used. A thermoanemometer probe was placed in the section x = 3 mm in the place along the transversal coordinate Y, where the maximum of the integral pulsation intensity was observed. Then selecting the powers and vibrations of the sound approximately equal levels of the structure were established at certain superposition of acoustic and vibrational influences. Then, there changed the phase γ of the vibrator feeding signal as related to the initial phase, and the amplitude of the corresponding coherent structure was measured. The results given in Fig. 19 show that in fact, under definite correlations of phases between the sound and vibrations an attenuation or amplification of the structures in the wake can be observed, i.e. a structure can be annihilated or doubled (in the figure along the axes of coordinates the relation of total structure amplitude to the amplitude of the structure appearing under the acoustic influence is given). A similar phenomenon is discovered in [40] for a laminar boundary layer.

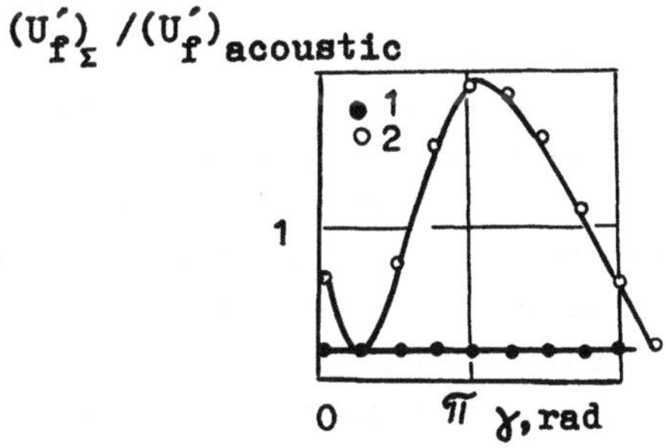

Fig. 19: Dependence of the velocity pulsation amplitude in the wake on the difference of phases between acoustic and vibrational influence. x=3 mm, y=0.5 mm. 1 - turbulent phone intensity; 2 - velocity pulsation intensity at the combined effect of sound and vibration

According to the results of the given study one can conclude that at a rather strong vibration coherent structures also can be generated, and these structures do not differ in their characteristics from the structures appearing under acoustic influence. Thus, for generation of the structures the oscillations of the flow near the hind edge as well as rather strong oscillations of the hind edge relative to the flow give the same result. With the combined effect of the sound and vibrations depending on the phase correlation between them a subtraction or addition of structures is possible. This fact is important to determine the receptivity of the turbulent flow in the wake to the acoustic disturbances of different structures. One of the possible applications of this results can be the use of the combined effect of acoustics and vibrations in a turbulent wake and, possibly, in other flow types.

3. Coherent structures in flows with reattachment of a boundary layer

The above-described phenomena take place under the effect of the acoustic field on the flow in turbulent wakes, i.e. on the free flows. The study of the acoustic field influence on the separated flow with adjusting of boundary layer is of a certain interest.

[41] treats separation of turbulent boundary layer behind a step, which is a classic, much-studied flow. Thereat there was noted that the major interest of the previous studies was devoted to the mean velocity distribution and turbulence intensity in the recirculating domain behind the step, and there was established the length of recirculating domain and the distances from the step where the turbulent boundary layer relaxed to the equilibrium turbulent boundary layer on the plane plate. Lately, due to the interest towards coherent structures observed for the first time in free shear layers, in the studies of the flow behind the step it was also tried to discover them in the adjusting shear layer. But the results we have got up to now do not allow to estimate precisely the conditions of the structures onset as well as the character of amplitudes and frequencies of such a flow.

In [34,37,42] the object was to show that coherent structures in the flow behind the step can appear under acoustic radiation of the domain under study. The step was formed by means of acrylic plastic plate 785 mm long, 700 mm wide and 20 mm high which was fixed on the lateral side of the working part of the tunnel. At the inlet of the working part the boundary layer was artificially turbulized by means of sand roughness. The scheme of the experiment is shown in Fig. 20. The experiments were carried out at a velocity of 15.5 m/s.

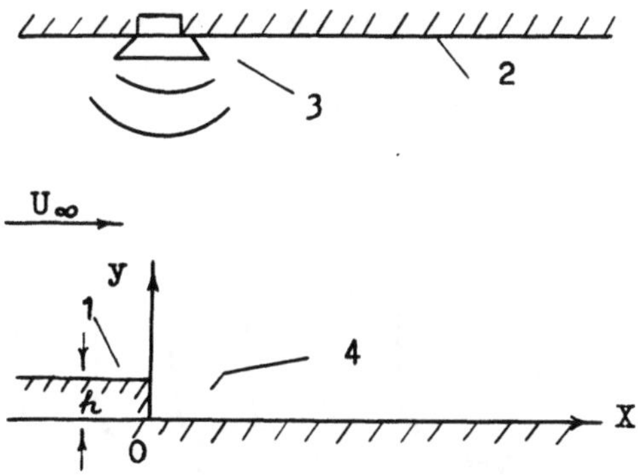

Fig. 20: Scheme of the experiment. 1 - step; 2 - working part wall; 3 - loudspeaker; 4 - thermoanemometer probe

The measurements carried out by means of a combined probe in the centre of the working part of the wind tunnel showed that there is no longitudinal pressure gradient at such a relatively small height of the step. The mean velocity distribution in the boundary layer of the plate measured by thermoanemometer corresponds to the law of the degree 1/7 with a high accuracy, which shows a well-developed turbulent layer on the plate. The ratio of the boundary layer thickness before the separation to the height of the step is approximately equal 1. The mean velocity profiles and those of the velocity pulsations integral intensity above the wall with the step are given in Fig. 21. Thereat it is to be noted that in the separation zone itself the profiles only qualitatively reflect the true picture because of the signal deformation caused by the insensitivity of thermoanemometers to reversal of the flow direction. When constructing the velocity profiles in this domain, it was taken into account that the mean velocity has a negative sign and it was allowed for at their construction the same as was done in [43]. The mean velocity profiles and the profiles of integral intensity of the velocity pulsations shown in the picture are typical for flow over a step and are in good agreement with data obtained by other scientists [44-47]. For the given case G/h - 1 the thermoanemometrical measurements give the reattachment length of the separated boundary layer $X_R/h = 8$. The average flow in the region of the step domain is characterized by a high spatial inhomogeneity and the presence of flexion points in the profiles. The maximum of integral disturbance of the flow behind the step grows with the longitudinal coordinate, and its location approximately coincides with inflexion point locations in this region.

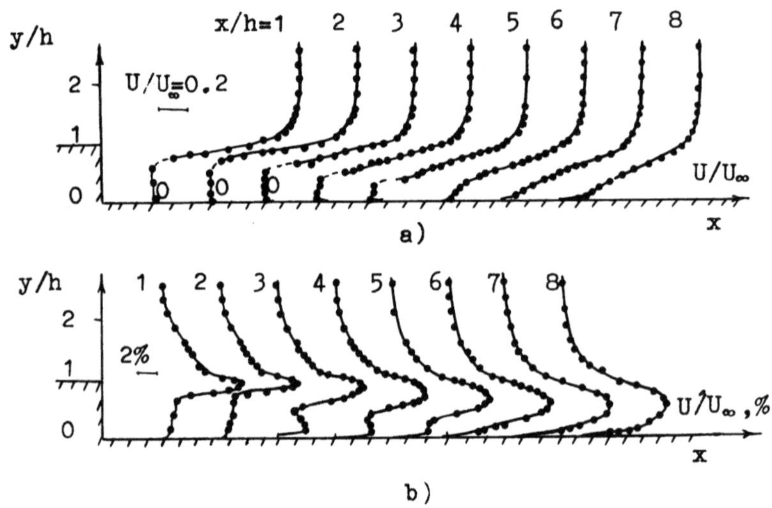

Fig. 21: Distribution of mean velocity (a) and integral intensity of velocity pulsations (b) in flow behind the step

When superposing a monochromatic acoustic field in the spectrum of the signal from the probe located behind the edge of the step, there arises a clear peak at the frequency of the acoustic disturbances. Thereat the influence over integral characteristics of the flow far from the step as well as near it at the intensity of the acoustic field equal to 116 dB was not observed.

The distribution and increment of the amplitude of velocity pulsations caused by an artificially generated coherent structure is shown in Fig. 22. This increment was determined as a square root from the difference $(U_f')^2/(U_\infty)^2$ with and without acoustics. In the figure one can see that a coherent structure appears and develops in a comparatively narrow shear layer. Limiting with the separation region, the maximal increment is being observed in the region of the mean velocity profile inflexion. The proceeding of the results of phase measurements showed that the velocity of travel of these disturbances constitutes ≈ 0.4 of the external flow velocity, which shows their eddy nature, since for the sound of the given frequency the phase velocity of spreading is one order greater.

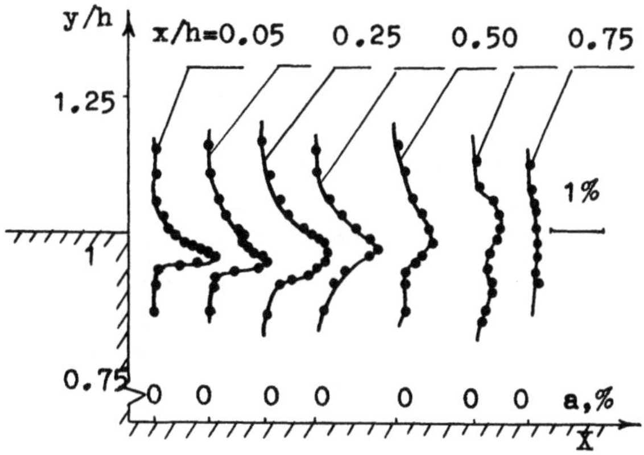

Fig. 22: Increment distribution of velocity disturbances amplitude appearing under acoustic influence over the flow behind the step. f=1072 Hz

In Fig. 23 is given a frequency spectrum of coherent structures disturbances under acoustic influence. It turns out that the shear layer appearing at the turbulent separation off the step reacts differently at different disturbance frequencies at equal intensity level of a monochromatic wave. With the decrease of velocity the maximum of the receptivity curve shifts to the left.

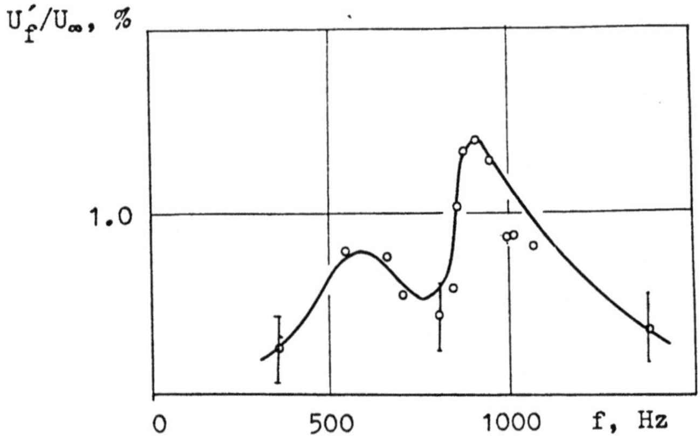

Fig. 23: The curve of receptivity of flow behind the step to acoustic disturbances

Thus, the studies conducted showed that acoustic oscillations are intensively transformed in eddy disturbances in a shear layer which arises at the separation of the turbulent shear layer behind the step, similar to what happens in laminar flows [23]. It was found out that a shear layer possesses a favourable range of disturbed frequencies which shows a susceptibility of the given flow to acoustic disturbances of definite frequencies.

One more type of the flow (in addition to those studies above) to which the problem on the receptivity to external disturbances can be applied is the flow behind a jet in a transverse flow.

The question whether coherent structures exist in such flows was not considered earlier. But it can be supposed that when jets flow out in the boundary layer in the boundaries of separated regions the formation of coherent structures is possible.

It is to be noted that the flow of a solitary round jet is rather complex and three-dimensional. Probably, at the given stage it is better to study a simple case of the development of a rectangular jet. In this case one can neglect three-dimensional effects and a two-dimensional picture of the jet's flowing out in the boundary layer.

In [48] the object was to study the onset of coherent structures under acoustic effect on the flow behind a two-dimensional jet entering in the turbulent boundary layer normal to the flow.

The scheme of the experiment is given in Fig. 24. In the wall of the working part of the tube a slot was made 114 mm long and 0.5 mm wide. The mean velocity of the jet flow was 7.5 m/s.

Acoustic disturbances were generated by a loudspeaker which was placed in a diffusor of the tube. At the superposition of the acoustic field with the frequency 883 Hz and intensity 120 dB a coherent structure was observed to appear. In Fig. 25 there are given the distributions of the velocity pulsation intensity along the transversal coordinate y in 4 Hz band for the given frequency in different sections along a longitudinal coordinate

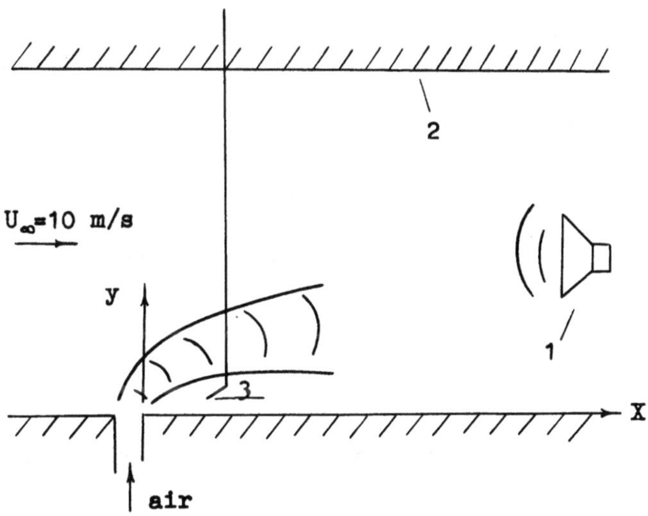

Fig. 24: The scheme of the experiment. 1 - loudspeaker; 2 - wall of the working part; 3 - thermoanemometer probe

X. It can be seen that when an acoustic field is superposed on the flow the pulsation intensity increases in comparison with the undisturbed flow. The disturbances arise in the region of the slot, increase up to the value 2.8 % from the velocity of an undisturbed flow in the section 3 mm and then attenuate. The maxima of these disturbances are located in the region of the maximum of integral intensity of the velocity pulsations.

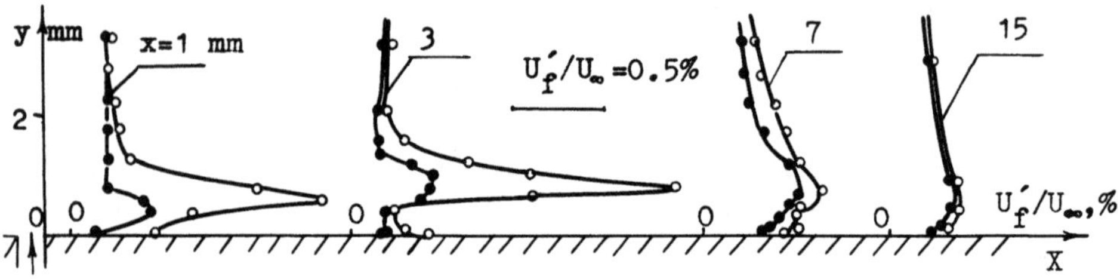

Fig. 25: Distribution of the amplitude of velocity disturbances in a 4 Hz band. f=883 Hz. 1 - without acoustic effect; 2 - with acoustic influence

The disturbances arising as a result of an acoustic effect are of an eddy nature. The spreading phase velocity of these disturbance or coherent structures are equal 0.4 - 0.5 of the velocity of an undisturbed flow.

Thus, as a result of the studies conducted it was found out that coherent structures appear when the acoustic field is superposed on the flow behind the turbulent jet flowing out in the turbulent boundary layer.

The given review of several experimental results concerning the onset and development of coherent structures in turbulent shear flows shows that the study of coherent structures can be based on the theory of hydrodynamic stability.

References

[1] Keller, L.V.; Friedmann, A.A.; Proc. 1st Intern. Congr. Appl. Mech. Delft, 1924.

[2] Kolmogoroff, A.N.: DAN SSSR, 30 299-303, 1941; 32, 19-21, 1941.

[3] Batchelor, G.K.: Cambridge University Press, 1953.

[4] Townsend, A.A.: Cambridge University Press, 1956.

[5] Hinze, J.O.: McGraw-Hill Book Comp., New York, 1959.

[6] Rotta, J.C.: Progress in Aeronautical Sciences, 2, 1962, 1-219, Pergamon Press, Oxford.

[7] Monin, A.S.; Yaglom, A.M.: "Nauka", Moskva, 1966 and 1976.

[8] Struminski, V.V.: DAN SSSR, 1985, t. 280, 3, 570-574.

[9] Struminski, V.V.: DAN SSSR, 280, 4, 1985, 820-826.

[10] Liepmann, H.W.: American Scientist, 67, 1979, 221-228.

[11] Rabinovich, M.I.; Sushchik, M.M.: In: Nelineinie volni, Moskva, 1983, 56-85.

[12] Goldshtic, M.A.; Shtern, V.N.: Novosibirsk: 1981, 48 s. (Preprint/AN SSSR. Sib. otdelenie. Inst. teplophiziki; 74-81).

[13] Reynolds, W.C.: J. Fluid Mech. 54, 1972, 481-488.

[14] Kovasznay, L.S.G.: In: Lect. Notes Phys. 75, 1978, 1-18.

[15] Brown, G.L.; Roshko, A.: J. Fluid Mech. 64, 1974, 775-816.

[16] Rotta, J.C.: Ing.-Arch. 24, 1956, 258-281.

[17] Ginevski, A.S.; Vlasov, E.V.: In: Modeli mekhaniki sploshnoi sredi. Novosibirsk, 1983, 91-117.

[18] Hussain, A.K.M.F.: Univ. of Houston, USA, Rep. FM-17, 1983, 92.

[19] Vlasov, E.V.; Ginevski, A.S.: Izvestiya AN SSSR. Mekhanika zhidkosti i gaza, 4, 1967, 133-138.

[20] Chandrsuda, C.; Mehta, R.L.; Weir, A.D., Bradshaw, P.: J. Fluid Mech., 85, 4, 1978, 693-704.

[21] Bradshaw, P.: In: Turbulent shear flows. 2, Springer-Verlag, 1980.

[22] Morkovin, M.V.: AFFDLTR, 1968, 68-149.

[23] Kachanov, Yu.S.; Kozlov, V.V.; Levchenko, V.Ya: Novosibirsk: Nauka, 1982, 151 s.

[24] Tam, C.K.W.: Proc. Symp. on Mechanics of Sound Generation in Fluids, Springer, 1971, 41-47.

[25] Hussain, A.K.M.F.; Reynolds, W.C.: J. Fluid Mech. 41, 1970, 241-258.

[26] Oster, D.; Wygnanski, J.: J. Fluid Mech., 123, 1982, 91-130.

[27] Sokolov, M.; Kleis, S.J.; Hussain, A.K.M.F.: AIAA J., 19, 8, 1981, 1000-1008.

[28] Furletov, V.N.; Kleis, S.J.; Hussain, A.K.M.F.: AIAA J., 19, 8, 1981, 1000-1008.

[29] Navosznov, O.I.; Pavelyev, A.A.: Isvestiya AN SSSR, Mekhanika zhidkosti i gaza, 4, 1980, 18-24.

[30] Chevray, R.; Kovasznay, L.S.G.: AIAA J., 8, 1969, 1641-1643.

[31] Sekundov, A.N.; Yakovlevski, O.V.: Izvestiya AN SSSR, Mekhanika zhidkosti i gaza, 6, 1970, 131-139.

[32] Ramaprian, B.R.; Patel, V.C.; Sastry, M.S.: AIAA J., 20, 9, 1982, 1228-1235.

[33] Yanenko, N.N.; Bardakhanov, S.P.; Kozlov, V.V.: DAN SSSR, 274, 1, 1984, 50-53.

[34] Yanenko, N.N.; Bardakhanov, S.P.; Kozlov, V.V.: In: Proc. of IUTAM Symp. of Turbulence and Chaotic Phenomena in Fluids, Kyoto, 1983, North-Holland Publ., 1984, 427-432.

[35] Bardakhanov, S.P.; Kozlov, V.V.: Novosibirsk, 1984, 29 s. (Preprint/AN SSSR. Sib. otdelenie. Inst. teor. i. prikl. mekhaniki; 14-84).

[36] Klebanoff, P.S.: NACA TN, No. 3178.

[37] Yanenko, N.N., Bardakhanov, S.P.; Kozlov, V.V.: In: Neustoichivost do- i sverhzvokovikh techni. Novosibirsk, 1982, 83-106.

[38] Bardakhanov, S.P.; Kozlov, V.V.; Yanenko, N.N.: Ing.-fiz. zhurnal, 47, 4, 1984, 533-536.

[39] Bardakhanov, S.P.; Kozlov, V.V.: In: Turbulentnie struinie techniya, Tallin, 1985, 94-99.

[40] Gedney, C.J.: Massachusetts Inst. Technology, Cambridge, Massachusetts, 1983, Rept. No. 83560-3.

[41] Eaton, J.K.; Johnston, J.P.: AIAA J., 19, 9, 1981, 1093-1100.

[42] Bardakhanov, S.P.; Kozlov, V.V.: Izvestiya SO AN SSSR. Ser. tekhn. nauk, 10, 2, 1985, 120-123.

[43] Sinha, S.N.; Gupta, A.K.; Oberai, M.M.: AIAA J., 19, 12, 1981, 1527-1530.

[44] Tani, I.; Juchi, M.; Komoda, H.: Aeronautical Research Institute; Univ. of Tokyo, 1961, Rept. 364.

[45] Etheridge, D.W.; Kemp, P.H.: J. Fluid Mech., 86, 1978, 545-566.

[46] Kim, J.; Kline, S.J.; Johnston, J.P.: Transaction of the ASME J. of Basic Eng., 102, 3, 1980.

[47] Eaton, J.K.; Johnston, J.P.: Stanford Univ., Rept. MD-39, 1980, 164.

[48] Bardakhanov, S.P.: In: Gidrodinamika i akustika odno- i dvukhfaznikh potokov. Novosibirsk, 1983, 43-47.

Energy Events in the Atmospheric Boundary Layer

Roddam NARASIMHA* and Sudarsh V KAILAS

Indian Institute of Science, Bangalore 560 012
*and National Aeronautical Laboratory, Bangalore 560 017

0. SUMMARY

Turbulent velocity fluctuations in the atmospheric boundary layer are analysed using the VITA technique, with a view to understand the relation of the special events detected by the technique with turbulent bursts at high Reynolds numbers. Utilising data acquired under conditions of neutral stability at the Boulder Atmospheric Observatory, it is found that the "energy events" so detected are qualitatively similar in many respects to the turbulent bursts observed at much lower Reynolds numbers in the laboratory, but last longer and occur less frequently, in a manner not inconsistent with outer scaling. It is concluded that high Reynolds number flows reveal bursts tending to scale on outer variables - a conclusion that is considered particularly significant as wall parameters are of the same order in the atmospheric boundary layer as in the laboratory flows that have been widely studied.

1. INTRODUCTION

Ever since the pioneering work of Kline et al (1967) showed that a key role in the process of turbulent energy production in a boundary layer is played by a phenomenon that has come to be known as bursting, a variety of experimental studies have been carried out by many workers in an attempt at elucidating the nature of the phenomenon. An intriguing aspect of all this work is that although there is complete agreement on the presence of a series of events associated with such a burst, there are still differences of opinion regarding their scaling. The problem of scaling is crucial in understanding the mechanics of the bursts, as there is no satisfactory theoretical picture yet of the phenomenon; decisive evidence on scaling can therefore provide strong clues to the nature of the process. It may be worthwhile to recount briefly some of these arguments, to set the stage for the present investigation.

The studies of Kline and others were followed by a series of measurements reported by Narahari Rao, Narasimha and Narayan (1971, to be referred to below as 3N) which suggested that the frequency of these brusts, even close to the surface, scaled on outer variables characterising the boundary layer, such as free-stream velocity and boundary layer thickness. Although there has been much discussion of the difficulties associated with measurement of burst frequencies (e.g. Offen and Kline 1974), the view that they scaled on outer variables came to be generally accepted. We may cite the experiments of Ueda and Hinze (1975), Willmarth (1975) and others, among the many that examined the problem. In the 3N study a significant piece of evidence had come from the measurements of Tu and Willmarth (1966), who later (Lu and Willmarth 1973) withdrew their high Reynolds number points because of uncertainties associated with their data processing. Fresh measurements reported by Willmarth (1975), however, agreed with the 3N conclusions. At the end of a comprehensive review, Cantwell (1981) concluded that the outer scaling proposed in 3N seemed well established.

Almost immediately thereafter, however, a study by Blackwelder and Haritonidis (1983) contested this conclusion, and found that their new experiments suggested inner scaling. These authors attributed the earlir conclusions to use of hot wires that were too long and had inadequate resolution. This conclusion has in turn been questioned (Narasimha 1983, Shah et al 1983, Rajagopalan and Antonia 1984). Furthermore, Johansson and Alfredsson (1982), using the variable interval time averaging (VITA) technique of Blackwelder and Kaplan (1976), have suggested an intermediate scaling (proportional to the geometric mean of the inner and outer scales) for the bursting rates.

In a detailed commentary on the work of Blackwelder and Haritonidis, Narasimha (1983) has examined the question from a more general viewpoint; it is worthwhile to present a gist of these considerations here.

The work of Grass (1971) shows that the dynamics of the boundary layer away from the surface, and in particular the series of events associated with the bursts, remain essentially the same in rough wall boundary layers as on smooth ones. As, however, the structure of the wall layer is vastly different in the two cases, this suggests that the flow at the wall cannot be thought to exert a decisive influence over the structure of the boundary layer as a whole. Consider now a neutrally stable atmospheric boundary layer, whose Reynolds number based on boundary layer thickness can be of order 10^8 or more, which is about 10^4 times larger than in laboratory studies. The wall scale (given by ν/U_*, where ν is the kinematic viscosity and U_* the friction velocity) remains of the order of 1 mm in the atmosphere, as in the familiar laboratory boundary layer. If burst parameters scaled **entirely** on wall variables, we would have to conclude that the turbulent energy production in the atmospheric boundary layer is the same as in a (rough wall) laboratory boundary layer, and in particular that the **same** bursts can sustain turbulence in boundary layers with thickness ranging from a few centimetres to a kilometre. This would be very hard to explain in any simple way. If on the other hand bursts scaled with outer variables, the **frequency** of bursting in the atmospheric boundary layer will be only about 10^{-4} of that in the laboratory, and one would now have to explain how so few bursts provide the turbulent energy in such a thick boundary layer. The possibility exists, of course, that in the latter case bursts are much longer, in the sense that their duration is so long that a smaller number of them produce all the energy required: this would indeed be consistent with outer scaling.

Now the Reynolds numbers in most academic wind tunnels, where much of the work on turbulent bursts has till now been carried out, are too low to provide decisive evidence on these questions, as inner and outer scales are not too well separated. For example at $U = 20$ m/s, $U_* = 1$ m/s, $\delta = 50$ mm, $50\ \nu/U_*^2$ and δ/U are both about 3-4 ms. It was therefore urged that a detailed examination of atmospheric boundary layers be undertaken, where the wall scale remains the same but the outer time scale is now of order 50 s for $= 1$ km.

There have been earlier investigations of energy production in geophysical flows. Thus, Gordon (1974) found very close agreement between the distributions of the Reynolds stresses from events in a tidal estuary and those in the laboratory;

using the quadrant method of Lu and Willmarth (1973) he found reasonable agreement with the results of 3N (Gordon 1975). Heathershaw (1974) could clearly detect the ejection and sweep phases of the bursts as seen in the laboratory in his studies of near-bottom turbulence in the Irish Sea, and found that the duration of an ejection or a sweep was of the order of 5 to 10 seconds, and intervals between them 20 to 100 seconds: these numbers are clearly inconsistent with inner scaling. Jackson (1976), after comparing various alluvial flow studies with his own work in the lower Wabash River, Illinois, suggested that the bursting process as seen in the laboratory might be responsible for the boils and kolks seen on the surface of rivers. It may be recalled that Grass (1971) had already connected the boils on the water surface of his laboratory flows to the strongly ejected fluid from a burst interacting with the water surface. Jackson found that the frequency of boils on the river surface scaled as in the 3N studies, irrespective of the bed surface characteristics.

It must be admitted, however, that these geophysical studies have generally not received the attention we believe they merit from laboratory fluid dynamicists. The reason may in part be that the methods used in such geophysical studies have often been very different from those used in the laboratory.

We have therefore set out here to study the atmospheric boundary layer using the well-known VITA technique (Blackwelder and Kaplan 1976) which has been widely used in laboratory studies, although it is realised that there continues to be a basic problem of **recognising** a burst using only velocity traces, when it is not possible to visualise flow. Such an investigation should be particularly decisive because of the identity of the wall scales in laboratory and geophysical flows in air that we have already noted. Apart from the light that may thus be thrown on the bursting phenomenon, a clearer picture of such events in the atmosphere is of value in itself for such problems as pollutant dispersion and mixing, evaporation in the maritime boundary layer and the transfer of heat across geophysical boundaries.

2. THE DATA

2.1 Data Source

The data that form the subject of this study were taken at the Boulder Atmospheric Observatory at Boulder, Colorado, and were kindly supplied by Dr J C Kaimal. This observatory (Kaimal and Gaynor 1983) has a 300 m instrumented tower with a large variety of atmospheric sensors, located on gently rolling terrain 25 km east of the foothills of the Colorado Rockies. The steepest gradient (7 per cent) is to the south. The tower has eight levels of instrumentation, which consists of sonic and propeller-vane anemometers, and platinum wire and quartz thermometers. The height of the tower "ensures that the instruments extend above the nocturnal boundary layer nearly all the time and to at least 25 per cent of the daytime convection layer in the mid-afternoon on most days" (Kaimal and Gaynor 1983).

The data are on two tapes that we shall call I and II, both

recorded for a continuous stretch of four hours between 1600 hours and 2000 hours Mountain Standard Time (local time at Boulder) on 11 September 1978. The stability conditions in the atmospheric boundary layer during this period were very close to neutral, as we shall demonstrate below. As shown in **Table 1**, Tape I contains wind speed (in m/s) and direction (in degrees clockwise from north) from three-axis sonic anemometers at all eight levels on the tower. These anemometers are pulsed synchronously 200 times per second; 20 point non-overlapping averages of the readings (every 0.1 s) are recorded. For more details of the instrumentation see Kaimal and Gaynor (1983). Tape II contains vertical velocity and temperature readings for the same four hours and recorded similarly.

2.2 Nature of traces

We begin with a comparison of the qualitative nature of the fluctuations in the atmosphere and the laboratory.

Figures 1a, 1b and 1c show typical traces of the fluctuations of horizontal east-velocity (u'), vertical velocity (w'), and instantaneous product ($u'w'$) respectively. The data are for 1602 hours to 1604 hours local time, and are plotted for all the available levels. **Figure 1d** is a reproduction from Blackwelder and Kaplan (1976) showing the time series traces of streamwise velocity at different distances from the wall in a neutral turbulent boundary layer flow generated in a laboratory. It is important to note here that the time axis in the two plots is on vastly different scales. Also, while the laboratory traces are all quite close to the wall ($z_+ \leq 100$), the atmospheric traces correspond to the height range $0.7 \times 10^6 \leq z_+ \leq 21 \times 10^6$. (The value of the friction velocity U_* used in obtaining $z_+ = zU_*/\nu$, where z is the height above the surface, is discussed in Section 3.) As expected, there is a marked decrease in the turbulence intensity at large z.

It will be observed that Figures 1a and 1d show distinct similarities. The sudden increase in the horizontal velocity seen at ~ 1603 hours (level 1) is similar to what can be expected during the sweep stage of a burst. In Figure 1b one notices a sharp increase in the w′ signal at ~ 1603 hours, which is characteristic of the ejection phase in the burst cycle. The sweep with the largest amplitude in this stretch occurs a little before 1603 hours (see level 2, Figure 1a) while the most intense ejection occurs a little after 1603 hours (Figure 1b). Figure 1c is a similar plot of the flux u′w′ for the same time stretch. One observes here the clearest indications of a burstiness: distinct periods of turbulent activity alternate with periods of quiet.

2.3 Selection of data sets

By examining five-minute averages of the horizontal wind speed and direction for the full four hours (samples of which are shown in **Figure 2**), the following three stretches of nearly stationary data are selected for further analysis:

Data set A: 1600 hours to 1715 hours,

B: 1745 hours to 1820 hours,

C: 1900 hours to 2000 hours.

3. MEAN FLOW PROPERTIES

Profiles of five-minute averages of horizontal speed, plotted on semi-logarithmic paper (**Figure 3**), show that the log law is satisfied. The friction velocity (U_*) may therefore be estimated from the expression

$$U_* = K \frac{d\ u}{d \log z},$$

assuming the rough-wall velocity profile

$$\frac{u}{U_*} = \frac{1}{K} \ln \frac{z}{z_o}$$

where K is the Karman constant (here taken as 0.41) and z_o is the roughness height.

Table 2 lists values of the friction velocity so determined for the three data sets analysed here, and shows that U_* is of the order of 1 m/s, about the same as in laboratory boundary layers. The wall unit for time is 1.46×10^{-5} s when $U_* = 1$ m/s, taking $\nu = 1.46 \times 10^{-5}\ m^2/s$ at 18^oC, which was the mean temperature of the atmospheric surface layer at Boulder.

Figure 4 shows the mean temperature profiles in data set A. The lapse rate is fairly constant in the region $10m < z < 300m$ and is seen to be around 10^oC per km which is very close to the lapse rate for the neutrally stable atmospheric boundary layer.

Table 3 shows a list of gradient Richardson numbers for the data sets A, B, and C, defined (see e.g. Dutton and Panofsky

1984) as

$$Ri = \frac{(g/T)\ (\gamma_d - \gamma)}{(dU/dz)^2},$$

where g is the acceleration due to gravity, γ_d the dry adiabatic lapse rate, γ the actual lapse rate and U the mean wind speed. It is seen that the value of Ri upto level 5 is generally low, showing that the flow is neutrally stable (Panofsky and Dutton 1984).

4. ANALYSIS OF ENERGY EVENTS

4.1 VITA technique

It was decided to apply the VITA technique to find out what periods of intense turbulent activity could be detected in the turbulent signals discussed in Section 2.3. The main consideration for the adoption of this technique has been its wide use in laboratory flow studies. As developed by Blackwelder and Kaplan (1976), the technique exploits the intermittent nature of the short-time variance D, defined for any fluctuating turbulent quantity p′ as

$$D(p', t, t_{av}) = (1/t_{av}) \int_{t - t_{av}/2}^{t + t_{av}/2} p'^2(s)ds - \left((1/t_{av}) \int_{t - t_{av}/2}^{t + t_{av}/2} p'(s)\ ds \right)^2$$

where t_{av} is a prescribed time interval of averaging. The mean square value of p′, defined as

$$\hat{p}^2 = \lim_{t_{av} \to \infty} D(p';\ t,\ t_{av}),$$

is of course independent of t in steady flow.

When a peak in the short-time variance signal D exceeds a certain discriminator level $k\hat{p}^2$, where k is any prescribed threshold, we shall say that a "special event" of intensity k has occurred. Clearly the number of such events will depend in general on t_{av} and k. In order not to prejudge the issue of their relation with 'bursts', we shall call "energy events" all those detected in the above way when the signal p′ is the streamwise velocity fluctuation u′.

To verify the VITA algorithm used here the variance of a test signal devised by Johansson and Alfredsson was calculated as a function of t_{av}; comparison with the result of Johansson and Alfredsson (1982; Figure 2) showed excellent agreement.

4.2 Special events as a function of the threshold and averaging time

To select the optimum length of data for analysis, it was first necessary to determine the averaging time that yields stable averages. For this purpose the cumulative averages of various quantities were studied. Results for the vertical velocity, the quantity with the largest high frequency content (Panofsky and Dutton 1984), showed that after around 15 minutes, the cumulative means are fairly steady. Thus an interval of 10-15 minutes seems large enough to obtain relative stationarity.

From Figure 2 it can be seen that the winds during the entire period covered by the present data were predominantly in the east-west direction, hence the east-velocity component is selected for analysis.

We first display results from a VITA analysis of Data Set A. The first 8000 values of the horizontal east-velocity component (positive easterly) at level 1, corresponding to the time interval 1600 hours to a little over 1613 hours, was used for this purpose.

The short-time variance of the u′ signal at level 1, determined for t_{av} ranging from 0.3 to 10 s, is shown in **Figure 5.** It is seen that while the fluctuations are fairly even for short averaging times, intense special events stand out conspicuously especially when $t_{av} > 1$ s. This averaging time corresponds to approximately 66 x 10^3 wall units, orders of magnitude larger than the value usually adopted (approximately 10) in laboratory studies. Comparison with similar plots from Johansson and Alfredsson (1982, Figure 3) show correspondingly vast differences in time scales; further, events detected in the laboratory seem to have a much smaller variation in intensity than in the atmosphere, where the most intense events stand out clearly against the medley of smaller ones.

In **Figure 6** is plotted the number of events per second versus the time interval of averaging for different values of threshold. Comparison with Johansson and Alfredsson (1982, Figure 8) shows the low threshold at which event frequencies in the atmosphere are comparable to those in laboratory flows. For example while Johansson and Alfredsson (1982) report a maximum event frequency around 1 s^{-1} for k = 0.7 (t_{av} ~ 1 s), in the atmosphere at the same k and t_{av} the event frequency is only

10^{-2} s^{-1} , orders of magnitude less. Yet the dependence on threshold is very similar, indicating the broad similarity of the structures detected by VITA and their character in both laboratory and atmosphere.

In **Figure 7** is plotted the event frequency versus the threshold for different values of t_{av}. Comparison with a similar plot in Johansson and Alfredsson (1982, Figure 7) shows that the trends in both flows are very similar.

However, neither of these studies reveals any method of determining a unique event frequency, confirming the conclusions of various earlier studies.

The above procedure was rerun for a larger range of k and t_{av} for the u′ signal (1602 hours to 1614 hours) at level 1, and the resultant variation of the event frequency with k and t_{av} is shown in **Figures 8 and 9.** It is seen that while, once again, no unique number for special event frequency emerges from this analysis, the larger events stand out nearly independently of the threshold over a certain range of high k.

Similar plots were obtained when the north-velocity, vertical velocity or temperature were used as detector signals, for different levels and data sets. The qualitative variation of the number of detected events with k and t_{av} was always found to be essentially the same at the other heights (level 8, level 7, level 2; 1600 hours to 1612 hours) for which the analysis was repeated, and also for other stretches of data (1609 hours to 1621 hours and 1700 hours to 1712 hours) at level 1 (see Narasimha and Kailas 1986).

Figures 9a, b show the number of energy events detected on a (k, t_{av}) plane for the u′ and w′ signals (for level 1 and the stretch 1602 hours - 1614 hours) respectively. These figures show that at lower t_{av}, the events detected from the u′ signal are of relatively lower intensities than those from the w′ signal. One recalls here the band pass filter characteristics of VITA (pointed out by Johansson and Alfredsson 1982), removing events with period upto approximately 1.3 times the value of t_{av} chosen. The more intense u′ events therefore seem to be of rather longer time scales than the sharper w′ events.

4.3 Conditional averages

To compare the structure of the energy events so detected in the atmosphere with similar events in laboratory flows, a study was made of the conditional averages of different numbers of events detected. To do this, the 'centre' of each event was first determined as the middle point ($\tau = 0$) of the time interval during which the variance signal is greater than the prescribed threshold. This definition is illustrated in Figure 5.

The VITA technique was now modified to detect events in order of intensity, say the most intense n events in a given data stretch, instead of the generally used method of prescribing a threshold value and averaging over all events so detected. This was accomplished simply by increasing the k value from zero in small steps and noting the number of events detected for a given t_{av}. When the specified number of events was detected, the k value and the centres of the events were determined.

Figure 10 shows the conditional averages of the horizontal east-velocity at level 1 (1602 hours to 1612 hours) over the ensemble of the n 'biggest' events so detected, with n = 1, 4 and 29: the averages are struck over the ensemble at fixed τ. (A 'bigger' event has a higher peak value of D.) The conditional average among the 30 biggest events does not show any strong fluctuation in the signal around the centre ($\tau = 0$) of the event. However, for the smaller ensembles, the signal shows clear signs of a retardation followed by an acceleration at $\tau = 0$, as seen in the laboratory conditional averages (Johansson and Alfredsson 1982, Thomas 1977): for comparison see Figure 11 in Johansson and Alfredsson (1982). The similarity between the laboratory and atmospheric signals is certainly striking. The reason that the 29-event ensemble shows no ordered event is presumably that there is too much 'jitter' in the events. The above study clearly shows the difficulty in determining a unique number of detections in a given stretch of data; it also shows that at low thresholds the VITA technique detects spurious sharp fluctuations in the flow, as Johansson and Alfredsson (1982) have also found.

Figure 11 is a similar plot from the present study of the conditional averages for the vertical velocity, which picks out the sharp increases characteristic of the ejection phase of the burst cycle. We note the conditional averages among w' events are not well defined, especially at the higher t_{av}, unlike the ensemble averages among u′ events. It therefore appears that w′ events detected at lower t_{av} indicate more clearly an ejection phase, whereas in contrast the u′ events detected at higher t_{av} show clearer sweep-like characteristics. Keeping in mind the band pass character of the VITA technique noted earlier, this observation indicates a relatively long sweep followed by a relatively sharp ejection.

5. CONCLUSIONS

From the above studies it is clear that the events detected by the VITA technique in the atmospheric boundary layer are very similar to those seen in the laboratory. It is equally clear that we cannot really fix a unique number of special events by just running the VITA test on a stretch of data: evidently events of widely varying intensity are present. This variation may in part be because events pass the sensor at varying stages of their lives, and at different spanwise distances from their respective centres.

Are these events detected in the atmosphere the same as the bursts detected at much lower Reynolds numbers in laboratory studies? The nature of both events appears to be the same. If however bursts scaled on inner variables, as has been suggested e.g. by Blackwelder and Harotinidis (1982), the present data will not detect them, as their time resolution (0.1 s) is not adequate. In this case the present analysis must be considered to reveal the existence of remarkably similar bursts on much longer time scales. An alternative, and to us more plausible, explanation is that there is only one kind of burst, and that it in fact occurs on much longer time scales than inner scaling would imply. In a separate study (to be published) we show that 10 events in a period of 12 minutes account for 90 per cent of the momentum flux in the boundary layer. The characteristic time for burst period is therefore about 70 s. To compare this with outer time scales, we note that according to Kaimal and Gaynor (1983), the total boundary layer thickness varies from < 300 m to about 1200 m, the figure being closer to the latter perhaps at the time of measurement. If we take the log law as extending to about $0.2\,\delta$ (Panofsky and Dutton 1984), and note that at 300 m there is no noticeable departure from it (Figure 3), we would estimate δ as about 1500 m. With $U \simeq 10$ m/s and $\delta \sim 1000$ m, the outer time scale is thus of order 100 s. This is indeed of the same order as the observed event frequencies in the present data.

A similar analysis of flux events is being carried out, and should throw further light on the bursting phenomenon.

Acknowledgement

It is a pleasure to thank Dr J C Kaimal for his kindness in supplying the appropriate data for a neutrally stable boundary layer from the excellent facility at Boulder. This work has been supported in part by a grant from the Department of Science and Technology.

R.N. is grateful for this opportunity to pay tribute to Prof J C Rotta. It is a pleasure to acknowledge how often it has proved profitable to appeal to a well-thumbed personal copy (now nearly in tatters) of Volume 2 of Progress in Aerospace Sciences, which contained Prof Rotta's long and remarkable essay on the turbulent boundary layer.

References

/1/ Blackwelder, R.F.; Kaplan, R.E.: On the wall structure of the turbulent boundary layer. J. Fluid Mech. (1976) 76:89-112.

/2/ Blackwelder, R.F.; Haritonidis, J.H.: Scaling of the bursting frequency in turbulent boundary layers. J. Fluid Mech. (1983) 132:87-104.

/3/ Cantwell, B.J.: Organised motion in turbulent flow. Ann. Rev. Fluid Mech. (1981) 13:457-515.

/4/ Gordon, C.M.: Intermittent momentum transport in a geophysical boundary layer. Nature (1974) 248:392-393.

/5/ Gordon, C.M.: Period between bursts at high Reynolds number. Phys. Fluids (1975) 18:141.

/6/ Grass, A.J.: Structural features of turbulent flow over smooth and rough boundaries. J. Fluid Mech. (1971) 50:233-255.

/7/ Heathershaw, A.D.: Bursting phenomena in the sea. Nature (1974) 248:394-395.

/8/ Jackson, R.G.: Sedimentological and fluid-dynamic implications of the turbulent bursting phenomenon in geophysical flows. J. Fluid Mech. (1976) 77:531-560.

/9/ Johansson, A.V.; Alfredsson P.H.: On the structure of turbulent channel flow. J. Fluid Mech. (1982) 122:295-314.

/10/ Kaimal, J.C.; Gaynor, J.E.: Boulder Atmospheric Observatory. J. Appl. Met. (1983) 22:863-880.

/11/ Kline, S.J.; Reynolds, W.C.; Schraub, F.A.; Runstadler, P.W.: The structure of turbulent boundary layers. J. Fluid Mech. (1967) 30:741-773.

/12/ Lu, S.S.; Willmarth, W.W.: Measurements of the structure of the Reynolds stress in a turbulent boundary layer. J. Fluid Mech. (1973) 60:481-511.

/13/ Narasimha, R.: Comments on Blackwelder & Haritonidis - Scaling of the bursting frequency in turbulent boundary layers (1983).

/14/ Narasimha, R.; Kailas, S.V.: Energy events in the atmospheric boundary layer. Report 86 AS 8, Indian Inst. Sci. (1986).

/15/ Offen, C.R.; Kline, S.J.: Combined dye-streak and hydrogen-bubble visual observations of a turbulent boundary layer. J. Fluid Mech. (1974) 62:223-239.

/16/ Panofsky, H.A.; Dutton, J.A.: Atmospheric turbulence: Models and methods for enginering applications. Wiley Inter-Science Publications, New York (1984).

/17/ Rajagopalan, S.; Antonia, R.A.: Conditional averages associated with the fine structure in a turbulent boundary layer. Phys. Fluids (1984) 27:1966.

/18/ Rao, K.N.; Narasimha, R.; Narayanan, M.A.B.: The 'bursting' phenomenon in a turbulent boundary lyer. J. Fluid Mech. (1971) 48:339-352.

/19/ Shah, D.A.; Chambers, A.J.; Antonia, R.A.: Report TN FM 83/7, Dept. Mech. Engg., University of Newcastle (1983).

/20/ Thomas, A.S.W.: Some observations of the structure of the turbulent boundary layer. AGARD CP271. Paper 26 (1977).

/21/ Tu, B.J.; Willmarth, W.W.: An experimental study of turbulence near the wall through correlation measurements in a thick turbulent boundary layer. Report 02920-3-T, Dept. Aerospace Engg., Uni. Mich. College Engg., Ann Arbor (1966).

/22/ Ueda, H.; Hinze, J.O.: Fine-structure turbulence in the wall region of a turbulent boundary layer. J. Fluid Mech. (1975) 67:125-143.

/23/ Willmarth, W.W.: Structure of turbulence in boundary layers. Adv. Appl. Mech. (1975) 15:159.

Table 1: Data on Boulder tapes

Time: 1600 - 2000 hours Mountain Standard Time

Day : 11 September 1978

Level	Height (m)	Parameters Recorded Tape I	Tape II	Remarks
1	10			
2	22			
3	50	Horizontal	Vertical	
4	100	wind speed	velocity	Level 4
5	150	and	and	missing
6	200	direction	temperature	
7	250			
8	300			

Table 2: Friction velocity and roughness parameter in atmospheric boundary layer data analysed

Code	Time (hours)	U_* (m/s)
A_1	1600 - 1605	1.15
	1605 - 1610	0.909
	1610 - 1615	1.2
A_2	1640 - 1645	1.45
	1645 - 1650	0.709
	1650 - 1655	0.729
A_3	1700 - 1705	1.04
	1705 - 1710	1.00
	1710 - 1715	0.92

Table 3: Gradient Richardson numbers in the atmospheric boundary layer

Height range, levels	Data set A	B	C
1 to 2	-0.027	-0.019	0.033
2 to 3	-0.025	0.066	0.077
3 to 5	-0.085	0.104	0.157
Mean temperature (°C)	18.16	17.51	16.87

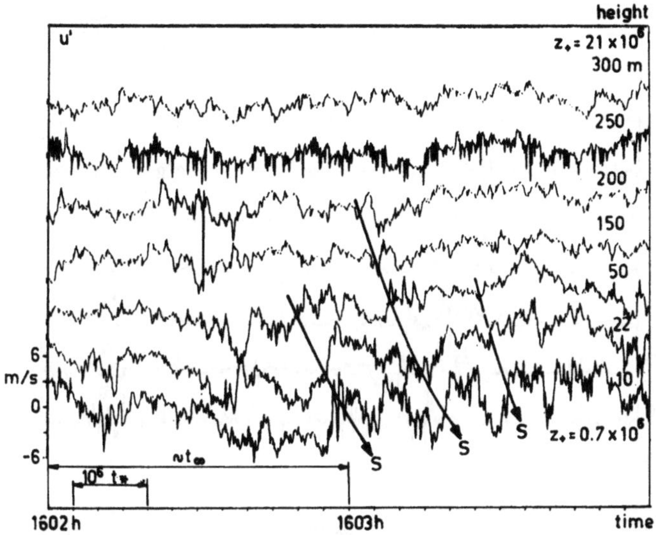

Fig. 1a. Simultaneous time-series traces of horizontal velocity fluctuations (u′) from 7 levels of the tower. The ′sweep′ (S) stage of bursting can be located especially in the lower levels.

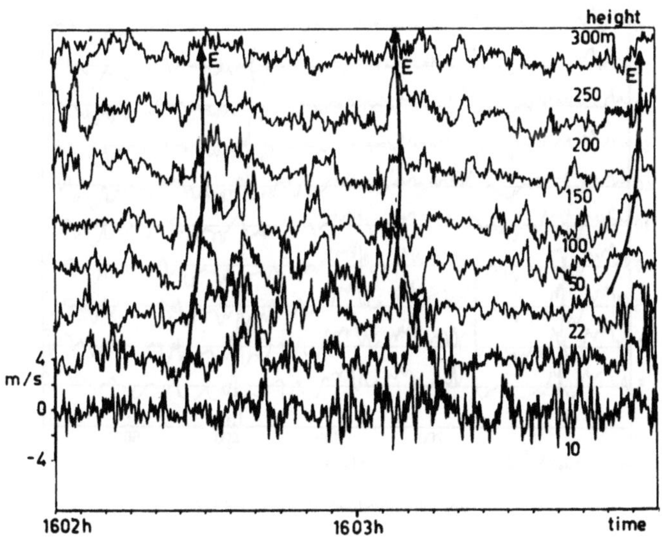

Fig. 1b. Simultaneous traces of the vertical velocity fluctuations (w′) from 8 levels of the tower. The sharp ′ejection′ (E) stage of bursting can be noted here.

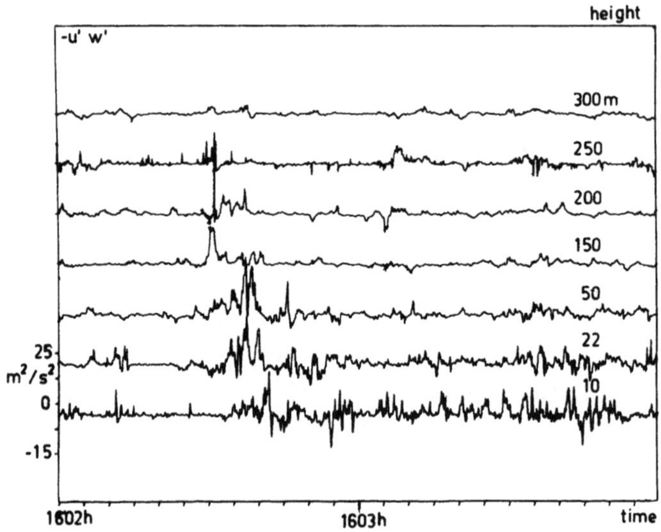

Fig. 1c. Simultaneous traces of the instantaneous product - u'w' from 7 levels of the tower. The intermittent nature of the flux - u'w' is easily seen.

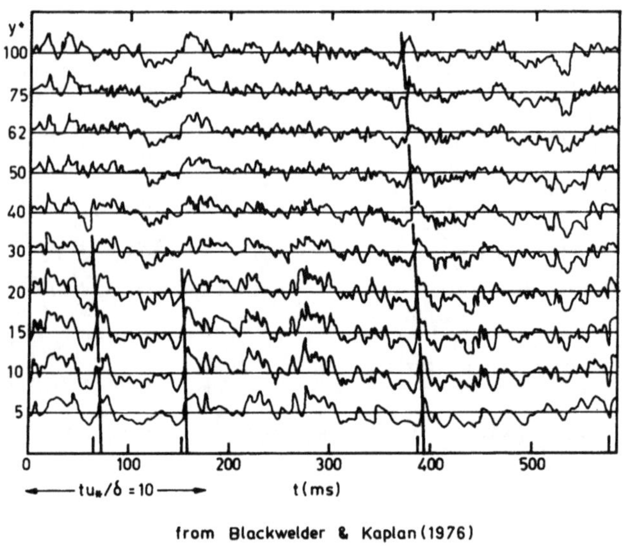

Fig. 1d. Simultaneous traces of the streamwise velocity from various levels of a boundary layer flow in the laboratory (Blackwelder and Kaplan 1976). The 'sweep' stage is marked in the figure.

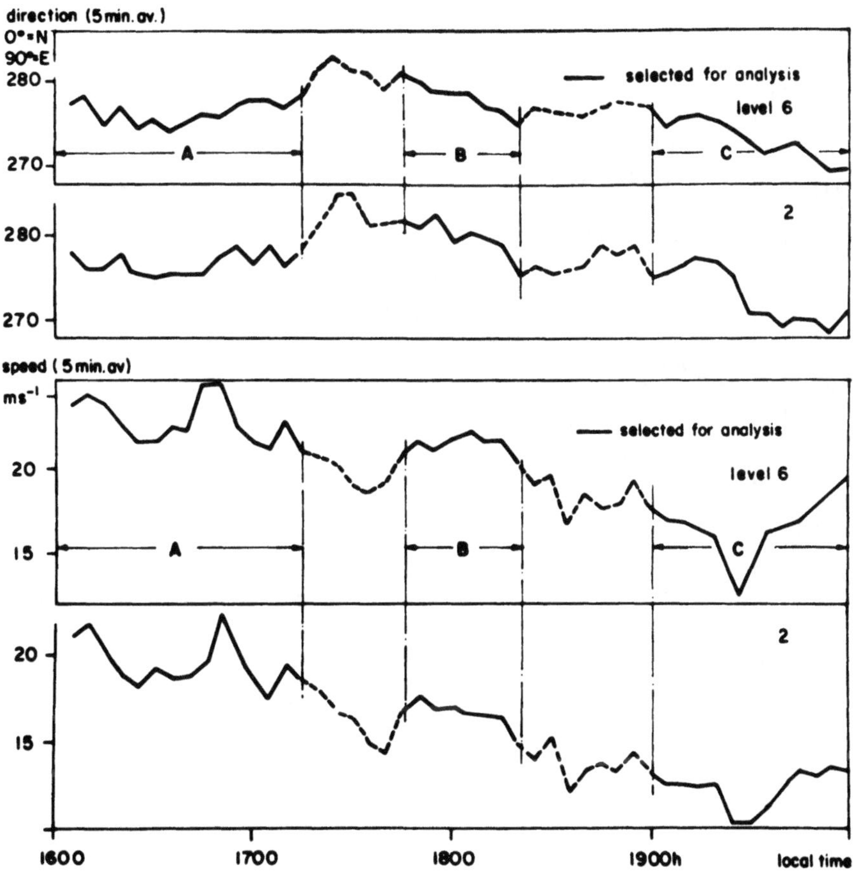

Fig. 2. Five-minute averages of wind speed and direction at two levels of the tower showing the subsets of data selected for analysis.

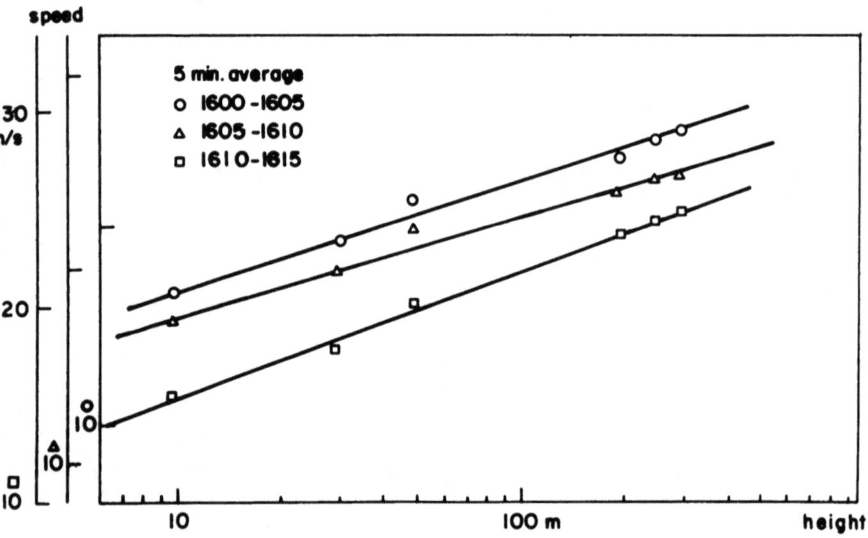

Fig. 3. Five-minute averages of wind speed showing the log-law profile.

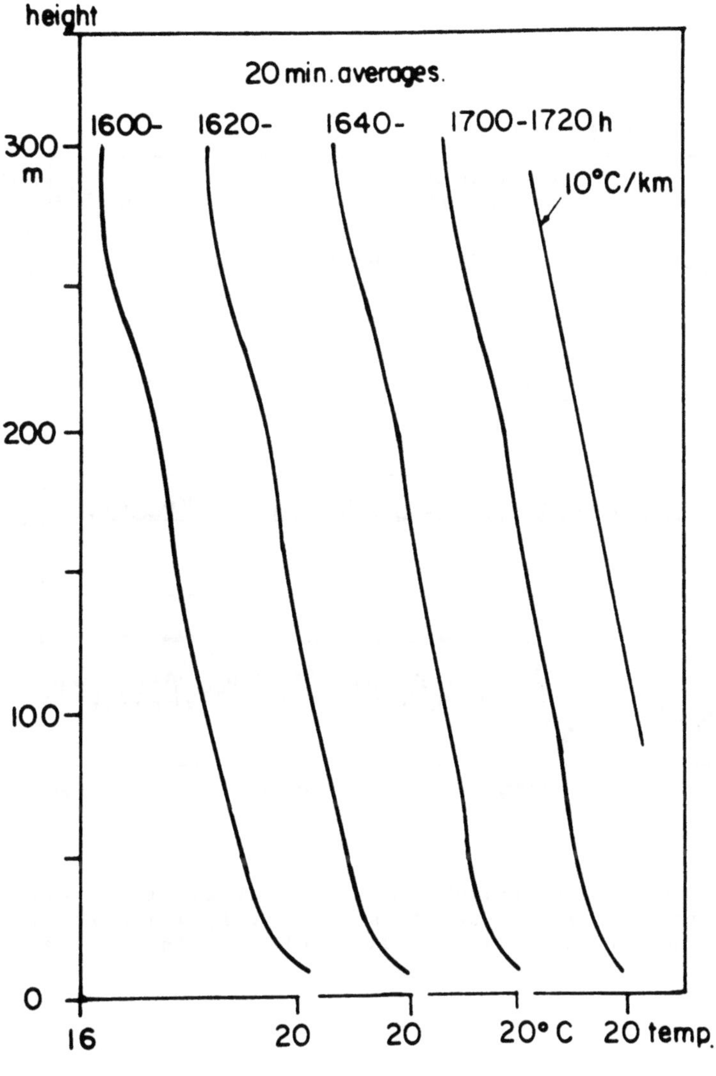

Fig.4. Mean temperature profile. The lapse rate in this region is around 10°C per km.

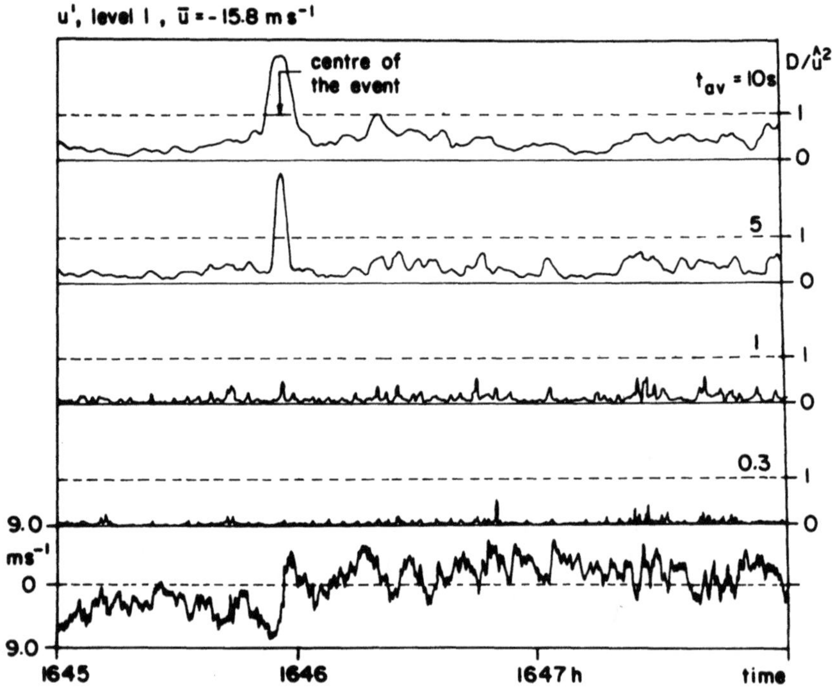

Fig.5. A trace of the atmospheric u′ signal and the short-time variance signals at averaging times of 0.3, 1, 5 and 10 s. Just before 1646 hours an event is seen clearly in the variance signal.

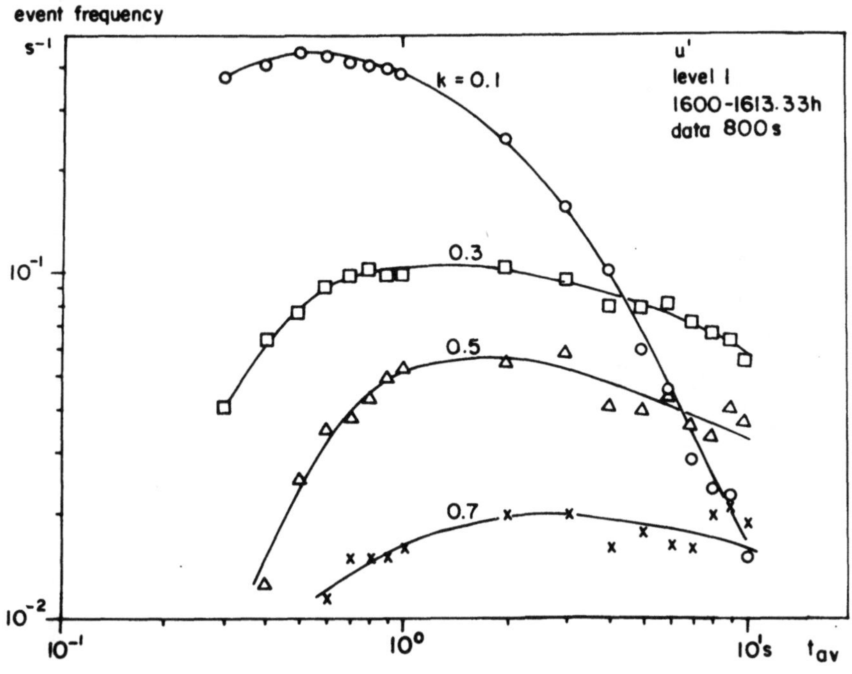

Fig. 6. The event frequency as a function of averaging time at various thresholds.

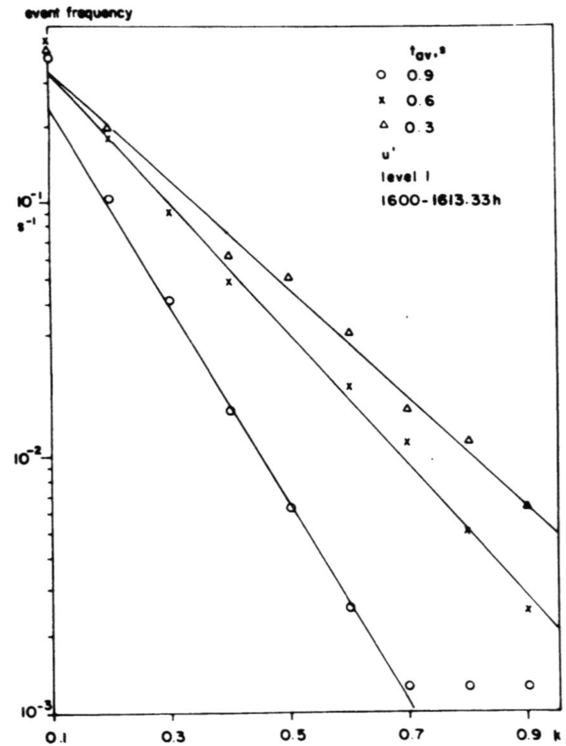

Fig. 7a. Frequency of occurrence of energy events of intensity k, at different averaging times.

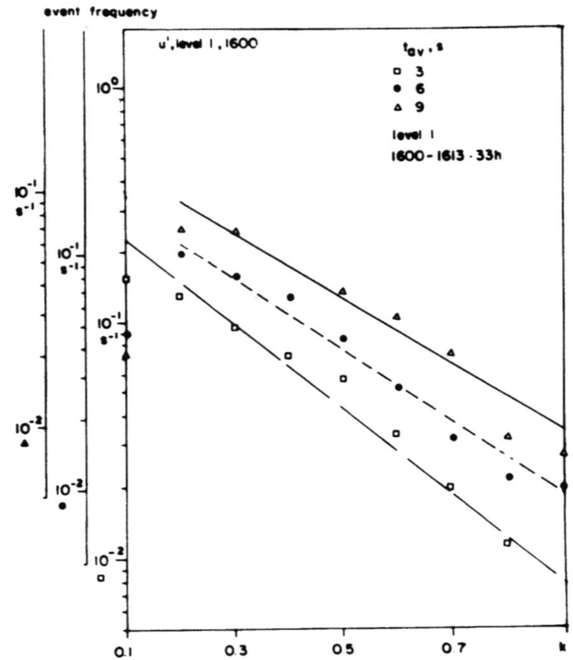

Fig. 7b. As in Fig. 7a, but at larger averaging times.

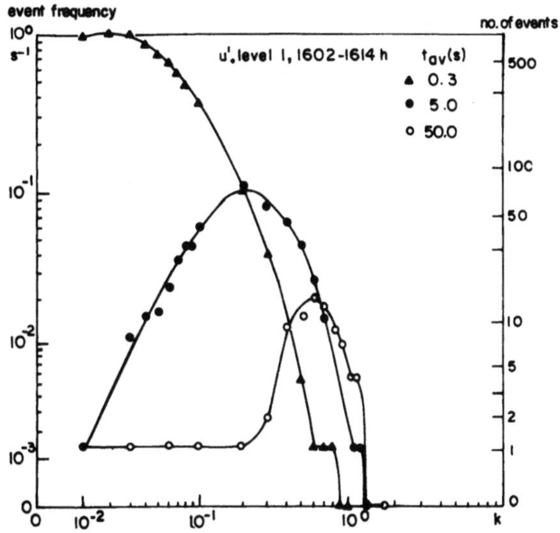

Fig. 8a. Frequency of energy events of different intensity in the atmosphere, at different averaging times.

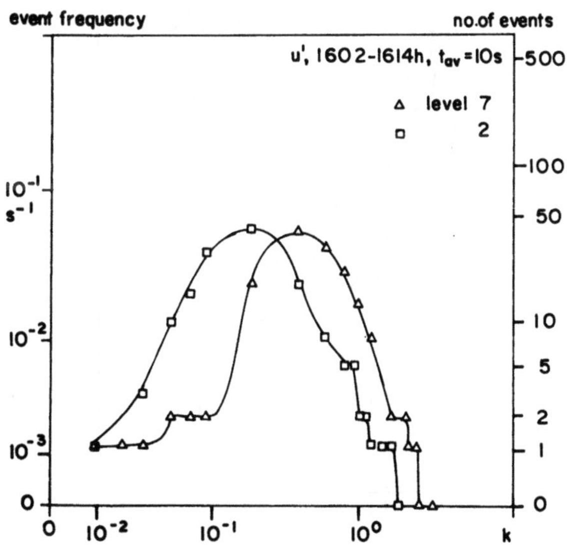

Fig. 8b. Comparison of detected event frequencies from the atmospheric data at two levels (2 and 7) at t_{av} = 0.3 s.

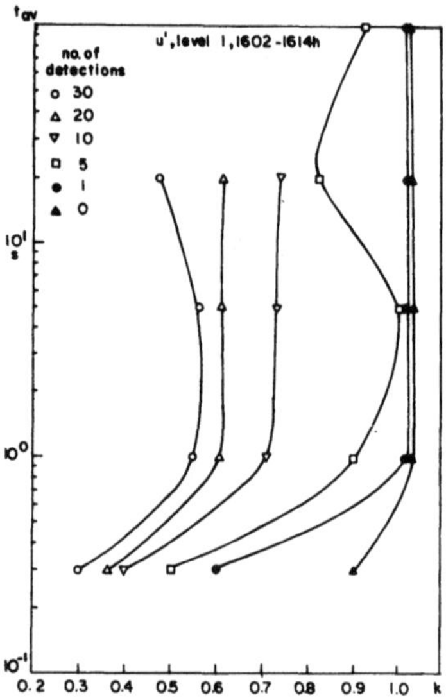

Fig. 9a. Contours of equal number of events detected from atmospheric data of horizontal velocity plotted on a (k, t_{av}) plane.

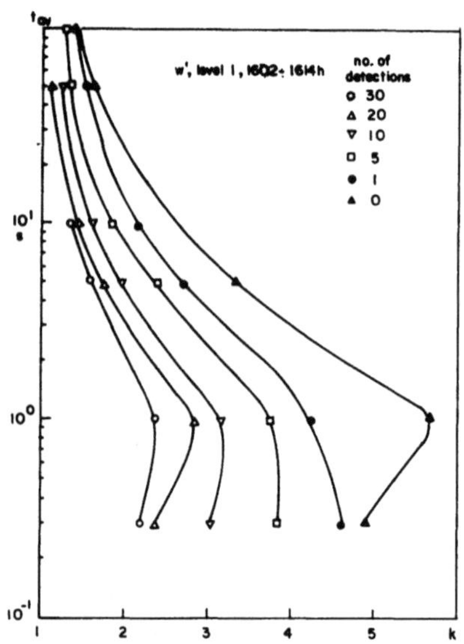

Fig. 9b. As in Fig. 9a, but from data on vertical velocity. Comparing with Fig. 9a, it is seen that w′ events are more intense than u′ events at low averaging times, and less intense at high averaging times.

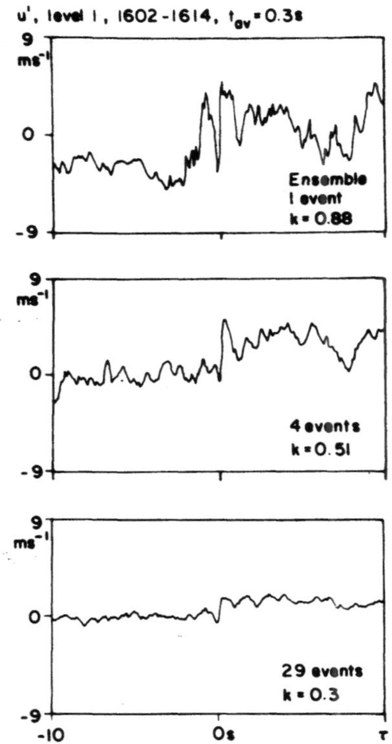

Fig. 10a. Conditional averages of events from ensembles of different sizes from the horizontal velocity signal from the atmosphere at t_{av} = 0.3 s.

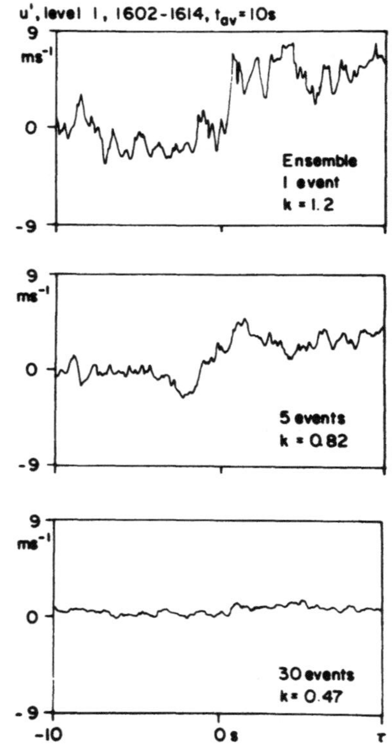

Fig. 10b. Conditional averages of events from ensembles of different sizes from the horizontal velocity signal from the atmosphere at t_{av} = 10 s.

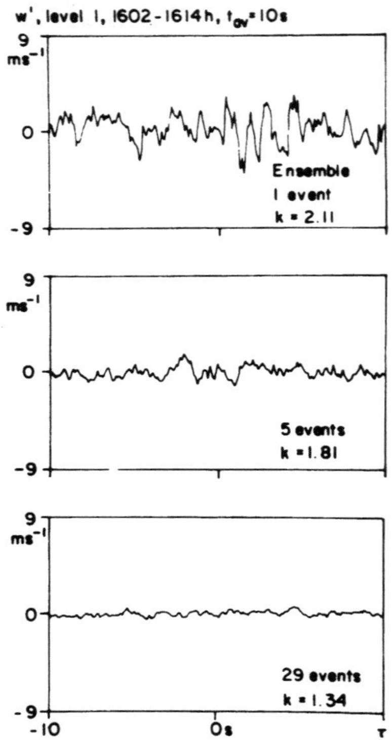

Fig. 11a. Conditional averages of events from ensembles of different sizes from the atmospheric vertical velocity signal at t_{av} = 0.3 s.

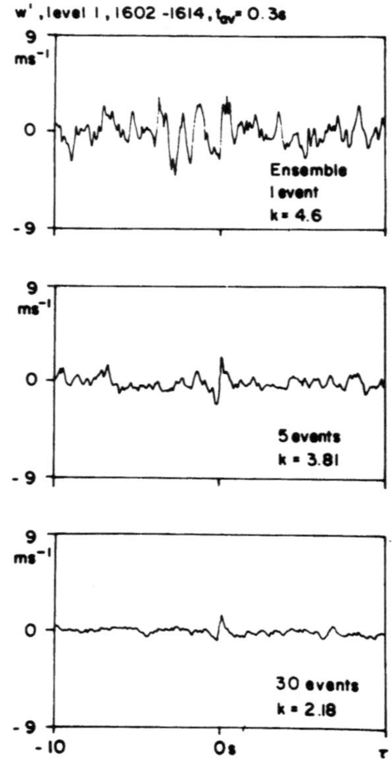

Fig. 11b. Conditional averages of events from ensembles of different sizes from the atmospheric vertical velocity signal at t_{av} = 10 s.

Turbulent Boundary Layer Development over Rough Surfaces

I. Tani

National Aerospace Laboratory
Jindaiji, Chofu, Tokyo, Japan

Summary

Using Coles additive law of the wall and law of the wake, with constant shift allowed for the effect of roughness, an integral method of analysis is put forward for evaluating the characteristic parameters (friction, wake and roughness parameters) from experimental mean velocity data of rough-wall turbulent boundary layers in zero pressure gradient. Emphasis is laid on the consistent determination of characteristic parameters and the interpretation of the effects of roughness in the context of these parameters. Re-evaluation of the published experimental data is made for three examples of the k-type roughness (the sand grain, sphere and circular rod roughnesses) and, furthermore, for the groove roughness of the d-type. A special interest attaches to the last example as it exhibits equilibrium behavior.

Introduction

The framework of our understanding of the turbulent flow in rough pipes was established by Nikuradse [1], who carried out extensive measurements in pipes roughened with closely packed sand grains. He found that with increasing Reynolds number the skin friction deviates from that of the smooth pipe and becomes dependent on the relative scale of roughness as well as the Reynolds number. At higher Reynolds numbers the skin friction becomes independent of viscosity and depends solely on the relative roughness scale, bringing about a regime called 'fully rough'. It was also confirmed by Nikuradse that the velocity defect law is valid except very close to the wall, no matter whether the wall be smooth or rough. Shortly later, Schlichting [2] investigated the effect of other types of roughness geometry in a duct of rectangular cross section and presented the result of experiments in terms of the equivalent sand grain roughness, which is the scale of the sand grain producing the same skin friction as the given roughness geometry in fully rough regime.

The extension to boundary layer flows was first made by Prandtl and Schlichting [3], who calculated the skin friction of a flat plate roughened with sand grains in zero pressure gradient, on a tacit assumption that the velocity profiles are the same as those measured by Nikuradse in rough pipes. The difference in velocity profile between the flow in a pipe and the boundary layer on a flat plate seems to have passed unnoticed until the investigation of Wieghardt [4] on the boundary layer developing over a smooth flat plate. On the other hand, Clauser [5] and Hama [6] demonstrated the validity of the same velocity defect law for boundary layer flows on both smooth and rough plates. It was also shown that the effect of roughness is represented by a constant shift of the logarithmic portion of the velocity profile by an amount characteristic of the roughness. Although the velocity defect profile is universal for both smooth-wall and rough-wall flat plate boundary layers, there is no universality in velocity defect profile for the pipe flow and the flat plate boundary layer flow. The difference between the two profiles is none other than that caused by the wake component of the additive law of the wall and law of the wake due to Coles [7][8][9].

Based on these earlier investigations, it is now possible to express the mean velocity profile in the form

$$\frac{U}{u_\tau} = \frac{1}{\kappa} \ln \frac{u_\tau y}{\nu} + C - \frac{\Delta U}{u_\tau} + \frac{\Pi}{\kappa} w \qquad (1)$$

for the turbulent boundary layer on a rough wall, where U is the mean velocity in the x-direction, x and y are coordinates parallel and normal to the wall, respectively, u_τ is the friction velocity (square root of the wall shear stress divided by density), ν is the kinematic viscosity, κ is the Karman constant (= 0.41), C is the smooth-wall constant (= 5.0), $\Delta U/u_\tau$ is the roughness shift, Π is the Coles wake parameter and w is the universal function of y/δ, δ being the thickness of the boundary layer. The effect of the viscous sublayer is considered negligibly small except for smooth-wall boundary layers at low Reynolds numbers. For smooth-wall boundary layers ($\Delta U/u_\tau = 0$), it is possible to evaluate the skin friction coefficient $c_f = 2(u_\tau/U_o)^2$ (U_o is the local free-stream velocity) and the wake parameter Π by fitting the measured velocity profile to the analytical expression (1). For rough-wall boundary layers, however, introduction of an additional parameter $\Delta U/u_\tau$ makes the evaluation imprecise, unless one of the three parameters is known by some other means. If the roughness of the same geometry were tested as rough pipe flow, it would be possible to transfer the roughness shift determined in pipe flow to boundary layer flow. Unfortunately, however, most of the recent roughness investigations have been made in rough-wall turbulent boundary layers in zero pressure gradient. As a natural result, the evaluation of the characteristic parameters and their correlations with other data have remained not entirely satisfactory. For example, the roughness shift has been determined by simply subtracting the rough-wall velocity profile from the smooth-wall velocity profile, without allowing for the difference in wake component for smooth-wall and rough-wall boundary layers (Bettermann [10]), or by assuming a certain correction for the simple subtraction in a way of crudely allowing for the wake component (Furuya and Fujita [11], Furuya, Fujita and Nakashima [12]). Otherwise, there has been no alternative but to blindly accept the Schlichting's value of equivalent sand grain roughness (Healzer, Moffat and Kays [13], Pimenta, Moffat and Kays [14]).

Negligence of the role of wake component does not seem to have been confined to the evaluation of roughness shift. It has been noted by Liu, Kline and Johnston [15], Healzer et al. [13] and Pimenta et al. [14] that the measured skin friction consistently deviates from the Prandtl-Schlichting calculation mentioned before. The discrepancy is attributed by Mills and Hang [16] to the neglect of the wake component in the renowned calculation.

The wake parameter Π is generally regarded as dependent on x. In the event that Π is independent of x, the velocity defect ratio $(U_o - U)/u_\tau$ turns out to be function of y/δ alone, which is the property assigned to equilibrium flow by Clauser [5]. Strictly speaking, however, $(U_o - U)/u_\tau$ is also a weak function of c_f, and this dependence precludes the possibility of even approximate equilibrium for smooth-wall boundary layers (Tani and Motohashi [17]). On the other hand, rough-wall boundary layers possess a characteristic feature of viscosity independent skin friction in fully rough regime. There is therefore a possibility for exact equilibrium to be realized with a certain type of rough-wall boundary layer (Rotta [18], Tani [19]).

This paper describes the author's recent attempt to develop an integral method of analysis for evaluating the characteristic parameters (friction, wake and roughness parameters) from experimental mean velocity data of rough-wall turbulent boundary layers. Emphasis is laid on the consistent determination of characteristic parameters and the interpretation of the effects of roughness in the context of these parameters.

Integral method of analysis

Wake function: The universal wake function suggested by Coles [9], $w(y/\delta) = 1 - \cos(\pi y/\delta)$, is incapable of yielding zero velocity gradient at the edge of the boundary layer, $y = \delta$. The deficiency is made up by using a quartic polynomial (Lewkowicz [20])

$$w(\eta) = 2\eta^2(3 - 2\eta) - \frac{1}{\Pi}\eta^2(1 - \eta)(1 - 2\eta), \qquad (2)$$

where $\eta = y/\delta$, although the second term makes it inevitable to correct the wall component by $-\kappa^{-1}\eta^2(1 - \eta)(1 - 2\eta)$. A quartic is chosen for w in preference to a cubic by reason of the smallness in magnitude of this correction.

Integral thicknesses: With the equations (1) and (2), the velocity defect $U_0 - U$ is expressed in the form

$$\frac{\kappa}{u_\tau}(U_0 - U) = -\ln\eta + \eta^2(1 - \eta)(1 - 2\eta) + 2\Pi(1 - \eta)^2(1 + 2\eta). \qquad (3)$$

The displacement and momentum thicknesses of the boundary layer are then given by

$$\delta^* = \alpha\frac{\delta}{z}, \quad \theta = (\alpha - \frac{\beta}{z})\frac{\delta}{z}, \qquad (4)$$

respectively, where

$$z = \kappa \frac{U_o}{u_\tau} = \kappa\sqrt{\frac{2}{c_f}}, \tag{5}$$

$$\alpha = \frac{59}{60} + \Pi, \quad \beta = \frac{8437}{4200} + \frac{667}{210}\Pi + \frac{52}{35}\Pi^2, \tag{6}$$

the friction parameter z thus being inversely proportional to the square root of the conventional skin friction coefficient c_f. Elimination of δ between the two equations of (4) gives

$$\frac{8437}{4200} - \frac{59}{60}\kappa G + \left(\frac{667}{210} - \kappa G\right)\Pi + \frac{52}{35}\Pi^2 = 0, \tag{7}$$

where

$$G = \frac{z}{\kappa}\left(1 - \frac{\theta}{\delta*}\right) \tag{8}$$

is the Clauser parameter. The equations (4) to (7) are all independent of roughness, although the flow may be affected by roughness through the boundary conditions. For further analysis it is convenient to introduce a non-dimensional quantity

$$s = z - 2\Pi - \kappa C. \tag{9}$$

Smooth wall: Evaluating the velocity profile (1), with $\Delta U/u_\tau = 0$, at $y = \delta$, we have

$$\delta = z\,\frac{e^s}{\kappa}\frac{\nu}{U_o}. \tag{10}$$

Substitution of (10) into (4) yields

$$\delta* = \alpha\,\frac{e^s}{\kappa}\frac{\nu}{U_o}, \quad \theta = \left(\alpha - \frac{\beta}{z}\right)\frac{e^s}{\kappa}\frac{\nu}{U_o}. \tag{11}$$

The equation (7), along with the first equation of (11), affords the means for evaluating the friction parameter z and the wake parameter Π from the measurable quantities $\delta*$ and θ. For low Reynolds numbers, however, it is necessary to account for the effect of viscous sublayer by using the corrected thicknesses

$$\delta*_c = \delta* - 50.63\,\frac{\nu}{U_o}, \quad \theta_c = \theta + \left(50.63 - \frac{136.47}{\kappa z}\right)\frac{\nu}{U_o} \tag{12}$$

in place of the measured thicknesses $\delta*$ and θ in the equations (8) and (11)(Tani and Motohashi [17]).

k-type roughness: Two types of roughness are considered, k-type and d-type according to the terminology of Perry, Schofield and Joubert [21]. The k-type roughness, typified by a smooth wall roughened with closely packed sand grains, or with sparsely spaced spanwise rods, has the roughness shift of the form

$$\frac{\Delta U}{u_\tau} = \frac{1}{\kappa} \ln \frac{u_\tau h}{\nu} + K, \tag{13}$$

where h is the height scale of roughness and K is a function of $u_\tau h/\nu$, tending to a constant value, characteristic of roughness geometry, for large values of $u_\tau h/\nu$ in fully rough regime. Evaluating the velocity profile (1), with $\Delta U/u_\tau$ given by (13), at $y = \delta$, we have

$$\delta = h\, e^{s + \kappa K}. \tag{14}$$

Substitution of (14) into (4) yields

$$\delta^* = \alpha \frac{h}{z} e^{s + \kappa K}, \qquad \theta = (\alpha - \frac{\beta}{z}) \frac{h}{z} e^{s + \kappa K}. \tag{15}$$

d-type roughness: The second type of roughness is typified by a smooth wall containing sparsely spaced narrow spanwise grooves. The roughness shift is expressed in the form independent of the roughness scale, such as the depth h of the groove,

$$\frac{\Delta U}{u_\tau} = \frac{1}{\kappa} \ln \frac{u_\tau \delta}{\nu} + D, \tag{16}$$

where D is a constant, characteristic of the roughness geometry. Evaluating the velocity profile (1), with $\Delta U/u_\tau$ given by (16), at $y = \delta$, we have

$$s = -\kappa D. \tag{17}$$

When the constant K or D is known from other source, the equation (7), together with the first equation of (15) or equation (17), would afford the means for evaluating z and Π from the measured values of δ^* and θ. When K or D is also to be known, an auxiliary equation must be sought for evaluation, as will be seen later.

Momentum integral equation: In the following we assume that the flow is in fully rough regime and that the constants K and D are independent of x. Substituting the integral thicknesses obtained in the preceding section into the momentum integral equation

$$\frac{d\theta}{dx} + \frac{\delta^* + 2\theta}{U_o} \frac{dU_o}{dx} = \left(\frac{u_\tau}{U_o}\right)^2, \tag{18}$$

we have

$$(\alpha z - \beta + \frac{\beta_c}{z})\, z \frac{dz}{dx} - [(2\alpha - \dot{\alpha})z - (2\beta - \dot{\beta})]\, z \frac{d\Pi}{dx} + (2\alpha z - \beta_c) \frac{z}{U_o} \frac{dU_o}{dx} = \frac{U_o}{\nu} \kappa^3 e^{-s}. \tag{19}$$

for the smooth-wall boundary layer,

$$[\alpha z - (\alpha + \beta) + \frac{2\beta}{z}]\,h\frac{dz}{dx} - [(2\alpha - \dot{\alpha})z - (2\beta - \dot{\beta})]\,h\frac{d\Pi}{dx}$$
$$+ (\alpha z - \beta)\frac{dh}{dx} + (3\alpha z - 2\beta)\frac{h}{U_0}\frac{dU_0}{dx} = \kappa^2 e^{-(s + \kappa K)} \quad (20)$$

for the k-type rough-wall boundary layer, and

$$[\dot{\alpha}z - (2\alpha + \dot{\beta}) + \frac{4\beta}{z}]\,\frac{\delta}{2}\frac{dz}{dx} + (\alpha z - \beta)\frac{d\delta}{dx}$$
$$+ (3\alpha z - 2\beta)\frac{\delta}{U_0}\frac{dU_0}{dx} = \kappa^2 \quad (21)$$

for the d-type rough-wall boundary layer, respectively, where the dot above α and β denotes differentiation with respect to Π, and $\beta_c = \beta - 136.47\,e^{-s}$.

Possibility of equilibrium: Each of the equations, (19), (20) and (21), provides a differential equation for z and Π (or δ) as functions of x. Owing to the closure difficulty, the equation is incapable of solution without being supplemented by an auxiliary equation. None the less, the differential equation serves to find the conditions under which the equilibrium solution could exist. It is readily seen (Tani [19]) that the equilibrium would be possible for boundary layers in favorable pressure gradient ($dU_0/dx > 0$) over smooth as well as k-type rough surface. If it is allowed for the roughness to increase in height linearly with x, equilibrium would be possible even in zero pressure gradient. If the roughness is of d-type, equilibrium would exist for a certain range of pressure gradient, from favorable to adverse. All these possibilities, except the d-type rough-wall flow, are those anticipated by Rotta [18] on dimensional grounds. It is intended in this paper to demonstrate the existence of equilibrium on the basis of evaluation of z and Π from available experimental data.

Sand grain roughness

Nikuradse's [1] experimental results for sand grain rough pipes have been reproduced in most textbooks absolutely intact for more than fifty years. The roughness is of stochastically distributed geometry. The microscopic photograph of the rough wall reveals the existence of interspace between the neighboring sand grains, which seems to have been considered as requisite by Nikuradse for the scale of the grain to equal the hydrodynamically effective scale of roughness. This geometry gives k-type roughness shift, and the values of κ, C and K found by Nikuradse are 0.40, 5.5 and -3.0, respectively. Following Coles [8][9] we take $\kappa = 0.41$ and $C = 5.0$, which make it desirable for consistency to re-evaluate K from the original data.

Nikuradse measured mean velocity at 17 positions along the radius of the pipe. He determined the effective radius by weighing the pipe filled with fluid (water), but did not explicitly define the origin of the normal coordinate y. Re-evaluation of the original data has recently been made by Grigson [22], who determined the origin of coordinate with the condition that would give the logarithmic distribution of velocity with $\kappa = 0.41$ in the neighborhood of the wall. The fact that only a small correction of origin is needed has been interpreted as disproving the frequently repeated idea of dependence of the Karman constant on Reynolds number. Unfortunately, however, the velocity profiles, even corrected, cannot be used to produce sufficiently accurate values of the integral thicknesses required for analysis, because of the insufficient number of measurement positions close enough to the wall. On the other hand, the pipe flow has a characteristic advantage that the constant K can be determined without recourse to the calculation of the integral thicknesses.

Assuming the velocity profile (1) with w and $\Delta U/u_\tau$ given by (2) and (13) respectively, and with δ replaced by the pipe radius r, we have

$$\frac{U_o}{u_\tau} = \frac{1}{\kappa}\ln\frac{r}{h} + C - K + \frac{2\Pi}{\kappa} \qquad (22)$$

on the pipe axis $y = r$. We also have

$$\frac{U_m}{u_\tau} = \frac{1}{\pi r^2}\int_0^r \frac{U}{u_\tau} 2\pi(r - y)dy = \frac{U_o}{u_\tau} - \frac{1}{\kappa}\left(\frac{3}{2} + \frac{7}{5}\Pi\right) \qquad (23)$$

for the rate of discharge U_m. Elimination of the wake parameter Π between (22) and (23) gives

$$K = C + \frac{1}{\kappa}\left(\ln\frac{r}{h} - \frac{15}{7}\right) + \frac{3}{7}\frac{U_o}{u_\tau} - \frac{10}{7}\frac{U_m}{u_\tau}, \qquad (24)$$

which makes it possible to evaluate K from the measured values of U_o, U_m and u_τ. Seeing that Nikuradse determined U_m by directly measuring the discharge for the bulk Reynolds numbers $2rU_m/\nu$ up to 3×10^5, but by integrating the velocity profiles for higher Reynolds numbers, this method of analysis would yield evaluation free from the effect of error in origin at least for lower Reynolds numbers. Results of calculation are shown in Fig. 1, where K is plotted against the roughness Reynolds number $u_\tau h/\nu$. The results for different values of r/h, covering different ranges of $u_\tau h/\nu$, form roughly into a single curve. This would suffice to dissipate the remaining doubt as to the effect of error in origin for higher Reynolds numbers. We thus obtain K = -3.6 for sufficiently large values of $u_\tau h/\nu$.

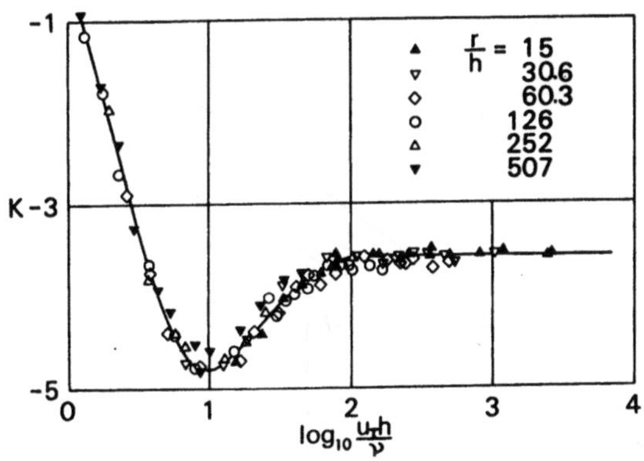

Fig. 1. Roughness shift constant re-evaluated from experimental data of sand grain rough-pipe flow [1].

Turning to boundary layer flows, we pose a problem to evaluate all the three characteristic parameters, z, Π and K from the mean velocity profiles measured on the flat plate roughened with sand grains (Furuya and Fujita [11], Bandyopadhyay [23]). There is a possibility that z can be guessed from the experimental data, for example, by the aid of the momentum integral equation (18) for zero pressure gradient,

$$\frac{d\theta}{dx} = \left(\frac{\kappa}{z}\right)^2, \tag{25}$$

which, however, provides a reliable basis only when the flow is two-dimensional to a considerable degree. Instead, therefore, we begin by regarding δ^* and θ as most reliable but z as susceptible to error and to be adjusted such that the equations (7) and (15) are satisfied. This procedure is justified on the ground that most of the experimental data of smooth-wall boundary layer flows could be made consistent by adjusting z by at most 2 to 3 percent (Tani and Motohashi [17]).

For rough-wall boundary layers, however, we have an additional parameter K, which makes the number of equations less than that of unknowns. The deficiency may be covered by making use of the integrated form of (25),

$$\theta - \theta_o = \kappa^2 \int_{x_o}^{x} \frac{1}{z^2}\, dx\,, \tag{26}$$

at this stage of calculation, where θ_o is the momentum thickness at a conveniently selected station $x = x_o$.

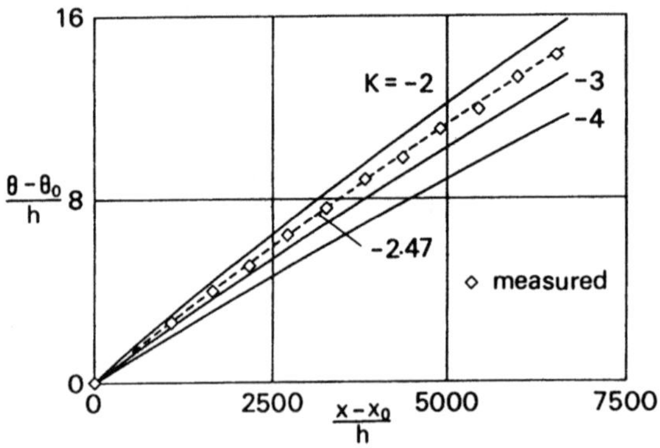

Fig. 2. Analysis of experimental data of sand grain rough-wall boundary layer flow [23]. x_o = 1219 mm, θ_o = 3.86 mm, h = 0.56 mm, U_o/ν = 2.67 x 10^3 mm^{-1}, $u_\tau h/\nu$ = 62-75.

The process of analyzing the Bandyopadhyay's [23] experimental data for fully rough regime is shown in Fig. 2, where values of $(\theta - \theta_o)/h$ calculated by (26) for three assumed values of K are compared with those obtained directly from measurements, h being 0.56 mm, the average roughness scale, and $U_o h/\nu$ = 1.5 x 10^3. We determine by comparison K = - 2.47 as the most probable value for the set of measurements. Similar analysis of the measurements of Furuya and Fujita [11] for h = 0.90 mm ($U_o h/\nu$ = 1.3 x 10^3) yields K = - 2.15. The difference between the two values of K is a little too large to disregard, and we are inclined to imagine that the Bandyopadhyay's wall might not exactly be similar in geometry to the Furuya and Fujita's wall, though both walls are roughened with aluminum oxide grits. It might further be imagined that neither of the two walls is similar to the Nikuradse's sand grain roughness, which possesses the value of K = -3.6 as already mentioned. It thus appears that the sand grain roughness cannot be uniquely defined without introducing at least an additional parameter, which might possibly be related to grain density.

Values of z and Π calculated from the two sets of measurements [11][23] are shown as functions of x/h in Figs. 3 and 4, respectively. x is the distance measured from the leading edge of the flat plate, where the turbulent boundary layer can safely be as-

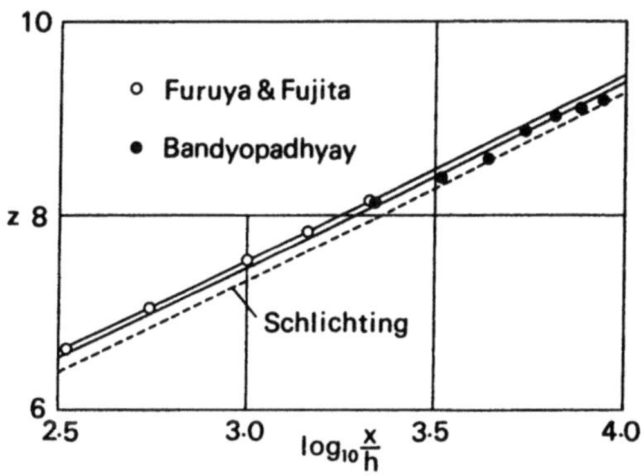

Fig. 3. Friction parameter evaluated from experimental data of sand grain rough-wall boundary layer flows [11] [23].

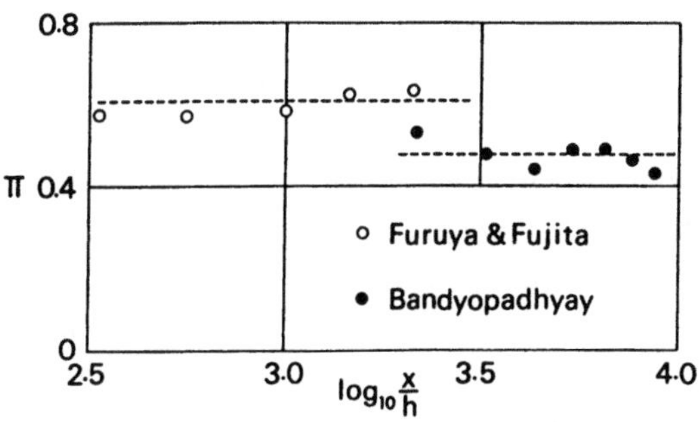

Fig. 4. Wake parameter evaluated from experimental data of sand grain rough-wall boundary layer flows [11][23].

sumed to begin. Values of z for the two experiments do not exactly overlap each other, and this is accounted for by the difference in K as well as Π. In Fig. 3 are also included the theoretical lines corresponding to the two experiments and a dotted line representing Schlichting's [2] interpolation formula. The theoretical lines are obtained by assuming U_o, h and Π as independent of x in the equation (20) and integrating it to the form

$$\frac{\kappa^2 x}{h} \exp[2\Pi + \kappa(C - K)] = \alpha[(z - 1)e^z + 1] - (\alpha + \beta)(e^z - 1) + 2\beta \int_0^z \frac{e^z}{z} dz, \quad (27)$$

where Π is assumed to be 0.61 and 0.48, respectively, as suggested by the dotted lines of Fig. 4. The theoretical prediction is in reasonable agreement with the evaluation from the mean velocity measurements.

Schlichting's interpolation formula, derived from the Prandtl-Schlichting [3] calculation for fully rough regime, is given by

$$z = \kappa\sqrt{\frac{2}{c_f}} = 0.580\,(2.87 + 1.58\log_{10}\frac{x}{h})^{1.25} \quad (28)$$

for the range of x/h from 2×10^2 to 10^6. It is seen from Fig. 3 that the formula gives result about 3 percent lower than the experimental values of z. Frequent mention has been made of the formula as predicting z as much as 10 percent lower (skin friction coefficient c_f 20 percent higher) than the experimental data (Liu et al. [15], Healzer et al. [13], Pimenta et al. [14], Mills et al. [16]). On reflection, however, this does not appear to stand to reason. It is to be noted that the formula has never been checked directly with the skin friction measured on the sand grain rough-wall boundary layer flow. What has been done is simply the comparison of the formula with the skin friction measured on the wall roughened with regularly spaced square rods (Liu et al. [15]), or on the wall roughened with closely packed array of uniform spheres (Healzer et al. [13], Pimenta et al. [14]), the scale of the roughness having been converted to that of the sand grain roughness by means of the equivalence concept of limited validity. Further comments on this subject will be made in the following sections.

Sphere roughness

We then evaluate the characteristic parameters from the velocity profiles measured on a flat plate roughened with coplanar uniform spheres packed in the most dense array. This geometry also gives the k-type roughness shift, and has early been investigated by Schlichting [2], who calculated the equivalent sand grain roughness as equal to 0.63 times the diameter of the sphere. Unfortunately, however, Schlichting's measurements were made in a duct of rectangular cross section of aspect ratio 4.2, only one of the broader walls having been roughened. The complicated geometry makes the accurate determination of skin friction rather difficult. In addition, the origin of the y-coordinate was determined by the condition of mass flow conservation, but not by the condition that would give the logarithmic distribution of velocity near the wall. Re-evaluation of Schlichting's original data has thus been made by Coleman, Hodge and Taylor [24], bringing about the equivalent sand grain roughness as equal to 0.38 times the diameter of the sphere.

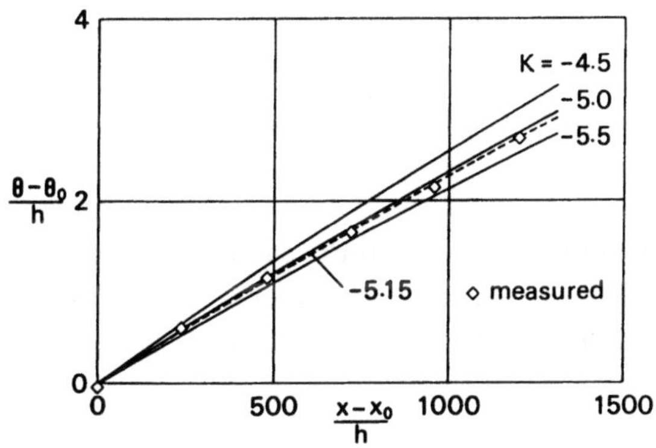

Fig. 5. Analysis of experimental data of uniform sphere rough-wall boundary layer flow [14]. x_o = 660 mm, θ_o = 2.29 mm, h = 1.27 mm, U_o/ν = 2.66 x 10^3 mm^{-1}, $u_\tau h/\nu$ = 160-172.

Measurements of velocity and temperature profiles have been made by several investigators on the same uniform sphere rough wall at Stanford University under a variety of conditions of free-stream velocity, mass and heat transfer, and pressure gradient (Healzer, Moffat and Kays [13], Pimenta, Moffat and Kays [14], Coleman,

Moffat and Kays [25], Ligrani, Kays and Moffat [26]). We consider only the velocity measurements for zero transfer and zero pressure gradient. A typical example of the calculation for $U_0/\nu = 2.66 \times 10^3\ mm^{-1}$ is shown in Fig. 5, where values of $(\theta - \theta_0)/h$ calculated by (26) are compared with those obtained directly from experimental data, h being 1.27 mm, diameter of sphere. We take $K = -5.15$ as the most probable value. Results obtained for different unit Reynolds numbers U_0/ν are summarized in Fig. 6, where K is plotted against the mean value of roughness Reynolds number $u_\tau h/\nu$. K increases with the increase in $u_\tau h/\nu$, tending to a constant value of about -4.8 for sufficiently large values of $u_\tau h/\nu$.

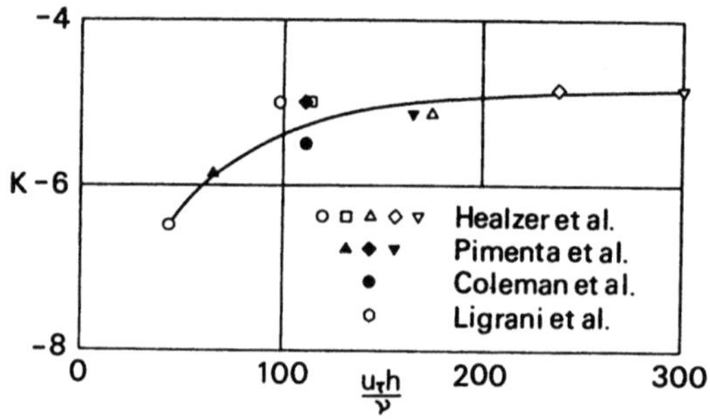

Fig. 6. k-type rough-wall constant evaluated from experimental data of uniform sphere rough-wall boundary layer flows [13][14][25][26]. h = 1.27 mm.

Obviously, the concept of equivalent sand grain roughness is effective only for flow in pipes, where the wake component of mean velocity profile is negligibly small. With $\Pi = 0$ in (22), the equality of U_0/u_τ gives

$$\frac{h_s}{h} = \exp\left[\,\kappa(K - K_s)\right], \qquad (29)$$

where h_s is the scale of equivalent sand grain roughness and K_s is the sand grain rough-wall constant. If we take $K = -4.8$ and $K_s = (-2.47 - 2.15)/2 = -2.3$, we have $h_s/h = 0.36$, which nearly coincides with the re-evaluation due to Coleman et al. With $K_s = -3.6$, instead of -2.3, then we would have $h_s/h = 0.61$, which is closer to Schlichting's original value. It is thus seen that the equivalent sand grain roughness critically depends on the choice of K and K_s.

Values of z and Π calculated from the experimental data for fully rough regime are shown in Figs. 7 and 8, respectively, as functions of θ/h. The results appear almost independent of the unit Reynolds number. Scatter of data points is accounted for by the dependence of θ/h on both z and Π as expected from (15), accentuated by the slow response of Π in downstream development.

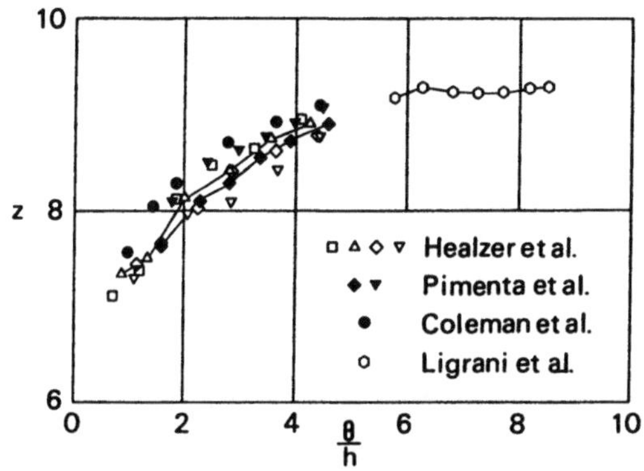

Fig. 7. Friction parameter evaluated from experimental data of uniform sphere rough-wall boundary layer flows [13][14][25][26]. h = 1.27 mm.

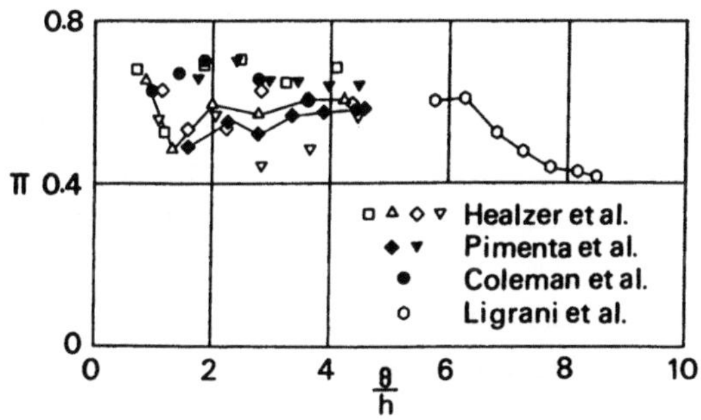

Fig. 8. Wake parameter evaluated from experimental data of uniform sphere rough-wall boundary layer flows [13] [14][25][26]. h = 1.27 mm.

Evaluation of the characteristic parameters is extended to the measurements made on an artificially thickened boundary layer developing over the same rough wall (Ligrani et al. [26]). The results obtained are included in Figs. 7 and 8, and generally indicate the artificial boundary layer to behave similarly to the natural boundary layer. Some sort of dissimilarity observed in the variation of z and Π might suggest the artificial boundary layer is not yet sufficiently settled in downstream development.

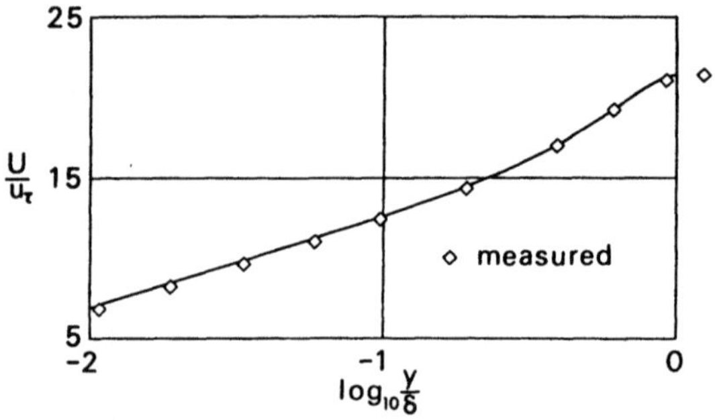

Fig. 9. Typical mean velocity profile. Comparison of analytical expression (1) with measured values [14]. $h = 1.27$ mm, $U_0/\nu = 2.66 \times 10^3$ mm^{-1}, $\delta = 34.9$ mm.

En passant, brief mention is to be made of the accuracy of the evaluation of characteristic parameters. Values of z determined by the integral method of analysis (and shown in Fig. 7) differ from those estimated by the measured Reynolds shear stress near the wall (Pimenta et al. [14], Coleman et al. [25]) by less than 4 percent. Also shown in Fig. 9 is a typical mean velocity profile, for which the analytical expression (1) with the evaluated parameters, $z = 8.77$, $\Pi = 0.65$ and $K = -5.15$, is compared with measurements (Pimenta et al. [14]). The agreement is tolerable from the practical viewpoint, but not entirely satisfactory when compared to the case of smooth-wall boundary layer flows (Tani and Motohashi [17]). This suggests the need of refining the formulation of roughness shift for more advanced purposes.

Circular rod roughness

Another example of the k-type roughness is afforded by the wall roughened with regularly spaced spanwise wires or circular rods. Measurements of mean velocity profiles have been made by Furuya, Fujita and Nakashima [12] on a flat plate of this roughness geometry for the ratio L/h ranging from 1 to 64, where L and h are the spacing and diameter of the rod, respectively. Results of analysis of experimental data are shown in Figs. 10 and 11. It is interesting to find in Fig. 10 that the roughness shift constant K attains a maximum for L/h = 8, in which condition the flow separated over the rod is considered to reattach just prior to encountering the succeeding rod. It is also seen that values of K originally reported by Furuya et al. are not much different from those re-evaluated by integral method by Matsumoto [27], implying that the assumption of different velocity defect profiles for smooth and rough walls is not very far from the truth. On the other hand, Fig. 11 serves to indicate the general trend of the change in downstream development of the parameters z and Π with the increase of the ratio L/h. It is seen that the change of the value of K is directly related to that of the level of z and that the downstream variation of Π changes from moderate decrease to monotonic increase as L/h exceeds 32. It may be noted parenthetically that a similar trend is observed in the behavior of the smooth-wall turbulent boundary layer perturbed by a small circular rod (Marumo, Suzuki and Sato [28], Tsiolakis [29]). The downstream variation of Π exhibits monotonic decrease or increase according as the rod is placed close to or away from the wall.

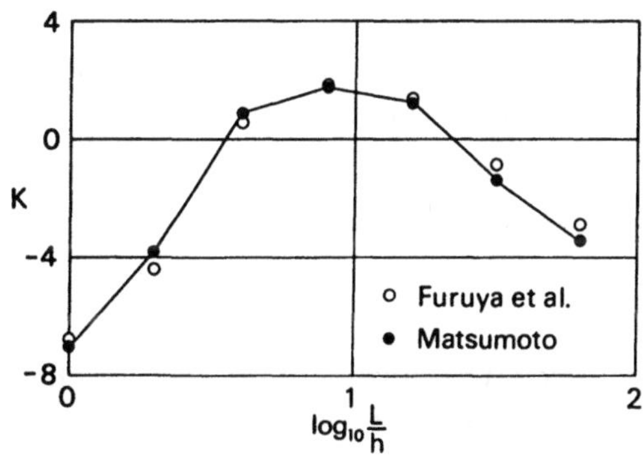

Fig. 10. k-type rough-wall constant evaluated from experimental data of circular rod rough-wall boundary layer flows [12]. h = 2 mm, $U_o/\nu = 1.45 \times 10^3\ mm^{-1}$.

In addition to the change of flow pattern just mentioned, the increased spacing causes the effect of periodic geometry of roughness to penetrate through the thickness of the boundary layer. This occurs for L/h ≥ 16 in the present example. For smaller L/h the mean velocity profile retains the same form except very close to the crest of the roughness element, irrespective of the location of the measurement station relative to the element.

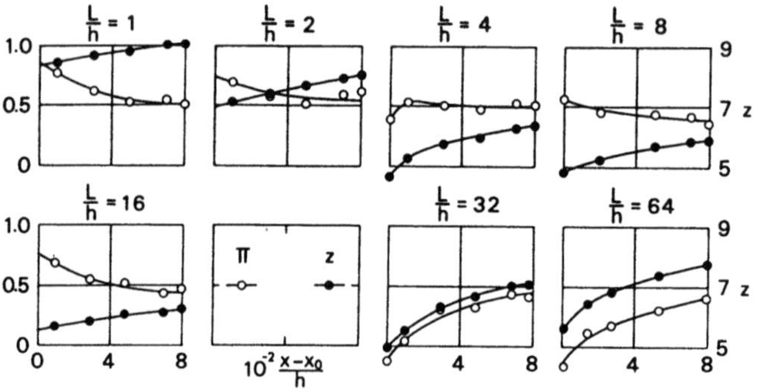

Fig. 11. Friction and wake parameters evaluated from experimental data of circular rod rough-wall boundary layer flows [12]. h = 2 mm, $U_0/\nu = 1.45 \times 10^3$ mm^{-1}.

The roughness composed of regularly spaced square rods, instead of circular rods, has most often been used in friction and heat transfer experiments as representative of periodically repeated geometry (Moore [30], Liu, Kline and Johnston [15], Bettermann [10]). Unfortunately, however, none of the available data for this type of roughness is sufficient for the purpose of the present study, owing to the scarcity in number either of the measurement station or of the spacing to roughness height ratio. Broadly speaking, the effect of spacing would be similar to that observed for circular rod roughness, but there might be some differences in quantitative details.

Equivalent roughness

As might be surmised from the arguments made in the preceding sections, we interpret the results of evaluation as suggesting that the wake parameter Π is seriously affected by the upstream conditions and only slowly relaxing, while the friction parameter z is perturbed to a lesser extent and more rapid in relaxation. Thus, the variation of Π in the streamwise direction might serve as nothing but an indication of the distortion. On the other hand, a legitimate inference may be drawn from the streamwise variation of z, provided that due consideration is given to the secondary effect of Π. Take, as an example, Figs. 7 and 8. It is useful in Fig. 7 to imagine a 'standard' curve on which Π remains constant in the course of downstream development.

Based on these considerations, an attempt is made to extend the Schlichting's concept of equivalence to the k-type rough-wall boundary layer flows. The equivalent roughness is defined as the scale of the reference roughness producing the same friction and wake parameters as the given roughness on the flat plate in fully rough regime. In view of (27) we then have

$$\frac{h_s}{h} = \exp\left[\,\kappa(K - K_s)\right], \qquad (30)$$

which is the same as (29), but in which h_s is the scale of the equivalent reference roughness and K_s is the reference rough-wall constant. For reasons already stated, we feel a little reluctant to take the sand grain roughness as reference. Tentatively, the circular rod roughness (Furuya, Fujita and Nakashima [12]) with spacing ratio L/h = 8 is adopted as reference for the simple reason that the constant K is the maximum of those evaluated so far. Thus we take K_s = 1.74, and calculate the values of h_s/h as listed in Table 1.

Table 1

roughness [reference]	K	h_s/h
sand grain [1]	- 3.58	0.113
sand grain [23]	- 2.47	0.174
sand grain [11]	- 2.15	0.203
sphere [13][14][25][26]	- 4.85	0.067
circular rod [12] L/h = 1	- 7.09	0.027
2	- 3.83	0.102
4	0.84	0.692
8	1.74	1.000
16	1.17	0.792
32	- 1.36	0.281
64	- 3.43	0.120

According to Table 1, Furuya and Fujita's sand grain roughness [11] has the equivalent roughness 14 percent larger than that of Bandyopadhyay's sand grain roughness [23]. On the other hand, the reverse is true of the comparison shown in Fig. 3, in which Furuya and Fujita's roughness is 8 to 9 percent smaller in scale than Bandyopadhyay's roughness. The apparent discrepancy is due to the effect of Π, which is assumed to take the same value for defining the equivalent roughness, but different values, 0.61 and 0.48, for calculating the theoretical curves of Fig. 3. It is to be borne in mind that the extended definition of equivalent roughness is based on the comparison of roughness scale for producing the same friction under the fictitious condition of equilibrium, in which Π is maintained constant.

d-type roughness

The analysis is also applied to the mean velocity profiles measured on a flat plate with d-type roughness, which is composed of regularly spaced spanwise grooves of narrow rectangular section (Perry, Schofield and Joubert [21], Wood and Antonia [31], Osaka Nakamura and Kageyama [32], Bandyopadhyay [23], Matsumoto, Munakata and Abe [33], Sakamoto and Osaka [34]). For this type of roughness a fairly steady vortex appears to occupy the main portion of the groove, making the momentum exchange across the opening of the groove almost independent of the groove depth (roughness scale).

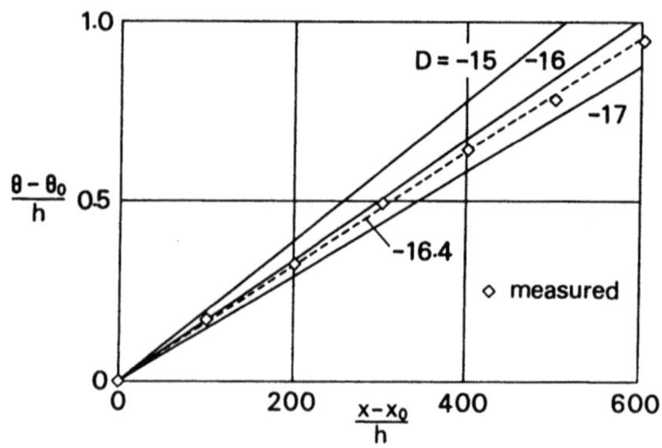

Fig. 12. Analysis of experimental data of d-type groove rough-wall boundary layer flow [33]. x_o = 1000 mm, θ_o = 2.47 mm, h = 5 mm, U_o/ν = 0.90 x 10^3 mm^{-1}.

We have now three equations, (7), (17) and (26), available for determining the friction parameter z, the wake parameter Π and the d-type rough-wall constant D. The procedure of determination is exactly similar to that for k-type rough-wall boundary layer flows. A typical example is shown in Fig. 12, where the values of $(\theta - \theta_o)/h$ calculated by (26) are compared with those obtained directly from experimental data of Matsumoto et al. [33], h being 5 mm, the depth of the square groove repeated at a spacing of 50 mm. We obtain D = - 16.4 as the most probable value. Then the values of z and Π are determined as shown in Fig. 13. It is seen that both z and Π are nearly constant with the streamwise distance x. The thicknesses δ*, θ and δ are proportional to each other, all increasing linearly with x.

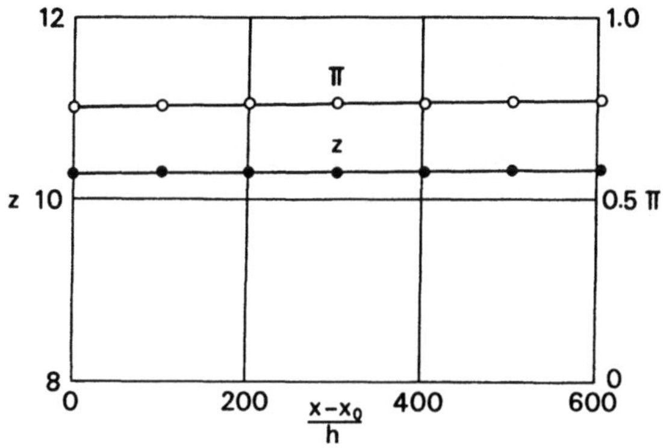

Fig. 13. Friction and wake parameters evaluated from experimental data of d-type groove rough-wall boundary layer flow [33]. x_o = 1000 mm, θ_o = 2.47 mm, h = 5 mm, U_o/ν = 0.90 x 10^3 mm^{-1}.

Evaluation from other sets of experimental data also bears the same characteristics as just mentioned, although the value of D is not the same and the constancy of z and Π in the streamwise direction is not so perfect for some data. The results are summarized in Table 2, and also depicted in Fig. 14, where values

Table 2

Reference	h	b	L	D	z	Π	$u_\tau h/\nu$
[21]	3.1	2.8	28.0	- 13.7	9.00	0.67	250
[23]	0.33	0.23	0.53	- 15.0	9.56	0.68	26
[31]	3.2	3.2	6.4	- 15.5	9.74	0.67	60
[32]	3.0	3.0	6.0	- 14.7	9.29	0.60	132
[33]	5.0	5.0	50.0	- 16.4	10.30	0.76	180

h, b and L are depth, breadth and spacing in millimeter of rectangular groove, respectively.

of z and Π are plotted as functions of D. Since the roughness employed in the experiments is not geometrically similar to each other, the feature exhibited in Fig. 14 should be taken as simply suggesting the broad outline. In spite of this restriction, it is evident that z increases with the decrease in D, namely, with the approach toward the smooth-wall flow. The relation between z and Π can be understood by reference to (17).

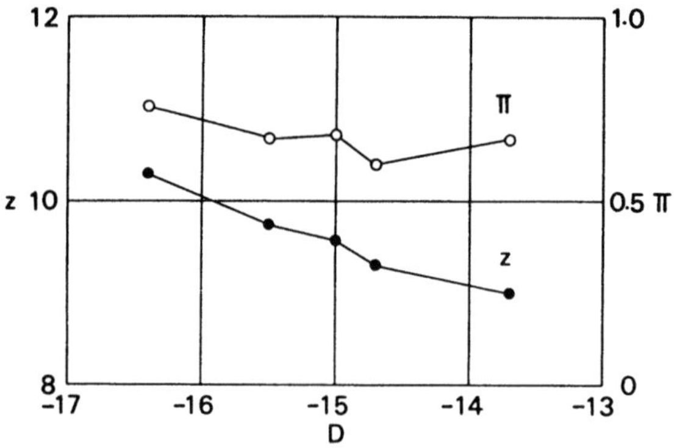

Fig. 14. Friction and wake parameters evaluated from experimental data of d-type groove rough-wall boundary layer flows [21][23][31][32][33].

As mentioned before, theoretical consideration leads to the prediction that the equilibrium solution $dz/dx = 0$, $d\Pi/dx = 0$ could exist with zero pressure gradient for d-type rough-wall boundary layer flows. Results of evaluation of z and Π from velocity profile data certainly afford evidence in support of the prediction. Some of the evaluation indicate values of z and Π more or less distant from the equilibrium value, presumably affected by the initial conditions,but approaching the equilibrium value as the flow develops downstream. This is worthy of mention since the tendency of recovery is implicit in the definition of equilibrium.

Recently, additional evidence for equilibrium has been furnished by perturbation experiments, independently carried out by Matsumoto, Munakata and Abe [33] and Sakamoto and Osaka [34]. The turbulent boundary layer over a flat plate roughened with d-type groove was perturbed by a circular rod of small diameter d, which was attached to the mid-crest of the roughness in the former experiment [33], while suspended above the mid-crest in the latter [34]. Values of z and Π calculated from velocity profile measurements [33] are shown in Figs. 15 and 16, respectively. Also shown in the figures are the results for unperturbed rough plate (d = 0), in which the thickness of the boundary layer at the rod location (x = 1000 mm) is 21 mm. It is seen that both z and Π increase by perturbation, but tend to approach new equilibrium values, different from those of the unperturbed case, as the flow develops downstream. This is the kind of behaviour that might naturally be expected from the equation (21). In other words, the like behaviour cannot be noted in similarly perturbed smooth-wall as well as k-type uniform rough-wall boundary layer flows.

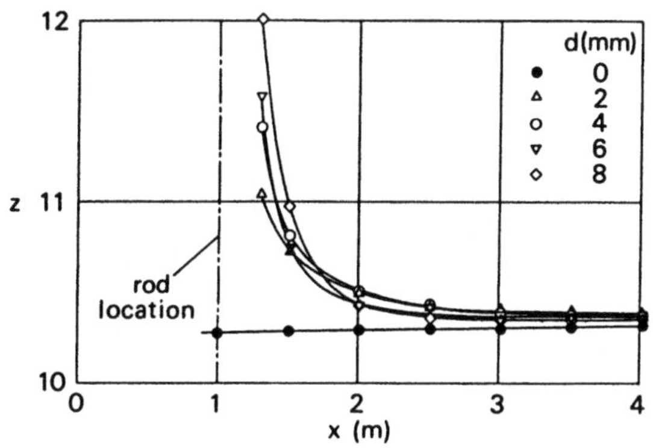

Fig. 15. Friction parameter evaluated from experimental data of d-type groove rough-wall boundary layer flow perturbed by a circular rod of diameter d at x = 1000 mm, where the unperturbed boundary layer has a thickness of 21 mm [33].

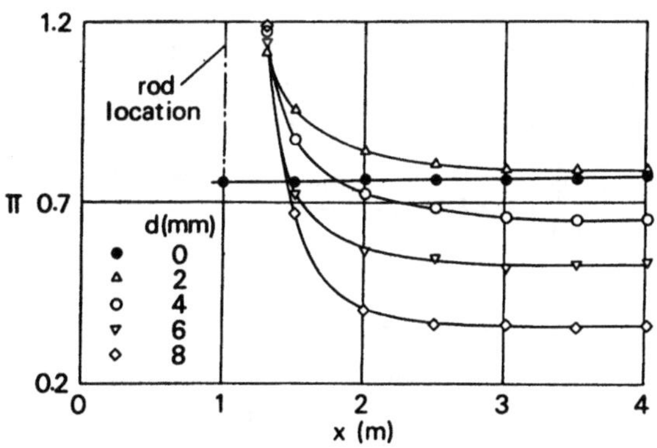

Fig. 16. Wake parameter evaluated from experimental data of d-type groove rough-wall boundary layer flow perturbed by a circular rod of diameter d at x = 1000 mm, where the unperturbed boundary layer has a thickness of 21 mm [33].

Perturbation experiment has been carried out by Matsumoto, Munakata and Abe [33] also for smooth-wall boundary layer flow. The characteristic parameters evaluated from the measured velocity profiles are shown in Figs. 17 and 18. It seems that the effect of perturbation persists for an exceedingly long distance downstream without any sign of recovery to equilibrium. This behavior of smooth-wall boundary layer flows, which forms the subject of the author's previous paper (Tani and Motohashi [17]), offers a striking contrast with that of the d-type groove rough-wall boundary layer flows.

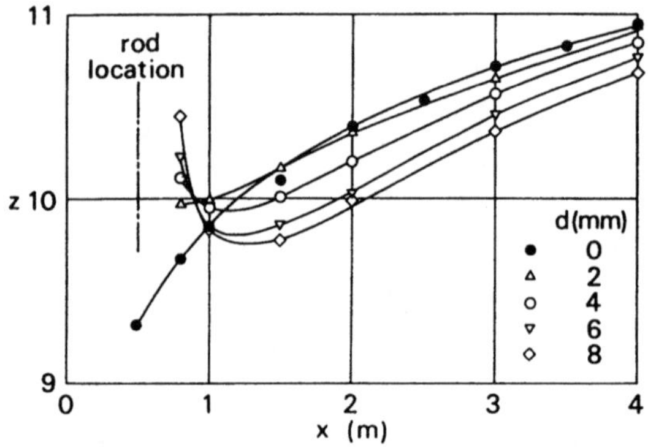

Fig. 17. Friction parameter evaluated from experimental data of smooth-wall boundary layer flow perturbed by a circular rod of diameter d at x = 500 mm, where the unperturbed boundary layer has a thickness of 16 mm [33].

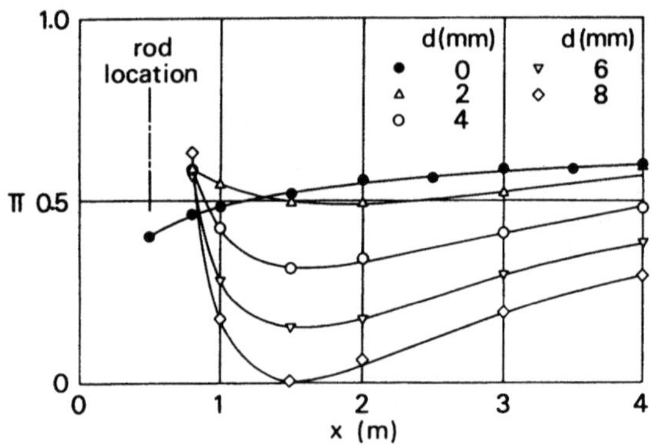

Fig. 18. Wake parameter evaluated from experimental data of smooth-wall boundary layer flow perturbed by a circular rod of diameter d at x = 500 mm, where the unperturbed boundary layer has a thickness of 16 mm [33].

In the experiments of d-type roughness quoted above [21][23][31][32][33], the crests of the roughness were aligned with the upstream smooth (but tripped) surface. Since little interaction appears to occur between the mean flow of the boundary layer and the quasi-stable flow within the groove, there are features similar to those of a smooth-wall boundary layer. In particular, the mean velocity and Reynolds stress profiles are closely similar to those observed on a smooth wall except for the region immediately above the grooves (Wood et al. [31]). The friction parameter is not much different from that of the smooth-wall boundary layer. The value $z = 10.3$ given by Fig. 13 is even slightly higher (skin friction c_f lower) than the corresponding smooth-wall value for the range of x up to 1.8 m according to Fig. 17.

Decreasing the spacing to depth ratio L/h of the grooves, we have the value of z reduced with the increase in D, as seen in Table 2 and Fig. 14. This brings about approach toward the k-type roughness. As a matter of fact, it is generally agreed that a line of demarcation between k-type and d-type roughness may be drawn at about $L/h = 4$ for the roughness consisting of regularly spaced square grooves (Osaka et al. [32]). Care must be taken, however, to distinguish such secondary effects as leading to a change in overall flow patterns. In one of the experiments carried out by Liu, Kline and Johnston [15], the bottoms of the square grooves spaced at the ratio $L/h = 2$ were aligned with the upstream smooth (but tripped) surface. The geometry of roughness may thus be referred to as that composed of regularly spaced square rods, and has actually been dealt with by Liu et al. as displaying k-type roughness shift. The calculation by integral method of analysis yields $K = -6.5$, which is not much different from the corresponding value for the circular rod roughness (Fig. 10). On the other hand, however, it is interesting to note that the same experimental data can also be analyzed by assuming d-type roughness shift. The results are $z = 8.0$, $\Pi = 0.5$ and $D = -12.0$, which do not appear to be wholly inconsistent with the values of Table 2 and Fig. 14. This example might be regarded as a borderline case.

References

1. Nikuradse, J.: Strömungsgesetze in rauhen Rohren. VDI Forschungsheft No. 361 (1933).
2. Schlichting, H.: Experimentelle Untersuchungen zum Rauhigkeitsproblem. Ing. Arch. 7 (1936) 1-34.
3. Prandtl, L.; Schlichting, H.: Das Widerstandsgesetz rauher Platten. Werft Reederei Hafen 15 (1934) 1-4.
4. Wieghardt, K.: Ueber die turbulente Strömung im Rohr und längs einer Platte. Z. Angew. Math. Mech. 24 (1944) 294-296.
5. Clauser, F. H.: Turbulent boundary layers in adverse pressure gradients. J. Aeron. Sci. 21 (1954) 91-108.
6. Hama, F. R.: Boundary-layer characteristics for smooth and rough surfaces. Trans. Soc. Nav. Arch. Mar. Engrs. 62 (1954) 338-358.
7. Coles, D.: The law of the wake in the turbulent boundary layer. J. Fluid Mech. 1 (1956) 191-226.
8. Coles, D.: The turbulent boundary layer in a compressible fluid. Rand Corp., Rep. R-403-PR (1962).
9. Coles, D.: The young person's guide to data. Proc. 1968 AFOSR-IFP-Stanford Conference on Computation of Turbulent Boundary Layers, 2 (1969) 1-45.
10. Bettermann, D.: Contribution à l'étude de la convection forcée turbulente le long de plaques rugueuses. Int. J. Heat Mass Transfer 9 (1966) 153-164.
11. Furuya, Y.; Fujita, H.: Turbulent boundary layers on wire-screen roughness (in Japanese). Trans. Jpn. Soc. Mech. Engrs. 32 (1966) 725-733.
12. Furuya, Y.; Fujita, H.; Nakashima, H.: Turbulent boundary layers on plates roughened by wires in equal intervals (in Japanese). Trans. Jpn. Soc. Mech. Engrs. 33 (1967) 939-946.
13. Healzer, J. M.; Moffat, R. J.; Kays, W. M.: The turbulent boundary layer on a rough, porous plate: experimental heat transfer with uniform blowing. Thermosc. Div., Mech. Eng. Dept., Stanford Univ., Rep. HMT-18 (1974).
14. Pimenta, M. M.; Moffat, R. J.; Kays, W. M.: The turbulent boundary layer: an experimental study of the transport of momentum and heat with the effect of roughness. Thermosc. Div., Mech. Eng. Dept., Stanford Univ., Rep. HMT-21 (1975).
15. Liu, C. K.; Kline, S. J.; Johnston, J. P.: An experimental study of turbulent boundary layer on rough walls. Thermosc. Div., Mech. Eng. Dept., Stanford Univ., Rep. MD-15 (1966).
16. Mills, A. F.; Hang, Xu: On the skin friction coefficient for a fully rough flat plate. ASME J. Fluids Eng. 105 (1983) 364-365.
17. Tani, I.; Motohashi, T.: Non-equilibrium behavior of turbulent boundary layer flows. Proc. Jpn. Acad. B 61 (1985) 333-340.

18. Rotta, J. C.: Turbulent boundary layers in incompressible flow. Progr. Aeron. Sci. 2 (1962) 1-219.

19. Tani, I.: Some equilibrium turbulent boundary layers. Fluid Dyn. Res. 1 (1986) 49-58.

20. Lewkowicz, A. K.: An improved universal wake function for turbulent boundary layers and some of its consequences. Z. Flugwiss. Weltraumforsch. 6 (1982) 261-266.

21. Perry, A. E.; Schofield, W. H.; Joubert, P. N.: Rough-wall turbulent boundary layers. J. Fluid Mech. 37 (1969) 383-413.

22. Grigson, C. W. B.: Nikuradse's experiment. AIAA J. 22 (1984) 999-1001.

23. Bandyopadhyay, P. R.: The performance of smooth-wall drag reducing outer-layer devices in rough-wall boundary layers. AIAA Paper 85-0558 (1985).

24. Coleman, H. W.; Hodge, B. K.; Taylor, R. P.: A re-evaluation of Schlichting's surface roughness experiment. ASME J. Fluids Eng. 106 (1984) 60-65.

25. Coleman, H. W.; Moffat, R. J.; Kays, W. M.: Momentum and energy transport in the accelerated fully rough turbulent boundary layer. Thermosc. Div., Mech. Eng. Dept., Stanford Univ., Rep. HMT-24 (1976).

26. Ligrani, P. M.; Kays, W. M.; Moffat, R. J.: The thermal and hydrodynamic behavior of thick, rough-wall, turbulent boundary layers. Thermosc. Div., Mech. Eng. Dept., Stanford Univ., Rep. HMT-29 (1979).

27. Matsumoto, A.: To be published.

28. Marumo, E.; Suzuki, K.; Sato, T.: A turbulent boundary layer disturbed by a cylinder. J. Fluid Mech. 87 (1978) 121-141.

29. Tsiolakis, E. P.: Geschwindigkeiten und Reynoldssche Spannungen in einer mit einem Kreiszylinder gestörten turbulenten Plattengrenzschicht. Dissertation, Technische Hochschule Aachen (1982).

30. Moore, W. F.: An experimental investigation of the boundary layer development along a rough surface. Dissertation, Iowa State University (1951).

31. Wood, D. H.; Antonia, R. A.: Measurements in a turbulent boundary layer over a d-type surface roughness. ASME J. Appl. Mech. 42 (1975) 591-597.

32. Osaka, H.; Nakamura, I.; Kageyama, Y.: Time averaged quantities of a turbulent boundary layer over a d-type rough surface (in Japanese). Trans. Jpn. Soc. Mech. Engrs. B 50 (1984) 2299-2306.

33. Matsumoto, A.; Munakata, H.; Abe, K.: Relaxation behavior of smooth-wall and rough-wall turbulent boundary layers perturbed by a circular rod (in Japanese). Proc. 18th Turbulence Symposium, Tokyo, July 1986.

34. Sakamoto, M.; Osaka, H.: Equilibrium behavior of disturbed rough-wall turbulent boundary layer (in Japanese). Proc. 18th Turbulence Symposium, Tokyo, July 1986.

The Role of Free Flight Experiments in the Study of Three-dimensional Shear Layers

Arild BERTELRUD
FFA The Aeronautical Research Institute of Sweden,
S-161 11 BROMMA, Sweden

Summary

In past decades the structure of three-dimensional shear layers has been studied extensively – both experimentally and analytically, as well as through use of numerical methods. While most experimental studies have been performed in wind tunnels or under other laboratory conditions, there are several remaining questions concerning how model tests can be used to simulate the shear layers on full-scale vehicles in flight. The present paper will be concentrated on what role free flight experiments may play in the study of turbulent shear layers, and in particular the study of three-dimensional boundary layers on swept wings.

It is shown that numerous aerodynamic studies in flight were performed in the quite early research stage, and that several factors today indicate that aerodynamic flight testing may become of increasing importance in the future.

Reference is made to several earlier studies, but in order to illustrate the type of information in question, the main part of the paper concerns analysis of recent flight tests on a swept wing aircraft in Sweden.

Introduction

In the present paper shear layers concerns the boundary layer flow on a flight vehicle, but also free wakes, separated areas as well as wing-body junctions with embedded vortices. To analyze three-dimensional turbulent shear flows it is essential to combine theoretical and experimental approaches, involving extensive detail in the description of the flow properties (POLHEM/1/, ROTTA/2/). To describe the viscous flow fields relevant to flight vehicles, it is necessary to do experiments in laboratory facilities (wind tunnels etc.) as well as in flight. There are several reasons why most of the information is gained in laboratory tests:

- Costs. There is in general an order of magnitude difference between wind tunnel and flight tests to obtain the same amount of information.
- Repeatability. In a wind tunnel the flow conditions may be repeated from day to day – this is obviously not possible in flight tests.
- Accuracy. This has been a problem in flight tests, but today airworthy instrumentation of high quality is available.
- Independent parameter variation is not possible in flight. In a wind tunnel the angle attack is always a free variable, and in a pressurized tunnel also Mach and Reynolds numbers may be varied independently. In flight there is always a coupling between these parameters, and although various techniques exist to separate them, no satisfactory solution is possible.

However, it is clear that flight data is required for a variety of purposes; any model result obtained or computation performed is the subject of interpretation and extrapolation, while a properly performed flight experiment is a direct documentation of reality. Some of the factors creating problems in the use of wind tunnel data is:

- Tunnel turbulence, and its influence on transition.
- Wall interference.
- Reynolds number; in most facilities this is too low.
- Model imperfections, either simplicity or surface finish; this may be a severe problem in pressurized tunnels due to the small model dimensions. It also includes the rigidity of the model.

One of the problems is also the inability to simulate unsteady phenomena encountered in maneouvers. In general the maneouvers are sufficiently moderate and slow to allow quasi- stationary analysis, but there are conditions when a strong coupling exists between the aerodynamic, flight mechanical and structural properties.

To show the role of flight tests in the past, an account is given of some selected work. As the space does not permit a comprehensive description of each experiment, the reader is referred to the publications listed. To illustrate the type of information that may be obtained in flight tests, an investigation of the flow on a swept wing aircraft is described in more detail.

When the experiment was designed, it was decided to obtain information essential to the development of future codes rather than present production or experimental codes. A proper description of experiments of this complexity, whether it is a flight or wind tunnel test, is not suitable for traditional report format. Instead the reports contain sufficient information to characterize the flow and the experiment, and the actual data is retreivable in a computerized database. In the present case such a database is available from the FFA. It includes the experimental data as well as the computer code required to access the data easily.

Background

Historically, shear flows have been analyzed experimentally in carefully conducted laboratory experiments. Much like the theoretical analysis the initial experiments were simple but crucial, aimed at establishment of the main guidelines governing viscous flows. The Wright brothers conducted their wind tunnel development of wing profiles, followed by glider experiments to verify their hypotheses. As new aircraft were developed the flight tests generally followed closely the wind tunnel tests – aircraft were still reasonably simple and fast to build, and guidelines on extrapolation from wind tunnel to flight were generally not very reliable. Also the information required in flight was rather limited. Generally the emphasis was put on exploration of the aircraft's performance as well as potentially dangerous conditions that might be experienced with the aircraft, and the interest for comprehensive studies of shear layers was rather small. In the 1920s reducing parasite drag involved removing protruding parts and in general "cleaning up" the aircraft; the wing profile drag was a minor item that was not considered worthwhile of too much effort.

From 1930 and onwards, this changed, and a series of advanced experiments were performed in flight. Several reviews have been written on the various aspects of flight tests, and therefore only selected references will be included in the present paper. For example, SALTZMAN and AYERS/3/(1981) did a survey of the wind tunnel/flight correlations to date, and HOLMES, OBARA and YIP/4/(1984) made a survey of experiments regarding laminar flow conditions in flight. POISSON-QUINTON/5/ examined wind tunnel and flight tests over the entire speed regime from low speed to hypersonics regarding their value in the development of aircraft, and concluded that differences in Reynolds number was the primary reason for discrepancies.

While on-line data reduction and analysis of aerodynamic flow in flight is considered a desirable (but not always obtainable) feature today, the flight tests of the early days were often conducted in a very straight-forward and direct manner, allowing direct observations of the main phenomena.

One experiment of particular interest was performed by STÜPER/6/(1934), who did the first free flight transition experiment. He made measurements of the pressure distributions and boundary layer development on the wing of a Klemm L26 aircraft. The measurements were performed with a traversing pressure rake located at 18 different chordwise positions of the wing, as shown in Fig. 1. Reynolds numbers based on the wing chord ranged from 2.82 to 4.88 millions, and the lift coefficient was C_L=0.31 to 0.91.

In Fig. 2 the experimental pressure distributions for Stüper's four test cases are compared with inviscid profile calculations using the panel method found in KUETHE and CHOW/7/. Stüper also performed boundary layer calculations using the method of GRUSCHWITZ/8/, and the results agree quite well with current finite difference calculations using the method of BRADSHAW et al/9/. One may appreciate the care and accuracy involved in obtaining the flight data in this early experiment.

Generally the investigations in this time period were concentrated on two-dimensional flows, but are included here as they not only gave valuable information for the study of turbulent shear flows in general, but also because they demonstrate the instrumentation development.

In the late thirties several flight experiments were performed to explore the drag characteristics of various airfoils, JONES /10/ developed a method to evaluate the profile drag from surveys of the impact- and static pressures in the wake. Transition measurements were performed in England (Cambridge and Farnborough), and were reported by JONES/10/(1938). While Stüper's experiment gave complete velocity profiles at a number of chordwise stations, the latter experiment employed only up to 5 pressure tubes in the boundary layer, but instead these were movable along the surface and capable of yielding details on the chordwise development of the flow.

During this time aerodynamic flight tests were being peformed in several countries. In U.S.A., for example, GOETT and BICKNELL/11/(1939) measured the profile drag on a Fairchild N-22 and ZALOVCIK/12/(1945) examined the flow on and behind the wing of a P-47.

In the late fourties, the increasing flight Mach numbers resulted in the introduction of swept wings with their particular problems. In several countries the new conditions were explored utilizing existing, large aircraft as testbeds for swept wings mounted vertically on the fuselage. In other cases existing wing profiles were modified though the use of gloves covering a part of the span. BROWN and CLOUSING/13/ made a rather extensive documentation of pressure distributions on a P-80 jet aircraft, describing an aileron buzz problem encountered in certain flight regimes. Their results have been used in comparisons with recent Navier-Stokes computations, where it has been found that some these codes are capable of predicting both the occurence and the frequency of the oscillations.

A few years later, with the rapidly growing commercial airline traffic, the interest in laminar flow profiles and aircraft prompted another series of flight experiments and even some flying testbeds. This concerned both natural laminar flow as well as laminar flow control.

However, the main effort regarding flight tests, and in particular the aerodynamic aspects, was the series of experimental aircraft built to explore a variety of design ideas. These projects have been reviewed by SMITH/14/(1978). The series X-1 to X-15 covered the increase in Mach number from transonic up to Mach 3 (X-3) and with the X-15 up to hypersonic conditions, and also included many experimental aircraft for particular configuration studies. After the cancellation of the X-15 program in 1968, work continued with the XB-70 Mach 3 cruise delta wing aircraft and later with the lifting-body vehicles intended for reentry configurations.

Fortunately, during these years the experience in extrapolating wind tunnel results to flight increased, since the cost of performing flight tests increased tremendously. Also the computational techniques developed to a stage where it was meaningful to rely on them for guidance in the initial development of a new aircraft or concept, and it was clear that the future would yield steadily increased advantages for computational techniques.

The development of these techniques required a variety of experiments to supply the empirical basis for computational models, and it has been a general assumption that flight tests of aerodynamic flows is neither sufficiently accurate nor repeatable and hence not worthwhile performing. Generally measurements of lift and drag for aircraft is determined from the general characteristics as measured under specific flight conditions (BOWES)/15/ or maneouvers (ARNAIZ)/16/, or more recently also from computational methods (SLOOF)/17/. The thrust is estimated, and although this type of flight testing is essential for the development and certification of new aircraft, the aerodynamic information obtained is generally limited to flow diagnostics in local problem areas. Thus little information of general use in defining the local flow behaviour on the vehicle is obtained. While several experiments were performed in the period 1960-75 (ERLICH)/18/, these mainly consisted of local measurements in one or two locations. Skin friction laws were established and extrapolated to higher Reynolds numbers than before, but as there was virtually no up- or down-stream history documented, very few general conclusions could be drawn. Although lessons could be learned concerning the boundary layer properties at very high Reynolds numbers or under extreme conditions, very limited new information pertaining to shear layers was obtained – this was expecially true for three-dimensional shear layers. During this period most of the information required for aircraft development and even basic research could be more efficiently obtained in wind tunnel tests.

Concerning particular issues, though, a number of experiments have been performed, which highlights particular problems in flow conditions in general and drag reduction in particular. One example is the investigation concerning the effects of cabin pressurization on fuselage drag, performed by GEORGE-FALVY/19/, or the "sweep-rake" used by GEORGE-FALVY/20/ to reduce the areas of separated flow on engine nacelles. COULOMB, LEDOUX and LERAT/21/ performed a comparative measurement of the flow on a leading edge slat of a Nord 2501 aircraft in flight as well as under tunnel conditions. They used static pressure taps extensively and also surface hot film gauges. GOECKE POWERS/22/ used a YF-12 aircraft to obtain information on the flow at aft-facing steps for transonic and supersonic speeds up to Mach 2.8.

One of the most popular configurations used for comparative tests of transonic wind tunnel performance, has been a 10-degree cone. The transition position has been estimated, and rightfully or not has been considered a measure of the tunnel quality. FISHER and DOUGHERTY/23/ performed a flight test to correlate the tunnel results with real flight conditions.

For small aircraft, sailplanes etc., it has been found reasonable to operate flying testbeds for a variety of flow properties as well as profile development. QUAST and HORSTMANN/24/ describes one aircraft of this type, JENKINS/25/ another.

Three-dimensional experiments

Although a large number of flight tests have been performed to investigate more or less pronounced three-dimensional viscous effects, few detailed investigations have been carried out.

Flight tests have been pursued primarily in cases where transition location, wall interference or wind tunnel turbulence have been considered critical to obtain a valid result for a configuration. Essential in the use of three-dimensional flight experiments is the value of wind tunnel/flight correlations. While a 2D comparison can be obtained through the use of a wing glove etc. and not necessarily be related to actual flight conditions or problems for the "host" aircraft, SALTZMAN and AYERS/3/(1981), describe various aircraft, the discrepancies between model- and flight tests, and their apparent causes. The proceedings from the Mini-Workshop on Wind Tunnel/Flight Correlations at Langley 1981, edited by McKINNEY and BAALS/26/ gives the relation of the flight test results to the high Reynolds number data obtainable in the cryogenic NTF wind tunnel. The report is illuminating, in the sense that it contains the stories of development of various projects and the wind tunnel/flight correlations of aerodynamic properties of a series of different aircraft, but very limited information on detailed flow documentation in flight is presented.

The extrapolation technique developed by CAHILL and CONNOR/27/ based on the C-141 flight test data by PATERSON et al/28/ may be useful as an empirical formula in many cases, but it provides no further physical understanding of the phenomena encountered in flight concerning shock/boundary layer interaction.

Buffeting has been one of the most important areas of study in flight tests in high subsonic or transonic speeds. MABEY/29/ (1973) reviewed the problem, and concluded that the inability to produce the correct flows in ground facilities, was connected to the low Reynolds numbers of most wind tunnels.

The first comprehensive series of experiments concerning the boundary layer on a fully three-dimensional configuration were performed some thirty years ago at Cranfield, when GRAY/30/ (1952) mounted a 45 degree swept wing on top of the fuselage of a Lancaster bomber. The intention was to investigate the flow on the wing, with particular emphasis on the transition region. Flow visualizations were performed with particular interest in the instabiltiy of the flow.

Later BURROWS/31/ re-examined the transition data, and documented the boundary layer development for angles of incidence up to 10 degrees, using pitot combs. Oil flow visualizations were attempted, and tufts were used comprehensively. However, the conclusion in both these experiments was that the boundary layer was turbulent almost everywhere, due to the high wing sweep; this was an encounter with the so-called attachment-line contamination.

Some years later a 12.5% thick RAE profile with slitted surface for suction was mounted on the Lancaster bomber, and areas of laminar flow obtained were investigated using pitot tubes and hot wires (LANDERYOU and TRAYFORD/32/) as well as hot films (GASTER/33/ and /34/). The wing sweep was 43 degrees, and a large portion of the experiment concerned the state of the attachment line flow and how to avoid the attachment line contamination, in this case through the introduction of a bump at the leading edge (GASTER/35/). It may be of interest to note, though, that the bump was developed through wind tunnel tests and only in a couple of cases verified during tests in flight under real conditions. Apart from the aerodynamic achievements in these experiments, it may be worthwile noting that the main portion of the 50 flights performed were allocated to proving the aircraft and instrumentation and the development of the instrumentation, and the last 10 flights were most productive concerning aerodynamic results. The wing was tested in a low speed wind tunnel for comparison, and later remounted on a Lincoln bomber (LANDERYOU and PORTER/36/1966).

A large number of flight tests of various types were performed at Cranfield. For example, aerodynamic mapping of the pressure distribution on the wing of a supersonic delta wing aircraft, the Fairey Delta 2, was done by NICHOLAS/37/(1960), utilizing Scanivalves and a total of 180 pressure taps that were built into the wing during manufacture. Experiments with pressure belts were also done, and later WEBB/38/ made measurements of the pressure fluctuations at one location on the wing to gain experience for the supersonic transport to be built, presumably Concorde. Later, BROWNE et al/39/ carried out a rather complete pressure mapping on the wing of a VC10 over a span of subsonic Mach numbers. It was found, however, that the rather large pressure belts (5 mm tube diameter) in some cases affected the local flow measurement as well as the general flow pattern on the wing. The use of pre-fabricated, tape-on pressure belts is advantageous in many cases, and to verify their performance relative to pressure taps drilled in the wing skin. MONTOYA and LUX/40/ performed a flight test in the Mach region 0.50 to 0.97 using an F-8 with supercritical wing. Belts with 2.5 mm outer diameter were used, and the results showed a good agreement with the flush-mounted taps. However, it was again observed that the presence of the belt at critical positions of the wing might create a shape change for the airfoil of sufficient magnitude to change the wing flow characteristics.

One problem area particularly feasible for flight tests is the development of vortices over and behind aircraft configurations. FENNELL/41/ describes rather comprehensive tests with the HP115 delta wing aircraft to document vortex breakdown, and presents the results in the form of a correlation between vortex breakdown position and the angle of attack for the aircraft.

In the late 1960s and early 70s flight test interest was focussed on supercritical wing technology, as a series of wind tunnel investigations had indicated that large gains in the form of drag-rise Mach number should be obtainable. Extensive tests were performed using new, supercritical wings applied to a North American T-2C and an F-8 for evaluation (FISCHEL et al/42/). Pressure distributions were obtained using static pressure taps and movable wake rakes as well as traversing boundary layer rakes. Although the general agreement between wind tunnel and flight tests was good, various differences were observed and also it became clear that improvement of the testing technique at transonic speeds was needed to obtain the required accuracy.

A supercritical wing was developed for the F-111; it was flight tested both on an experimental aircraft (TACT) as well as in a smaller drone configuration (ECKSTROM/43/). Measurement of pressure distributions along with boundary layer and wake properties (LUX/44/) yielded a rather comprehensive picture of the most important flow features on the wing, and the data may be used as a test case for computational codes. If aerodynamic measurements are performed during a maneouver, it is desirable to have a technique for optimum performance and repeatability. DUKE and LUX/45/ described the use of an autopilot to perform the desired maneouvers.

A supercritical wing was also developed for the Dornier Alpha Jet, and here an investigation was performed to examine the results on performance as well as the local flow properties, see JACOB et al./46/ and BUERS et al/47/, who also made comparisons between flight and wind tunnel results as well as computations.

In the experiments discussed above, a limited amount of data was obtained for an aircraft or a configuraton of particular interest. In the flight experiment reported here a different approach was taken. Complete (or nearly so) data was obtained for present and future computer code verification. While supercritical, three-dimensional wings have complicated geometrical descriptions and at off-design flight conditions may perform rather oddly, it was assumed that a classical wing profile would experience basically the same physical phenomena, albeit at other Mach numbers or altitudes.

Lansen Experiment

This experiment concerns the viscous flow on swept wings at subsonic speeds, but some of the information is relevant even for delta wings if the sweep angle is not too large. The aircraft chosen for the experiment was a SAAB 32A Lansen, with a 39 degrees leading edge sweep, 10% thick classical NACA 64A010 profile normal to the 25% chord line – i.e. a symmetrical profile with zero twist. It was a common problem in many research institutes that the aircraft under investigation was either novel, and hence under constant pressure to be transferred to other tasks, or a prototype with all operational/spare parts problems. In contrast the Lansen used in present series of experiments was an ordinary serial aircraft being phased out from the Swedish air force, and an aircraft was made available to the FFA for aerodynamic investigations with very limited requirements on the time scale of the experiment.

Figure 3 illustrates the aircraft and the flight envelope performed each flight; in addition to the fixed flight conditions indicated, a number of maneouvers and special tests were performed each flight. A total of 130 flights were flown over a time period of more than 5 years, and a large variety of probes and instrumentation types were utilized, as shown in Fig. 4:

- Static pressure taps in
  - leading edge (7 spanwise stations)
  - instrumented segment
  - wing tip
  - fuselage/inlet
  - wing/body junction (fillet)
  - stabilizer.

  The static pressure taps were drilled in the surface itself, as the use of pressure belts would have caused disturbances to the three-dimensional flow in these areas. As the wing could not be opened, the tubes were routed inside the wing only a limited distance away from the measurement region.

- Boundary layer rakes
  - pressure, yaw probes
  - pressure
  - hot wires
  - rotating hot wires
  - split film probes.

  As a large variety of boundary layer thicknesses were experienced depending on location or flight conditions, it was necessary to build a variety of pressure rakes; the smallest was 6 mm high and the largest 150 mm. In many cases measurements were repeated with different probes at the same location to improve resolution. The hot wire rakes were either single, X-wire or split-film, and whenever possible they were mounted in connection with pressure rakes to benefit from the knowledge of velocity magnitude as a check on hot wire calibration.

- Local skin friction probes
  - modified Preston tubes (PRESTON/48/, BERTELRUD/49/, BRADSHAW & UNSWORTH/50/)
  - Stanton tubes (razor blades) (STANTON/51/)
  - hot films (LUDWIEG/52/, McCROSKEY/53/, MATHEWS and POLL/54/)
  - piezo-electric foil (NITSCHE/55/).

  The modified Preston tubes were easy to move to different parts of the aircraft, allowing a mapping of the flow properties. The main concern was to ensure that a non-interfering probe pattern was obtained each flight. The Stanton tube technique was used the following way: one flight was performed with the normal static pressure tap open, then another flight was performed with the razor blade normal to the flight direction. If directional information was desired, which was generally the case on the wing tip, flights were flown with the razor blade at yaw angles to the flow. The hot film and piezo sensors were flush-mounted, but could not be moved from flight to flight.

- Flow diagnostics techniques
  - tufts
  - liquid injection/video camera
  - liquid crystals.

  Simple, light nylon tufts were used, and movies were taken of the wing pattern from the rear seat and of the wing/body junction and the fuselage from another aircraft flying in formation. The liquid was injected through pressure taps in the leading edge region, and photographed using a miniature (25 mm cube) video camera clamped to the pitot tube. Liquid crystals in the form of self-adhesive tape was tested to explore possible effects of UV-radiation.

The flight envelope of the aircraft covers the entire subsonic Mach regime, and hence it was possible to study a wide variety of flow problems:

- Leading edge flow; i.e. separation bubbles
- massive part-span separations
- shock/boundary layer interaction
- buffeting

As can be seen, the instrumentation provides most of the information needed to describe turbulent shear layers, and the flow phenomena cover most problems encountered on other flight vehicles, be it at a different Mach number, angle of attack etc.

In addition to the measurements performed on the aircraft itself, several concepts for flow manipulation were evaluated:

- LEBU (Large Eddy BreakUp) devices for drag reduction. Flat and profiled tandem devices were tested, showing reduction in local skin friction downstream, and alteration of the turbulence structure through the entire flight regime. The feasability of using LEBUs in flight was shown.

- Passive shock control through surface perforation. A high-Reynolds number verification of the concept was done, the drag-rise Mach number was increased.

- Attachment line transition trips. Wind-tunnel criteria were confirmed.

The measurement system, as well as some of the flow properties have been reported previously; see /49/ and /56/ to /59/. In the present context the main information may be found in the AGARD database definition and in a recent paper on the computerized database system used to handle the vast amount of data available.

In the present paper some of the flow problems handled in the flight tests will be discussed in more detail, using appropriate parts of the database. In the AGARD test case selection, four different flight conditions were chosen, and for consistency the same flight conditions will be used in the present paper:

Case	Mach	Altitude
A	0.89	10 km
B	0.89	7 km
C	0.8	7 km
D	0.4	7 km

The four test cases cover a variety of flow problems. In cases A and B there is a strong shock on the outer part of the wing; case A is close to shock-induced separation due to the lower dynamic pressure (higher angle of attack) than case B. Test case C is a high subsonic Mach number without serious problems regarding shocks or separation bubbles, while test case D has a rather highly loaded bubble over the majority of the span. For lower Mach numbers, the bubble breaks up, starting at the tip and progressing inwards as the flight Mach number decreases past 0.3. It may be worth noting that this aircraft has a very stable stall progression, and hence it was possible to obtain valid data all the way down to stall. In flight the stall was defined as the condition when the aircraft could not be kept at constant altitude despite full afterburner power – this distinction between flight and standard wind tunnel investigations should be kept in mind when comparisons are made.

Two different flow problems have been chosen for special attention in the present paper; shock/boundary layer interaction and wing/body junction flows. In both these cases the large physical scale and the sensitivity to unsteadiness in the reference conditions make them suitable for discussion, as well as the need to utilize several independent measurements to obtain a proper description of the flow.

General Flow Description

On a swept wing one of the main areas of interest is the leading edge region, where some of the properties may be deduced from two-dimensional analyses, whereas others are sensitive to the three-dimensionality of the flow.

The attachment line, which is the three-dimensional equivalent to the two-dimensional stagnation point, is subject of considerable interest. Figure 5 shows how the pressure distribution changes in the present case at one spanwise position if the Mach number is varied with the flight altitude kept constant. One may first note that there is a considerable region of turbulent flow all the way down to a flight Mach number of 0.4, and while the pressure gradients normal to the leading edge are small at transonic speeds, a suction peak starts to build up as the angle of attack is increased. At angles larger than 4-5 degrees this particular wing has a short separation bubble. For the present flight altitude there is a complex interaction between the bubble and a shock on top of it, and at M=0.4 the bubble fails to close and a complete redistribution occurs for the entire pressure distribution on the wing.

In Fig. 5 the attachment line is indicated, and if the pressure distributions are translated into the more familiar parameter C* used to determine attachment line transition, it can be seen that laminar attachment line flow is possible at this location only for intermediate Mach numbers. This is illustrated in Fig. 6, where the experimentally determined pressure distributions have been used to evaluate C*, and the actual transition condition for the attachment line boundary layer is determined from hot film gages and a miniature hot wire rake located immediately next to the attachment line itself. One point of interest here, is that the transitional region as found by CUMPSTY and HEAD/60/, does not appear. It may be argued that this is a results of the three-dimensionality of the wing, i.e. the variation in spanwise direction. While the attachment line conditions may be instable at an inboard station, indicating a turbulent flow, the local pressure distribution a little further outboard might indicate a transitional flow.

The fact that the laminar bubble bursts at low Mach number is an indication that the flow has relaminarized from its turbulent attachment line condition – and using the LAUNDER-JONES/61/ criterion, further support for the assumption of relaminarization can be found. However, the hot-film gages indicated a strong unsteadiness, as can be expected with a laminar viscous layer growing underneath the already turbulent layer.

Close to the wing tip the separation bubble becomes more three dimensional, i.e. its length size increases with spanwise posi tion and the spanwise mass flow within the bubble increases considerably. Figure 7 shows how the bubble changes from short to long with a corresponding change in pressure distribution, lowering the suction peak considerably. The liquid flow visualization was intended to explore this problem, and in Fig. 7 samples of pictures in a 30 minutes long video movie are given. The liquid was injected into the separation bubble, and the highly unsteady character of the flow, as well as the variation with time and flight conditions of the spanwise mass flow was evident. Also the wing tip was instrumented, and in Fig. 8 a clear correlation between the changes in bubble flow and the wing tip flow itself can be seen.

In many cases the wing tip flow may be very critical to how the start of stall occurs for the aircraft, and in the figure the pressure distributions at several chordwise positions are shown for high and low Mach numbers respectively.

The pressure distribution in the leading edge region differs vastly with flight condition; Figure 9 can be seen as an example of this, with a highly loaded separation bubble present in test case D. The stall fence influences the amount of loading on the separation bubble in- and outboard, as illustrated in Fig. 10. These two figures illustrate the resolution required to determine the pressure distributions properly in the leading edge region; the segment of Fig. 9 had 16 taps within 2% chord, while the two other spanwise stations of Fig. 10 had 8 each - yet both of the distributions seem far from complete.

In Fig. 11 it is shown how other types of instrumentation like surface hot films can be used to augment the flow information. It is seen how as few as three sensors allow a good interpretation of what is occurring for various flight conditions if they are used together with the pressure information. It also clear, however, that hot film sensors alone would have been very hard to interpret. This is one of the keys to flight tests of aerodynamic properties: a variety of instrumentation yielding partly redundant information.

The flow on the stabilizer and its tip is very much dependent on whether or not the aircraft is flown trimmed out, and also on the flight Mach number. The aircraft had an all-movable stabilizer, free to change between positive and negative angles of attack, and the stabilizer had similar bubble patterns as the main wing.

While the information above to a large extent concerned the pressure distributions on various parts of the wing, it may be seen as a representation of the time-averaged properties of the flow on the aircraft. However, if results obtained from instrumentation with a higher frequency response is examined, one may find that superimposed on the steady state results, unsteady or quasi-steady variations associated with aircraft movement or structure may be identified. For example, a fast pressure transducer in the boundary layer close to the leading edge demonstrates clearly (see Fig. 12), the pressure variations associated with control surface deflections etc. The response is quasi-steady, as no phase shift is detectable. However, the wind bending frequency of 8 Hz produces a coupled response from the pressure variation; this has been exemplified in latter part of the figure. Note that this type of unsteadiness should not be confused with the variations occuring due to the turbulence, rather the quasi-steady variations should be seen as an equivalent to the models used for parameter identification in flight mechanics. (ILIFF and MAINE/62/, HAMEL/63/).

The instrumentation used in the present experiment was capable of documenting even rapid changes in local flow pattern occuring during maneouvers. Figure 13 shows a local separation lasting approximately 0.3 seconds when the aircraft was rolling and a rapid pith movement was introduced in the middle of the roll. The data is taken from the miniature hot wire rake, and it is shown how the local separation forms, but is unable to grow.

The major information of interest for analysis of three- dimensional shear layers consists of time-averaged distributions of pressure and local skin friction. Figures 14 to 16 show the pressure distributions and the local skin friction distributions for the entire wing at the four chosen conditions. It may be seen that as the angle of attack is increased the stall fence starts to influence the flow, lowering the suction peak close to the leading edge. On the other hand it is clear that a rather strong shock is formed on the outer part of the wing at high Mach numbers.

With the resolution available in the present case, it was possible to determine both pressure and skin friction distributions through use of razor blades mounted on several pressure taps. The data in Fig. 17 was obtained as the aircraft was climbing, and the region without data corresponds to an area of pressure belts. In this particular case, the instrumentation was used to verify whether or not the pressure belts had any adverse local effects.

One of the experiments performed on the Lansen aircraft, was an attempt to see whether or not use of thin transverse ribbons (LEBUs) in the outer part of the boundary layer were capable of reducing skin friction through turbulence modification. Figure 18 shows the configuration used, and the reductions in local skin friction as verified using the modified Preston tubes in a diagonal array. Wind tunnel information was confirmed, even in the region of local supersonic flow and with highly swept ribbons. Hot wire measurements performed far downstream showed that the vertical convective velocity was drastically reduced for the LEBU cases, and it was also found that it was possible to tension the ribbons properly to avoid breaking as well as fluttering.

Shock/Boundary Layer Interaction

It is important to have a well defined static pressure distribution close to the shock; both in front of it and in the very vicinity of the pressure rise. This is generally difficult to obtain in wind tunnel tests, due to the small scale of transonic models. In the present case an overall, but not detailed, description of the flow was already available from static pressure taps in the surface and the large number of modified Preston tubes already employed on the outer part of the swept wing. Figure 19 serves to illustrate the flow properties from pressure and skin friction distributions. The figure shows how an increase in angle of attack causes the shock to move forward. The boundary layer characteristics was also determined from pressure rakes (Fig. 20). As the shock was known to move between 50-65% of the chord at one of the outer stations, it was decided to make pressure taps every 0.5% in this region to be able to monitor how the pressure rise took place. This, of course, put strict requirements on the data handling and also on the performance on the pilot; the shock is by itself unsteady in many cases, and partly this unsteadiness had to be checked, partly it was necessary to ascertain that the conditions were indeed repeated when another type of instrumentation was employed.

In Fig. 20 it can be seen how the shock changes when the Mach number is increased for a specific altitude; in this case the angle of attack is closely constant. The boundary layer increases in thickness at a station close behind the shock, and at a certain Mach number there is a local separation at the shock itself. Initially the flow reattaches in front of the measurement station, but if the Mach number is increased just a little more there is a massive separation extending all the way to the aileron.

In several recent wind tunnel studies the concept of passive shock control through ventilation of the surface close to the shock has been investigated (NAGAMATSU et al/64/, KROGMANN /65/). In these model tests it has been shown that the drag-rise Mach number can be raised – the strong shock causing the separation has been split into several weaker ones.

In the present flight tests, various combinations of porosity was investigated, drilling holes in the wing skin upstream as well as at the shock. The percentage porosity corresponded to Krogmann's wind tunnel tests. In order to evaluate the drag of the configuration, a pressure rake was located downstream of the shock, close in front of the aileron. Using the measured momentum thickness, shape factor as well as the local C_p, SQUIRE-YOUNG's /66/ method can be used to determine the effects on the profile drag.

It was found that while porosity far upstream in the supersonic region would increase the drag, porosity underneath the shock would split it, and while no effects could be noted below the drag-rise Mach number, that Mach number was increased significantly, as is illustrated in Fig. 21.

Wing/Body Junction Flow

When an object protrudes from the wall into a turbulent boundary layer, horseshoe vortices are created at the intersection, trailing downstream along the body. This is the case also with a wing extending from the fuselage; the vorticity in the boundary layer of the fuselage is rolled up into a strong vortex. In the present case one half runs underneath the wing and the other one in the junction of the wing and fuselage. The past years unfilleted junctions have been studied in low speed laboratory conditions, see SHABAKA/67/ for example, where idealized geometries were used. In the present case the interest is focused upon the conditions in a filleted junction in flight at Mach numbers from transonic down to stall. The stall pattern of the aircraft was favourable in this case, with the separation progressing slowly inward on the wing as the Mach number was decreased, allowing the wing conditions for the junction flow to be progressively more severe without causing vortex breakdown.

For several reasons, measurements in the wing/body junction are significantly different from the main portion of the flight tests:

- layers were thicker, requiring larger rakes,
- the flow is not a boundary layer, i.e. the static pressure varies through the viscous layer,
- crossflow in the vicinity of the vortices is small, making tuft visualization unsuitable, but meaning that useful data could indeed be obtained with two-dimensional instrumentation.

For example, the total pressure coefficient measured would equal that of the surrounding static pressue, but because of the static pressure variation through the layer, the dynamic pressure (and hence the local Mach number) in the vortex might be equal to the flight reference dynamic pressure. Moreover, the number and location of vortices in the corner is unknown initially. In the present case a test series in several steps was initiated.

To determine the extent of junction influence on the wing, chordwise pressure and skin friction distributions were obtained, as seen in Fig. 22. While the pressure distributions appear relatively undisturbed, the skin friction distributions are more and more severely distorted as the fuselage is approached, and it is clear that the resolution in the experiment is insufficient.

Thus, the major information has been obtained in spanwise surveys of good resolution, as indicated in Fig. 23. This figure also shows the effects of angle of attack and Mach number on the pressure distribution: in the former case the local distribution is changed, while the level of pressure is affected when Mach number is varied.

Spanwise distributions of static pressure and local skin friction were obtained at several chordwise positions to identify the region of interest. Figure 24 shows the resulting static pressure distributions at three chordwise positions. The Preston tubes were used in several spanwise locations, superimposing results obtained at different flights to obtain the desired spanwise resolution. In general three flights were used, before the probes were moved back to the next chordwise measurement station.

From Fig. 24 a rather consistent picture of the flow pattern is seen as it moves downstream in the junction, and the skin friction distribution shown in Fig. 25 confirms this development. It may be noted here that as the vortices proceed downstream, they may move between the wing and the contoured part of the fillet or also up onto the fuselage.

Pressure rakes were used in the same manner as mentioned for the modified Preston tubes, and in Fig. 26 typical total pressure profiles are shown. As the same flight conditions were flown each time, it was possible to accumulate data even for the total pressure distributions in the junction region.

Figure 27 shows isobars of total pressure obtained in the wing/ body junction at several flight conditions. It is clear that there is an immense change in magnitude as well as location of the vortex as the Mach number is increased (with increasing angle of attack).

To measure the three-dimensional flow in the vicinity of the vortex, a rake with 5 rotating X-wires was employed, see Fig. 28 for the general outline of the instrumentation. This rake was mounted at positions were total pressure profiles were already available, and the deduced velocity from the rake was used as an in-flight calibration of the magnitude (and hence the sensitivity) of each hot wire. The sensors were rotated through 12 angular positions from 0 to 360 degrees at 30 degree interval. This then produces redundant information on the velocity vector, making it possible to determine the angular calibration characteristics for each particular X-wire at each flow condition.

The probes remained at each position for 3.2 seconds, allowing turbulence data to be gathered as well. While the time-averaged part of the data is included in the computerized database, the turbulence data is only available on magnetic tape and has not yet been analyzed.

Conclusions

In the present paper a description has been given of aerodynamic flight tests performed over several decades, and it can be seen from the publications that a large body of good quality work was performed in early tests. For several decades the flight tests were used for verification of model concepts, with no need to document the flow extensively. However, with the advent of complex computational methods and the need to verify these methods with flight data rather than simply aircraft design verification, it is now necessary perform flight tests with a comprehensive, if not "complete", documentation of the viscous flow. For certain model geometries comprehensive investigations have been performed in wind tunnels. The investigation of the flow around a prolate spheroid by MEIER et al/68/ is an example of this. Results for this type of experiment on a swept wing aircraft at subsonic speeds in flight have been given here, to illustrate the complexity of the measurements, and also to indicate the insight that may be gained. This means that it will be possible to go back to the experimental data over and over again to examine new additions to models. This also means that the database system has to be completely computerized, with experimental data and all codes required to reduce it treated as a self-contained system.

Of course, this experiment has to be repeated in various ways – in many areas it may raise more questions than it gives answers. But it may serve as a reminder of the amount of data and the complexity of a flight measurement designed to yield information on three-dimensional turbulent shear layers. In terms of modelling one may safely assume that many surprises are in store, especially when unsteady effects due to flight mechanics or structural interaction are to be included. Even a three-dimensional full Navier-Stokes solutions will be invalid if the boundary conditions are not known or met.

The previously common assumption that flight data was too inaccurate for proper analysis of the results, is no longer true. The data of WALSH/69/, obtained to evaluate a few percent skin friction change on the fuselage of a Lear Jet aircraft is an example of this.

To obtain the required information concerning the flow on aircraft it is necessary to develop instrumentation for quantitative as well as qualitative documentations. Two recent examples of this development are the vapor screen technique used in flight by LAMAR/70/, and the techniques utilized by HOLMES et al/71/. The measurement of shear stresses (ROSE and OTTEN/72/) and large-scale coherent structures (CHOU et al/73/) are feasible, and it may be expected that an increasing amount of reliable turbulence data may be obtained in flight.

References

1 Polhem, C.: "The merging of theories and Experiment in Mechanics". In Swedish. Vetenskapsakademiens handlingar, Stockholm, Sweden, 1774, pp. 149-166.

2 Rotta, J.C.: "Turbulente Strömungen. Eine Einführung in die Theorie und ihre Anwendung". B.K. Teubner, Stuttgart, 1972.

3 Saltzman, E.J. & Ayers, T.G.: "A Review of Flight to Wind Tunnel Correlation". AIAA Paper 81-2475, Nov. 1981.

4 Holmes, B.J., Obara, C.J. & Yip, L.P.: "Natural laminar flow experiments on modern airplane surfaces". NASA TP 2256, June 1984.

5 Poisson-Quinton, Ph.: "From Wind Tunnel to Flight, the Role of the Laboratory in Aerospace Design". Journal of Aircraft, Vol. 5, No. 3, 1968, pp. 193-214.

6 Stüper, J.: "Untersuchung von Reibungsschichten am fliegenden Flugzeug". Lufo, Band 11, No. 1. Also see: "Investigation of Boundary Layers on an Airplane in Free Flight". NASA TM 751, 1934.

7 Kuethe, A.M. & Chow, C.Y.: "Foundations of Aerodynamics. Bases of Aerodynamic Design". 4th edition, John Wiley & Sons, 1986.

8 Gruschwitz, E.: "Turbulente Reibungsschicht in Ebener Strömung bei Druckabfall und Druckansteig". Ingenieur-Archiv 11. Band, 1931, pp. 321-346.

9 Bradshaw, P., Mizner, G.A., & Unsworth, K.: "Calculation of Compressible Turbulent Boundary Layers with Heat Transfer on Straight-Tapered Swept Wings". Imperial College IC Aero Report 75-04, 1974. See also: AIAA J. Vol. 14, 1976, p. 399.

10 Jones, B.M.: "Flight Experiments on the Boundary Layer". First Wright Brother's Lecture. J Aero. Sci., Vol. 5, No. 3, 1938, pp. 81-101.

11 Goett, H.J., & Bicknell, J.: "Comparison of Profile-Drag and Boundary-Layer Measurements Obtained in Flight and in the Full-Scale Wind Tunnel". NACA TN 693, March 1939.

12 Zalovcik, J.A.: "Flight Investigation of Boundary-Layer and Profile-Drag Characteristics of Smooth Wing Sections of a P-47D Airplane". NASA Wartime Report L-86, Oct 1945.

13 Brown, H.H. & Clousing, L.A.: "Wing pressure measurements up to 0.866 Mach number in flight on a jet propelled airplane". NASA TN 1181, 1947.

14 Smith, R.H.: "Antecedents and Analogues – Experimental Aircraft". AIAA Paper 78-3008, In "Diamond Jubilee of Powered Flight", edited by J.D. Pinson, AIAA, 1978.

15 Bowes, G.M.: "Aircraft Lift and Drag Prediction and Measurement". AGARD Lecture Series No. 67, 1974.

16 Arnaiz, H.H.: "Flight-Measured Lift and Drag Characteristics of a Large, Flexible, High Supersonic Cruise Airplane". NASA TM X-3532, May 1977.

17 Sloof, J.W.: "Aircraft Drag Prediction and Reduction – Computational Drag Analyses and Minimization; Mission Impossible?" AGARD-R-723 Addendum 1, 1985.

18 Erlich, E.: "Sondage en vol de la couche limite sur l'Avion Supersonique Mirage IV". La Recherche Aerospatiale, Vol. 122, 1968, pp. 11-19.

19 Georgy-Falvy, D.: "Effect of Pressurization on Airplane Fuselage". J. Aircraft, Vol. 2, No. 6, Nov-Dec 1965, pp. 531-537.

20 George-Falvy, D.: "Applications of Flow Diagnostic Techniques to Aerodynamic Problem Solving". AIAA 9th Aerodynamic Testing Conference, 1976, p. 246.

21 Coulomb, J., Ledoux, M., and Lerat, J.: "Measures en vol et en soufflerie sur le bord d'attaque d'une aile de Nord 2501". L'Aeronautique et L'Astronautique, No. 69, 1978-2, pp. 19-28.

22 Goecke Powers, S.: "Flight-Measured Pressure Characteristics of AFT-Facing Steps in Thick Boundary Layer Flow for Transonic and Supersonic Mach Numbers". In NASA CP-2054, Volume 1, 1978.

23 Fisher, D.F., and Dougherty, Jr. N.S.: "Flight and Wind-Tunnel Correlation of Boundary Layer Transition on the AEDC transition cone". AGARD CP 339, 1982, Paper 5.

24 Quast, A., and Hostmann: "Private Communication". 1986.

25 Jenkins, M.W.M.: "Small Scale Free Flight Research at Lockheed-Georgia". Society of Flight Test Engineers 11th Aunnual Symposium, Atlanta, Georgia, Aug. 27-29, 1980, Paper 15.

26 McKinney, L.W. & Baals, D.D.: "Wind-Tunnel/Flight Correlation". NASA CP 2225, 1981.

27 Cahill, J.F. & Connor, P.C.: "Correlation of Data Related to Shock-Induced Trailing Edge Separation and Extrapolation to Flight Reynolds Number". NASA CR-3178, 1979.

28 Paterson, J.H., Blackerby, W.T., Schwanebeck, J.C. & Braddock, W.F.: "An Analysis of Flight Test Data on the C-141A Aircraft". NASA CR-1558, 1970.

29 Mabey, D.G.: "Beyond the Buffet Boundary". Aeronautical Journal, April 1973, pp. 201-215.

30 Gray, W.E.: "The effects of wing sweep on laminar flow". RAE TM Aero 255, 1952.

31 Burrows, F.M.: "A Theoretical and Experimental Study of the Boundary Layer Flow on a 45 degrees Swept Back Wing". College of Aeronautics Cranfield Report 109, 1956.

32 Landeryou, R.R. & Trayford, R.S.: "Flight tests of a laminar flow swept wing with boundary layer control by suction". College of Aeronalutics Cranfield Report Aero No. 174, 1964.

33 Gaster, M.: "The Application of Hot Film Gauges to the Detection of Transition in Flight". College of Aeronautics Cranfield, Report 189, 1965.

34 Gaster, M.: "On The Flow Along Swept Leading Edges". The Aeronautical Quarterly, May 1967, pp. 165-184.

35 Gaster, M.: "A Simple Device for Preventing Contamination on Swept Leading Edges". J. of the Royal Aero. Soc., Vol. 69, Nov 1965, p. 788.

36 Landeryou, R.R. & Porter, P.G.: "Further tests of a laminar flow swept wing with boundary layer control by suction". College of Aeronautics Cranfield Report Aero No. 192, 1966.

37 Nicholas, O.P.: "Pressure Plotting The Wing of the Fairey Delta 2". Flight Test Instrumentation. Proc. 1st International Symposium, Cranfield, 1960. (Ed. M.A. Perry, Pergamon Press, 1961).

38 Webb, D.R.: "Instrumentation for the Measurement of Pressures and Strains Arising From Fluctuations in The Turbulent Boundary Layer of a Fairey Delta 2 Aircraft". Flight Test Instrumentation, Vol. 2, Proc. 2nr International Symposium, 1962, ed. M.A. Perry, Pergamon Press, 1963.

39 Browne, G.C., Bateman, T.E.B., Pavitt, M., and Haines, A.B.: "A Comparison of Wing Pressure Distributions Measured in Flight and on a Wind Tunnel Model of the Super VC10". ARC R & M 3707, 1972.

40 Montoya, L.C., and Lux, D.P.: "Comparisons of Wing Pressure Distribution from Flight Tests of Flush and External Orifices for Mach Numbers from 0.50 to 0.97". NASA TM X-56032, 1975.

41 Fennell, L.J.: "Vortex Breakdown - Some Observations in Flight on the HP115 Aircraft". ARC R & M 3805, 1977.

42 Fischel, J., Bower, R.E., and Weil, J.: "Supercritical Wing Technology. A Progress Report on Flight Evaluations". NASA SP-301, 1972.

43 Eckstrom, C.V.: "Flight Measurements of Surface Pressures on a Flexible Supercritical Research Wing". NASA TP 2501, Dec 1985.

44 Lux, D.P.: "In-flight Three-Dimensional Boundary Layer and Wake Measurements from a Swept Supercritical Wing". NASA CP-2045, Part 2, 1978, pp. 643-655.

45 Duke. E.L., and Lux, D.P.: "The Application and Results of a New Flight Test Technique". AIAA Paper 83-2137, 1973.

46 Jacob, D., Welte, D. & Wonnenberg, H.: "Ground/flight Correlation on the Alpha-Jet Experimental Aircraft with a Transonic Wing. A Comparison between Wind Tunnel and Flight Results for Aerodynamic Performance". AGARD Paper 4A, 1982.

47 Buers, H., Schmitt, V. & Lerat, J.: "A Comparison of the Wing Pressure Distribution and Local Wake Survey from Analytical Wind Tunnel and Flight Results". See: Ground/Flight Test Techniques and Correlation, = AGARD-CP-339, 1982.

48 Preston, J.H.: "The Determination of Turbulent Skin Friction by Means of Pitot Tubes". JR. Ae. Soc. Vol. 58, 1954, p. 109.

49 Bertelrud, A.: "Total Head/Static Measurements of Skin Friction and Surface Pressure". AIAA Journal, March 1977.

50 Bradshaw, P. & Unsworth, K.: "A Note on Preston Tube Calibrations in Compressible Flow". Imperial College IC Aero Report 73-07, 1973.

51 Stanton, T.E., Marshall, O., and Bryant, C.N.: "On the Conditions at the Boundary of a Fluid in Turbulent Motion". Proc. Roy. Soc. Series A, 97, 1920, pp. 413-434.

52 Ludwieg, H.: "Ein Gerät Zur Messung der Wandschubspannung turbulenter Reibungsschichten". Ing. Archiv, Band XVIII, 1949, pp. 207-218.

53 McCroskey, W.J., and Durbin, E.J.: "Flow Angle and Shear Stress Measurements Using Heated Films and Wires". J. Basic Eng. Trans. of the ASME, March 1972, pp. 46-52.

54 Poll, D.I.A., and Matthews, J.R.: "On the measurement of turbulent skin friction by heated film gauge technique". Cranfield, College of Aeronautics, Report No. 8515, Sept. 1985.

55 Nitsche, W., Thünker, R., and Weiser, N.: "Piez-Folien als Sensor für instationäre Drücke am Beispiel von Wanddruck- und Pulsdruckmessung". Sensor '85, Transducer-Technik: Entwicklung v. Anwendung, Messegelände Karlsuhe, 1985.

56 Bertelrud, A.: "Instrumentation for Measurement of Flow Properties on a Swept Wing in Flight". DGLR Vortrag 81-034, 1981.

57 Bertelrud, A.: "Pressure Distribution on a Swept Wing Aircraft in Flight". In AGARD AR-138, Addendum, 1984.

58 Bertelrud, A.: "Full Scale Experiments with Two Techniques Suggested for Drag Reduction on Aircraft". AIAA Paper 85-1713, 1985.

59 Bertelrud, A. & Olsson, J.: "Method of Analysing Data on a Swept Wing Aircraft in Flight". ICAS Paper 86-1.9.3, in ICAS Proceedings 1986, 15th Congress of the International Council of the Aeronautical Sciences, London, England, Sept 1986.

60 Cumpsty, N.A., and Head, M.R.: "The calculation of the three-dimensional turbulent boundary layer. Part 3. Comparison of attachment-line calculations with experiment". Aeronautical Quarterly, Vol. 20, 1969, pp. 99-113.

61 Launder, B.E., and Jones, W.P.: "On the prediction of laminarization". ARC CP 1036, 1969.

62 Maine, R.E., and Iliff, K.W.: "Formulation and implementation of a practical algorithm for parameter estimation with process and measurement noise". AIAA Paper 80-1603, 1980.

63 Hamel, P.G.: "Status of methods for aircraft state and parameter identification". AGARD CP-187, 1976.

64 Nagamatsu, H.T., Dyer, R. & Ficarra, R.V.: "Supercritical Airfoil Drag Reduction by Passive Shock Wave/Boundary Layer Control in the Mach Number Range 0.75 to 0.90". AIAA Paper 85-0207.

65 Krogmann, P.: "Transonic Drag Reduction by Active/Passive Boundary Layer Control". VKI Lecture Series "Aircraft Drag Prediction and Reduction", May 1985.

66 Squire, M.G. & Young, A.D.: "The Calculation of the Profile Drag of Aerofoils". ARC R & M 1838, 1937.

67 Shabaka, I.M.M.A.: "Turbulent Flow in a Simulated Wing-Body Junction". PhD Thesis, Imperial College, London, 1978.

68 Meier, H.U., Kreplin, H.P., and Vollmers, H.: "Development of boundary layer and separationpatterns on a body of revolution at incidence". 2nd Symp. on Numerical and Physical Aspects of Aerodynamic Flows, Long Beach, 1983.

69 Walsh, M.: "Private Communication"., 1986.

70 Lamar, J.E.: "In-flight and wind tunnel leading-edge vortex study on the F-106B airplane". In NASA CP-2416, Vortex Flow Aerodynamics, Vol. 1, 1986.

71 Holmes, B.J., Croom, C.C., Gall, P.D., Manuel, G.S. & Carraway, D.L.: "Advanced Transition Measurement Methods for Flight Applications". AIAA paper 86-9786. AIAA/AHS/CASI/DGLR/IES/ISA/ITEA/SETP/SFTE 3rd Flight Testing Conference, Las Vegas, April 2-4, 1986.

72 Rose, W.C., and Otten III, L.J.: "High Reynolds-Number Shear-Stress Measurements". AIAA Paper 83-0053, 1983.

73 Chou, D.C., de Jonckheere, R.K., and Lynch, G.: "Large-Scale Coherent Structures in High Speed Turbulent Shear Flows". AIAA Paper 86-0028, 1986.

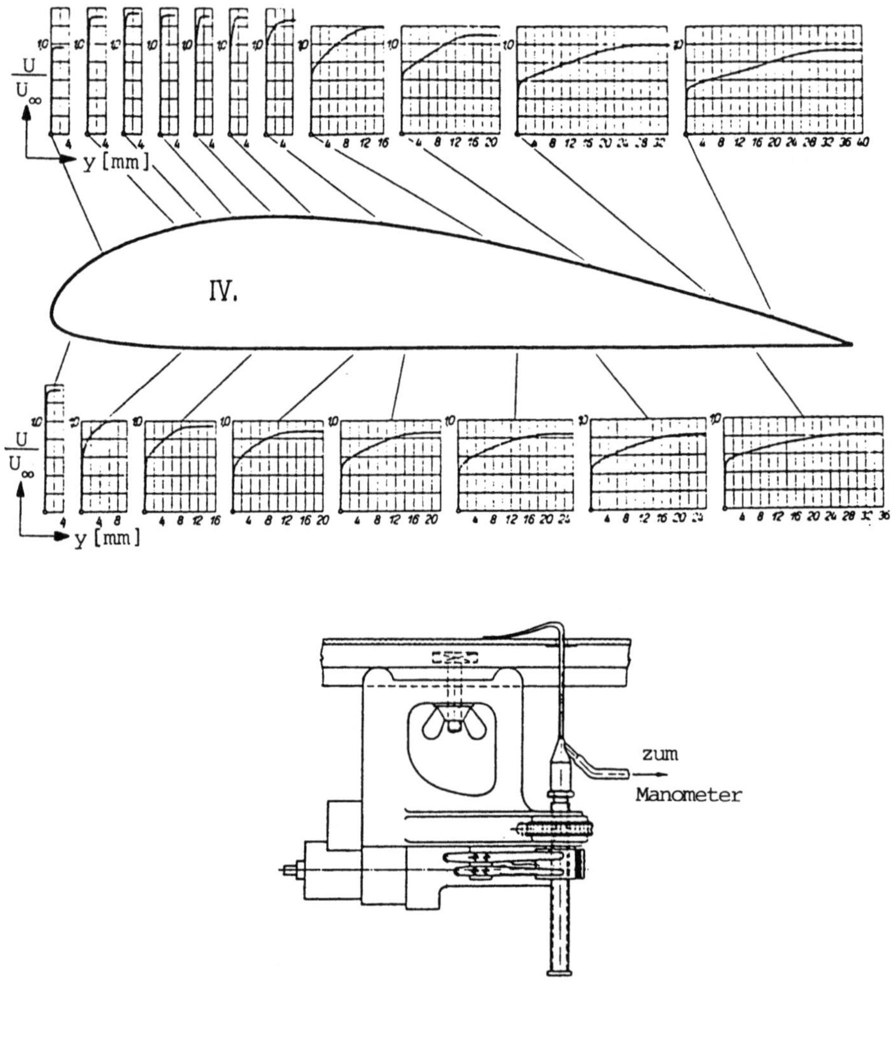

Fig. 1 Velocity profiles measured in flight, and traversing pressure probe. From Stüper /6/.

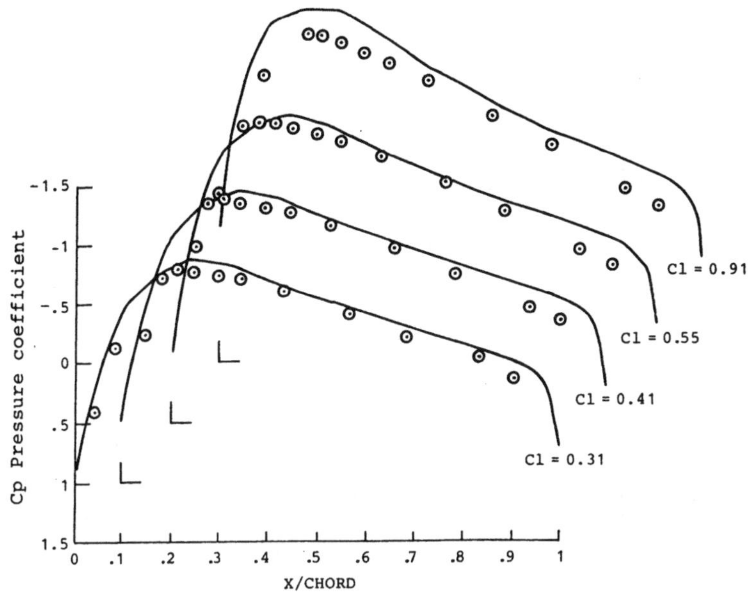

Fig. 3 Aircraft sketch and flight envelope with stationary conditions repeated each flight. Reynolds number is based on mean aerodynamic chord.

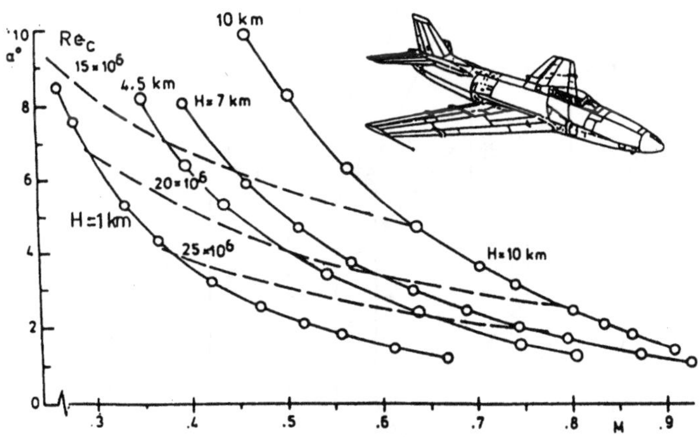

Fig. 2 Computed and experimental (From /6/) pressure distributions on suction side.

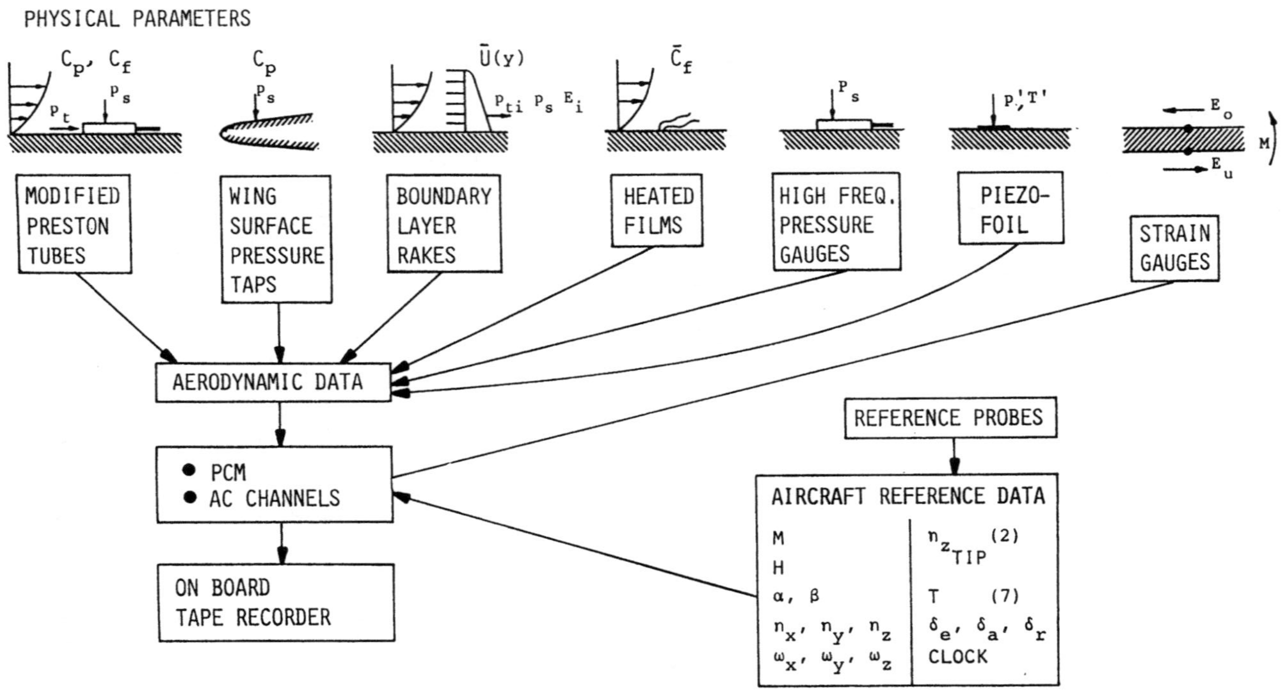

Fig. 4 Sensors and data acquisition system on 32-209 Lansen (N5468X).

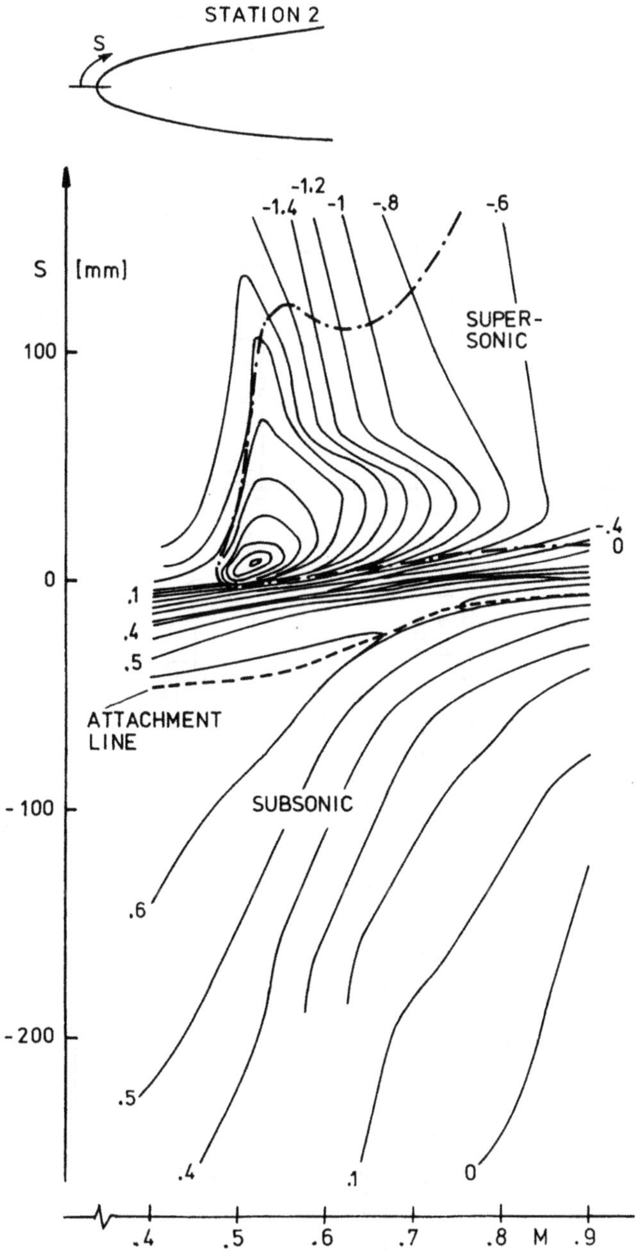

Fig. 5 Change in pressure distribution close to the leading edge for varying Mach number (i.e. also varying angle of attack), constant altitude.

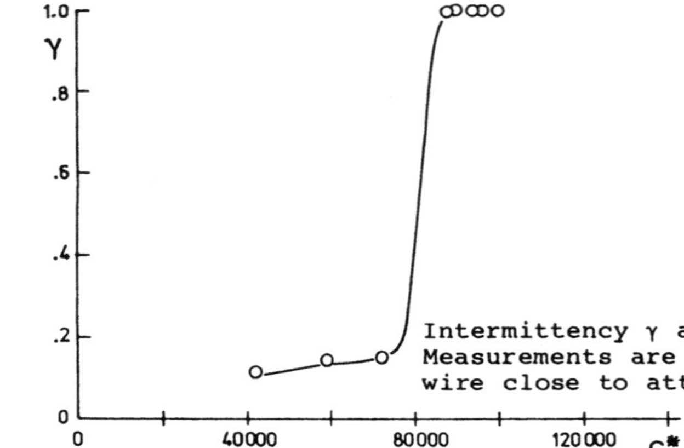

Intermittency γ as function of C*. Measurements are taken with a hot wire close to attachment line

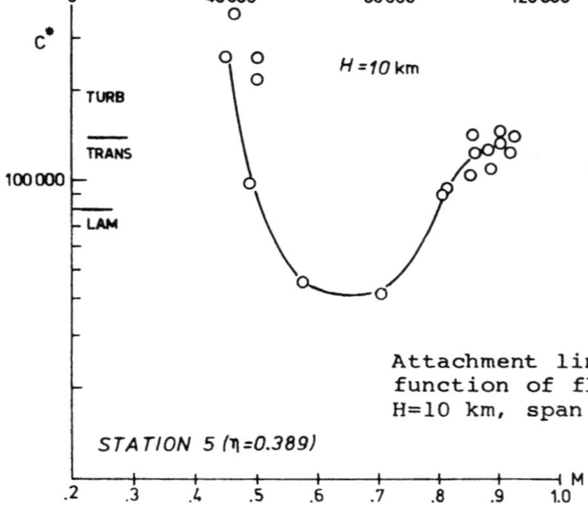

$$C^* = \frac{W_e^2}{\nu \cdot \left(\frac{dU_e}{ds}\right)_{s=0}}$$

Attachment line parameter C* as function of flight Mach number; H=10 km, span station 5

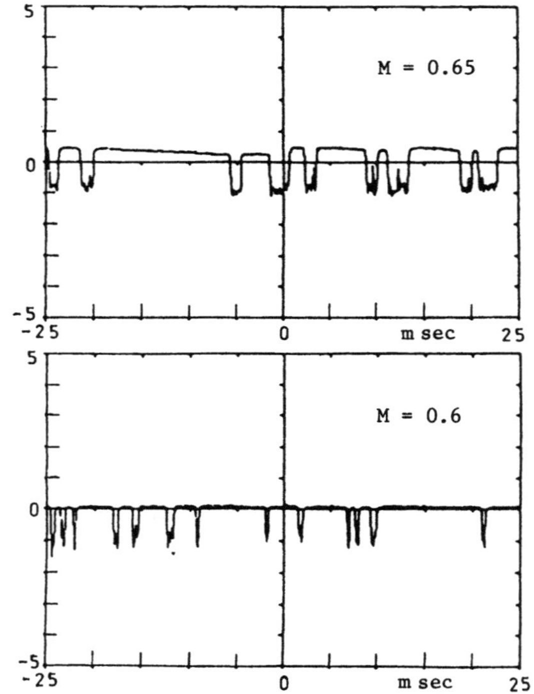

Intermittent hot wire signal close to attachment line

Figure 6

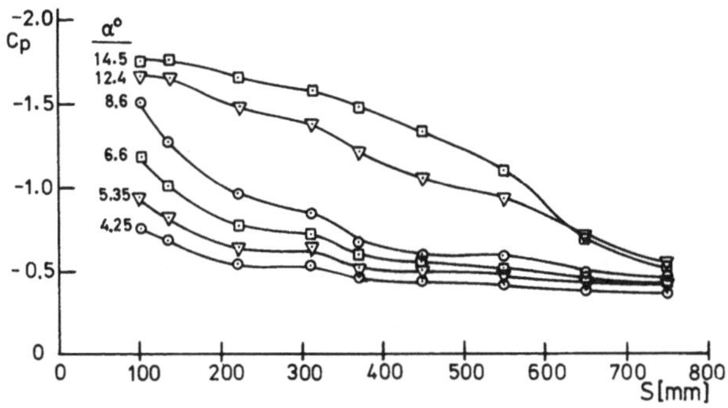

Fig. 7a) Pressure distribution in leading edge region for increasing angle of attack (decreasing Mach number). Long bubble evident at the two highest angles of attack.

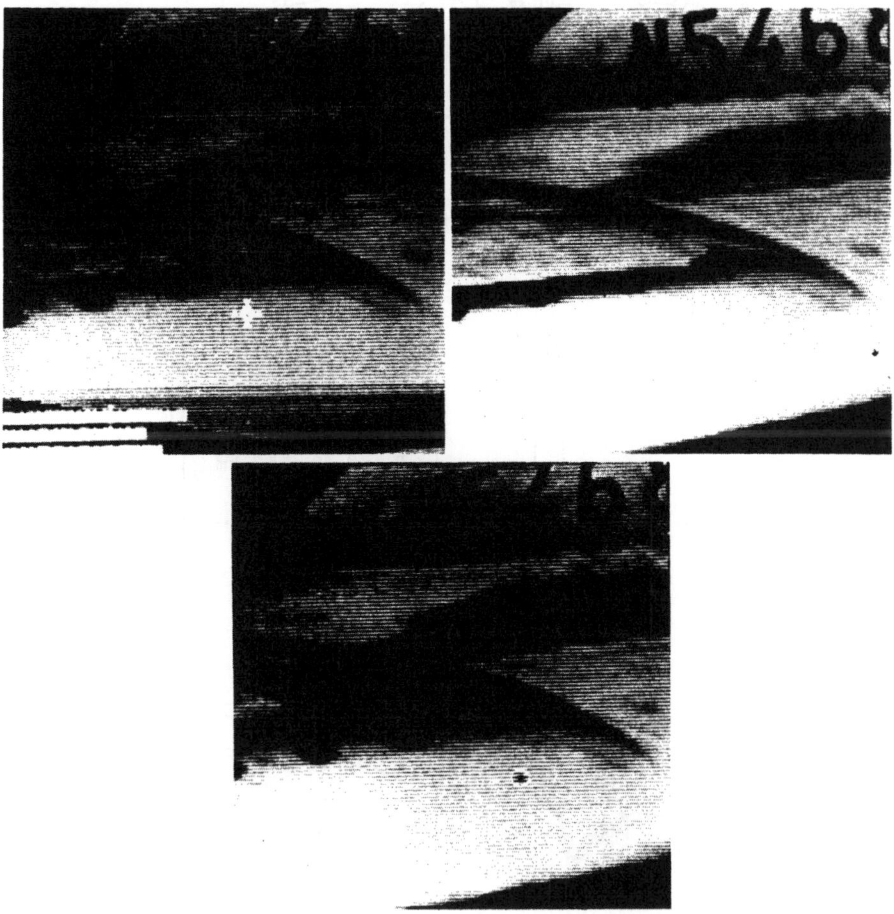

Fig. 7b) Samples from video movie on separation bubble flow.

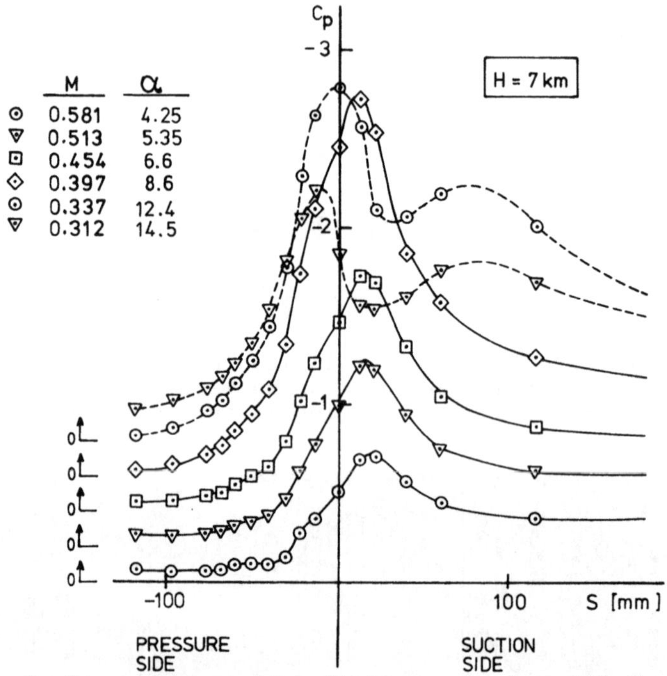

Fig. 8 Tip pressure distributions showing effect of change from short to long bubble.

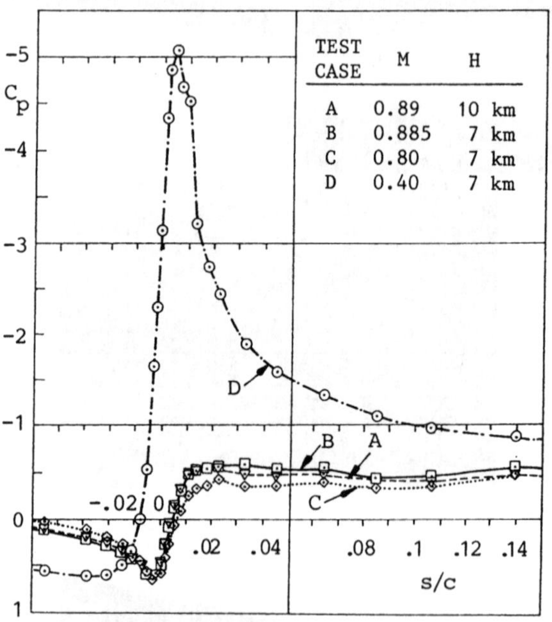

Fig. 9 Pressure distributions at instrumented segment for the four test cases.

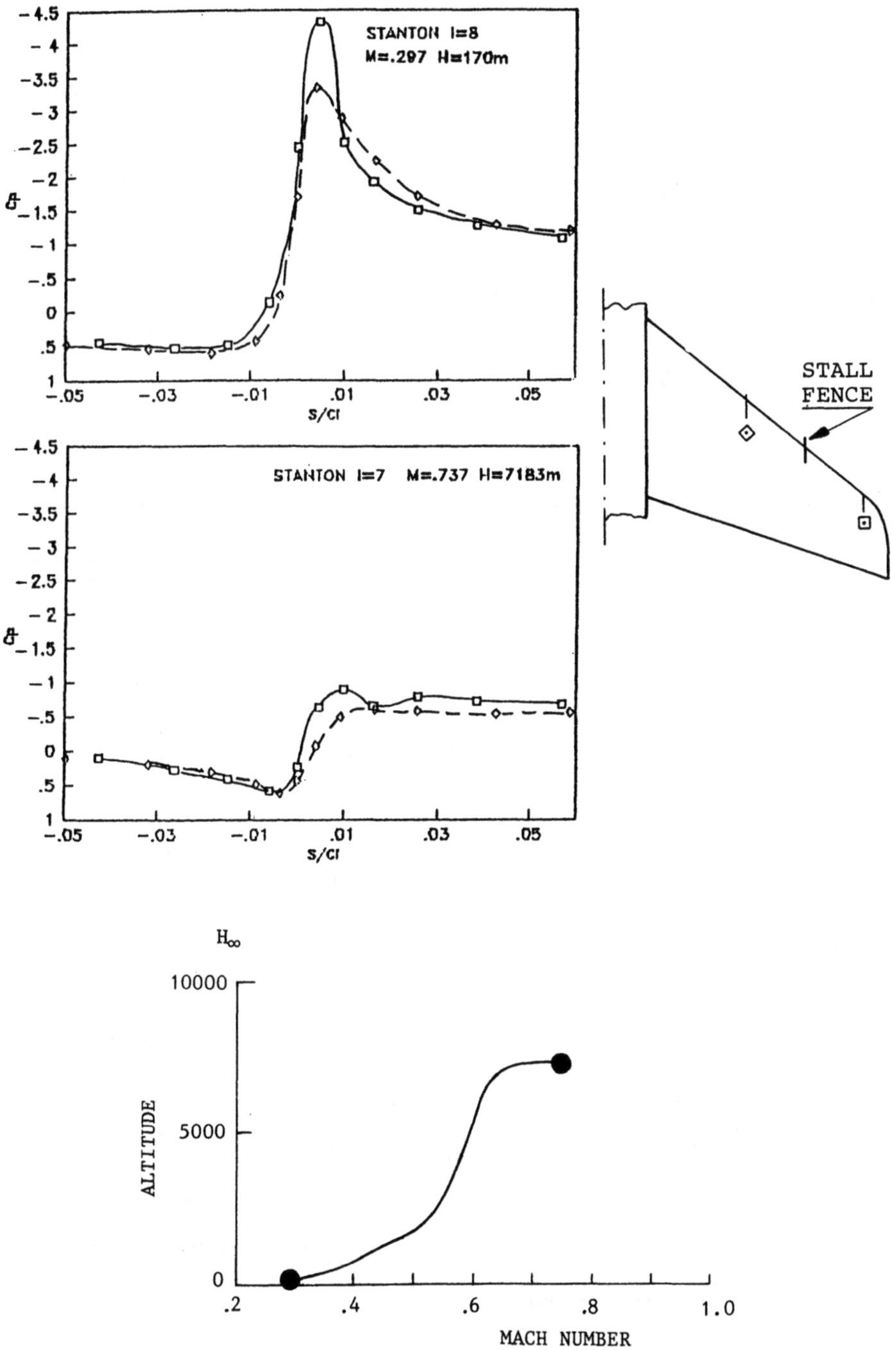

Fig. 10 Pressure distributions at two spanwise stations during climb.

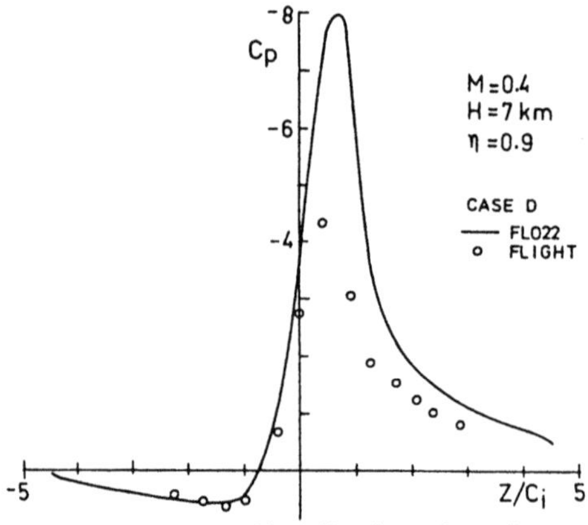

Pressure distribution plotted appropriately for leading edge suction evaluation. η=0.9

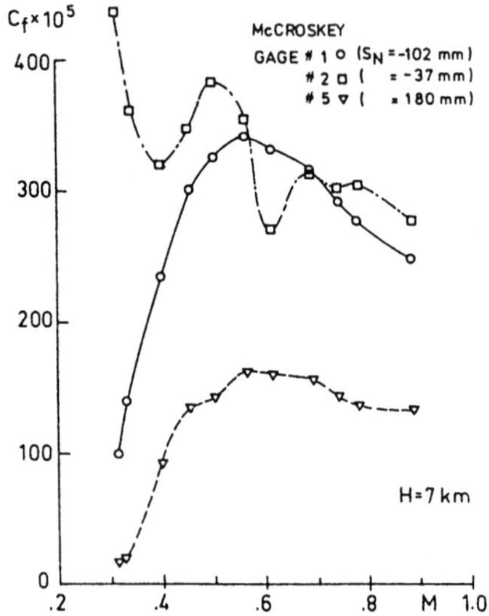

Local skin friction magnitude as function of flight Mach number

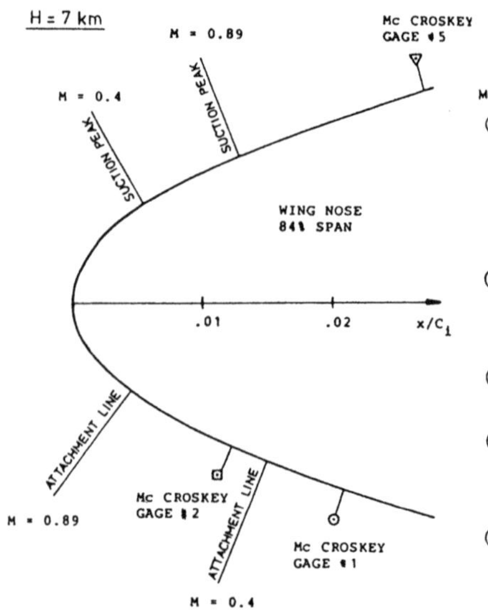

Location of attachment line and suction peak for a Mach number range; position of relevant McCroskey (heated film) gages

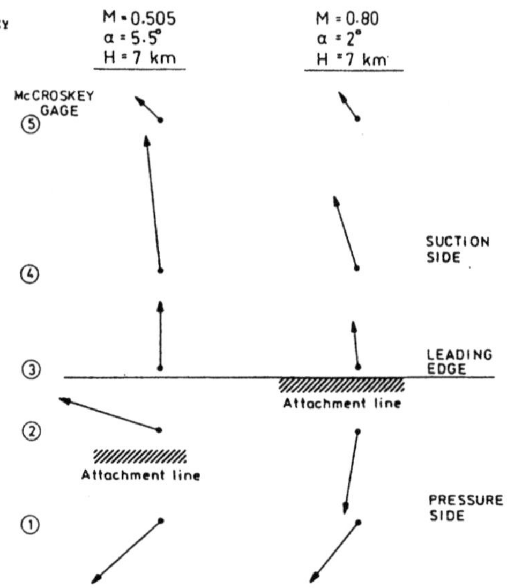

Local skin friction (magnitude and direction) at two flight conditions

Figure 11

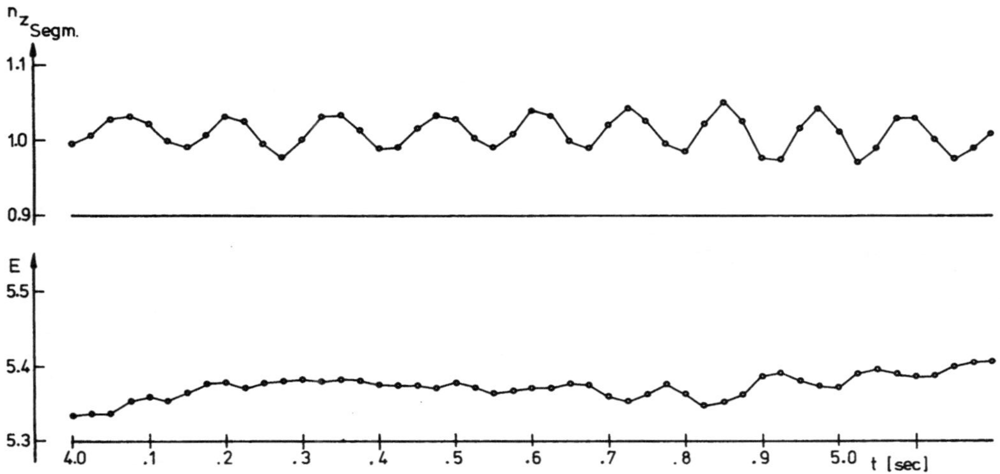

Fig. 12a) Local pressure history (full line) compared with predictor due to the elevator movement (dotted line).

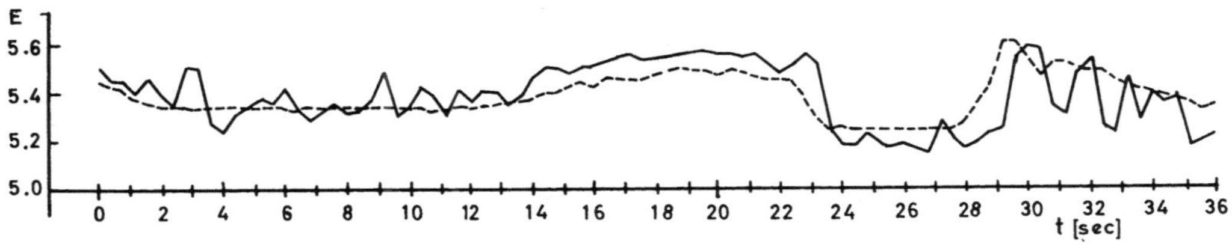

Fig. 12b) PCM datasampled at 40 Hz showing the correlation between the pressure fluctuations and an accelerometer at a nearly location. The wing bending frequency is seen clearby.

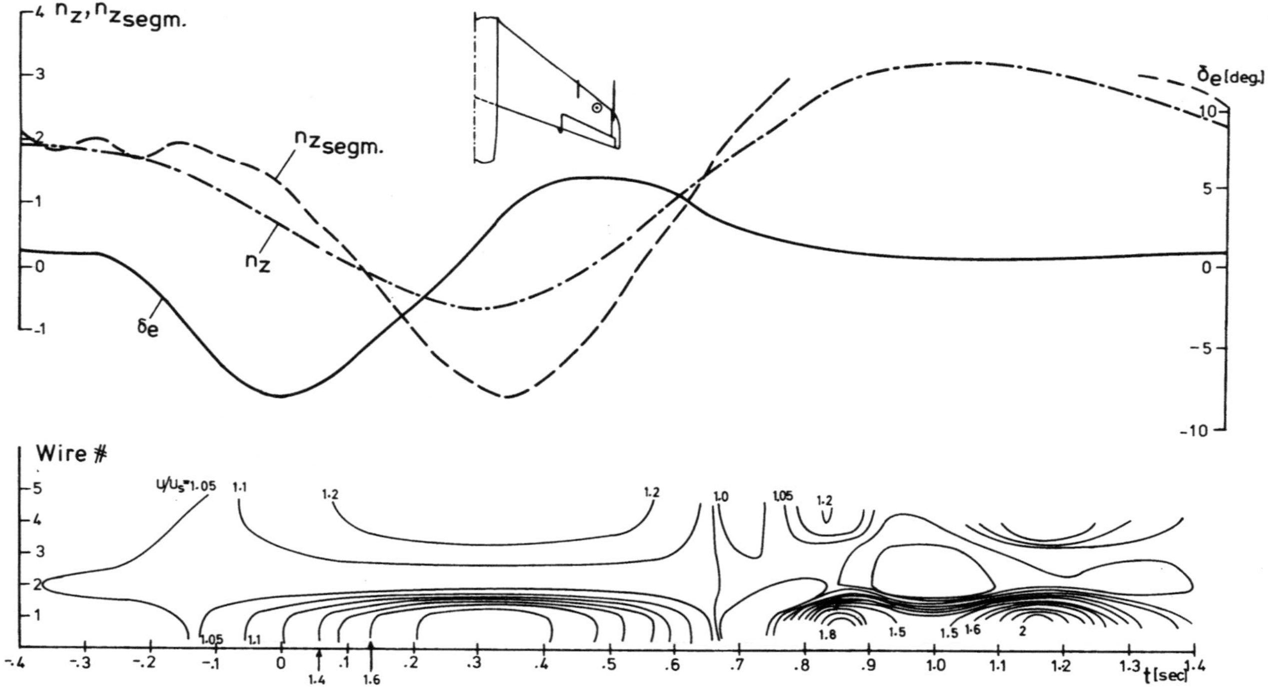

Fig. 13 Change in local flow close to the leading edge during a roll with superimposed pitch movement.

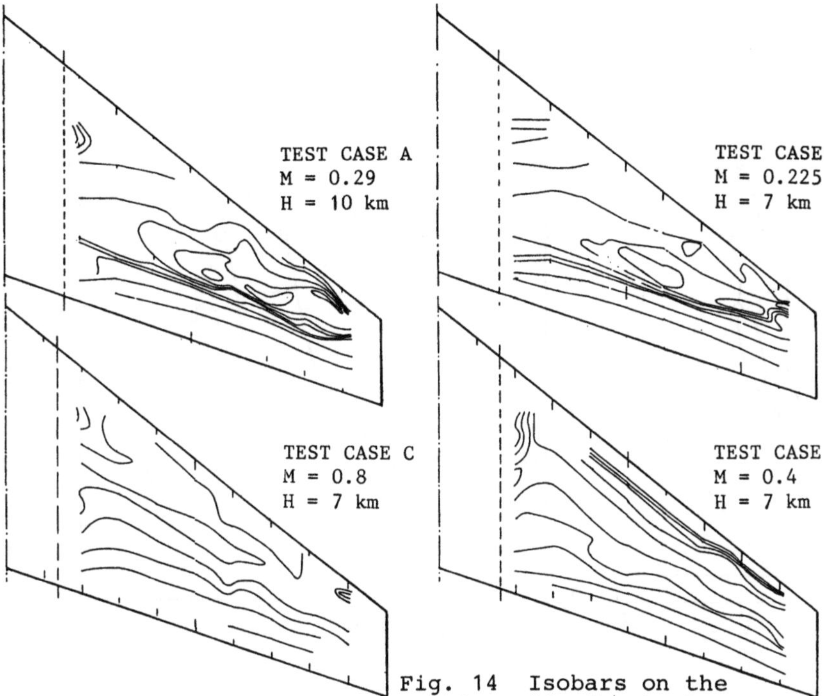

Fig. 14 Isobars on the suction side for test cases.

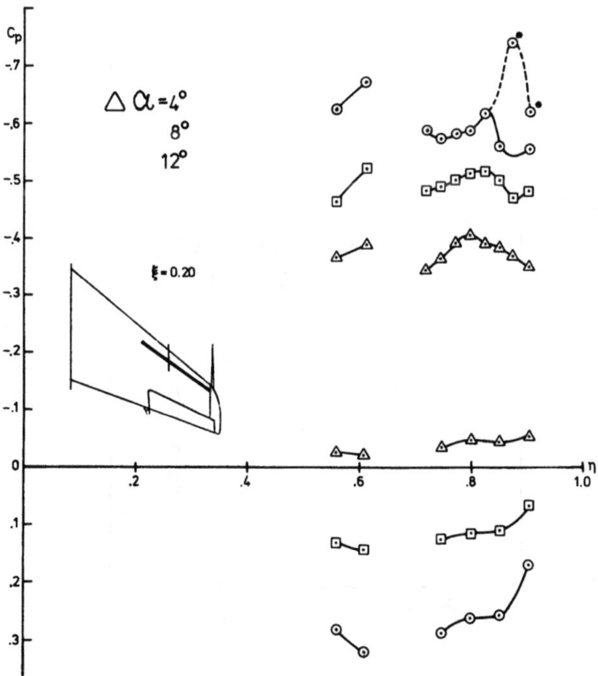

Fig. 15 Spanwise C_p in- and outboard of the stall fence.

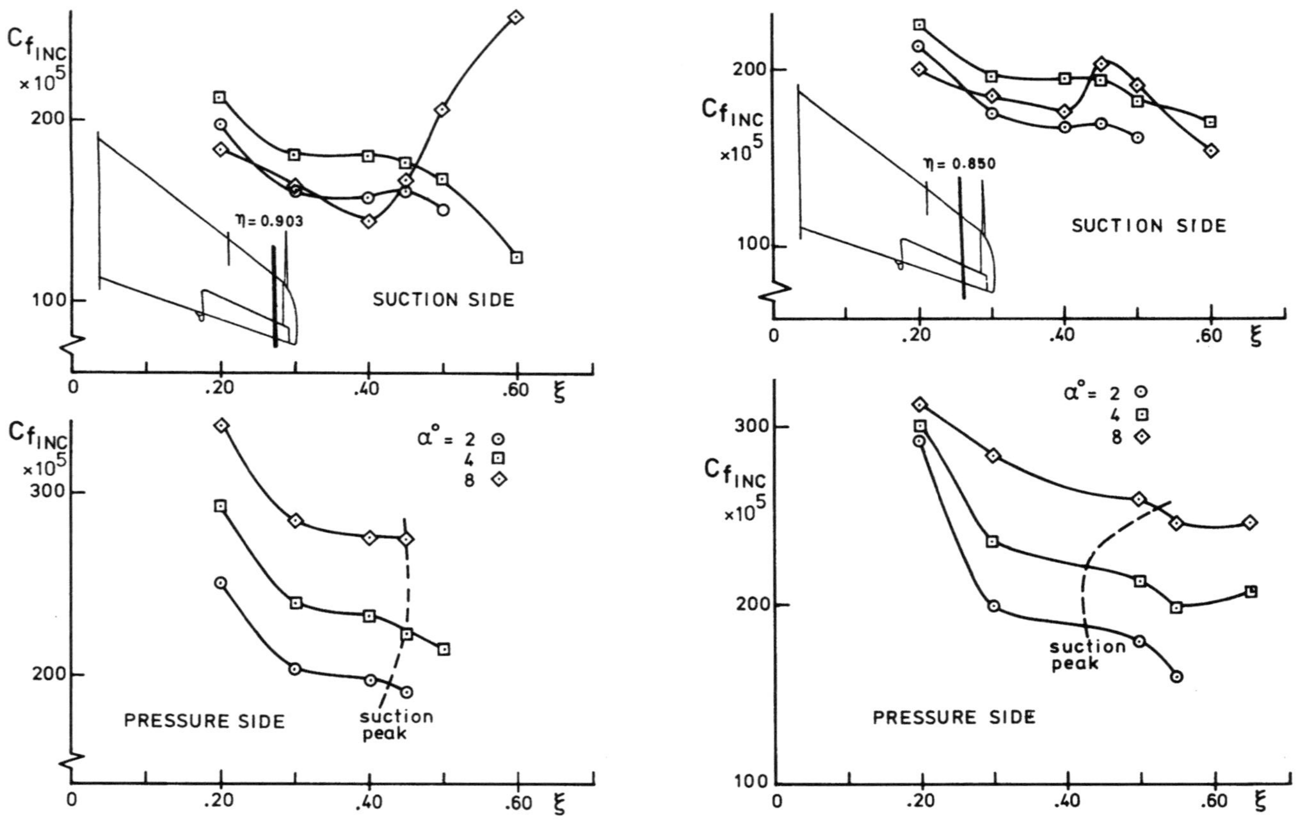

Fig. 16 Local skin friction coefficient $C_{f\ inc}$ (incompressible calibration) for two span stations.

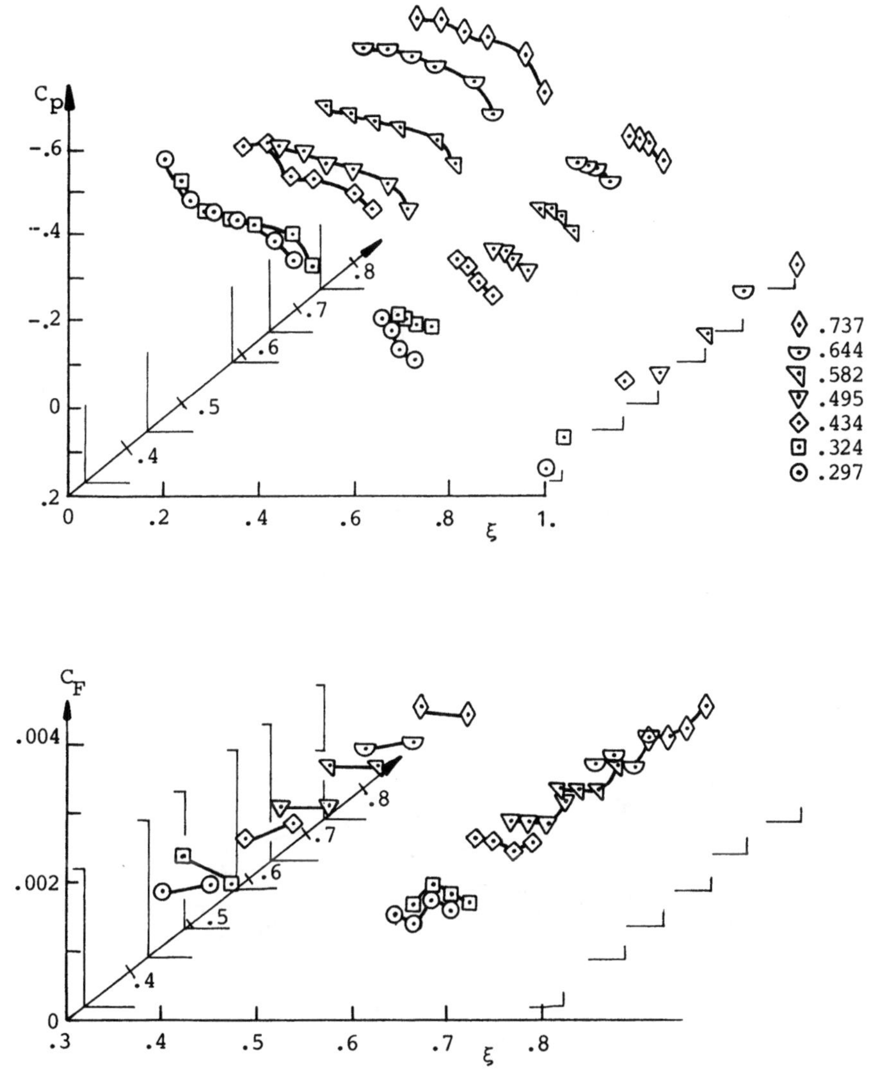

Fig. 17 C_p and C_f from static pressure taps with and without razor blades. (Stanton tubes).

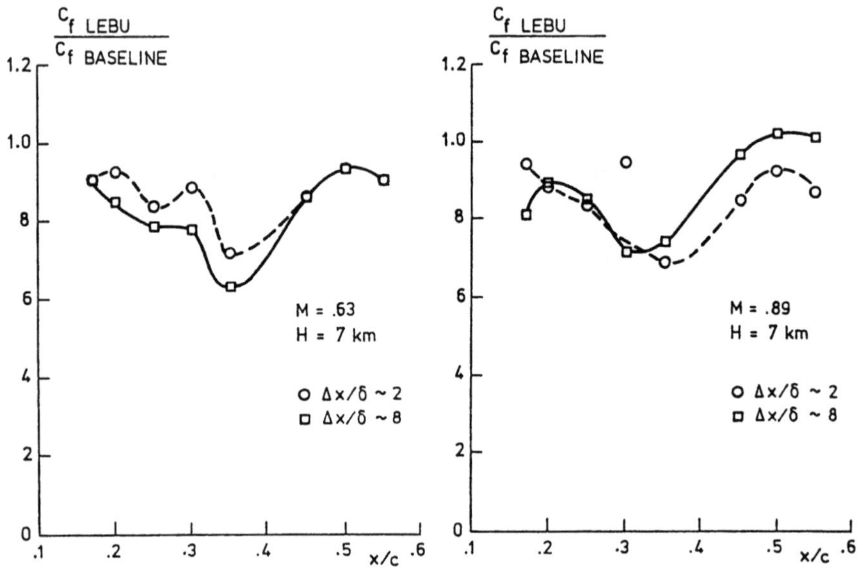

Reduction of local skin friction.

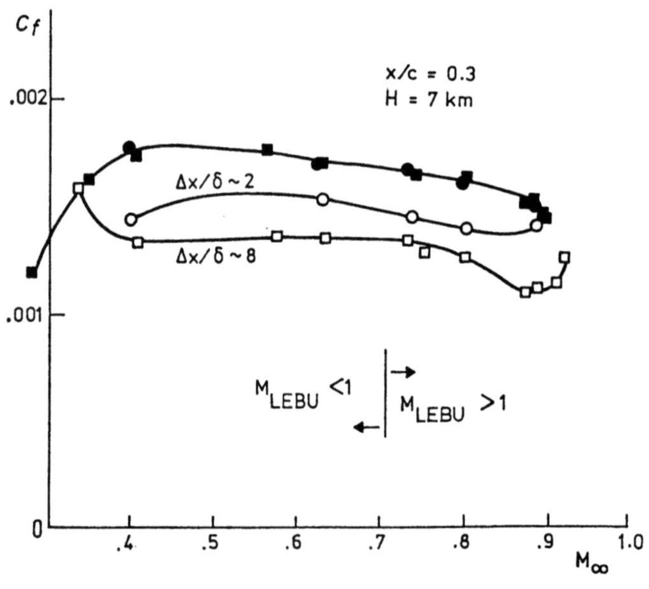

Effect of Mach number on C_D at one location.

Figure 18

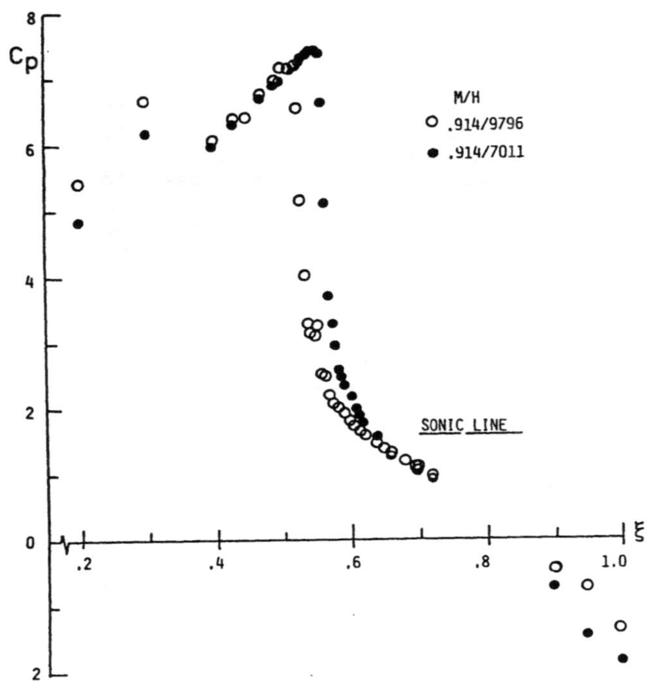

Fig. 19 Pressure distributions for M = 0.914 at two altitudes H (7 and 10 km) illustrating shock movement.

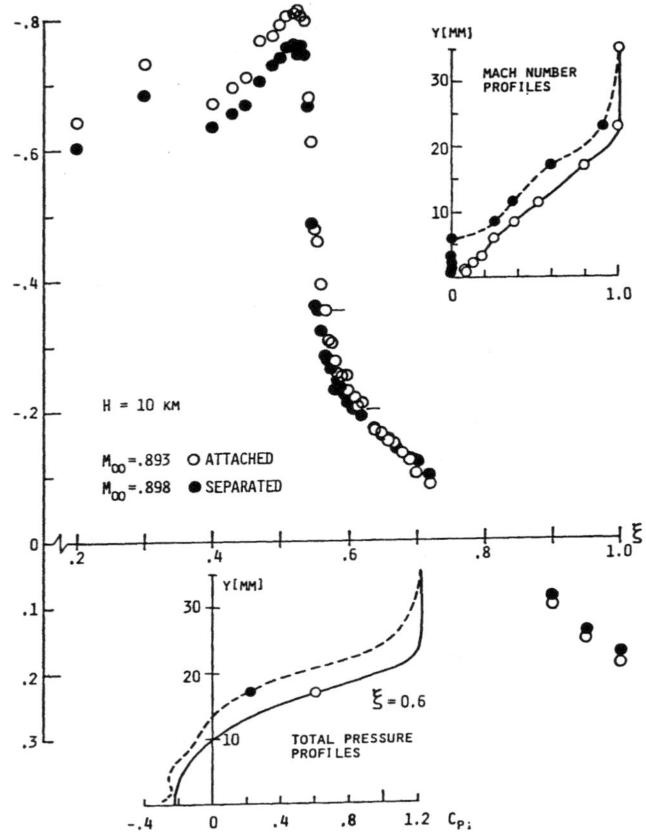

Fig. 20 Flip/flop effect of separation. Pressure and Mach number profiles.

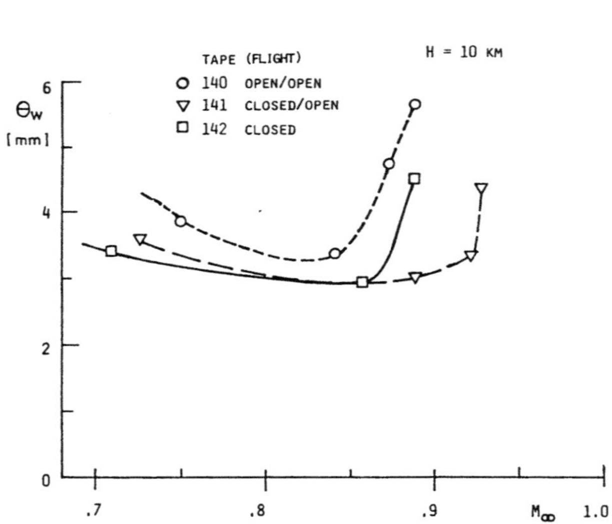

Fig. 21
Momentum deficit at ξ=0.6 transformed to wake position. Open corresponds to 3.1% perforation θ_w is momentum thickness transformed to undisturbed (far wake) conditions.

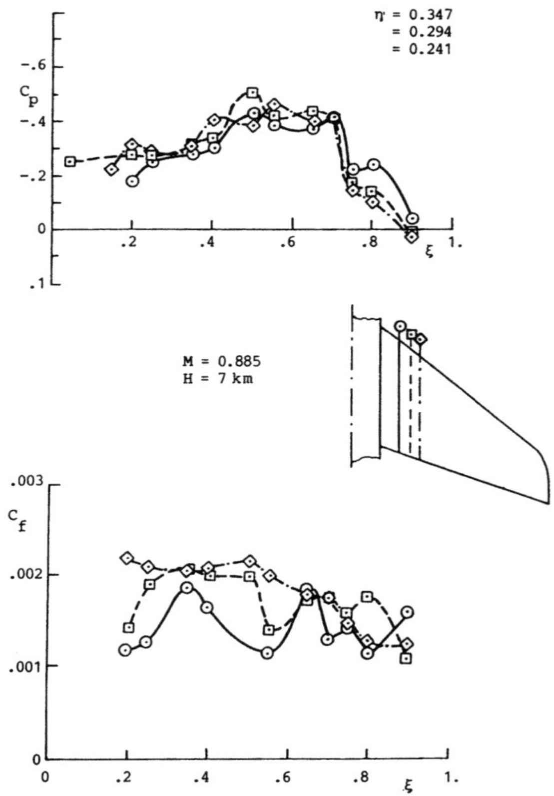

Fig. 22 Chorwise distribution of C_p and C_f on the inner part of wing.

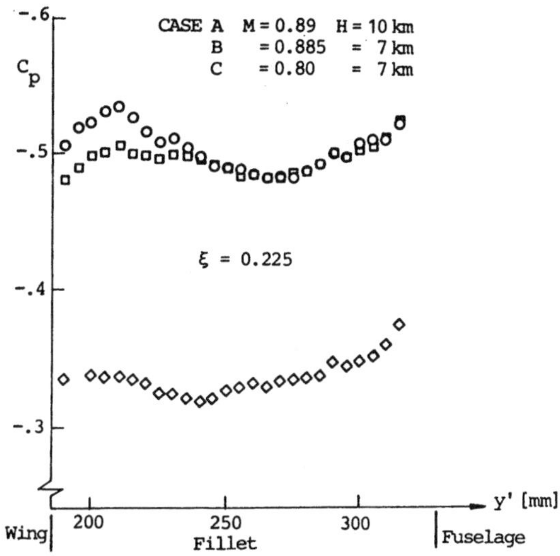

Fig. 23 Spanwise C_p distribution at 22.5% chord in wing/ body junction. y' is spanwise coordinate in mm. Three test cases.

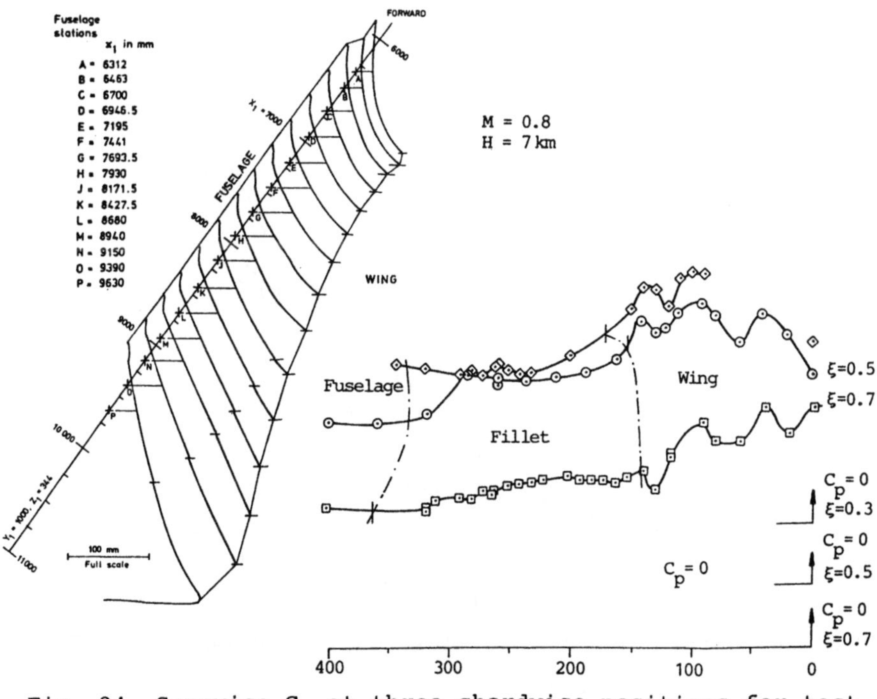

Fig. 24 Spanwise C_p at three chordwise positions for test case C. Sketch indicates shape of fillet; y' is spanwise coordinate in mm.

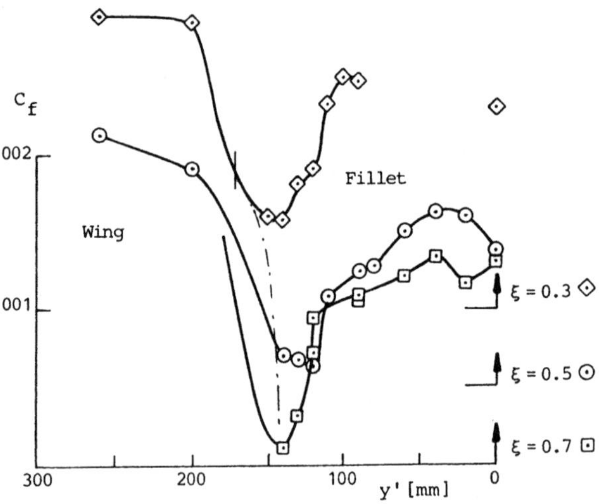

Fig. 25 Spanwise C_f at three chorwise positions.

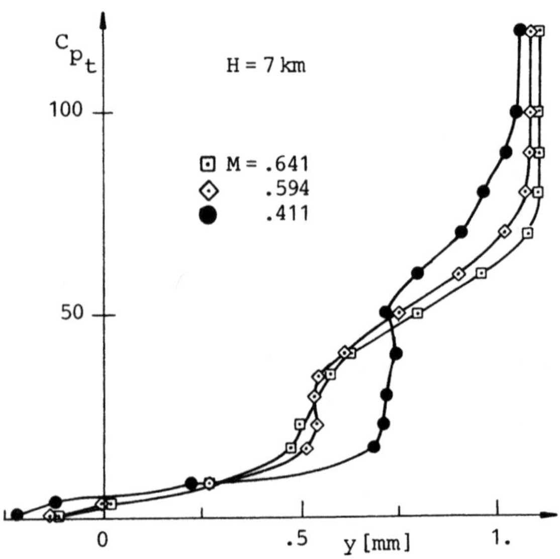

Fig. 26 Total pressure profiles, C_{p_t}, in wing/body junction.

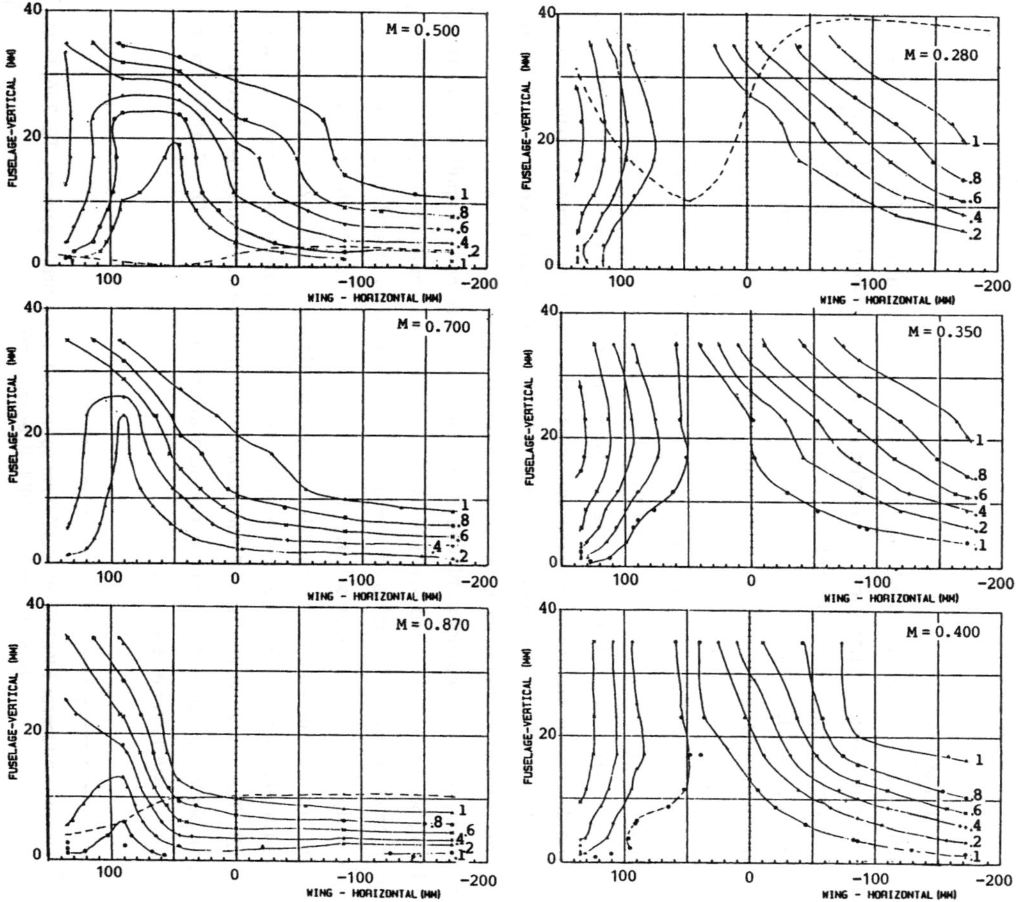

Fig. 27 Isobars, C_{p_t}, in wing/body junction for 6 Mach numbers,
Altitude H = 7 km.
y' is spanwise coordinate in mm.

Developments in Measuring Reynolds Stresses

U. R. MÜLLER

Aerodynamisches Institut der Rheinisch-Westfälischen Technischen Hochschule Aachen, D-5100 Aachen, Federal Republic of Germany *)

Summary

Single, dual and multi-sensor hot-wire anemometry for measuring instantaneous velocities as well as time-averaged statistics in incompressible turbulent flows is reviewed with emphasis on accuracy of calibration and data reduction schemes and on methods applicable to high-intensity and recirculating flows. Briefly some recent developments in Laser-Doppler anemometry like three-component measuring techniques are reported. Importance of turbulence measurements for guiding Reynolds stress closures is illustrated by measured balances of Reynolds stresses.

1. Introduction

Design of flow machinery and flight vehicles by computational fluid dynamics has received increasing attention by engineers and aerodynamicists, Kutler (1985). Supported by rapid development of large-scale and high-performance computers, computational fluid dynamics is an accepted design tool now, and beyond that has found manyfold applications in civil engineering. Advantages of cost-effective computations of complex flow fields at various Reynolds and Mach numbers are elucidated in Kutler's review as opposed to disadvantages of experimentation, yielding costly and sometimes wind-tunnel contaminated limited amounts of data. On the other hand, advance in the development of prediction methods for high-Reynolds-number viscous flows still has to be paralleled by progress in experimental research, with the link being turbulence modeling, which was listed by Kutler as one of the primary pacing items. Since direct numerical simulations of turbulent flows based on the time-dependent Navier-Stokes equations are not yet applicable to flows of engineering interest, solution of Reynolds-averaged equations together with an appropriate closure assumption represents the highest level of sophistication reached presently. Turbulence models currently being used do not apply universally and are often "tuned" for computing specific classes of flow problems, with the empirism advantageously being obtained from fundamental building-block experiments, Marvin (1982). The state of the art of turbulence modeling will certainly be discussed in other papers presented at this symposium intending to honour Dr. Rotta's contributions to turbulence research.

---

*) Present address: MBB GmbH UT-Transport-und Verkehrsflugzeuge, TE 234, Postfach 107845, D-2800 Bremen, FRG

The mutual promotion of experiment and theory has been acknowledged for a long time, Rotta (1962, 1972), Hinze (1975). The desire to verify theoretical concepts like Kolmogorov's hypotheses or to enhance understanding of turbulent mass and heat transfer has stimulated much experimental work in the past. Among others, appropriate techniques for measuring not only mean velocities but fluctuating ones as well have been developed , resulting in present methods for measuring the full Reynolds stress tensor. In the present paper, the instrument most widely being used for turbulence measurements, namely the hot-wire anemometer, will receive particular attention. Various techniques for measuring one or more velocity components instantaneously as well as those for measuring one-point mean statistics are reviewed with emphasis on incompressible isothermal flow. Trends of developing calibration and measuring techniques are discussed, including techniques proposed for complex flows with high turbulence intensity or recirculation. Comments on Laser-Doppler anemometry cover some recent developments like multi-component measurements. In a final chapter, experimental investigations into the balances of the Reynolds stresses are discussed on the basis of the transport equations.

2. Hot-Wire Calibration and Measuring Techniques

Fundamentals of hot-wire anemometry need not be repeated in the present context, readers are referred to the excellent reviews on the subject by Corrsin (1963), Kovasznay (1959), Bradshaw (1975), Comte-Bellot (1976), Bruun (1979), Vagt (1979) and Perry (1982). These experimenters all appreciate the wide-spread range of applicability of hot-wire anemometry, but as well they clearly point out the numerous problem areas associated with the use of a delicate instrument like a hot-wire. Knowledge of these papers is presumed, only some of the difficulties arising in hot-wire anemometry will be discussed inhere. Since actual developments of fluid machinery and flight vehicles require the understanding of more and more complex flows, experimenters are often pushed to use a particular measuring tool and its signal analysis to the very limits and sometimes even beyond. Several though interesting investigations published in the literature cannot be backed up by this writer and are therefore deliberately not discussed in the present context. Emphasis will be focused on the various techniques applicable for one-, two- or three-component measurements of turbulence. Methods for measuring instantaneous velocities as well as mean statistics will be covered. Depending on the quantity to be measured and on the flow to be investigated, a judicious choice of the instrument to be picked and the signal analysis technique to be applied will always be necessary. A summary of successfully used techniques will be given, and an attempt is made to appreciate more recent developments of hot-wire anemometry, in particular those being applicable to moderate or high-intensity or even recirculating flows.

Admittedly, after more than a decade of hot-wire measurements (which period is short compared to those of the experimenters quoted above!), it is still surprising to this writer to find new ingenious techniques to emerge and some to establish, e.g. the "pulsed-wire" and the "flying hot-wire" techniques, which will be discussed later. Also new calibration techniques arose, which improved accuracy or by using digital computers increased speed of turbulence measurements. Due to the inherent necessity of calibrating each hot wire to be used, this important area will be discussed first.

2.1 Calibration techniques for measuring instantaneous velocities

2.1.1. Velocity calibration relationships

In Tables I to III examples of calibration techniques are given, which were proposed in the literature for subsequently measuring one-, two- or three-velocity components instantaneously. The velocity calibration relationship has to be established first, Table I. With U_{N1} and U_T being the velocity components normal and tangential to the wire in the plane of the prongs and U_{N2} being binormal, Fig. 1, the cooling law is generally calibrated with the velocity vector being normal to the hot-wire axis, i.e. with $\alpha = \beta = 0$ the magnitude of the velocity vector $Q = |\vec{U}|$ is equal to U_{N1}. For a single normal wire probe, the angle γ between sensor and probe axis is equal to 90^0. With E being the voltage, one of the cooling laws

$$E^2 = A + BU^N \tag{1}$$

$$E^2 = A + BU^{1/2} + CU \tag{2}$$

as formulated by King (1914) or Siddal and Davies (1972), respectively, needs to be calibrated throughout the anticipated measuring range. Though the coefficients in Eq. (1) or (2) might vary with velocity U, Bruun (1971), a fixed set of coefficients is generally found to be sufficient for incompressible flows with $2\ m/s \le U \le 40\ m/s$. Equation (2) was least squares fitted for optimum values of A,B,N,C by Bruun and Tropea (1985), while Oster and Wygnanski (1982), Elsenaar and Boelsma (1974) or Thompson and Whitelaw (1984) described the velocity U by polynomials in E. The often tedious use of analogue linearizers yielding $E_{Lin} = S \cdot U$ is presently being replaced by using digital computers to evaluate the cooling law. With the use of Eq. (1) or (2), a one-component measurement of the instantaneous velocity U(t) is feasible by neglecting the V and W components.

2.1.2. Calibration of dual sensor probes

For measuring two velocity components instantaneously, a cross-wire probe having two slanted sensors with angles $\gamma_i = 45^0$ nominally is most widely being used.

Corresponding to the applicability of this probe for measuring Reynolds stresses $\overline{u_i u_j}$, many calibration and data reduction schemes have been developed, which take into account the yaw sensitivity of a hot-wire sensor, Table II. Let us consider a slanted hot-wire cooled by a velocity Q with components U and V, Fig. 1, with W being zero. An ideal, infinitely long sensor feels the normal velocity component U_{N1} only, Schubauer and Klebanoff (1946), but actual hot wires show deviations from this "cosine law", which have a non-negligible effect on the accuracy of Reynolds stress measurements, Champagne, Sleicher and Wehrmann (1967), Müller (1982b). Several expressions for the definition of an effective cooling velocity have been proposed, Table II.1.b, but most experimenters use the formulation

$$U_c = (U_{N1}^2 + K U_T^2)^{1/2} = Q(\cos^2\alpha + K\sin^2\alpha)^{1/2} \qquad (3)$$

because it conveniently expresses the cooling velocity in terms of orthogonal components U_{N1} und U_T. Since the work of Champagne et al., numerous calibrations of the yaw factor K have been carried out, more recent ones are those of Bruun and Tropea (1985), Samet and Einav (1985) or Müller (1982b). Axial cooling was found to depend, among others, on aspect ratio ℓ/d, wire Reynolds number and also type of probe. Normal or slanted, plated or unplated sensors of hot films or hot wires all yield different directional cooling sensitivities. So do small production faults, and finally prong and stem disturbances substantially contribute to the yaw factors. Compared to Champagne et al.'s calibrations yielding $K \approx 0.04$ for $|\alpha| \leq 60^0$, most other results show increasing K values up to 0.5 for $\alpha \to 0^0$, and some results might even be negative, see Bruun and Tropea (1985), Samet and Einav (1985) or Müller (1985). An example is given in Fig. 2, which shows the results for two sensors of an X-probe, taken from the latter work. Since calibration of K is subject to experimental inaccuracies itself, Müller (1982b), and might give relative errors of about 10 percent for a 1 percent error in U obtained from Eq. (1) or (2), the following more accurate procedure was proposed by Müller (1985), which does not require knowledge of an analytic velocity relationship: Given a velocity vector of magnitude Q_1 at a flow angle α, the cooling velocity U_{c1} or a corresponding voltage is measured. Then at $\alpha = 0$, the velocity Q_2 is searched for, which yields $U_{c2} \equiv U_{c1}$. The yaw factor is calculated from

$$K = [(Q_2/Q_1)^2 - \cos^2\alpha] / \sin^2\alpha \quad . \qquad (4)$$

The results, Fig. 2, obtained for $|\varepsilon| = |90^0 - \gamma - \alpha| < 25^0$ and $10\ m/s \leq Q \leq 30\ m/s$ were found to be independent of Q for the measuring range tested. This simplifying result is attributed to the use of gold-plated wire sensors in contrast to the use of unplated ones, Müller (1982b), showing velocity dependence of K as well.

Accuracy of turbulence measurements requires to account for each hot wire's yaw characteristics. Following the proposal of Müller and Krause (1979), the two unknowns Q and ε, the magnitude and direction of the velocity vector, are computed most accurately from the measurements by iteratively updating the yaw factors in the coefficient matrix. For Reynolds stress measurements, Andreopoulos (1981) and this author arrived at significant improvements of accuracy compared to simpler data reduction schemes.

For avoiding the explicit description of the yaw calibration by an effective cooling velocity, in particular for varying yaw factors, calibration in form of a "look-up table" is advantageous, since it includes the directional sensitivities. Willmarth and Bogar (1977) were among the first to develop a computer-aided calibration and measuring technique. Prescribing the velocity components U and V, which is equivalent to Q and ε, voltage pairs for the two sensors were measured. Data reduction of measurements was performed by linear interpolation between the calibration points. The variable angle method of Johnson and Eckelmann (1984) improved the aforementioned one by incorporating a second-order two-dimensional Taylor series expansion for interpolating between calibration points. Within the range $|\varepsilon| < 30^{0}$ and 0.02 m/s $\leq Q \leq 0.2$ m/s, Fig. 3, the required voltage pairs (E_1, E_2) as well as first and second partial derivatives of the voltages with respect to Q and ε were collected. With voltages E_{1j} and E_{2k} being closest to an actual measurement, an equation of fourth degree for the velocity difference $(Q-Q_j)$ was obtained from the expressions

$$E_i(Q,\varepsilon) = E_{i,jk} + (Q-Q_j)\left.\frac{\partial E_i}{\partial Q}\right|_{jk} + (\varepsilon-\varepsilon_k)\left.\frac{\partial E_i}{\partial \varepsilon}\right|_{jk} + \frac{1}{2}\left[(Q-Q_j)^2\left.\frac{\partial^2 E_i}{\partial Q^2}\right|_{jk} + 2(Q-Q_j)(\varepsilon-\varepsilon_k)\left.\frac{\partial^2 E_i}{\partial Q\partial \varepsilon}\right|_{jk} + (\varepsilon-\varepsilon_k)^2\left.\frac{\partial^2 E_i}{\partial \varepsilon^2}\right|_{jk}\right] \quad , \quad i=1,2 \qquad (5)$$

by eliminating the angular difference $(\varepsilon-\varepsilon_k)$. For solving the equation for $(Q-Q_j)$ with a Newton technique, an approximate solution was computed from Eq. (5) by neglecting all second derivatives. In fact, as stated by the authors, the approximate solution yielded results, the accuracy of which was already sufficient for many applications.

Oster and Wygnanski (1982) have also used a variable angle calibration method and curve fitted the results to obtain third-order surfaces for ε and Q in terms of the voltages E_1 and E_2. Since generally the polynomial coefficients varied for different ranges of velocity and angle, using various sets of coefficients improved the accuracy.

A major drawback of establishing the look-up tables or the curve fit mentioned seems to be the rather complex and tedious calibration procedure to be carried out before each measurement. Calibration drift is expected to yield serious problems. Therefore one might seek for separate calibrations of the yaw characteristics, which are constant sensor features and have to be evaluated only once, and of the velocity relationship Eq. (1) or (2), which always has to be established prior to a measurement. The following calibration procedure might be more convenient and less time consuming. Since the yaw factors, Fig. 2, of the gold-plated sensors of an X-wire probe were yaw dependent only, so was the ratio of the effective cooling velocities, regardless of the magnitude Q of the velocity vector. Corresponding to the multi-sensor procedure of Pailhas and Cousteix (1986), we have recast the same set of calibrations shown in Fig. 2 into the functions

$$F(\varepsilon) = (U_{c1}/U_{c2} - 1) / (U_{c1}/U_{c2} + 1)$$
$$H(\varepsilon) = (U_{c1}^2 + U_{c2}^2)^{1/2} / Q \qquad (6)$$

shown in Fig. 4. For $K_i = 0$ and $\gamma_i = 45^o$, F is equal to tan ε and H is equal to unity. Actually H varied slightly between $1.016 \le H \le 1.02$. Replacing analytic expressions for the cooling velocities U_{c1} and U_{c2} by the functions H = 1.018 and F(ε) as given in Fig. 4, this non-iterative time-dependent two-component method was applied to the X-wire measurements of Müller (1985). Compared to the previous iterative scheme, data reduction of identical sets of measurements was about seven times faster without loss in accuracy!

2.1.3. Calibration of multi-sensor probes

Calibration procedures for measuring three instantaneous velocity components, summarized in Table III, are straightforward extensions of those for two components. Jörgensen's (1971) definition of an effective cooling velocity

$$U_c = (U_{N1}^2 + K\, U_T^2 + h\, U_{N2}^2)^{1/2} \qquad (7)$$

can most conveniently be used for matrix inversion when using mutually orthogonal triple hot-wire sensors, Moffatt, Yavuzkurt and Crawford (1978), Andreopoulos (1983a), Andreopoulos and Rodi (1984), Müller (1983, 1985).

Examples for the use of look-up tables were proposed by Mathioudakis and Breugelmans (1985) and Pailhas and Cousteix (1986). In the former paper pitch, yaw and velocity-magnitude parameters were derived from calibration measurements within two quadrants, $|\psi| \leq 70^{o}$. In the data reduction, approximate hot-wire response equations were solved with $K_i = 0$ and estimated values of h_i being prescribed. Rectification was removed by two successive measurements with different probe orientations, thereby heavily relying on periodicity of the instantaneous velocity pattern in the flow examined.

Pailhas and Cousteix calibrated a four-sensor Kovasznay-type probe by means of the two angular functions $F(\psi)$ and $G(\varepsilon, \psi)$ and the velocity parameter $H(\varepsilon, \psi)$ being defined in Table III.2. The calibrations displayed in Fig. 5 preclude an explicit description of cooling velocities, and data reduction advantageously is non-iterative.

Curve-fitting calibration techniques for three-component measurements are still rarely being applied. Examples are published by Skinner and Rae (1984) and Huffmann (1980), who used least squares fittings of the velocity relationship and of the directional cooling law, Eq. (7), for each sensor. In Butler and Wagner's (1983) technique, the magnitude of the velocity vector as well as its angles with respect to the sensors were described by third-order surfaces in terms of the voltages E_i.

Presently this writer investigates an extension of the X-wire method of Table II.3. Calibrating the functions F(W/U), G(V/U) and H, see Table III.3, which for an ideal sensor with $K_i = 0$ and $h_i = 1$ would yield $F = \tan^{-1}(W/U)$, $G = \tan^{-1}(V/U)$ and $H = 1$, is expected to yield self-similar calibration curves. Slight dependences of F on V/U and G on W/U should be corrected for by functions $F = F_1(V/U, W/U)\ \tan^{-1}(W/U)$ and $G = G_1(V/U, W/U)\ \tan^{-1}(V/U)$, respectively. Compared to previous data reduction schemes, this non-iterative one is expected to substantially reduce the required computer time.

With calibrations of single, dual or triple sensor probes as described above, the measurement of one, two or three components of the instantaneous velocity vector is feasible. Since turbulence always is three-dimensional, only the latter measurement can be exact, while the others need to be simplified. From the time series of data, computation of mean velocities, Reynolds stresses and higher-order products of fluctuations is straightforward.

2.2 Measurement of mean velocities $\bar{U}_i$, Reynolds stresses $\overline{u_i u_j}$ and triple correlations $\overline{u_i u_j u_k}$

2.2.1. Low turbulence intensity, Tu << 1

For a slanted hot wire lying in the x-y-plane as displayed in Fig. 1, an instantaneous velocity vector $\vec{U} = (\bar{U}+u, v, w)$ generates the effective cooling velocity according to Eq. (7)

$$U_c = (U_{N1}^2 + K\,U_T^2 + h\,U_{N2}^2)^{1/2}$$
$$= \{[(\bar{U}+u)\sin\gamma - v\cos\gamma]^2 + K[(\bar{U}+u)\cos\gamma + v\sin\gamma]^2 + h\,w^2\}^{1/2} \quad , \qquad (8)$$

which expression may be abbreviated as

$$U_c = \{F^2(\bar{U}) + f(u_i) + g(u_i u_j)\}^{1/2} \qquad (9)$$

the terms F^2 and g contain double velocity correlations of mean and fluctuating velocities, f includes the linear fluctuation terms. Separating Eq. (9) into mean and fluctuating parts requires the binomial expansion

$$U_c = F + \frac{f+g}{2F} - \frac{f^2}{8F^3} + \text{h.o.t.} \qquad (10)$$

For turbulence levels $Tu = \sqrt{\overline{u^2}}/\bar{U} = u'/\bar{U} << 1$, the linearized hot-wire response equations

$$\bar{U}_c = F + \text{second-order terms}$$
$$u_c = U_c - \bar{U}_c = \frac{f}{2F} + \text{second-order terms} \qquad (11)$$
$$\overline{u_c^n} = \overline{\left(\frac{f}{2F}\right)^n} + \text{h.o.t.} \quad , \quad n \geq 2$$

are obtained, which relate the mean and fluctuating hot-wire signals to the mean and fluctuating velocities. Two comments seem to be appropriate concerning these equations, which form the basis for the majority of turbulence measurements published in the literature. Firstly, methods proposed to square and time average Eq. (9) yielding the exact relationship $\overline{U_c^2} = F^2 + \bar{g}$ cannot be used in practice, since even for turbulence levels up to 20 percent, say, the ratio $\bar{g}/F^2$ is about 0.04 and consequently the coefficient matrix needed for computing the Reynolds stress tensor is ill-conditioned, Tsiolakis, Krause and Müller (1983). Secondly, the sensitivity of voltage fluctuations (or u_c) with respect to velocity fluctuations has been analysed by several investigators and led to the discussion about the adequacy of the static calibrations described above as opposed to dynamic calibrations, see Perry and

Morrison (1971b), Morrison, Perry and Samuel (1972), Bruun (1976). Perry and co-workers developed a dynamic calibrator and obtained the sensitivity $\partial E/\partial U$ by shaking a probe in a known stream. Bruun traced back apparent discrepancies between static and dynamic calibration to the inappropriateness of Eq. (1) to cover a wide range of velocities. Presently the discussion seems to be settled. Provided an accurate velocity relationship, static calibration is accepted by the majority of experimenters. For a recent discussion on the subject, see Kühn and Dreßler (1985).

The approximate signal analysis according to Eq. (11) is valid for low local turbulence levels only, but applications up to $Tu \approx 0.2$ are in general admissible. In the linearized signal interpretation the influence of the lateral velocity component w has to be neglected at all. Turbulence measurements on the basis of Eq. (11) have been carried out with single slanted and cross-wire probes, as summarized in Table IV.1. The single-sensor method of Hoffmeister (1972) requires mean and rms measurements of $\bar{U}_c(\psi)$ and $\overline{u_c^2}(\psi)$ at least at six different roll angles ψ about the probe axis for evaluating the mean velocities $\bar{U}_i$ and the Reynolds stresses $\overline{u_i u_j}$. Attention has to be paid to the conditioning of the coefficient matrix of the response equations. The single-wire technique reduces aerodynamic interferences between flow and probe compared to an X-wire probe, and is therefore preferred for measurements in sensitive flows like the asymmetric corner flow, Kornilov and Kharitonov (1984). Use of a constantly rotating hot-wire probe was proposed by Fujita and Kovasznay (1968). This method also requires measurements of cooling velocities at various roll angles and applies least squares fitted data to the response equations. It seems that this method cannot easily be used to yield accurate measurements, therefore applications are rare.

The bulk of turbulence measurements in the past has been carried out with X-hot-wire probes being rotatable around the probe axis, which was placed in x-y plane. For ideal hot-wire response with $K_i = 0$ and $\gamma_i = 45^o$, the governing equations to evaluate particular turbulence quantities are given in Table IV.1.b. Two velocity components and their double and triple products were measured, for example, by Chandrsuda and Bradshaw (1981). All components of the Reynolds stress tensor were measured in three-dimensional turbulent boundary layers with the probe axis being aligned with the local mean flow direction by Elsenaar and Boelsma (1974), Fernholz and Vagt (1981) or Müller (1982a, 1982b). For minimizing disturbances induced by the probe, Fernholz and Vagt used specially designed low-interference probes with built-in fixed orientations of the X-wire plane. Using analogue electronic equipment for measuring all double and triple velocity correlations, numerous measurements of the squares and cubes of the fluctuating cooling velocities have to be carried out, Müller (1982b). When applying digital techniques, all required data are obtained from measurements in four hot-wire positions with $\Delta\psi = 45^o$. Even in nominally two-dimensional mean flow, it is recommended to measure all turbulence quantities obtainable for checking symmetry conditions.

It should be noted that accuracy of measured Reynolds shear stresses, in particular of $\overline{vw}$, can only be obtained under carefully controlled experimental conditions, even then uncertainties might be of the order of 15 percent. On the other hand, in most cases such accuracy is sufficient for comparison with prediction methods, which presently cannot do better for complex flows.

2.2.2 Moderate turbulence levels, Tu < 0.4

For increasing turbulence levels up to 0.4, say, a serious decrease of accuracy of turbulence measurements is expected when using the linearized signal interpretation, Eq. (11). A thorough and detailed estimate of the inherent errors due to rectification of the hot-wire signal, occuring for instantaneous flow reversal or, e.g., for angular fluctuations with $|\varepsilon| > 45^o$ when using an X-wire probe, and due to the neglect of the lateral velocity component w was computed by Tutu and Chevray (1975). These authors prescribed quasi-Gaussian probability density functions (pdf) P(u,v,w) with given central moments of the fluctuations and then calculated the moments, which would be measured by the X-hot-wire. The results indicated that errors for the shear stress $\overline{uv}$ might reach 28 percent for a turbulence intensity of 35 percent. Several authors have continued this approach of error estimate. From the comprehensive analysis of Castro and Cheun (1982) and Castro (1986) some interesting results arose. Assuming jointly normal distributed velocity fluctuations and a typical cross-correlation coefficient R = 0.3, the probability p of a velocity vector to lie outside of an elliptical acceptance cone with semi-angles $\varepsilon = \tan^{-1}(V/U)$ and $\varphi = \tan^{-1}(W/U)$, Fig. 6, was computed. For $\varepsilon = 70^o$ and $\varphi = 20^o$, representing the optimistically chosen upper limits for a single normal wire placed along the y-axis, $p(\sigma)$ with $\sigma = u'/\overline{U} = v'/\overline{U} = w'/\overline{U}$ rose rapidly for $\sigma > 0.10$ and yielded p = 0.25 for $\sigma = 0.3$. For $\varepsilon = \varphi = 20^o$, representing the bounds of the acceptance cone for an X-wire lying in the x-y-plane, p was about 0.3 for $\sigma = 0.25$ and 0.7 for $\sigma = 0.4$. Due to the long tails of the distribution functions, there even was a substantial probability of instantaneous reverse flow. Quantifying the results of Castro or Tutu and Chevray, correction factors for a given turbulence level can be given. Certainly such results give valueable hints concerning possible errors in hot-wire measurements, and where applicable, the corrections considerably improved the accuracy as found by Dengel, Fernholz and Vagt (1981) and Jaroch (1985) from comparisons with the pulsed-wire technique. However, as already acknowledged by Castro and Cheun or Jaroch, corrections might even worsen the accuracy of turbulence measurements in cases where the probability density function is not distributed normally. For a long time, the present writer was very much concerned about the applicability of any pdf-based correction method. Firstly, our own measurements generally do not show the long tails of a Gaussian distribution, therefore the probalibility of a velocity vector to lie outside of an acceptance cone is reduced drastically. Examples, obtained in the turbulent shear flow sketched in Fig. 7, are shown in Fig. 8. The pdf's of the angles ε and φ were measured by triple hot-wire anemometry. At $x/D = 70$, with turbulence levels below 0.15, and at $x/D = 20$, where turbulence levels reach 0.25 and the upper limit for standard X-wire techniques is approached, acceptance cones with $\varepsilon = \varphi = 25^0$ or $\varepsilon = \varphi = 40^0$, respectively, cover 100 percent of the measurements; neither excursions nor rectifications were observed. Secondly, the joint pdf of the u and v fluctuations often shows substantial departures from normality, Wallace and Brodkey (1977), Johnson and Eckelmann (1984) or Durst, Jovanovic and Kanevce (1985). In separated flows, Owen (1983) even obtained bimodal distribution functions. Skewed pdf's, measured by Müller (1986b), Fig. 9, show negative skewness of the v-fluctuation and positive one for u at $x/D = 20$ and $y = 8$ mm, while in the outer layer the signs of the skewness factors were opposite. Obviously a single correction would not apply across the boundary layer.

On the same reason, improvement of the linearized standard signal analysis, Eq. (11), by inclusion of third-order velocity correlations for measuring Reynolds stresses

$$\begin{aligned} \bar{U}_c &= F + \frac{\bar{g}}{2F} - \frac{\overline{f^2}}{8F^3} + \text{h.o.t.} \\ u_c &= \frac{f + g - \bar{g}}{2F} - \frac{f^2 - \overline{f^2}}{8F^3} + \text{h.o.t.} \\ \overline{u_c^2} &= \overline{\left(\frac{f}{2F}\right)^2} + \frac{\overline{2fg}}{4F^2} - \frac{\overline{f^3}}{8F^4} \end{aligned} \tag{12}$$

in general cannot be based on corrections by triple products based on assumed quasi-Gaussian pdf's, Heskestad (1965), Vagt (1979). Instead, Müller (1982b) proposed to measure all third-order terms simultaneously with the second-order ones, and even if the accuracy of the triple products decreases with increasing turbulence level, the signs and orders of magnitude of the corrections take into account the actual structure of the turbulence field. Using a digital measuring and processing technique, standard and improved techniques can easily be compared with each other, thereby checking the accuracy of Reynolds stress measurements.

An extension of the scope of standard X-wire anemometry was proposed by Kawall, Shokr and Keffer (1983) and Legg, Coppin and Raupach (1984). With a new hot-wire probe having three in-plane wires, Table IV.2.b., the lateral velocity intensity w' is taken into account and need not be neglected. The sign of w(t) cannot be resolved with this probe. Based on a Gaussian pdf of the fluctuations, Kawall et al. expected a high degree of accuracy for Reynolds stress measurements up to turbulence levels of 0.4. According to Legg et al., however, the intensity w' is difficult to measure and might even come out negative. In case w' would be known (or prescribed), an error analysis using skewed non-Gaussian pdf's showed, that for turbulence intensities below 0.5 the relative errors of the measured intensities u' and v' are below 10 percent. The same accuracy is expected to be obtained for the shear stress $-\overline{uv}$ for $Tu < 0.31$.

A most promising development for turbulence measurements consists in advanced applications of triple sensor hot-wire probes. Following the early work of Spencer (1970), who used a custombuilt probe, and Lakshminarayana and Poncet (1974), who placed an X-wire and a single-sensor probe closely together, three-component measurements were carried out e.g. by Fabris (1978), Moffatt, Yavuzkurt and Crawford (1978), Andreopoulos and Rodi (1984) or Müller (1985). Using a processor-aided measuring technique, instantaneous velocity vectors can be measured. Mean velocities, Reynolds stresses and higher-order correlations of the fluctuations are conveniently obtained from one set of measurements. Optionally, time-domain analysis is applicable as well. The completeness of one-point statistics obtained by far outweighs the calibration effort.

From the three hot-wire signals measured, the three unknown instantaneous velocity components are computable from the exact hot-wire response equations. Moffatt, Yavuzkurt and Crawford (1978) used an analogue computer, the other investigators mentioned digitally inverted three equations according to the cooling velocity defined in Eq. (7). Andreopoulos (1983a) and Andreopoulos and Rodi (1984) iteratively took into account the calibrated yaw and pitch factors, Müller (1985) did the same for the yaw sensitivity. For mutually orthogonal sensors as used by the latter experimenters, accurate measurements are expected for turbulence levels up to 0.3 according to the statistical error analysis of Andreopoulos (1983b).

One should keep in mind that only for orthogonal hot-wire arrangements signal interpretation by matrix inversion is unique throughout the acceptance range, i.e. the octant spaned by the three sensors. Independently Chang, Adrian and Jones (1982) and Müller (1983) showed that non-orthogonal wire probes, e.g. Fabris (1978), have a rather limited uniqueness range giving single-valued results. Outside of this domain, up to four real solutions can be found from one triplet of measurements. Occurance of multiple solutions may easily be explained geometrically from the intersection of cylinders with radii U_{ci} aligned with the sensors, Willmarth (1985).

2.2.3. High turbulence levels, Tu > 0.4

In flow with high turbulence levels, instantaneous flow directions are likely to lie outside of the acceptance range of a particular hot-wire probe and might even be reversed. Rectification of the signal does not allow for unique signal interpretation and velocity vectors lying outside the acceptance domain are misinterpreted as being inside. An example of observed rectification is given by Castro (1986) for the measured pdf of the angle of incidence ε of a velocity vector in the x-y-plane; a similar, even more pronounced example measured with a triple-sensor probe in the flow field of Fig. 7 at x/D = 10 is shown in Fig. 10. Velocity vectors with $\varepsilon > 45^0$ and $\varepsilon < -45^0$ are rectified and therefore yield additional peaks near the bounds of the acceptance range instead of monotonic decreasing tails of the pdf. For removing the sign ambiguity of the instantaneous velocity, two measuring principles have been proposed in the literature for applying hot-wire anemometry in high-intensity flow, Table IV.3: In the "flying hot-wire" technique a standard probe is moved through the flow field that fast that relative to the probe no reverse flow occurs; the other technique detects the direction of a thermal wake ("thermal tuft") of a heated sensor, the most advanced technique of this kind being the "pulsed hot-wire" anemometer. Both techniques will be discussed next.

Flying hot-wire anemometry was successfully applied by Coles and Wadcock (1979) in a separated trailing-edge flow of an airfoil. Cantwell and Coles (1983) reported cross-wire measurements of phased-averaged mean velocities and Reynolds stresses in the near-wake of a circular cylinder. In these experiments the probe was mounted on a whirling arm, and electrical signals were passed through low-noise mercury slip rings. The novelty of eliminating the slip rings was introduced by Walker and Maxey (1985). An anemometer bridge was placed inside the hub of the whirling system, and the signals were optically transmitted by a light emitting diode to a fixed receiver. A linearly traversed sled for catapulting the probe through the flow was developed by Watmuff, Perry and Chong (1983). Phase-averaged mean velocities, an example is given in Fig. 11, and Reynolds stresses were measured downstream of a stationary and an oscillating ellipsoid. Thompson and Whitelaw (1984) studied the trailing-edge separation of airfoil-like flow. Moving the probe (single slant or crosswire) along a closed circuit trajectory enabled the application of otherwise standard measuring techniques to obtain four components of the Reynolds stress tensor.

In general, aerodynamic disturbances of the flow field are expected to be acceptable or small for a probe moving upstream. The investigations mentioned proved the flying wire technique to be applicable to complex flows not accessible for standard hot-wire anemometry. The experimental arrangement, however, requires enormous efforts, in particular for the traversing mechanism. As well the data acquisition and processing system, being digital necessarily, for simultaneously monitoring hot-wire signals and probe speed and location is rather complex. Many passages through the flow are required to obtain the mean turbulence statistics. Since external flows can be investigated exclusively and extension to three-dimensional flow is expected to be complicated, other instrumentation is needed for flows not accessible to a flying hot wire.

Thermal wake detecting probes have been employed for a long time, Kovasznay (1948). Surveys of applications intending to remove rectification of hot-wire signals are summarized by Ligrani, Gyles, Mathioudakis and Breugelmans (1983) and Müller (1983). Examples of three-element probes are those of Carr and McCroskey (1979) and Eaton et al. (1979), who used a heated wire with two parallel temperature detectors on either side. In their near-wall measurements, instantaneous flow direction was indicated by one of the detector sensors, the magnitude of the velocity was obtained from the center wire. For measurements away from solid surfaces, the pulsed-wire technique was developed by Bradbury and Castro (1971) and Bradbury (1976). A central wire being normal to and between two temperature-wake detectors, Fig. 12, is heated by a voltage pulse of about 5 μs duration and the time of flight of the wake to reach a detector is measured. The velocity component normal to the sensors is computed from $U = s/\Delta t$. Ideally U is equal to $U_0 \cos \Phi$, where U_0 is the instantaneous velocity vector and Φ its angle with respect to the sensor normals; s is half the distance between the sensor wires, Δt is the measured time of flight. From actual calibrations the relation $U = A/\Delta t + B/\Delta t^3$ was found to be more accurate, Castro (1986). Deviations from the ideal cosine law response orginally being accounted for by $U(\Phi) = U_0(\cos \Phi + K \sin \Phi)$ were largely eliminated in parallel developments by Castro (1986) and Jaroch (1985), who rotated the plane of the detector sensors by 45^0 around the stem axis, Fig. 13. By closely spacing the sensors, the limits of the acceptance cone given by $\Phi_{max} = \tan^{-1} l/2s$ were increased to about 80^0, thereby minimizing the zone of blindness. Yaw and pitch characteristics of a pulsed-wire probe were carefully analysed by Jaroch (1985) for developing probes with large symmetrical measuring cones in which the velocity is not disturbed by the wake of a prong.

Though generally the pulsed-wire anemometer cannot be used in flows with velocities larger than 15 m/s or with turbulence levels below 0.1, the experiments of Castro and Cheun (1982), Dengel, Fernholz and Vagt (1981), Jaroch (1985) and Ruderich and Fernholz (1986) demonstrated the applicability of this technique for measurements in separated flows behind bluff bodies or in the high-intensity flow regime of turbulent jets. Several theoretical and experimental tests gave confidence into the technique. Bradbury (1976) computed the rectified signal to be expected for a standard hot-wire measurement and found reasonable agreement compared to actual single-sensor measurements. Dengel et al. and Jaroch compared standard and pulsed-wire measurements, and in many cases they found the discrepancies to be reduced substantially when correcting the conventionally obtained data according to the estimates of Tutu and Chevray (1975), Fig. 14. In the recent work of Castro et al. and Fernholz et al. the former one component measurements were extended to turbulence measurements of u', v' and $\overline{uv}$. Similar to the single slanted wire technique, three measurements at probe orientations with 0^0 and $\pm 45^0$ relative to the y-axis were necessary; meanwhile Castro has been able to measure the intensity w' also. Though repeated measurements with the same probe may show considerable scatter, in particular for v' and $\overline{uv}$, Jaroch (1985), the state of the art of the pulsed-wire technique is best being described by Castro and Cheun's (1982) conclusion: "In flows with such high intensity that hot wires would be useless, measurements of all Reynolds stresses can be made with an accuracy probably better than 30 % (for v') or even 15% (for u' and $\overline{uv}$). In the medium intensity range (10-30%, say), provided the yaw response extends to large enough angles, pulsed-wire measurements can be as accurate as hot-wire measurements."

A new approach of using the thermal-tuft principle, aiming at three-component measurements in complex flows, has been proposed by Müller (1983). With a thermal-wake detector, wire no. 2 in Fig. 15, being added to a triple sensor probe, velocity vectors lying within the range $|\varepsilon|, |\varphi| \lessapprox 20^0$ produce a small ratio of the cooling velocities U_{c2}/U_{c1}. Thereby the approximate flow direction is identified and rectification is avoided. Fig. 16 shows the actual probe design. The gold-plated velocity sensors are mutually orthogonal and symmetrical to the stem, their active parts of 5 μm in diameter and 1 mm in length produce a measuring volume of about 1.5 mm in diameter. The detector sensor, being separated from the heated wire by about 90 μm, is operated by CTA with overheat ratio of about 1.3. In the digital conditional sampling technique used, explained in Fig. 17, the criterion $U_{c2}/U_{c1} < 0.4$ is used to detect instantaneously acceptable velocity vectors. The components are computed exactly from the signals of the three velocity sensors as outlined above. Tracing the time development of velocity vectors, they are accepted as long as they are within a prescribed domain with angular bounds S_i. Rotating the probe as ideally sketched in Fig. 15 enables to measure the conditional flow field statistics of further acceptance ranges. With ΔN being the number of samples accepted out of a total of N, for i measuring positions the sum $\Sigma(\Delta N/N)_i$ should ideally be unity; a flow quantity $\bar{A}$ to be measured is obtained from the weighted ensemble averages

$$\bar{A} = \Sigma_i (\Delta\bar{A} \cdot \Delta N / N)_i \tag{13}$$

Preliminary tests of the technique have been completed successfully, Müller (1986c). At x/D = 20 in the flow of Fig. 7, where standard X- and triple wire measurements are still applicable, four measuring positions with yaw and pitch angles of the probe of $\pm 15^0$ were used to measure velocities within the domains $V \lessgtr 0$, $W \lessgtr 0$ with U being positive. Good agreement for streamwise mean velocity, Reynolds stresses and triple correlations was achieved. At x/D = 10, where the vortices shed from the cylinder still produce high angular fluctuations of the velocity vector with $|\varepsilon|$, $|\varphi| < 55^0$, conventional anemometry is not applicable as demonstrated by Fig. 18, which shows the conditionally sampled results for the Reynolds stress $-\overline{uv}$ compared to rectified X-wire measurements. Following the preliminary qualification experiments, the technique described is planed to be tested in recirculating flow.

3. Remarks on Further Aspects of Hot-Wire Anemometry

3.1 Temperature variation

A large variety of applications of hot-wire anemometry beyond one-point ensemble-averaged Reynolds stress measurements, e.g. space-time correlations, multi-point structural analysis, or short-time averaging techniques, has to be skipped in this context for brevity. Some remarks, however, should be added to the previous discussion of hot-wire techniques. Calibrations as discussed above are valid for constant ambient temperature. Variations of temperature during measurements are among the most serious sources of error, Perry (1982), and are negligible only for slight changes below 0.5^0 C, otherwise the voltage E^2 according to the velocity relationship, Eq. (1), changes proportional to temperature, Bremhorst (1985). Temperature drift should be accounted for by either measuring velocity and temperature simultaneously, or by special anemometer circuitry, Chevray and Tutu (1972), Bremhorst (1985). For a large temperature range from -55 to 220^{0} C and velocities up to 25 m/s, Nitsche and Haberland (1984) proposed a new calibration technique, which evaluates the ratio of heat loss to the prongs and convected heat transfer versus wire Reynolds number.

Temperature has to be accounted for in supersonic flows, to which hot wires have frequently been applied until the late seventies. In compressible turbulent flow, the sensor responds to heat conduction due to mass flow and total temperature fluctuations, Kovasznay (1950), Morkovin (1956). Using a single wire, the variances of both quantities as well as their correlation can be determined by taking several readings at various operating temperatures of the constant current anemometer. Following Morkovin's hypothesis for negligible pressure fluctuations for non-hypersonic flows, results for velocity and density and their correlation can be evaluated, for further discussion on the modal analysis technique see Demitriades (1973). Applications of hot-wire anemometry to hypersonic flows were reported by Laderman and Demitriades (1974) and several papers following, and by Mikulla and Horstman (1975), who were able to measure total heat flux with a special three-sensor probe. Even more difficult are measurements in transonic flows, since the hot-wire calibration additionally is Mach number dependent, Horstman and Rose (1977). Recently Smits et al. (1983, 1984) discussed the problem of applying constant temperature anemometry to supersonic flow, but it seems fair to say that turbulence measurements in compressible flow have now nearly entirely been replaced by Laser-Doppler anemometry.

Corollaries from Morkovin's hypothesis have been discussed by Bradshaw (1977). Neglect of density fluctuations is valid for many compressible boundary layer flows, and the structure of turbulence when scaled properly virtually coincides with that of low-speed flows. Therefore most recent work of measuring velocity and temperature fields concentrated on incompressible flow using temperature as a passive contamination. With the temperature disturbance being a few degrees above ambient at most, hot-wire anemometry for measuring velocity and temperature simultaneously can conveniently be applied. The temperature sensor with a diameter of 1 μm or less has to be run as a "cold wire" with careful verification of proper frequency response, Perry and Morrison (1971a), Antonia, Browne and Chambers (1981). In numerous papers by Antonia and colleagues, e.g. Browne and Antonia

(1986), temperature and velocity variances, heat fluxes as well as spatial derivatives were investigated. Two velocity components plus temperature were also measured instantaneously by La Rue, Libby and Seshadri (1981), Dekeyser and Launder (1983), Gibson and Verriopoulos (1984) or Nagano and Hishida (1985). Fabris (1978, 1983) used a special four-sensor probe to obtain all three velocity components and the temperature instantaneously and evaluated all velocity-temperature correlations up to third order. Such experimental data provide a basis for developing turbulence models for heat and mass transport. For example, the Reynolds analogy implying equal diffusivities for momentum and heat transport as opposed to the use of a turbulent Prandtl number being different from unity, the turn-over time scales for fluctuating velocity or fluctuating scalar fields, Elghobashi and Launder (1983), or closure assumptions for the turbulent transport terms $\overline{u_i \theta}$ and $\overline{u_i u_j \theta}$, occurring in the transport equations for scalar heat flux $\overline{u_i \theta}$, Launder (1978), or the mean square temperature variance $\overline{\theta^2}$, respectively, need to be examined experimentally. Velocity and temperature fluctuations are denoted by u_i and θ. Closure of the transport equations is advantageously guided by comparison with measured balances of the quantities involved, Nagano and Hishida (1985).

3.2. Wall proximity

Special interest in near-wall measurements including buffer and viscous sublayers of turbulent boundary layers arises for understanding the time-dependent structural behaviour of near-wall turbulence and for developing low-turbulence-Reynolds-number models. Unique measurements were reported by Kreplin and Eckelmann (1979) for the near-wall region of oil flow in a channel with a viscous sublayer thickness of about 3 mm. In air flow, measurements within a few tenth of a millimeter off the wall are subject to serious errors arising from inaccurately measured distance from the wall, inaccurate calibration for the expected low-speed range, aerodynamic disturbance of the flow by the probe, and additional heat loss to the wall substrate. A comprehensive analysis of theoretical and experimental corrections to single-wire near-wall measurements was reported by Azad (1983), who summarized results from the literature in a diagram reprinted in Fig. 19. Deviations from the velocity profile $u^+ = y^+$, plotted in the form $\Delta u^+ = f(y^+)$, suggest that corrections are essential for wall distances $y^+ \leq 2$. Considerable experimental scatter is traced back by Azad to aerodynamic disturbances of the near-wall flow by the various hot-wire probes used. A corresponding conclusion was obtained by Hartmann (1982), who experimentally investigated the influence of particular probes and their orientations with respect to the wall on the measured mean and fluctuating velocities.

3.3. Small scale turbulence

Measurement of small scale turbulence with hot-wire anemometry necessarily requires small scale measuring volumes for sufficient spatial and temporal resolution. Development of miniature probes resulted in sensors with lengths down to 25 μm, the diameter being 0.5 μm, Willmarth and Sharma (1984). A major objective of recent small scale measurements is to obtain conditionally sampled Reynolds stresses for detecting coherent structures by means of short-time averaging, Blackwelder and Kaplan (1976), or quadrant analysis, Wallace, Eckelmann and Brodkey (1972). Other investigations are associated with verification tests of Kolmogorov's (1941) concept of local isotropy, which proved to be one of the most important contributions to the theory of turbulence. Measurements for assessing local isotropy were carried out, e.g., by Champagne (1978) and recently by Antonia, Anselmet and Chambers (1986). For accurate measurements of high wave number spectral components, hot-wire sensors with length below 2 or 3 times the Kolmogorov length $(\nu^3/\varepsilon)^{1/4}$ have to be used according to the error estimates of Wyngaard (1968); ν is the kinematic viscosity, ε is the dissipation of kinetic turbulence energy. Possible sources of errors arising in interpreting fine-scale measurements are tabulated by Antonia et al. (1986).

In the present contex of discussing turbulence measurements being relevant to turbulence modeling, the estimation of dissipation of kinetic energy is of major importance. Measurements of spatial derivatives of velocity fluctuations are rare and do not yet exist for all components of the dissipation tensor. Most estimates are based on Kolmogorov's second hypothesis, and isotropic dissipation is derived from the $k^{-5/3}$ law, k being the wave number, and additionally assuming Taylor's hypothesis to apply. Surveys and discussions of measurements can be found in Rotta (1962, 1972) and Hinze (1975). Recent detailed measurements were reported by Azad and Kassab (1985a), who thoroughly applied various techniques and compared the results with those derived from the turbulence energy budget, Azad and Kassab (1985b).

4. Comments on Laser-Doppler Anemometry

In the proceeding chapters, hot-wire techniques for measuring turbulence have been discussed extensively to appreciate decades of experimental research, which has substantially contributed to the understanding of turbulence. In this writer's opinion, hot-wire anemometry will remain in force as a major tool for turbulence research in the future. The only other established technique for local velocity measurements is the Laser-Doppler Anemometer (LDA), which has successfully been applied to flow not easily, if at all, accessible to hot wires, in particular to flows in liquids, high-speed and recirculating flows, or those in hostile enviroments like in combustion engines. The relevance of the advantageously non-intrusive LDA to turbulence measurements cannot be overestimated, and the length of the present few remarks is certainly inverse proportional to it!

LDA measurements have particularly contributed to the understanding of separated flows. In his excellent review on two-dimensional turbulent separated flow, Simpson (1985) discussed the experimentally-gained insight into the structure of detached steady and unsteady cases. For example, transonic separated trailing-edge flow was investigated by Seegmiller, Marvin and Levy (1978); Fig. 20 shows their results for mean velocity, Reynolds shear stress and kinetic energy of u and v-velocities in the regime beyond detachment. Following that work, series of experiments in transonic separated flows were conducted at NASA Ames Research Center, a recent one being the work of Bachalo and Johnson (1986). A comprehensive study of mean-flow and turbulence quantities measured in incompressible pressure-induced separated flow was reported by Simpson, Chew and Shivaprasad (1981a, b). Durst and Schmitt (1985), Driver and Seegmiller (1985) and Adams and Eaton (1985) applied the LDA to the backstep flow. Examples for measurements in reciprocating engine chambers are the investigations by Vafidis and Whitelaw (1984) and Durst and Krebs (1986), who developed miniature optical heads connected with the system by fiber optics.

In the literature, a widespread variety of LDA systems has been reported. The fundamentals of the various components, i.e. laser, transmitting and receiving optics, particle size and concentration, photomultiplier, band-pass filter, signal processing and data analysis, are described in detail by Durst, Melling and Whitelaw (1981). Inherent uncertainties of optics and electronics as well as problems associated with finite-transit time or statistical biasing, Buchhave, George and Lumley (1979), Lehmann (1982), Castro (1986), can be reduced to tolerable amounts in most cases. In general, specific measuring applications require specific optimized systems. Therefore catagorizing LDAs applied in the literature seems difficult, in particular, since most of the components are still in a stage of rapid development and refinement. However, many LDAs employed presently are based on dual-beam fringe-type interferometry, apply compact optics and high-powered lasers for backscattering and Bragg-cell driven frequency shift for removing sign ambiguity of the velocity. Beat frequency of Doppler bursts is advantageously measured and validated by means of counters, which however require large signal-to-noise ratios. For measuring two instantaneous velocity components in the plane perpendicular to the optical axis, a second set of optical components and processing electronics is required; orthogonal polarizations of the beams or two different wave lengths (colours) may be used, with the latter method being preferable for backscatter systems. Simpler and less expensive, basically one-component equipment was utilized for two-component measurements by Adler, Menn and Kalekin (1986) and Bahnen and Koeller (1984). The former authors applied a new optical method with a marked non-parallel fringe system, the latter applied an electro-optical modulator (Pockels cell) for switching polarization of the incident laser beam during transit of a particle through the measuring volume. Thereby passage of the beam through one of the two orthogonal beam splitters was switched to the other, Fig. 21, and two velocity components were measured from one Doppler burst.

The scope of the LDA is presently being extended to long-distance measurements in large-scale industrial and research wind-tunnels, Boutier and Canu (1982), Hunter, Honaker and Gartrell (1983), Bütefisch and Sauerland (1985), Durst, Ernst and Völklein (1985), or even in the earth's lower atmosphere, Durst and Richter (1983a, b). Durst et al. (1985) used special zoom optics, which kept the beam angle constant for focal lengths up to 4.5 m. Natural aerosol particles were found to be sufficient when applying the photon correlation technique described by Durst and Richter. In cases where photomultipliers still produce analogue signals, but the signal-to-noise ratio is too low for accurate application of counters, acquisition of Doppler bursts with a fast transient recorder and subsequent frequency analysis by software Fast Fourier Transform is preferable, Lehmann, Helbig and Simon (1984), Pallek (1985). With Lehmann's technique, presently up to 1000 Doppler bursts can be processed per second.

Similar to developments in hot-wire anemometry, intensive efforts are devoted to the development of three-component LDA systems, Orloff and Snyder (1982), Boutier et al. (1984, 1985), Bütefisch and Sauerland (1985), Driver and Hebbar (1985). Possible arrangements using separation of channels by polarization, colour or frequency shift or combinations thereof are discussed by Boutier et al. (1984). Three orthogonal channels are most advantageously used to yield good signal quality and accuracy, but require corresponding access to the flow by windows. With two transmitting optics for forward scattering, Fig. 22, Boutier et al. (1985) investigated the accuracy of the measured on-axis component for non-orthogonal channels. Two separate argon lasers were used to achieve equal intensities of the three beams. Varying the coupling angle between the two optical axes, test measurements indicated that accuracy for the on-axis component requires angles of at least 60°. A corresponding system was also used by Driver and Hebbar (1985).

Developments like application of a LDA for flow field scanning, e.g. Durst, Lehmann and Tropea (1981), are likely to support future turbulence research. An important step in this direction was recently reported by Chehroudi and Simpson (1985), who for the first time measured transverse space-time correlations in a separating turbulent boundary layer with a rapidly scanning LDA. Possibly high-speed anemometry using Smeets' (1984) spectrometric velocity measuring technique, Lehmann (private communication), will provide the means for future optical multipoint flow-field measurements.

5. Measured Balances of Reynolds Stresses

In the seventies, turbulence measurements with analogue equipment required tedious efforts for obtaining the mean flow field and the Reynolds stresses. In three-dimensional turbulent boundary layers, experimental sets of data at most contained twenty measuring stations, Müller (1982a). Therefore turbulence modeling, in particular for Reynolds stress closures, was hampered by lack of experimental data in engineering-type flows. Most comparisons of computations with experiment were based exclusively on measured profiles of mean velocities and components of the Reynolds stress tensor. With the emergence of powerful digital data acquisition and processing techniques, nowadays full sets of detailed measurements including all components of the mean velocity vector and of the double and triple correlation tensors of the fluctuations can be taken with reasonable effort. Such data are most useful for developing and testing turbulence models, in particular full Reynolds stress closure. From the Reynolds stress measurements, structural parameters like the ratios of the intensities or the shear stress-turbulent energy ratio $-\overline{uv}/\overline{k}$ with $\overline{k} = \overline{u_i u_i}/2$ can be checked for their validity in a particular flow. From the measured triple products, the adequacy of gradient-diffusion assumptions for turbulent convection can be tested, and in fact, closures are often found not to represent the experimental data sufficiently well, Chandrsuda and Bradshaw (1981), Driver and Seegmiller (1985), Müller (1986d). Moreover, complete sets of data enable analysis of extra terms in the boundary layer equations describing complex shear layers as opposed to equilibrium thin boundary layers, e.g. normal-stress diffusion $-\partial \overline{u^2}/\partial x$ of momentum or normal-stress production $(\overline{u^2}-\overline{v^2})\,\partial \overline{U}/\partial x$, or longitudinal diffusion $\partial \overline{ku}/\partial x$ of kinetic energy. Derivation of balances of the Reynolds stresses according to the transport equations is considered mandatory by this writer for achieving progress in turbulence modeling by enhanced understanding of the energy transfer among the fluctuating velocity components. Reynolds stress models should be analysed according to the balances first before recommending less sophisticated closures. Since measuring techniques for the pressure-strain correlation, besides some limited attempts, do not exist, and those for the dissipation of kinetic energy are restricted by the assumptions of isotropy and Taylor's hypothesis, the measurements of mean velocities, Reynolds stresses and triple-order products as well as their spatial gradients need to be particularly accurate to derive the non-measurable terms as closing entries to the transport equations. The pressure-strain correlations, though obtained indirectly with some experimental uncertainty, are expected to substantially aid the numerous efforts to model this term adequately, Rotta (1951), Launder, Reece and Rodi (1975), Higuchi and Rubesin (1981) or Gibson and Younis (1986). Details are omitted here, very likely experts on that subject will discuss the state of the art at the present symposium.

Balances for kinetic energy and shear stress, derived from two-component velocity measurements, were presented, e.g., by Chandrsuda and Bradshaw (1981) and Driver and Seegmiller (1985). The relaxation of their reattached flows bares much in common with the relaxing boundary layer of Fig. 7 studied by Müller (1986d), in which investigation balances for all non-zero components of the Reynolds stress tensor were evaluated from X-wire and triple-sensor hot-wire measurements. Examples from that work are given in Figs. 23a-c for kinetic energy $\overline{k}$, its individual components $\overline{u_i^2}$, and the shear stress $\overline{uv}$. The energy dissipation derived as closing entry to the $\overline{k}$-equation generally did not agree with the attempts of measuring it with various methods, in constrast to the favourable agreement Azad and Kassab (1985b) achieved in conical diffuser flow. Assuming isotropy of dissipation, the balances of the normal stresses, Fig. 23b, yielded an estimate for the pressure-strain correlations. In the disturbed flow, x/D = 40, the enlarged eddy structure enabled the results of the pressure-strain term of the $\overline{v^2}$ fluctuation to show the "splatting effect" close to wall, theoretically predicted by Moin and Kim (1982) by Large Eddy Simulation of a channel flow. Due to the constraints imposed on the v-fluctuation by the solid wall, energy is transferred from the u-component to the w-component directly. The balance of the Reynolds shear stress $\overline{uv}$ showed equally large production and pressure-strain terms, with mean and turbulent convection being considerably smaller. A similar result was obtained by Azad and Kassab (1985b). It should be noted, that the sum of the evaluated pressure-strain correlations $\overline{p\,\partial u_i/\partial x_i}$, Fig. 23b, in fact was close to zero, as well the diffusion terms integrated to zero, thereby giving some confidence into the accuracy of the results.

For the relaxing turbulent boundary layer discussed, balances of Reynolds stresses reduced from the measurements according to the transport equations, as well as analysis of structural parameters, Müller (1986b), have already served as a basis for some comparison calculations employing the Hanjalić and Launder (1976) three-equation model, Müller (1986a). One of the major discrepancies encountered was the inadequacy of a gradient-diffusion assumption for the turbulent convection terms. The computed pressure-strain correlations of the $\overline{uv}$-transport equation approximately balanced the production term throughout the flow relaxation in reasonable agreement with the experimental data.

Evaluation of balances in three-dimensional flows are still rare. Andreopoulos and Rodi (1984) derived the kinetic energy balance in a jet-to-crossflow experiment and discussed the production term of the turbulent shear stress $\overline{vw}$ of the lateral direction. A three-dimensional turbulent boundary layer along a spinning cylinder (zero azimuthal gradients) entering a fixed portion of the body was investigated by Driver and Hebbar (1985) by three-component LDA measurements. Their results for the $\overline{vw}$ balances obtained at the first measuring station on the fixed part, displayed in Fig. 24, are comparable to the $\overline{uv}$ balances discussed above. With the isotropic dissipation rate assumed to be small, production and pressure-strain correlation were the dominant terms.

6. Concluding Remarks and Perspectives

Hot-wire and Laser-Doppler techniques for measuring instantaneous velocities and time-averaged statistics in turbulent flows have been reviewed with emphasis on recent efforts to increase scope, accuracy and efficiency of turbulence measurements. From inherent limitations of present methods, requirements for improved ones immediately follow. Some perspectives of anticipated future measuring techniques will be discussed below after summarizing the present state of the art.

The majority of turbulence measurements has been and still is being performed by standard single and dual sensor hot-wire anemometry. Besides measurements of mean velocities and Reynolds stresses, a variety of applications including measurements of spectra, space-time correlations and temperature has been reported in the literature. Hot-wire anemometry is particularly suited for low-speed, low intensity, steady or unsteady turbulent flows being wall bounded or free. There exist, however, a number of well-known limitations on the use of hot wires. One of the most serious problems is temperature drift of calibration, which should be avoided by carefully controlling the fluid temperature or may be accounted for by temperature corrections. Flow disturbances by probe and its support as well as spatial resolution problems, in particular in wall proximity, generally limit the accuracy of measurements, but both problems can be reduced by the ever continuing efforts to design low-interference or miniaturized probes. The restriction of hot-wire measurements to flows with low local turbulence levels justifies linearized hot-wire response equations to be used for conventional signal interpretation. A general drawback of hot-wire anemometry consists in time-consuming, tedious calibration of each sensor to be used. Increase of required accuracy of measurements necessitates increasing efforts for calibrating the sensors with respect to both the magnitude and the direction of the velocity vector. Since most present data reduction schemes for turbulence measurements iteratively include the directional characteristics, a time-dependent mode of signal analysis costs extensive computer time.

Some of the limitations mentioned could be relaxed by recent developments of specific probes and techniques as well. The emergence of triple-sensor hot-wire anemometry was particularly attractive to experimenters. Firstly, from time-dependent three-component measurements, the mean velocities as well as the central moments of the fluctuations are advantageously obtained from a single set of measurement. Secondly, solution of exact response equations eliminates the necessity of a priori assumptions about the flow pattern and yields all components of the instantaneous velocity vector. As long as flow reversal and rectification are insignificant, the range of applicability extends to flows with moderate turbulence levels up to 0.4, say. For turbulence measurements in flows with even higher intensities or with regions of recirculation, the "flying hot wire" and the "pulsed hot wire" techniques have been developed. Both have been applied for investigating two-dimensional separated flows; with the former technique, all components of the Reynolds stress tensor are measurable by X-wire anemometry, while in recent years the scope of the pulsed-wire technique has been extended to include measurements of the normal velocity fluctuation v and the cross correlation $\overline{uv}$.

From the discussion of both, the state of the art of present standard hot-wire techniques as well as recent developments intending to relax inherent limitations and restrictions, some perspectives of future hot-wire anemometry immediately arise. Excluding instantaneous and mean flow reversal, the future standard technique anticipated by this writer should be a fast, accurate, time-dependent, three-component one applicable to two- and three-dimensional mean flows. The first feature, namely speed, calls for digital computers to aid not only calibration and measurement but data reduction and graphical display of results as well. Three-dimensional calibration of each hot-wire sensor throughout the anticipated measuring range presently is rather time consuming and should advantageously be automated. Results of calibrations have to be prepared in a form suitable for subsequent incorporation in the signal reduction scheme. Use of a computer enables high data rates obtained from a software-controlled multi-channel analogue to digital converter to be shifted to a mass storage like disc or tape. Subsequent off-line processing of data substantially helps to save wind-tunnel time, during the measurements a "quick look" should in general be sufficient to ensure the acquisition system to work well. Computation of time-dependent velocity components from the data requires fast and therefore non-iterative data reduction schemes, the development of which is considered one of the key pacing items towards an efficient measuring technique. Inclusion of curve-fitted results of the directional calibrations in the data reduction method would yield an analytical, non-iterative scheme, which combines accuracy and speed. Such a scheme has successfully been applied to X-wire data, while the challenging extension to three-component measurements has just recently been attacked. Since data reduction of single and X-wire measurements has to be based on linearized response equations in order to separate mean and fluctuating parts of the signals, only triple-sensor probes are inherently capable to measure instantaneous velocity vectors correctly. Solving the exact response equations together with the equations describing the directional calibrations is expected to yield accurate turbulence measurements even at moderate intensities. In order to achieve relative errors of the measured Reynolds stresses below ten percent, which accuracy is often claimed but seldom proven, particular attention will have to be paid to the accuracy of calibration, measurement and data reduction as well. On the other hand, such data will be sufficiently accurate to guide future developments of prediction methods and turbulence models, especially for complex flows.

Another aspect of an anticipated measuring technique is the demand for a minimum level of completeness of data sets: Measurements not only of mean velocities and Reynolds stresses but also of triple velocity products is considered mandatory by this writer in order to evaluate momentum balances as well as balances of the Reynolds stresses according to the transport equations. Thereby non-measurable terms like the dissipation of kinetic turbulence energy and the pressure-strain correlations can be estimated as closing entries to the equations. Reynolds-stress closure models are expected to particularly benefit from such results. Obtaining the required data by means of single or X-hot-wire probes would necessitate several traverses through the flow at various probe orientations in order to get a sufficient number of measurements to evaluate all quantities desired, and therefore not only efficiency will suffer but accuracy as well. One-point, three-component statistics up to third order are most conveniently achievable from time-dependent triple-sensor measurements obtained from a single traverse. It should be mentioned that triple-wire measurements would be particularly well suited for three-dimensional mean flows: Their large angular acceptance range does not require the probe to be aligned with the local mean flow direction as is the case for single or X-wire measurements. Hopefully efficiency and accuracy of new techniques will encourage more experimenters than before to investigate the complicated nature of three-dimensional turbulent flow fields.

Time-dependent measurements of velocities, already mentioned several times above, do not only allow for computation of mean velocities and central moments of the fluctuations, but also enable time-series analyses yielding power spectra or correlation functions. As well joint probability density distributions are desired results for comparisons with Gaussian distributions. Last not least time-dependent data reduction offers the opportunity to apply conditional sampling techniques, which are used, e.g., to detect coherent or intermittent flow structures or are applied for phase-averaged measurements in oscillating flow fields.

From summarizing the features anticipated for a future standard hot-wire technique, namely a fast, accurate, time-dependent, three-component one being applicable to two- and three-dimensional mean flows, triple-sensor anemometry is found to fulfill the requirements best. With reasonable optimism the anticipated developments are conjectured to be achieved within the next decade or so. Concerning, however, hot-wire measuring techniques applicable to complex flows including high turbulence levels or separation, future perspectives must be regarded less optimistical. Though appropriate techniques have already been proposed and are still being refined, relative errors of Reynolds stresses will be of the order of twenty percent or even more, and unfortunately completeness of results will decrease with increasing flow complexity. Three-component statistics have not yet been measured in separated flow, but might be expected from recent developments.

Future turbulence measurements in complex flows including separation are expected to be covered by the Laser Doppler Anemometer (LDA), in particular the accuracy of measurements in small scale separated regions will benefit from non-intrusiveness of the optical method. As well hot-wire anemometry for compressible flows has nearly entirely been replaced by the LDA. Most present techniques employ compact, two-colour, backscattering optics and Bragg-cell driven frequency shift for two-component measurements of mean velocities, Reynolds stresses and triple velocity products. Present developments of refined optics and signal analysis techniques are proceeding rapidly. For instance, for traversing the measuring volume through the flow field, miniaturized optical heads, possibly connected to the incident laser beam or to the photomultiplier by fiber optics, or appropriate zoom optics have been tested successfully. In flows with high turbulence intensities or in long-distance measurements, the signal-to-noise ratio of the burst signal becomes to low for accurate frequency measurements by a counter. Use of a transient recorder and subsequent frequency analysis by a Fast Fourier Transform algorithm was shown to be advantageous, for very low signal-to-noise ratios the photoncorrelation technique has to be used .

Similar to hot-wire anemometry, developments towards efficient, accurate, time-dependent, three-component LDA techniques are being aimed at. Development of bias-free averaging techniques for turbulent flows will be one of the major pacing items. Possibly by highly seeding the flow, sufficient data rates can be obtained for time-accurately integrating mean and fluctuating velocities. From high data rates, efficiency would benefit as well. Development of three-component techniques remains a most challenging task. Orthogonal beam arrangement would yield best accuracy of measurements, but is often impracticable due to limited optical access to the flow field. On the other hand, some test measurements have shown, that coupling angles below 60^0 result in reduced accuracy of Reynolds stress measurements, in particular for those including the velocity component along the optical axis. Another problem associated with LDA systems, in particular for three-component ones, is the high price. If future systems combine multi-dimensionality with low-price and simple optics and electronics, they may be expected to become the instrumentation most widely being used for turbulence measurements.

Beyond one-point measurements, the advantages of the high-speed, optical LDA method are likely to be utilized for flow field scanning. Measurements of two- or multi-point statistics including correlation functions and length scales as well as investigations of coherent structures are among the options anticipated for future measuring techniques.

In the immediate future, Laser-Doppler and hot-wire anemometry will remain complementary tools for experimental turbulence research. Both have their specific advantages as well as peculiarities, and therefore their judicous and accurate application requires a "calibrated experimenter", Perry (1982). On the other hand, given a certain flow field to be investigated, the experienced experimenter is able to decide, which turbulence quantity can be measured with an anticipated accuracy as well as with reasonable effort by means of a particular instrument. Consequently, experimenters should feel requested to mutually contribute to design and performance of each others experimental investigation. Participation of theoreticians in experimental research programmes should be considered mandatory in order to pin-point the theoretical concepts and hypotheses to be explored. In fact, an important step in this direction has been made. A joint international cooperation, aiming at experimental investigations of three-dimensional turbulent boundary layers on a swept wing and within a curved duct, has been initiated by Professors G. Drougge and E. Krause, who headed a group of experimenters and theoreticians for designing the test series. Specialists have already reviewed suitability and applicability of instrumentation as well as techniques for data processing and handling, Bertelrud et al. (1983). Under the chairmanship of Professor I. Rhyming, preparation of the duct flow experiment proceeded fast at EPFL, Lausanne, and is finished presently, Humphreys (1986). Survey measurements and flow visualization have already been carried out, and the group of Professor H. Fernholz participated in measuring wall shear stresses, Truong et al. (1986). Hot-wire turbulence measurements are about to follow.

This most promising example for an international cooperation in turbulence research is expected not only to yield progress in developing and applying a particular measuring technique, but also to significantly enhance understanding of complex turbulent flows.

7. References

Adams, E.W., Eaton, J.K.: An LDA study of the backward-facing step flow, including the effects of velocity bias. Symposium on Laser Anemometry at the ASME Winter Annual Meeting (1985).

Adler, D., Menn, A., Kalekin, E.: Measurement of mean velocities, velocity fluctuations and Reynolds stresses in a turbulent jet with a modified laser interferometric technique. J. Phys. E.: Sci. Instrum. 19 (1986).

Andreopoulos, J.: Comparison test of the response to pitch angles of some digital hot wire anemometry techniques. Rev. Sci. Instrum. 52 (1981).

Andreopoulos, J.: Improvements of the performance of triple hot wire probes. Rev. Sci. Instrum. 54 (1983a).

Andreopoulos, J.: Statical errors associated with probe geometry and turbulence intensity in triple hot-wire anemometry. J. Phys. E.: Sci. Instrum. 16 (1983b).

Andreopoulos, J., Rodi. W.: Experimental investigation of jets in a crossflow. J. Fluid Mech. 138 (1984).

Antonia, R.A., Anselmet, F., Chambers, A.J.: Assessment of local isotropy using measurements in a turbulent plane jet. J. Fluid Mech. 163 (1986).

Antonia, R.A., Browne, L.W.B., Chambers, A.J.: Determination of time constants of cold wires. Rev. Sci. Instrum. 52 (1981).

Azad, R.S.: Corrections to measurements by hot wire anemometer in proximity of a wall. Report MET-7, Department of Mechanical Engineering, University of Manitoba (1983).

Azad, R.S., Kassab, S.Z.: Dissipation of energy in turbulent flow. Report METR-9, Department of Mechanical Engineering, University of Manitoba (1985a).

Azad, R.S., Kassab, S.Z.: Turbulent energy budget and shear stress equation in a conical diffuser. Report METR-11, Department of Mechanical Engineering, University of Manitoba (1985b).

Bachalo, W.D., Johnson, D.A.: Transonic, turbulent boundary-layer separation generated on an axisymmetric flow model. AIAA J. 24 (1986).

Bahnen, R.H., Koeller, K.H.: Laser-Doppler velocimeter for multicomponent measurements using an electro-optical modulator demonstrated for a two-component optical configuration. Rev. Sci. Instrum. 55 (1984).

Bertelrud, A., Alfredson, P.H., Bradshaw, P., Landahl, M.T., Pira, K.: EUROEXPT-Report of the specialist group for instrumentation. FFA TN 1983-56 (1983).

Blackwelder, R.F., Haritonidis, J.H.: Scaling of the bursting frequency in turbulent boundary layers. J. Fluid Mech. 132 (1983).

Blackwelder, R.F., Kaplan, R.E.: On the wall structure of the turbulent boundary layer. J. Fluid Mech. 76 (1976).

Boutier, A., Canu, M.: Application of laser velocimetry to large industrial wind-tunnels. International Symposium on Applications of Laser-Doppler Anemometry to Fluid Mechanics, Lisbon (1982).

Boutier, A., D'Humières, Ch., Soulevant, D.: Three-dimensional laser velocimetry: a review. 2nd International Symposium on Applications of Laser Anemometry to Fluid Mechanics, Lisbon (1984).

Boutier, A., Pagan, D., Soulevant, D.: Measurements accuracy with 3 D laser velocimetry. International Conference on Laser Anemometry - Advances and Application, Manchester (1985).

Bradbury, L.J.S.: Measurements with a pulsed-wire and a hot-wire anemometer in the highly turbulent wake of a flat plate. J. Fluid Mech. 77 (1976).

Bradbury, L.J.S., Castro, I.P.: A pulsed-wire technique for velocity measurements in highly turbulent flows. J. Fluid Mech. 49 (1971).

Bradshaw, P.: An introduction to turbulence and its measurement. 2nd Ed., Pergamon, Oxford (1975).

Bradshaw, P.: Compressible turbulent shear layers. Ann. Rev. Fluid Mech. 9 (1977).

Bremhorst, K.: Effect of fluid temperature on hot-wire anemometers and an improved method of temperature compensation and linearisation without use of small signal sensitivities. J. Phys. E: Sci. Instrum. 18 (1985).

Browne, L.W.B., Antonia, R.A.: Reynolds shear stress and heat flux measurements in cylinder wake. Phys. Fluids 29 (1986).

Bruun, H.H.: Interpretation of a hot-wire signal using a universal calibration law. J. Phys. E.: Sci. Instrum. 4 (1971).

Bruun, H.H.: A note on static and dynamic calibration of constant-temperature hot-wire probes. J. Fluid Mech. 76 (1976).

Bruun, H.H.: Interpretation of hot-wire probe signals in subsonic airflows. J. Phys. E.: Sci. Instrum. 12 (1979).

Bruun, H.H., Tropea, C.: The calibration of inclined hot-wire probes. J. Phys. E.: Sci. Instrum. 18 (1985).

Buchhave, P., George, W.K., Lumley, J.L.: The measurement of turbulence with the Laser-Doppler anemometer. Ann. Rev. Fluid Mech. 11 (1979).

Bütefisch, K.A., Sauerland, K.H.: A three component dual beam Laser-Doppler anemometer to be operated in large wind tunnels. ICIASF '85 Record, IEEE publication 85 CH2210-3 (1985).

Butler, T.L., Wagner, J.W.: Application of a three-sensor hot-wire probe for incompressible flow. AIAA J. 21 (1983).

Cantwell, B., Coles, D.: An experimental study of entrainment and transport in the turbulent near wake of a circular cylinder. J. Fluid Mech. 136 (1983).

Carr, L.W., McCroskey, W.J.: A directionally sensitive hot-wire probe for detection of flow reversal in highly unsteady flow. ICIASF '79 Record, IEEE Publication 79 CH 1500-8 AES (1983).

Castro, I.P.: The measurement of Reynolds stresses. In: Encyclopedia of Fluid Mechanics I, Ed. N.P. Cheremisinoft, Gulf (1986).

Castro, I.P., Cheun, B.S.: The measurement of Reynolds stresses with a pulsed-wire anemometer. J. Fluid Mech. 118 (1982).

Champagne, F.H.: The fine-scale structure of the turbulent velocity field. J. Fluid Mech. 86 (1978).

Champagne, F.H., Sleicher, C.A.: Turbulence measurements with inclined hot-wires, Part 2. Hot-wire response equations. J. Fluid Mech. 28 (1967).

Champagne, F.H., Sleicher, S.A., Wehrmann, O.H.: Turbulence measurements with inclined hot-wires, Part 1. Heat transfer experiments with inclined hot-wire. J. Fluid Mech. 28 (1967).

Chandrsuda, B., Bradshaw, P.: Turbulence structure of a reattaching mixing layer. J. Fluid Mech. 110 (1981).

Chang, P.H., Adrian, R.J., Jones, B.G.: Comparison between triple-wire and X-wire measurement techniques in high intensity turbulent shear flow. Eighth Biennial Symposium on Turbulence, Rolla, Missouri (1982).

Chehroudi, B., Simpson, R.L.: Space-time results for a separating turbulent boundary layer using a rapidly scanning laser anemometer. J. Fluid Mech. 160 (1985).

Chevray, R., Tutu, N.K.: Simultaneous measurements of temperature and velocity in heated flows. Rev. Sci. Instrum. 43 (1972).

Coles, D., Wadcock, A.J.: Flying hot-wire study of two-dimensional turbulent separation on a NACA 4412 airfoil at maximum lift. AIAA J. 17 (1979).

Comte-Bellot, G.: Hot-wire anemometry. Ann. Rev. Fluid Mech. 8 (1976).

Corrsin, S.: Turbulence: experimental methods. Handbook of Physics 8/2, Springer, Berlin (1963).

Dekeyser, I., Launder, B.E.: A comparison of triple-moment temperature-velocity correlations in the asymmetric heated jet with alternative closure models. Proc. 4th Intern. Symposium on Turbulent Shear Flows, Karlsruhe (1983).

Demetriades, A.: Modal analysis of turbulent correlations in compressible flow. J. Appl. Mech. 40 (1973).

Dengel, P., Fernholz, H. H., Vagt, J.-D.: Turbulent and mean flow measurements in an incompressible axisymmetric boundary layer with incipient separation. 3rd Intern. Symposium on Turbulent Shear Flows, Davis (1981).

Driver, D.M., Hebbar, S.K.: Experimental study of a three-dimensional, shear-driven, turbulent boundary layer using a three-dimensional Laser-Doppler velocimeter. AIAA-85-1610 (1985).

Driver, D.M., Seegmiller, H.L.: Features of a reattaching turbulent shear layer in diverging channel flow. AIAA J. 23 (1985).
Durst, F., Ernst, F., Völklein, J.: Laser-Doppler-Anemometer-System für lokale Geschwindigkeitsmessungen in Windkanälen, Teil I: Systemauslegung und Verifikation. LSTM 108/E/85, Lehrstuhl für Strömungsmechanik, Universität Erlangen-Nürnberg (1985).
Durst, F., Jovanovic, J., Kanevce, Lj.: Probability density distribution in turbulent wall boundary layer flows. Proc. 5th Intern. Symposium on Turbulent Shear Flows, Ithaca, New York (1985).
Durst, F., Krebs, H.: Adaptive optics for IC-engine measurements. Exp. Fluids 4 (1986).
Durst, F., Lehmann, B., Tropea, C.: Laser-Doppler system for rapid scanning of flow fields. Rev. Sci. Instrum. 52 (1981).
Durst, F., Melling, A., Whitelaw, J.H.: Principles and practice of Laser-Doppler anemometry. 2nd Ed., Academic Press, London (1981).
Durst, F., Richter, G.: Neuere Entwicklungen auf dem Gebiet der Laser-Doppler-Windgeschwindigkeitsmessungen, Teil 1: Theorie und Verifikationsexperimente. Z. Flugwiss. Weltraumforsch. 7 (1983a).
Durst, F., Richter, G.: Neuere Entwicklungen auf dem Gebiet der Laser-Doppler-Windgeschwindigkeitsmessungen, Teil 2: Meßsystem für Freilandmessungen und sein Einsatz. Z. Flugwiss. Weltraumforsch. 7 (1983b).
Durst, F., Schmitt, F.: Experimental studies of high Reynolds number backward-facing step flow. Proc. 5th Intern. Symposium on Turbulent Shear Flows, Ithaca, New York (1985).
Eaton, J.K., Jeans, A.H., Ashjaee, J., Johnston, J.P.: A wall-flow-direction probe for use in separating and reattaching flows. J. Fluids Eng. 101 (1979).
Elsenaar, A., Boelsma, S.H.: Measurements of the Reynolds stress tensor in a three-dimensional turbulent boundary layer under infinite swept wing conditions. NLR TR 74095 U (1974).
Elghobashi, S.E., Launder, B.E.: Turbulent time-scales and the dissipation rate of temperature variance in the thermal mixing layer. Phys. Fluids 26 (1983).
Fabris, G.: Probe and method for simultaneous measurements of true instantaneous temperature and velocity components in turbulent flow. Rev. Sci. Instrum. 49 (1978).
Fabris, G.: Third-order conditional transport correlations in the two-dimensional turbulent wake. Phys. Fluids 26 (1983).
Fernholz, H.-H., Vagt, J.-D.: Turbulence measurements in an adverse-pressure-gradient three-dimensional turbulent boundary layer along a circular cylinder. J. Fluid Mech. 111 (1981).
Friehe, C.A., Schwarz, W.H.: Deviations from the cosine law for yawed cylindrical anemometer sensors. J. Appl. Mech. 35 (1968).
Fujita, H., Kovasznay, L.S.G.: Measurements of Reynolds stress by a single rotated hot-wire anemometer. Rev. Sc. Instrum. 39 (1968).
Gibson, M.M., Younis, B.A.: Calculation of swirling jets with a Reynolds stress closure. Phys. Fluids 29 (1986).
Gibson, M.M., Verriopoulos, C.A.: Turbulent boundary layer on a mildly curved convex surface. Part 2. Temperature field measurements. Exp. Fluids 2 (1984).
Hartmann, U.: Wall interference effects on hot-wire probes in a nominally two-dimensional highly curved wall jet. J. Phys. E.: Sci. Instrum. 15 (1982).
Hanjalić, K., Launder, B.E.: Contribution towards a Reynolds-stress closure for low-Reynolds-number turbulence. J. Fluid Mech. 74 (1976).
Heskestad, G.: Hot-wire measurements in a plane turbulent jet. J. Appl. Mech. 32 (1965).
Higuchi, H., Rubesin, M.W.: An experimental and computational investigation of the transport of Reynolds stress in an axisymmetric swirling boundary layer. AIAA-81-0416 (1981).
Hinze, J.O.: Turbulence. 2nd Ed., McGraw Hill (1975).

Hoffmeister, M.: Using a single hot-wire probe in three-dimensional turbulent flow fields. DISA Inf. 13 (1972).

Horstman, C.C., Rose, W.C.: Hot-wire anemometry in transonic flow. AIAA J. 15 (1977).

Huffmann, G.D.: Calibration of triaxial hot-wire probes using a numerical search algorithm. J. Phys. E.: Sci. Instrum. 13 (1980).

Humphreys, D.A.: EUROVISC series of internal 3 D turbulent boundary layer flows: A report of the Lausanne meeting. FFAP-164 (1986).

Hunter, W.W., Honaker, W.C., Gartrell, L.R.: Application of laser anemometry to cryogenic wind tunnels. ICIASF '83 Record, IEEE Publication 83 CH 1954-7 (1983).

Jaroch, M.: Development and testing of pulsed-wire probes for measuring fluctuating quantities in highly turbulent flows. Exp. Fluids 3 (1985).

Jörgensen, F.E.: Directional sensitivity of wire and fiber-film probes. DISA Inf. 11 (1971).

Johnson, F.D., Eckelmann, H.: A variable angle method of calibration for X-probes applied to wall-bounded turbulent shear flow. Exp. Fluids 2 (1984).

Kawall, J.G., Shokr, M., Keffer, J.F.: A digital technique for the simultaneous measurement of streamwise and lateral velocities in turbulent flows. J. Fluid Mech. 133 (1983).

King, L.V.: On the convection of heat from small cylinders in a stream of fluid. Phil. Trans. R. Soc. A214 (1914).

Kolmogorov, A.N.: Dissipation of energy in locally isotropic turbulence. C.R. Acad. Sci. USSR 32 (1941).

Kovasznay, L.S.G.: Some improvements in hot-wire anemometry. Hung. Acta Phys. 1 (1948).

Kovasznay, L.S.G.: The hot-wire anemometer in supersonic flow. J. Aeron. Sc. 17, (1950).

Kovasznay, L.S.G.: Turbulence measurements. Appl. Mech. Rev. 12 (1959).

Kornilov, V.I., Kharitonov, A.M.: Investigation of the structure of turbulent flows in streamwise asymmetric corner configurations. Exp. Fluids 2 (1984).

Kreplin, H.-P., Eckelmann, H.: Behavior of the three fluctuating velocity components in the wall region of a turbulent channel flow. Phys. Fluids 22 (1979).

Kühn, W., Dreßler, B.: Experimental investigations on the dynamic behaviour of hot-wire probes. J. Phys. E.: Sci. Instrum. 18 (1985).

Kutler, P.: A perspective of theoretical and applied computational fluid dynamics. AIAA J. 23 (1985).

Laderman, A.J., Demetriades, A.: Hot-wire measurements of hypersonic boundary-layer turbulence. Phys. Fluids 11 (1973).

Lakshminarayana, B., Poncet, A.: A method of measuring three-dimensional wakes in turbomachinery. J. Fluids Eng. 96 (1974).

La Rue, J.C., Libby, P.A., Seshadri, D.V.R.: Further results on the thermal mixing layer downstream of a turbulence grid. Proc. 3rd Intern. Symposium on Turbulent Shear Flows, Davis, California (1981).

Launder, B.E.: Heat and mass transport. In: Topics in Applied Physics, Vol. 12: Turbulence; Ed.: P. Bradshaw, Springer (1978).

Launder, B.E., Reece, G.J., Rodi, W.: Progress in the development of a Reynolds stress turbulence closure. J. Fluid Mech. 68 (1975).

Legg, B.J., Coppin, P.A., Raupach, M.R.: A three-hot-wire anemometer for measuring two velocity components in high intensity turbulent boundary layers. J. Phys. E.: Sci. Instrum. 17 (1984).

Lehmann, B.O.: A spatially working model to correct the statistical biasing error of more directional one component LDA-measurements. International Symposium on Applications of Laser-Doppler Anemometry to Fluid Mechanics, Lisbon (1982).

Lehmann, B.O., Helbig, J., Simon, B.: LDA signal analysis by software FFT applied to the free shear layer of a circular jet. Ninth Symposium on Turbulence, Rolla, Missouri (1984).

Ligrani, P.M., Gyles, B.R., Mathioudakis, K., Breugelmans, F.A.E.: A sensor for flow measurements near the surface of a compressor blade. J. Phys. E.: Sci. Instrum. 16 (1983).

Marvin, J.G.: Turbulence modeling for computational aerodynamics. AIAA-82-0164 (1982).

Mathioudakis, K., Breugelmans, F.A.E.: Use of triple hot wires to measure unsteady flows with large direction changes. J. Phys. E.: Sci. Instrum. 18 (1985).

Mikulla, V., Horstman, C.C.: Turbulence stress measurements in a nonadiabatic hypersonic boundary layer. AIAA J. 13 (1975).

Moffat, R.J., Yavuzkurt, S., Crawford, M.E.: Real-time measurements of turbulence quantities with a triple hot-wire system. Proc. Dynamic Flow Conference, Marseille, Baltimore (1978).

Moin, P., Kim, J.: Numerical investigation of turbulent channel flow. J. Fluid Mech. 118 (1982).

Morkovin, M.V.: Fluctuations and hot-wire anemometry in compressible flows. AGARDograph 24 (1956).

Morrison, G.L., Perry, A.E., Samuel, A.E.: Dynamic calibration of inclined and crossed hot wires. J. Fluid Mech. 52 (1972).

Müller, U.R.: Measurements of the Reynolds stresses and the mean-flow field in a pressure-driven three-dimensional boundary layer. J. Fluid Mech. 119 (1982a).

Müller, U.R.: On the accuracy of turbulence measurements with inclined hot wires. J. Fluid Mech. 119 (1982b).

Müller, U.R.: A hot-wire method for high-intensity turbulent flows. ICIASF '83 Record, IEEE Publication 83 CH 1954-7 (1983).

Müller, U.R.: Comparison of turbulence measurements with single, X and triple hot-wire probes. Submitted to Exp. Fluids (1985).

Müller, U.R.: Berechnung einer relaxierenden Grenzschicht mit einem Reynolds-Stress-Modell. ZAMM, Band 67, Heft 4/5 (1986a).

Müller, U.R.: The structure of turbulence measured in a relaxing boundary layer. Proc. European Turbulence Conference, Lyon (1986b).

Müller, U.R.: Erprobung einer Hitzdrahtmeßtechnik für abgelöste Strömungen. 5th DGLR-Fachsymposium "Strömungen mit Ablösung", München (1986c).

Müller, U.R.: Measurements of mean flow and Reynolds stress dynamics in a relaxing boundary layer. To be published (1986d).

Müller, U.R., Krause, E.: Measurements of mean velocities and Reynolds stresses in an incompressible three-dimensional turbulent boundary layer. Proc. 2nd Intern. Symposium on Turbulent Shear Flows, London (1979).

Nagano, Y., Hishida, M.: Production and dissipation of turbulent velocity and temperature fluctuations in fully developed pipe flow. Proc. 5th Intern. Symposium on Turbulent Shear Flows, Ithaca, New York (1985).

Nitsche, W., Haberland, C.: Ein vereinfachtes Eichverfahren für Hitzdrahtanemometer (Einzelpunkteichung). Z. Flugwiss. Weltraumforsch. 8 (1984).

Orloff, K.L., Snyder, P.K.: Laser-Doppler anemometer measurements using non-orthogonal velocity components: error estimates. Applied Optics 21 (1982).

Oster, D., Wygnanski, I.: The forced mixing layer between parallel streams. J. Fluid Mech. 123 (1982).

Owen, F.K.: Application of laser velocimetry to unsteady flows in large scale, high speed tunnels. ICIASF '83 Record, IEEE Publication 83 CH 1954-7 (1983).

Pailhas, G., Cousteix, J.: Méthode d'exploitation des données d'une sonde anémométrique à 4 fils chauds. Article pour la Recherche Aérospatiale (1986).

Pallek, D.: Fast digital data acquisition and analysis of LDA signals by means of a transient recorder and an array processor. ICIASF '85 Record, IEEE publication 85 CH2210-3 (1985).

Perry, A.E.: Hot-wire anemometry. Clarendon Press, Oxford (1982).

Perry, A.E., Morrison, G.L.: A study of the constant-temperature hot-wire anemometer. J. Fluid Mech. 47 (1971a).

Perry, A.E., Morrison, G.L.: Static and dynamic calibrations of constant-temperature hot-wire system. J. Fluid Mech. 47 (1971b).

Rotta, J.C.: Statistische Theorie nichthomogener Turbulenz. Z. Phys. 129 (1951).
Rotta, J.C.: Turbulent boundary layers in incompressible flow. Prog. Aeron. Sci. 2 (1962).
Rotta, J.C.: Turbulente Strömungen. Teubner, Stuttgart (1972).
Ruderich, R., Fernholz, H.H.: An experimental investigation of a turbulent shear flow with separation, reverse flow, and reattachment. J. Fluid Mech. 163 (1986).
Samet, M., Einav, S.: Directional sensitivity of unplated normal-wire probes. Rev. Sci. Instrum. 56 (1985).
Schubauer, G.B., Klebanoff, P.S.: Theory and application of hot-wire instruments in the investigation of turbulent boundary layers. NACA WR-86 (1946).
Seegmiller, H.L., Marvin, J.G., Levy Jr., L.L.: Steady and unsteady transonic flow. AIAA J. 16 (1978).
Siddal, R.G., Davies, T.W.: An improved response equation for hot-wire anemometry. Int. J. Heat and Mass Transfer 15 (1972).
Simpson, R.L.: Two-dimensional turbulent separated flow. AGARDograph No. 287, Vol. I (1985).
Simpson, R.L., Chew, Y.-T., Shivaprasad, B.G.: The structure of a separating turbulent boundary layer: I. Mean flow and Reynolds stresses. J. Fluid Mech. 113 (1981a).
Simpson, R.L., Chew, Y.-T., Shivaprasad, B.G.: The structure of a separating turbulent boundary layer: II. Higher order turbulence results. J. Fluid Mech. 113 (1981b).
Skinner, G.T., Rae, W.J.: Calibration of a three-element hot-wire anemometer. Rev. Sci. Instrum. 55 (1984).
Smeets, G.: Laser-Doppler velocimetry with a Michelson spectrometer. Proc. Laser Anemometry in Fluid Mechanics, Lisbon (1984).
Smits, A.J., Hayakawa, K., Muck, K.C.: Constant temperature hot-wire anemometer practice in supersonic flows. Part 1: The normal wire. Exp. Fluids 1 (1983).
Smits, A.J., Muck, K.C.: Constant temperature hot-wire anemometer practice in supersonic flows. Part 2: The inclined wire. Exp. Fluids 2 (1984).
Spencer, B.W.: Statistical investigation of turbulent velocity and pressure fields in a two-stream mixing layer. Dissertation thesis, University of Illinois at Urban-Champaign (1970).
Thompson, B.E., Whitelaw, J.H.: Flying hot-wire anemometry. Exp. Fluids 2 (1984).
Truong, T.V., Dengel, P., Moreau, V., Nakkasyan, A., Drotz, A., Rhyming, I.L.: EUROEXPT-Interim Report No. 2 by Specialist Group 2. Ecole Polytechnique Federal de Lausanne, IMHEF Internal Report (1986).
Tsiolakis, E.P., Krause, E., Müller, U.R.: Turbulent boundary layer-wake interaction. Proc. 4th Intern. Symposium on Turbulent Shear Flows, Karlsruhe (1983).
Tutu, N.K., Chevray, R.: Cross-wire anemometry in high intensity turbulence. J. Fluid Mech. 71 (1975).
Vafidis, C., Whitelaw, J.H.: Valve and in-cylinder flow in reciprocating engines. Imperial College, Mech. Eng. Report, FS/84/21 (1984).
Vagt, J.-D.: Hot-wire probes in low speed flow. Progr. Aerospace Sci. 18 (1979).
Walker, M.D., Maxey, M.R.: A whirling hot-wire anemometer with optical data transmission. J. Phys. E.: Sci. Instrum. 18 (1985).
Wallace, J.M., Brodkey, R.S.: Reynolds stress and joint probability density distributions in the u-v plane of a turbulent channel flow. Phys. Fluids 20 (1977).
Wallace, J.M., Eckelmann, H., Brodkey, R.S.: Some properties of truncated turbulence signals in bounded shear flows. J. Fluid Mech. 63 (1972).
Watmuff, J.H., Perry, A.E., Chong, M.S.: A flying hot-wire system. Exp. Fluids 1 (1983).
Willmarth, W.W.: Geometric interpretation of the possible velocity vectors obtained with multiple-sensor probes. Phys. Fluids 28 (1985).
Willmarth, W.W., Bogar, T.J.: Survey and new measurements of turbulent structure near the wall. Phys. Fluids 20 (1977).

Willmarth, W.W., Sharma, L.K.: Study of turbulent structure with hot wires smaller than the viscous length. J. Fluid Mech. 142 (1984).
Wyngaard, J.C.: Measurements of small scale structure with hot wires. J. Phys. E.: Sci. Instrum. 1 (1968).

8. Tables

Table I: Velocity calibration relationships

$E^2 = A + B\,U^N$	King (1914)
$E^2 = A + B\,U^N + C\,U$	A,B,N,C fixed: Siddal and Davies (1972)
	A,B,N,C least squares fitted: Bruun and Tropea (1985)
$U = \sum_n a_n E^n$	n = 0,5: Elsenaar and Boelsma (1974)
	n = 0,4: Oster and Wygnanski (1982)
$U = E^{1/N} + E^{2/N} + E^{3/N}$	Thompson and Whitelaw (1984)

Table II: Calibration techniques for measuring two velocity components instantaneously

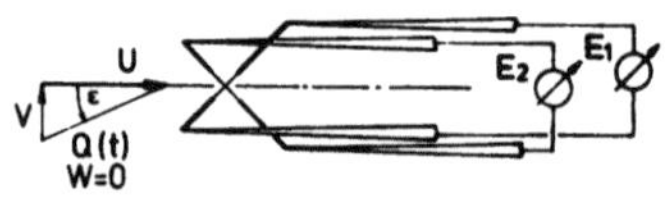

II.1. Effective cooling velocity U_c

II.1.a. $E_i(U_{N1})$ according to Table I, i = 1,2

II.1.b. Yaw dependence

$U_c = Q \cos\alpha$	Schubauer and Klebanoff (1946)
$U_c = Q\,(\cos^2\alpha + K\sin^2\alpha)^{1/2}$	Hinze (1975), Champagne and Sleicher (1967)
$U_c = Q(1 - b(1 - \cos^{1/2}\alpha))$	Friehe and Schwarz (1968)
$U_c = Q\cos\alpha_{eff}$	Bradshaw (1975)
$U_c = Q\cos^m\alpha$	Bruun (1979)

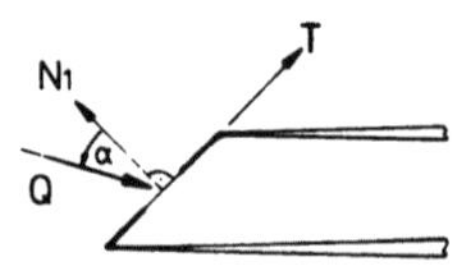

II.2. Look-up table

Willmarth and Bogar (1977): Calibration of $E_i = E_i\,(Q, \varepsilon)$, i = 1,2.

Measurement: Linear interpolation between calibration points.

Johnson and Eckelmann (1984): Calibration of

$$(E_i, \partial E_i/\partial Q, \partial E_i/\partial\varepsilon\ , \partial^2 E_i/\partial Q\,\partial\varepsilon\ , \partial^2 E_i/\partial\varepsilon\ \partial Q, \partial^2 E_i/\partial Q^2, \partial^2 E_i/\partial\varepsilon^2)$$

$$= f\,(Q, \varepsilon)\ , \quad i = 1,2$$

Measurement: Solution of fourth-order equations for the differences ΔQ and $\Delta\varepsilon$ with respect to the calibration point nearest to the measured voltage pair.

II.3. Curve fit

Oster and Wygnanski (1982), least squares fit:

$$Q(E_i, E_i^2, E_i^3) = \sum_j a_j E_1^m \; E_2^n \quad , j = 1{,}10;\; m + n \le 3;\; i = 1{,}2$$

$$\varepsilon\,(E_i, E_i^2, E_i^3) = \sum_j a_j E_1^m E_2^n \quad ; j = 1{,}10;\; m + n \le 3;\; i = 1{,}2$$

Müller (1985):

Algebraic approximations for calibrated functions F and H with

$$F(\varepsilon) = (U_{c1}/U_{c2} - 1) \,/\, (U_{c1}/U_{c2} + 1), \quad H(\varepsilon, Q) = |\vec{U}_c| \,/\, Q = (U_{c1}^2 + U_{c2}^2)^{1/2} \,/\, Q$$

Table III: Calibration techniques for measuring three velocity components instantaneously

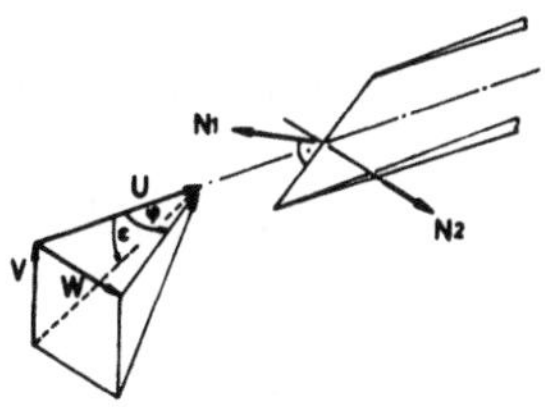

III.1. Effective cooling velocity U_c

III.1.a. $E_i(U_{N1})$ according to Table I, $i = 1{,}3$

III.1.b. Yaw and pitch dependence

$$U_c = (U_{N1}^2 + K\,U_T^2 + h\,U_{N2}^2)^{1/2}$$

Jörgensen (1971)

III.2. Look-up table

Pailhas and Cousteix (1986),

Kovasznay-type 4-sensor probe:

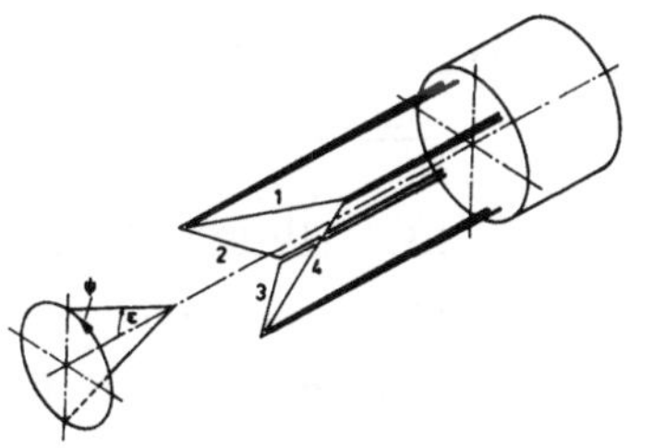

Interpolation between data points of calibrated functions F, G and H with

$$F(\varepsilon) = \tan^{-1} (U_{c2}^2 - U_{c4}^2) \,/\, (U_{c1}^2 - U_{c3}^3)$$

$$G(\varepsilon, \psi) = (|U_{c1} - U_{c3}| + |U_{c2} - U_{c4}|) \,/\, \Sigma U_{ci}$$

$$H(\varepsilon, \psi) = \Sigma U_{ci}^2 / (\Sigma U_{ci}^2\,(\varepsilon = 0))$$

III.3. Curve fit

Butler and Wagner (1983), least squares fit:

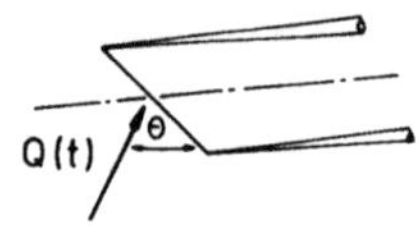

$$Q = f(E_i, E_i^2), \quad i = 1,3$$

$$\theta_j = f(E_i, E_i^2), \quad i = 1,3\ ;\ j = 1,3$$

Skinner and Rae (1984), Huffmann (1980):

$$E^2 = A + B\, U_c^N \big|_i$$

$$U_c = (U_{N1}^2 + K\, U_T^2 + h\, U_{N2}^2)^{1/2} \big|_i\ , \quad i = 1,3$$

$(A, B, N, K, h)_i$, $i = 1,3$: least squares fit

New proposal:

Algebraic approximations for the calibrated functions F, G, H with

$$F(W/U) = ((U_{c1}^2 + U_{c2}^2 - U_{c3}^2) \,/\, (U_{c2}^2 + U_{c3}^2 - U_{c1}^2))^{1/2}$$

$$G(V/U) = ((U_{c1}^2 - U_{c2}^2 + U_{c3}^2) \,/\, (U_{c2}^2 + U_{c3}^2 - U_{c1}^2))^{1/2}$$

$$H(|\vec{U}|) = ((U_{c1}^2 + U_{c2}^2 + U_{c3}^2) \,/\, (2\, |\vec{U}|^2))^{1/2}$$

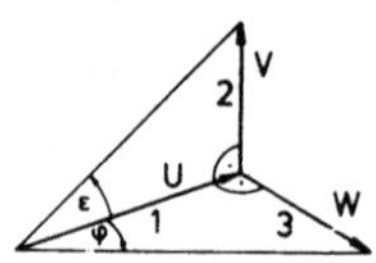

Table IV: Measurement of mean velocities and second and third-order products of fluctuations

IV.1. Low turbulence level, Tu«1: Linearized hot wire response equation

IV.1.a. Single slant wire

Hoffmeister (1972): fixed positions ψ_n

$$\left.\begin{aligned} \bar{U}_i &= f(\bar{U}_{cn}) \\ \overline{u_i u_j} &= f(\overline{u_{cn}^2}) \end{aligned}\right\} \quad n \geq 6$$

Fujita and Kovasznay (1968): rotating probe, $\psi(t)$;

least squares fitted response equations for $\bar{U}_i$, $\overline{u_i u_j}$

IV.1.b. Cross wire

	$\overline{U}_i$	$\overline{u_i u_j}$	$\overline{u_i u_j u_k}$	
Elsenaar and Boelsma (1974)	x	x		3DTBL
Fernholz and Vagt (1981)	x	x		3DTBL
Müller (1982a, b), (1986d)	x	x	x	3DTBL / 2DTBL
Azad and Kassab (1985b)	x	x	x	2DTBL

measurement	primary quantitity measured			
	$\psi = 0^o$	$\psi = 90^o$	$\psi = 45^o$	$\psi = -45^o$
$\overline{(u_{c1} + u_{c2})^2}$	$\overline{u^2}$	$\overline{u^2}$	$\overline{u^2}$	$\overline{u^2}$
$\overline{(u_{c1} - u_{c2})^2}$	$\overline{v^2}$	$\overline{w^2}$	$\overline{(v+w)^2}$	$\overline{(v-w)^2}$
$\overline{u^2_{c1}} - \overline{u^2_{c2}}$	$-\overline{uv}$	$-\overline{uw}$	$\overline{vw} - \sqrt{2}\,\overline{uw}$	$\overline{vw} + \sqrt{2}\,\overline{uw}$
$\overline{(u_{c1} + u_{c2})^3}$	$\overline{u^3}$	$\overline{u^3}$		
$\overline{(u_{c1} - u_{c2})^3}$	$\overline{v^3}$	$\overline{w^3}$		
etc.				

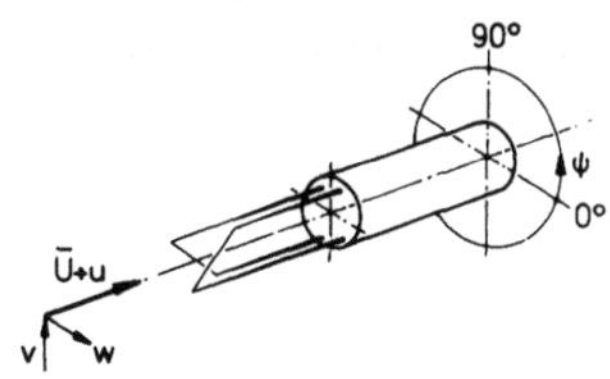

IV.2. <u>Moderate turbulence level, Tu < 0.4, no rectification</u>

IV.2.a. Correction of the linearized equations for $\overline{u_i u_j}$, Table IV.1.b., by third-order terms $\overline{u_i u_j u_k}$

Heskestad (1965), Vagt (1979):	$\overline{u_i u_j u_k}$ estimated from assumed pdf (u, v, w)
Müller (1982b, 1986d):	$\overline{u_i u_j u_k}$ measured

IV.2.b. Probe with three in-plane sensors to obtain U(t), V(t), | W(t) |

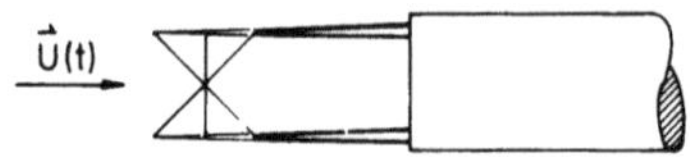

Kawall, Shokr and Keffer (1983): theoretical analysis

Legg, Coppin and Raupach (1984): theoretical analysis and test measurements

IV.2.c. Exact solution of the hot-wire response equations

Triple-sensor probe:

Moffatt, Yavuzkurt and Crawford (1978)
Andreopoulos (1983a)
Andreopoulos and Rodi (1984)
Müller (1983, 1985)

4-sensor Kovasznay-type probe:

Pailhas and Cousteix (1986)

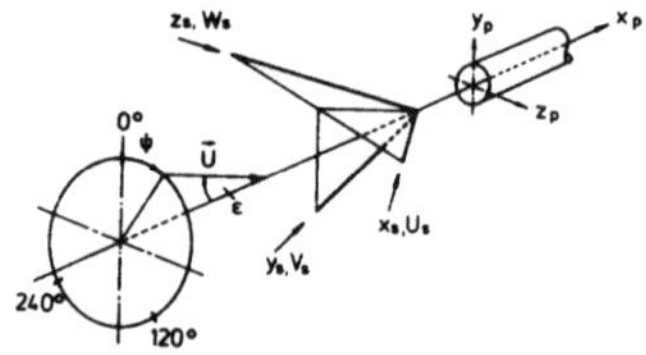

IV.3. <u>High turbulence levels, Tu > 0.4; large angular fluctuations</u>

IV.3.a. Flying hot wire

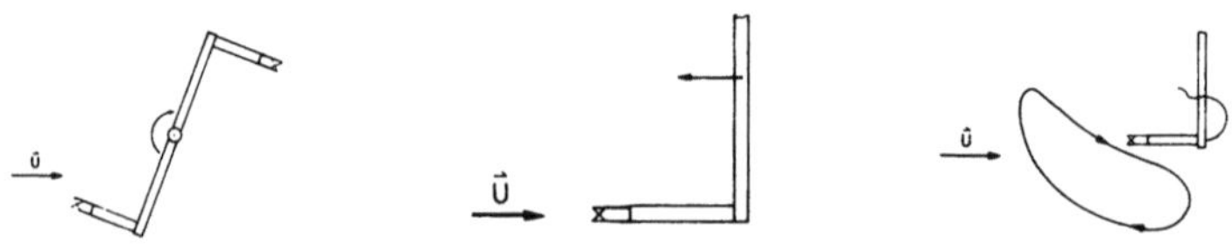

whirling probe:	catapulted probe:	probe moved along closed circuit:
Cantwell and Coles (1983)	Watmuff, Perry and Chong (1983)	Thompson and Whitelaw (1984)
Coles and Wadcock (1979)		

IV.3.b. Pulsed hot wire

$U = U_o \cos \Phi$

$U = \frac{s}{\Delta t}$, Δt = measured time of flight of temperature wake of pulsed wire

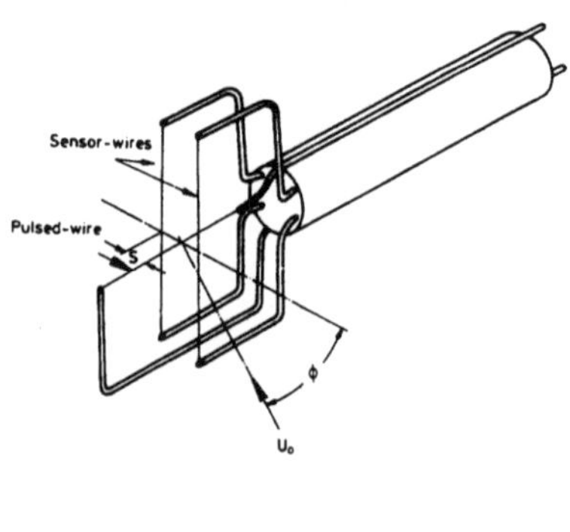

Bradbury and Castro (1971)
Bradbury (1976)
Dengel, Fernholz and Vagt (1981)
Castro and Cheun (1982)
Jaroch (1985)
Ruderich and Fernholz (1986)

IV.3.c. Thermal wake detection

Measurement of one velocity component:
Carr and McCroskey (1983)
Eaton et al. (1979)

Measurement of three velocity components:
Müller (1983, 1986c)

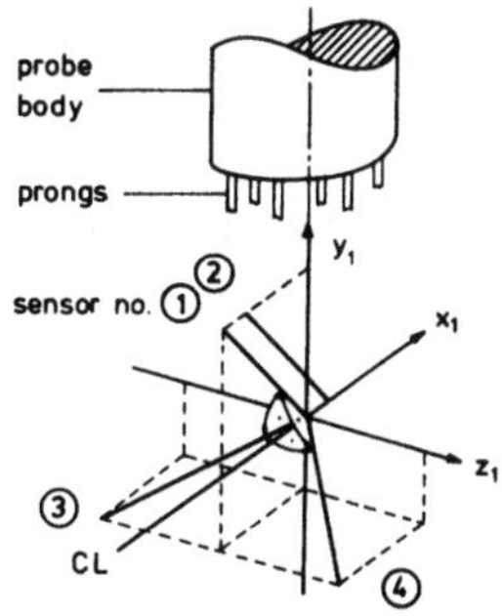

9. Figures

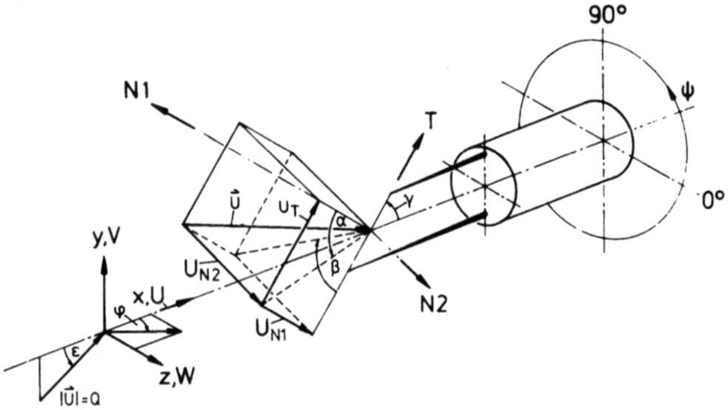

Fig. 1 Definition of reference and hot-wire fixed coordinate systems.

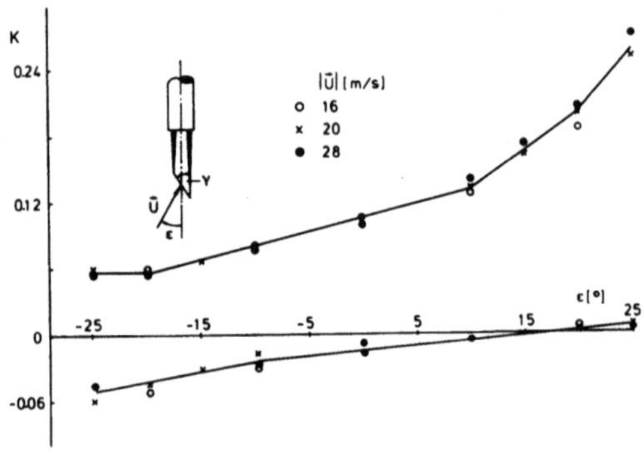

Fig. 2 Calibrated tangential sensitivities of X-wire probe.

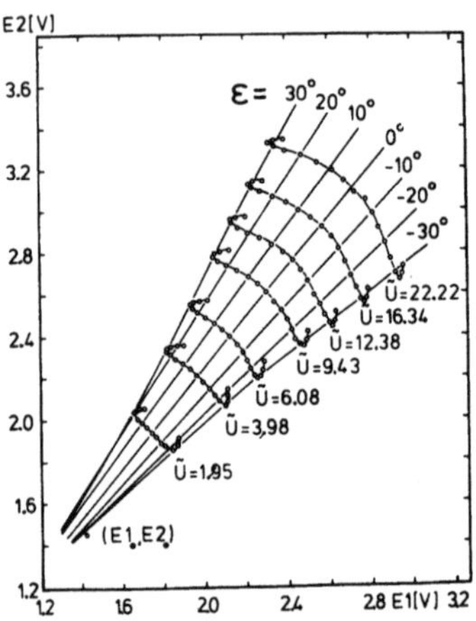

Fig. 3 Region of calibration as measured for a hot film X-probe. Speeds given in cm/s. From Johnson and Eckelmann (1984).

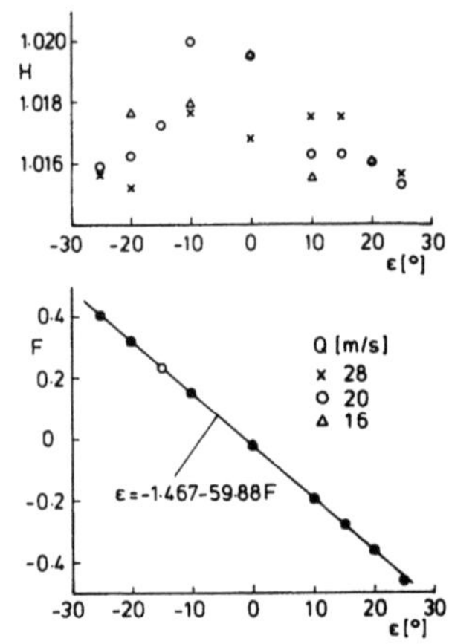

Fig. 4 Calibrated velocity parameter H and yaw function F (defined in Eq. 6) for an X-wire probe; equivalence to calibration given in Fig. 2.

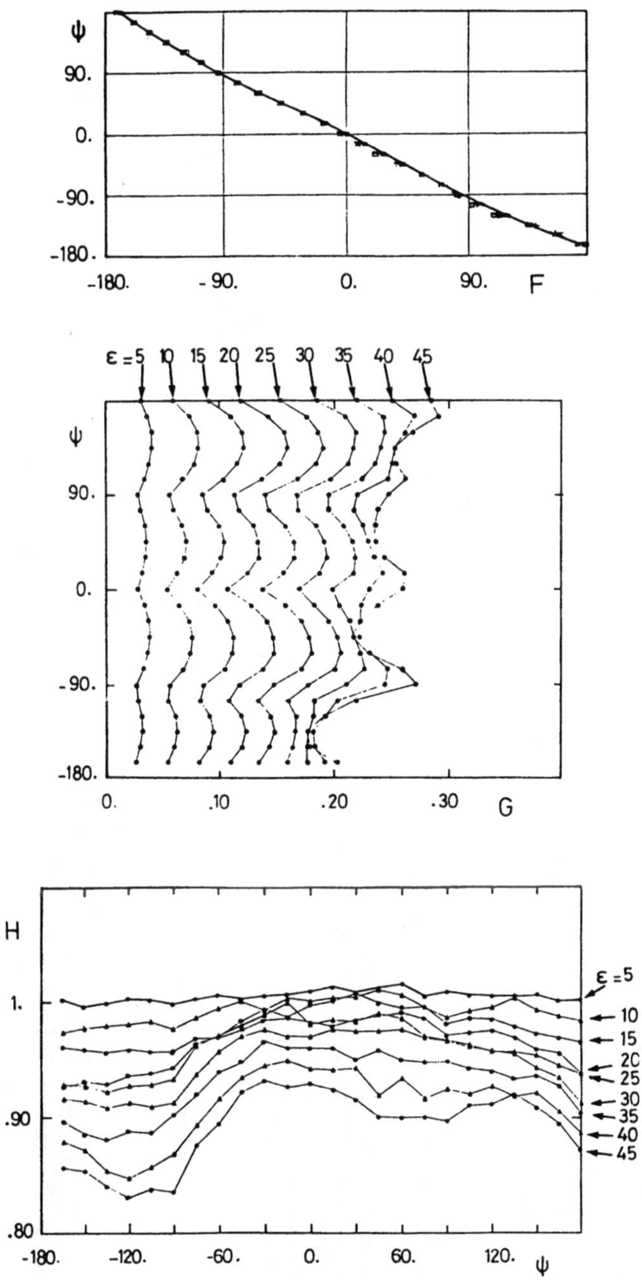

Fig. 5 Calibration of directional functions F and G and velocity function H (defined in Table III.2.) for a 4-sensor probe. From Pailhas and Cousteix (1986).

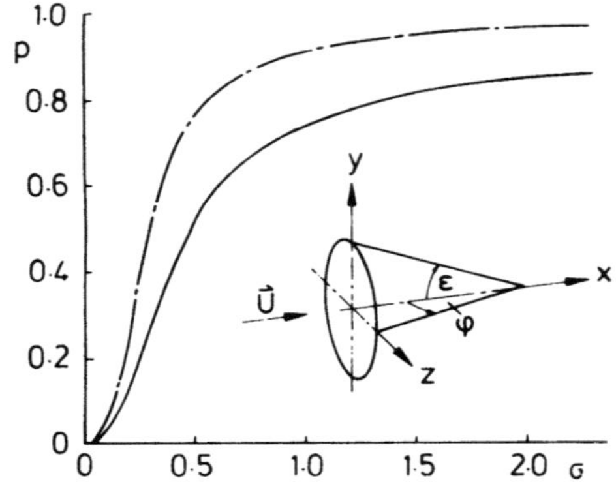

Fig. 6 Probability of velocity vectors to lie outside of an acceptance cone for a given turbulence level. From Castro (1986).
-·-·- single wire, $\varphi = 20^o$, $\varepsilon = 70^o$
—— X-wire, $\varphi = 20^o$, $\varepsilon = 20^o$

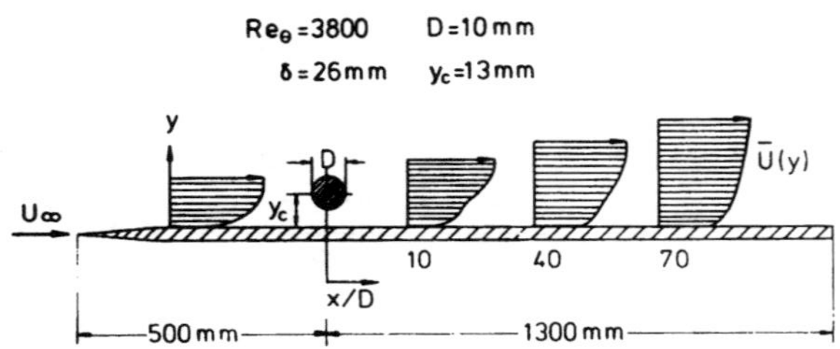

Fig. 7 Flow field investigated by Müller (1986 b, d).

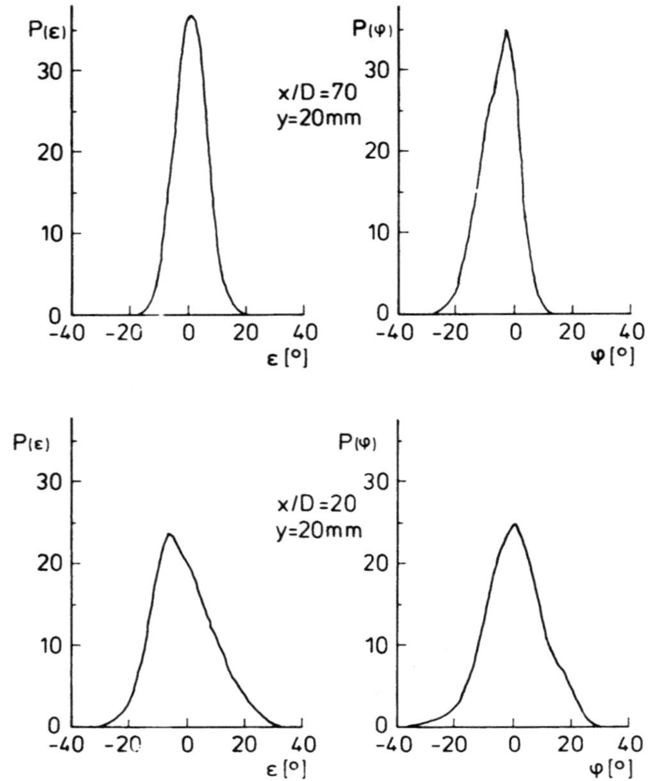

Fig. 8 Probability density functions of pitch and yaw angles measured in flow of Fig. 7.

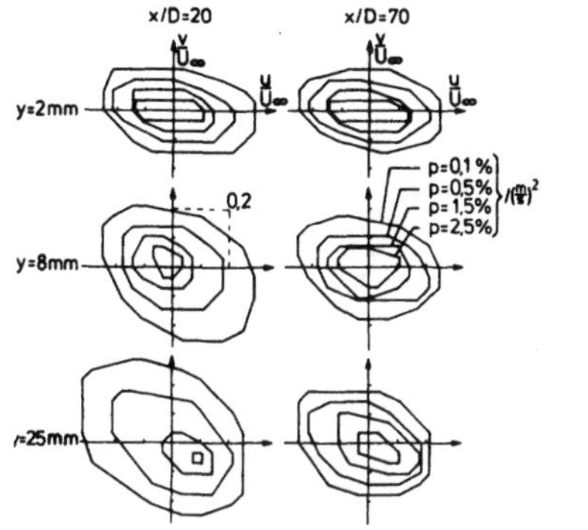

Fig. 9 Iso-contours of joint probability density functions of u- and v-fluctuations measured in flow of Fig. 7.

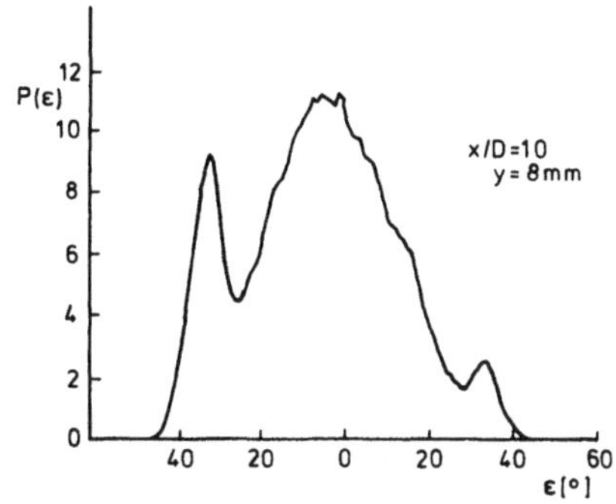

Fig. 10 Rectified probability density function of pitch angle measured in flow of Fig. 7.

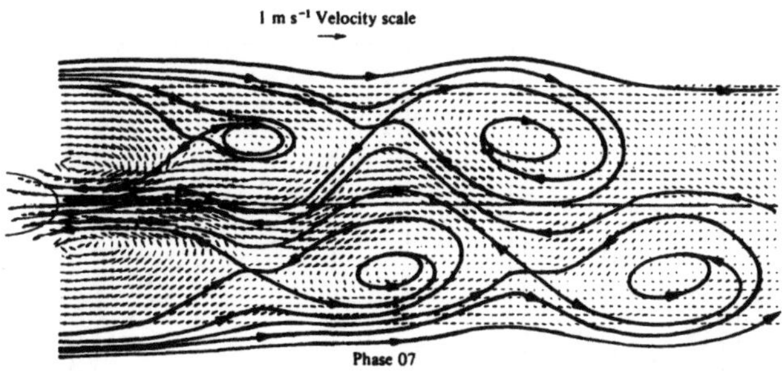

Fig. 11 Typical phase-averaged vector field measured by flying-wire anemometry, in the near wake behind an ellipsoidal body, $Re = 32 \cdot 10^3$. From Perry (1982).

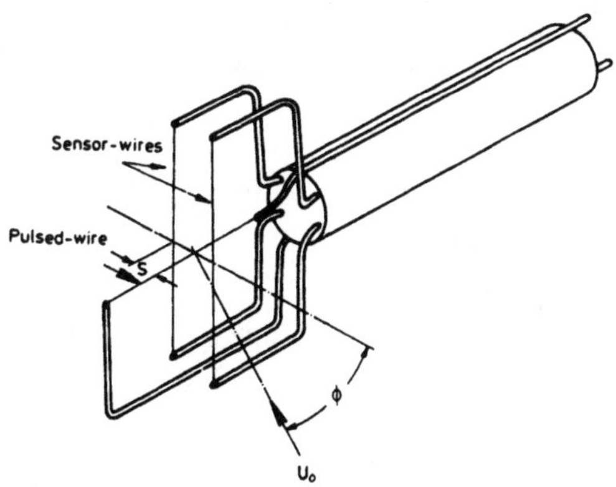

Fig. 12 Sketch of a pulsed-wire probe. From Dengel, Fernholz, Vagt (1981).

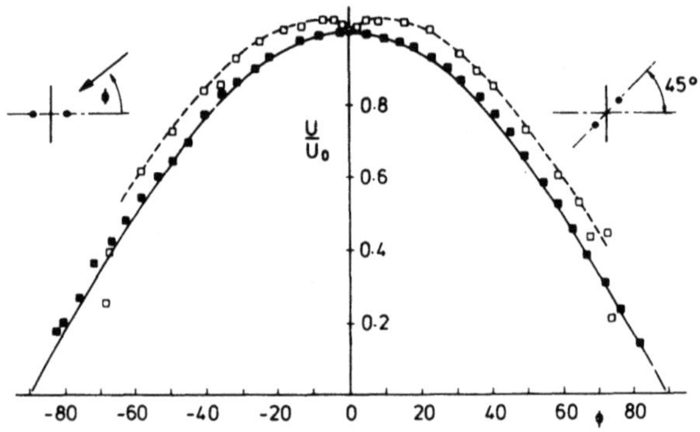

Fig. 13 Yaw response characteristics of pulsed-wire probes. From Castro (1986).
□, standard probe; ■, "offset" probe; —, cosine function.

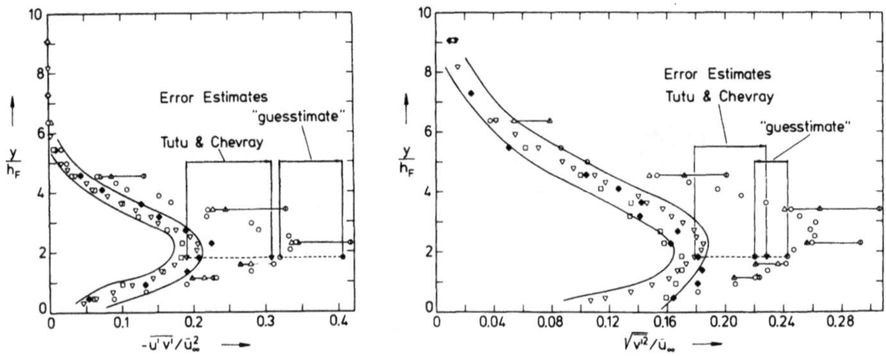

Fig. 14 Comparison of single, X- and pulsed-wire measurements of $\overline{uv}/U_\infty^2$ and v'/U_∞, reattached flow. From Jaroch (1985).
■, □, ◆, single wire; ▽, ▼, X-wire; ○, ●, △, ϕ, ⏀, pulsed-wire.

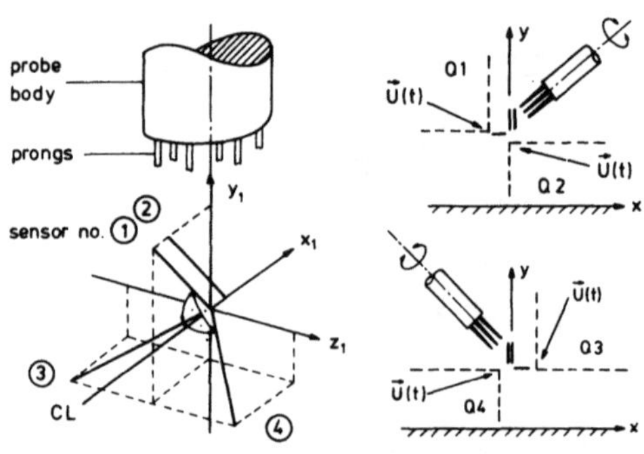

Fig. 15 Hot-wire configuration for measuring velocity vectors in high-intensity turbulent flows; proposed probe orientations.

Fig. 16 Four-sensor hot-wire probe for velocity measurements in high-intensity flows.

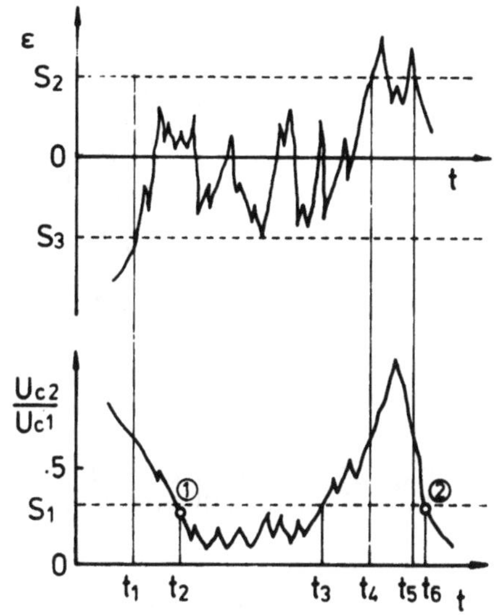

Fig. 17 Conditional sampling technique for velocity measurements in high-intensity flows.
Acceptance of data for

$U_{c2}/U_{c1} \leq S_1$ with $S_3 \leq \varepsilon \leq S_2$: $t_2 \leq t \leq t_3$

$U_{c2}/U_{c1} > S_1$ with $S_3 \leq \varepsilon \leq S_2$: $t_1 \leq t \leq t_2$, $t_3 \leq t \leq t_4$, $t_5 \leq t \leq t_6$

No acceptance for $\varepsilon \notin [S_2, S_3]$: $t_4 < t < t_5$

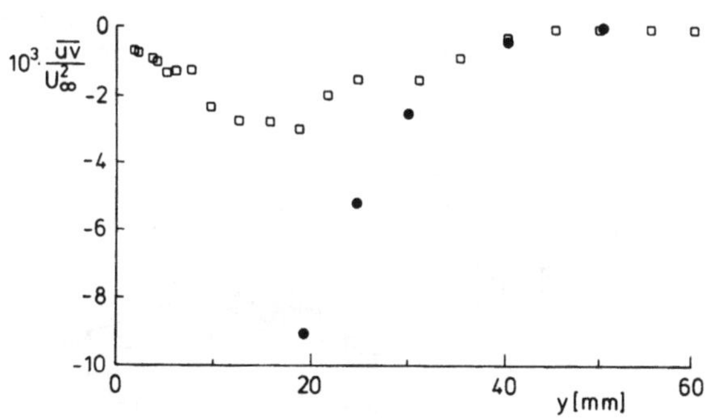

Fig. 18 Conditionally sampled four-wire measurements of Reynolds shear stress $\overline{uv}$ (●) compared with rectified x-wire data (□); x/D = 10 in flow of Fig. 7.

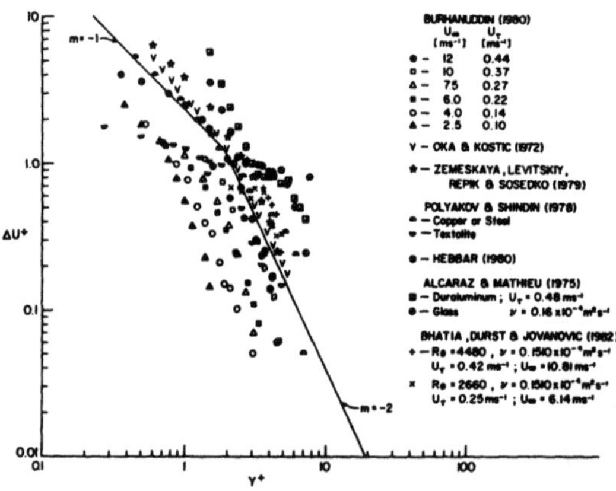

Fig. 19 The deviation of the measured velocity distribution from $u^{+} = y^{+}$ in wall proximity. From Azad (1983).

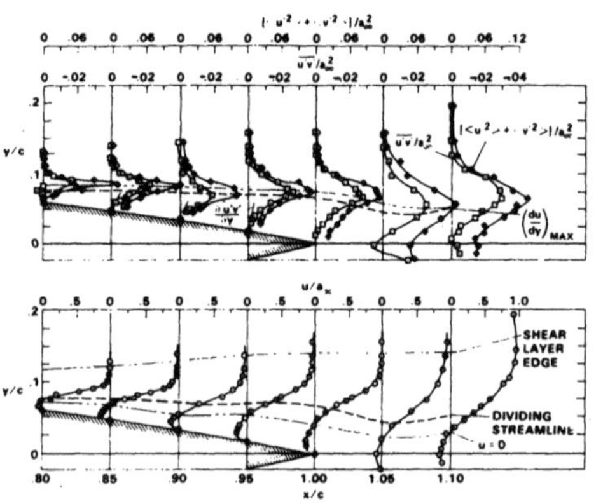

Fig. 20 Mean velocity, turbulent shear stress, and kinetic energy profiles in the shear layer downstream of separation, $M_{\infty} = 0.79$ and $Re_c = 11 \times 10^6$. From Seegmiller, Marvin, Levy (1978).

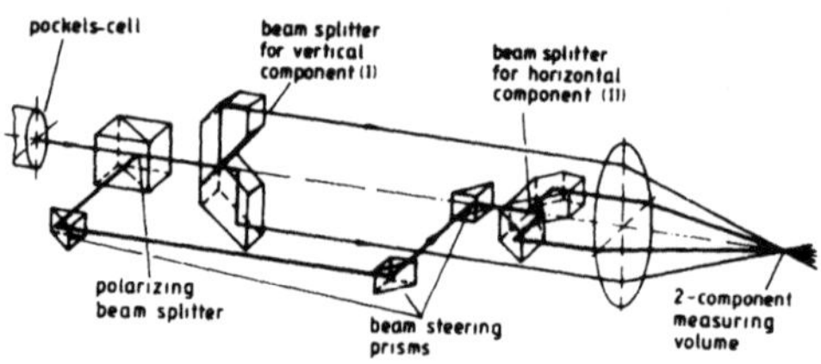

Fig. 21 Optical arrangement for a two-component velocity measurement using a Pockels cell. From Bahnen and Koeller (1984).

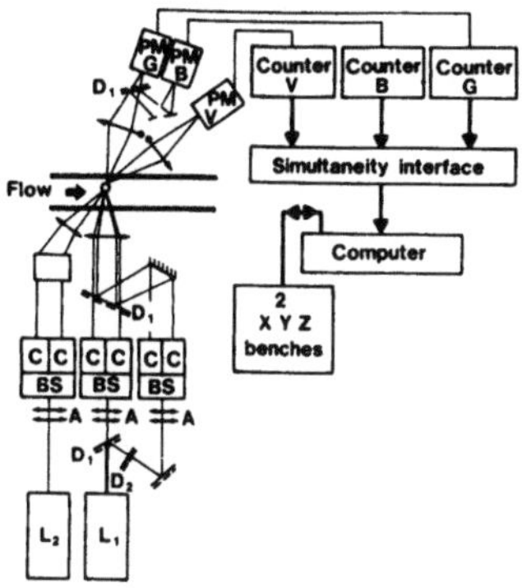

Fig. 22 Scheme of 3-D Laser-Doppler Velocimeter of ONERA. L, laser; D, dichroic plate; A, afocal telescope; BS, beam splitter; C, Bragg cell; PM, photomultiplier; G, green; B, blue; V, violett.
From Boutier et al. (1985).

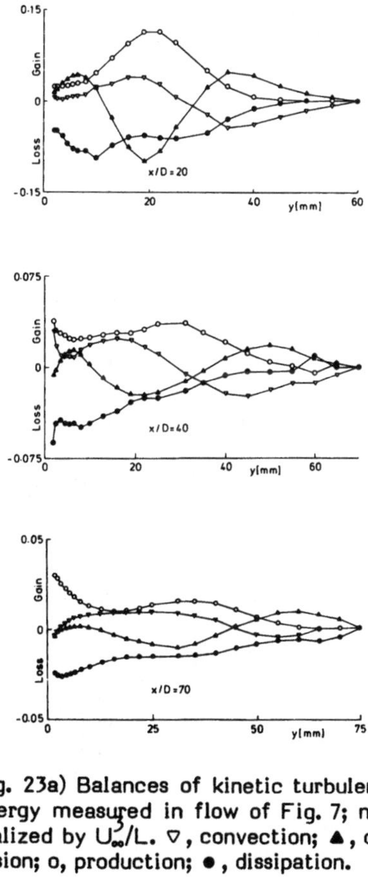

Fig. 23a) Balances of kinetic turbulence energy measured in flow of Fig. 7; normalized by U_∞^3/L. ▽, convection; ▲, diffusion; o, production; ●, dissipation.

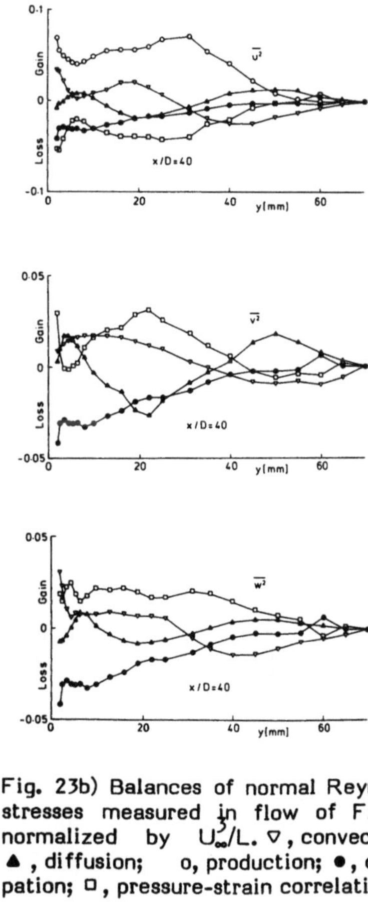

Fig. 23b) Balances of normal Reynolds stresses measured in flow of Fig. 7; normalized by U_∞^3/L. ▽, convection; ▲, diffusion; o, production; ●, dissipation; □, pressure-strain correlation.

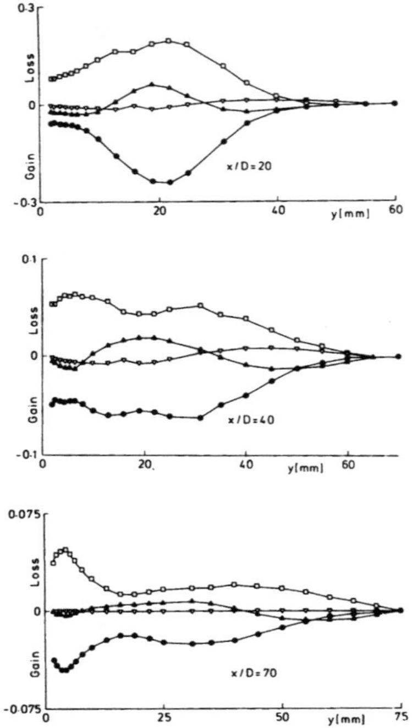

Fig. 23c) Balances of Reynolds shear stress measured in flow of Fig. 7; normalized by U_∞^3/L. ▽, convection; ▲, diffusion; ●, production; □, pressure-strain correlation.

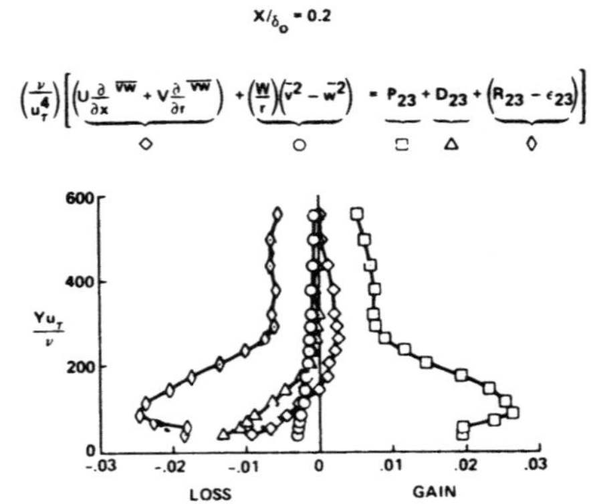

Fig. 24 Transport equation for $\overline{vw}$ Reynolds stress. From Driver and Hebbar (1985).

On the Theory of Turbulence for Incompressible Fluids

ZHOU (CHOU) PEI-YUAN

Peking University and Academia Sinica
Beijing, China

Abstract

The theory of turbulence, based upon the Reynolds equations of mean motion and the dynamical equations of the velocity correlations of successive orders, derived from the equations of the turbulent velocity fluctuation, by using the condition of pseudo-similarity and the hypotheses on the viscous dissipation terms in the correlation equations, is developed by a method of successive approximation. As examples in the first order approximation, we have solved the turbulent flows through a channel, in a plane wake and in jets by using the equations of mean motion and of the double correlation, while the terms in the triple correlation have been neglected. The agreements between the calculated values and the experiments are satisfactory.

In the present paper, the equations of the triple and quadruple correlations in addition to those used in the first order approximation are solved for the plane turbulent wake in the second order approximation by the method of substitution, starting from the solution of the first order approximation. Agreements between theory and existing experiments for the triple velocity correlation are also satisfactory. The theory has also yielded the components of the quadruple correlation which can be tested by experiment.

Introduction

The theory of turbulence for incompressible fluids, based upon the Navier-Stokes equations of motion, was first put forward by Reynolds in 1895. His important contribution consists of pointing out that a fully developed turbulent flow is composed of two parts, the mean motion and the turbulent fluctuation, and the derivation of the equations of mean motion by taking the average of the Navier-Stokes equations of motion. Due to the non-linearity of these latter equations, there appears the apparent stress, known as the Reynolds stress, in the equations of mean motion. On account of the Reynolds stress being unknown, the dynamical equations of mean motion thus derived are not closed. The subsequent mixture length theories of the momentum or vorticity transport proposed by Prandtl, Taylor and von Kármán in the early part of this centurty had the objective to relate the turbulent velocity fluctuations to the derivatives of the mean motion in order to make the set of equations of mean motion closed, while the equations of turbulent fluctuation which are the differences of the Navier-Stokes equations of motion and the Reynolds equations of mean motion were ignored since Reynolds' time.

The theory of turbulence, based upon the solutions of the dynamical equations of the velocity correlations of the double and triple orders derived from the equations of turbulent velocity fluctuation, and the equations of mean motion was first put forward in 1940(1) and then further developed in 1945(2). In the 1945 paper it was also pointed out that the direct approach to the solution of the general turbulence problem is to solve simultaneously the equations of mean motion and of turbulent fluctuation together with their corresponding equations of continuity. But this is a set of non-linear integro-partial differential equations and to seek their solutions for general shear turbulent motions of fluids is very difficult.

However, for simpler problems like the homogeneous isotropic turbulence and the homogenous shear turbulent flow, simultaneous solutions of the equations of mean motion and of turbulent fluctuation can be carried out. For homogeneous isotropic turbulence we introduce the condition of pseudo-similarity into the solution in order to choose the right kind of vortex to be the turbulence element. From the solutions of the turbulent velocity fluctuation we can then build the velocity correlations of any order to be compared with experimental measurements. For homogeneous turbulent shear flows special types of flow also satisfy this condition of pseudo-similarity.

For general shear turbulent flows, two methods to solve the problem have been developed without resorting to solve the equations of mean motion and turbulent fluctuation directly. The first method follows the line initiated in the above mentioned 1940 paper[1]. Due to the non-linearity of the equations of turbulent fluctuation, the dynamical equations of the correlations thus built are not closed, similar to the equations of mean motion, containing the unknown Reynolds stress which is a turbulent velocity correlation of the second order mentioned before. To overcome this difficulty in this method is to assume a relation between the velocity correlation of a given order with that of a lower one. Further hypotheses upon the pressure gradient and velocity fluctuation correlations and the terms in viscous decay have to be assumed in order to obtain definite solutions of the given turbulence problem. This line of approach was subsequently developed by Rotta (1951)[3] and a number of investigators in the sixties and seventies, especially after the computer was invented[4].

The second method of approach is to consider the derivation of the equations of mean motion, the building of the turbulent velocity correlations of successive orders and seeking the solutions of their corresponding dynamical equations as a method of successive approximation. This point of view was first brought out in 1945[2] and further developed in a recent paper[5]. Here we must point out that the condition of pseudo-similarity for the homogeneous isotropic turbulence has to be generalized for the general shear turbulent flows. The turbulent pressure gradient and velocity correlations and the dissipation terms in the dynamical equations of the double correlation have been treated before[2], while those for the equations of the triple and quadruple correlations can be treated in a similar way.

This second method of solution theoretically has the advantage over the first for being able to obtain velocity correlations of higher orders, if we carry out the process of successive approximation further.

As the first oder approximation we have used the equations of mean motion and the equations of the double velocity correlation to solve the problems of the flow in a channel, the plane wake[5], the plane and axial jets[6], while terms involving triple velocity correlations have been neglected. Agreements between the theoretical values of the mean velocities and double velocity correlations calculated with experiments are more satisfactory for free turbulence, the plane wake and jets, than the channel flow[5].

In the present paper we give the solution of the plane wake problem to the next order of approximation by using the dynamical equations for the velocity correlations of the triple and quadruple orders, while neglecting the fifth order correlation. Agreements between the calculations and the existing experimental data for the components of the triple velocity correlation are satisfactory. Since the present theory has given the values of those of the quadruple correlations which are not yet known, experiments can be carried out to test their validity.

I Equations of Motion

We first put down the equations of motion and start with the Navier-Stokes equations and the equation of continuity for incompressible fluids:

$$\frac{\partial u_i}{\partial t} + u^j u_{i,j} = -\frac{1}{\rho} p_{,i} + \nu \nabla^2 u_i,$$

$$u^j_{,j} = 0, \tag{1.1}$$

in which u_i is the velocity vector, p, the pressure, and ν, the kinematic coefficient of viscosity. Since we are using the rectangular system of coordinates x^i, the contravariat vector u^i is the same as its covariant form u_i and the comma sign followed by the coordinate x^j under u_i like $u_{i,j}$ denotes the covariant partial differentiation of u_i with respect to the coordinate x^j.

Following Reynolds, the vector u_i and pressure p in a fully developed turbulent flow can be separated into the mean motion and the turbulent fluctuation:

$$u_i = U_i + w_i, \qquad p = \bar{p} + \varpi, \tag{1.2}$$

in which U_i and $\bar{p}$ are the mean values of the velocity u_i and pressure p respectively, while w_i and ϖ are their turbulent fluctuations of which the mean values are both zero:

$$\overline{w_i} = 0\,, \qquad \overline{\varpi} = 0\,.$$

By introducing (1.2) into (1.1) and taking the average, we obtain the Reynolds equations and the equation of continuity for the mean motion,

$$\frac{\partial U_i}{\partial t} + U^j U_{i,j} = -\frac{1}{\rho}\bar{p}_{,i} + \frac{1}{\rho}\tau^j_{i,j} + \nu\nabla^2 U_i,$$

$$U^j_{,j} = 0, \tag{1.3}$$

with the Reynolds stress,

$$\tau^j_i = -\rho\overline{w_i w^j}.$$

By subtracting the Reynolds equations and the equations of continuity from the Navier-Stokes equations (1.1), we obtain the dynamical equations and the equation of continuity for the turbulent velocity fluctuation:

$$\frac{\partial w_i}{\partial t} + U^j w_{i,j} + w^j w_{i,j} + w^j U_{i,j}$$

$$= -\frac{1}{\rho}\varpi_{,i} - \frac{1}{\rho}\tau^j_{i,j} + \nu\nabla^2 w_i\,, \tag{1.4}$$

$$w^j_{,j} = 0.$$

From the equations of turbulent velocity fluctuation we can derive the dynamical equations of the double, triple and quadruple velocity correlations as follows:

$$\frac{\partial}{\partial t}\overline{w_i w_k} + U_{i,j}\overline{w^j w_k} + U_{k,j}\overline{w^j w_i} + U^j(\overline{w_i w_k})_{,j} + (\overline{w^j w_i w_k})_{,j}$$

$$= -\frac{1}{\rho}(\overline{\varpi_{,i} w_k} + \overline{\varpi_{,k} w_i}) + \upsilon \nabla^2 \overline{w_i w_k} - 2\upsilon g^{mn}\overline{w_{i,m} w_{k,n}}\,, \tag{1.5}$$

$$\frac{\partial}{\partial t}\overline{w_i w_k w_l} + U_{i,j}\overline{w^j w_k w_l} + U_{k,j}\overline{w^j w_l w_i} + U_{l,j}\overline{w^j w_i w_k}$$
$$+ U^j(\overline{w_i w_k w_l})_{,j} + (\overline{w^j w_i w_k w_l})_{,j}$$
$$= -\frac{1}{\rho}(\overline{\varpi_{,i} w_k w_l} + \overline{\varpi_{,k} w_l w_i} + \overline{\varpi_{,l} w_i w_k}) \tag{1.6}$$
$$+ (\overline{w^j w_i})_{,j}\overline{w_k w_l} + (\overline{w^j w_k})_{,j}\overline{w_l w_i} + (\overline{w^j w_l})_{,j}\overline{w_i w_k}$$
$$+ \upsilon g^{mn}(\overline{w_i w_k w_l})_{,mn} - 2\upsilon g^{mn}(\overline{w_{i,m} w_{k,n} w_l} + \overline{w_{k,m} w_{l,n} w_i}$$
$$+ \overline{w_{l,m} w_{i,n} w_k})\,,$$

$$\frac{\partial}{\partial t}\overline{w_i w_k w_l w_p} + U_{i,j}\overline{w^j w_k w_l w_p} + U_{k,j}\overline{w^j w_l w_p w_i} + U_{l,j}\overline{w^j w_p w_i w_k}$$
$$+ U_{p,j}\overline{w^j w_i w_k w_l} + U^j(\overline{w_i w_k w_l w_p})_{,j} + (\overline{w^j w_i w_k w_l w_p})_{,j}$$
$$= -\frac{1}{\rho}(\overline{\varpi_{,i} w_k w_l w_p} + \overline{\varpi_{,k} w_l w_p w_i} + \overline{\varpi_{,l} w_p w_i w_k} + \overline{\varpi_{,p} w_i w_k w_l})$$
$$+ (\overline{w^j w_i})_{,j}\overline{w_k w_l w_p} + (\overline{w^j w_k})_{,j}\overline{w_l w_p w_i} + (\overline{w^j w_l})_{,j}\overline{w_p w_i w_k}$$
$$+ (\overline{w^j w_p})_{,j}\overline{w_i w_k w_l} + \upsilon g^{mn}(\overline{w_i w_k w_l w_p})_{,mn} \tag{1.7}$$
$$- 2\upsilon g^{mn}(\overline{w_{i,m} w_{k,n} w_l w_p} + \overline{w_{k,m} w_{l,n} w_i w_p} + \overline{w_{l,m} w_{p,n} w_i w_k}$$
$$+ \overline{w_{i,m} w_{l,n} w_k w_p} + \overline{w_{k,m} w_{p,n} w_i w_l} + \overline{w_{i,m} w_{p,n} w_k w_l})\,.$$

In the above differential equations $\varpi_{,i}$ is the solution of the Poisson equation obtained before[2] and g_{ik} is the metric tensor. In the rectangular system of coordinates $g_{ik} = 1$, for $i = k$, and $g_{ik} = 0$, for $i \neq k$.

For the wall-bound and free turbulent flows like the channel, wakes and jets, we choose the solution of the equation of velocity turbulent fluctuation (1.4) of the type[5]:

$$
\begin{aligned}
& w_i = q\,\phi_i\left(\frac{x}{\lambda}, t\right) \\
& \text{with} \qquad q^2 = \overline{w_j w^j} = q^2(x,t), \qquad \lambda = \lambda(x,t)\,.
\end{aligned}
\tag{1.8}
$$

Here q is the magnitude of the velocity fluctuation and λ is the generalized Taylor's microscale of turbulence.

The velocity correlations of the second, third, and quadruple orders between two distinct points P and P' from (1.8) can be written as:

$$
\begin{aligned}
\overline{w'^n w_i} &= \overline{w^n w_i} + q^2 \phi_i^n(\eta, x, t)\,, \\
\overline{w'^m w'^n w_i} &= \overline{w^m w^n w_i} + q^3 \psi_i^{mn}(\eta, x, t) \\
\overline{w'^n w_i w_k} &= \overline{w^n w_i w_k} + q^3 \phi_{ik}^n(\eta, x, t) \\
\overline{w'^m w'^n w_i w_k} &= \overline{w^m w^n w_i w_k} + q^4 \psi_{ik}^{mn}(\eta, x, t)\,, \\
\overline{w'^n w_i w_k w_l} &= \overline{w^n w_i w_k w_l} + q^4 \phi_{ikl}^n(\eta, x, t)\,.
\end{aligned}
\tag{1.9}
$$

In the above equations we have

$$\xi^i = x'^i - x^i, \qquad \eta^i = \xi^i/\lambda . \tag{1.10}$$

The functions $\phi^n{}_i$, $\psi^{mn}{}_i$, $\phi^n{}_{ik}$, $\psi^{mn}{}_{ik}$ and $\phi^n{}_{ikl}$ all vary rapidly with ξ^i, the coordinate difference between the two points P and P', and vary slowly with x^i. Hence partial derivatives of the functions defined in (1.9) with respect to ξ^i are much greater than those with respect to x^i.

The turbulent pressure gradient and velocity fluctuation correlations, and the dissipation terms in the dynamical equations of correlations in (1.5), (1.6) and (1.7) are written below. Their calculations are explained in the Appendix.

For the pressure gradient and velocity fluctuation correlation we have

$$\frac{1}{\rho}(\overline{\varpi_{,i}w_k} + \overline{\varpi_{,k}w_i}) = q^2 a^{,n}_{mik}U^m_{,n} + q^2\lambda a^{,nr}_{mik}U^m_{,nr} + \frac{q^3}{\lambda}b'_{ik} \tag{1.11}$$

in which

$$a^{,n}_{mik} = \frac{1}{2\pi}\iiint[\phi^n_{i,mk} + \phi^n_{k,mi}]\frac{1}{r}dV' \tag{1.12}$$

$$a^{,nr}_{mik} = \frac{1}{2\pi\lambda}\iiint\{\phi^n_{i,m}\delta^r_k + \phi^n_{k,m}\delta^r_i + \xi^r[\phi^n_{i,mk} + \phi^n_{k,mi}]\}\frac{1}{r}dV' ,$$

$$b'_{ik} = \frac{\lambda}{4\pi}\iiint[\psi^{mn}_{i,mnk} + \psi^{mn}_{k,mni}]\frac{1}{r}dV' ,$$

$$\frac{1}{\rho}(\overline{\varpi_{,i}w_k w_l} + \overline{\varpi_{,k}w_l w_i} + \overline{\varpi_{,l}w_i w_k}) = q^3 b^{,n}_{mikl}U^m_{,n} + q^3\lambda b^{,nr}_{mikl}U^m_{,nr} + \frac{q^4}{\lambda}c'_{ikl} \tag{1.13}$$

in which

$$b'^{n}_{mikl} = \frac{1}{2\pi}\iiint[\,\phi^{n}_{ik,ml} + \phi^{n}_{kl,mi} + \phi^{n}_{li,mk}\,]\,\frac{1}{r}\,dV' \ ,$$

$$b'^{nr}_{mikl} = \frac{1}{2\pi\lambda}\iiint\{\phi^{n}_{ik,m}\delta^{r}_{l} + \phi^{n}_{kl,m}\delta^{r}_{i} + \phi^{n}_{li,m}\delta^{r}_{k} \tag{1.14}$$

$$+ \xi^{r}\,[\,\phi^{n}_{ik,ml} + \phi^{n}_{kl,mi} + \phi^{n}_{li,mk}\,]\}\,\frac{1}{r}\,dV' \ ,$$

$$c'_{ikl} = \frac{\lambda}{4\pi}\iiint[\,\psi^{mn}_{kl,mni} + \psi^{mn}_{li,mnk} + \psi^{mn}_{ik,mnl}\,]\,\frac{1}{r}\,dV' ;$$

$$\frac{1}{\rho}\,(\overline{\varpi_{,i}w_k w_l w_p} + \overline{\varpi_{,k}w_l w_p w_i} + \overline{\varpi_{,l}w_p w_i w_k} + \overline{\varpi_{,p}w_i w_k w_l}\,)$$

$$= q^4 c'^{n}_{miklp}U^{m}_{,n} + q^4\lambda\, c'^{nr}_{miklp}U^{m}_{,nr} + \frac{q^5}{\lambda}d'_{iklp} \ , \tag{1.15}$$

in which

$$c'^{n}_{miklp} = \frac{1}{2\pi}\iiint[\phi^{n}_{ikl,mp} + \phi^{n}_{klp,mi} + \phi^{n}_{lpi,mk} + \phi^{n}_{pik,ml}]\ \frac{1}{r}\ dV',$$

$$c'^{nr}_{miklp} = \frac{1}{2\pi\lambda}\iiint\{\,\phi^{n}_{ikl,m}\delta^{r}_{p} + \phi^{n}_{klp,m}\delta^{r}_{i} + \phi^{n}_{lpi,m}\delta^{r}_{k} + \phi^{n}_{pik,m}\delta^{r}_{l} \tag{1.16}$$

$$+ \xi^{r}\,[\phi^{n}_{ikl,mp} + \phi^{n}_{klp,mi} + \phi^{n}_{lpi,mk} + \phi^{n}_{pik,ml}\,]\}\ \frac{1}{r}\ dV'$$

$$d'_{iklp} = 0 \ .$$

All the partial differentiations under the above integrals are taken respect to the coordinate ξ^i and all the tensors a', b', c', d' with various ranks defined integrals (1.12) - (1.16) are approximately constants proved before[2].

The viscous dissipation terms in (1.5), (1.6) and (1.7) are given by (cp. Appendix)

$$2\nu g^{mn}\,\overline{w_{i,m}w_{k,n}} = -\frac{2\nu}{3\lambda^2}\,(k-5)\,q^2\delta_{ik} + \frac{2\nu k}{\lambda^2}\,\overline{w_i w_k}\,, \tag{1.17}$$

$$\begin{aligned} &2\nu g^{mn}(\overline{w_{i,m}w_{k,n}w_l} + \overline{w_{k,m}w_{l,n}w_i} + \overline{w_{l,m}w_{i,n}w_k}) \\ &= \frac{2\nu}{\lambda^2}\,[c_1(\overline{w_i w_j w^j}\delta_{kl} + \overline{w_k w_j w^j}\delta_{li} + \overline{w_l w_j w^j}\delta_{ik}) + c_2\overline{w_i w_k w_l}\,]\,, \end{aligned} \tag{1.18}$$

$$\begin{aligned} &2\nu g^{mn}\,(\overline{w_{i,m}w_{k,n}w_l w_p} + \overline{w_{k,m}w_{l,n}w_i w_p} + \overline{w_{l,m}w_{p,n}w_i w_k} \\ &\qquad + \overline{w_{i,m}w_{l,n}w_k w_p} + \overline{w_{k,m}w_{p,n}w_i w_l} + \overline{w_{i,m}w_{p,n}w_k w_l}) \end{aligned} \tag{1.19}$$

$$\begin{aligned} &= \frac{2\nu}{\lambda^2}\,[d_1(\overline{w_i w_k w_m w^m}\delta_{lp} + \overline{w_k w_l w_m w^m}\delta_{ip} + \overline{w_l w_p w_m w^m}\delta_{ik} \\ &\qquad + \overline{w_i w_l w_m w^m}\delta_{kp} + \overline{w_k w_p w_m w^m}\delta_{il} + \overline{w_i w_p w_m w^m}\delta_{kl}) + d_2\,\overline{w_i w_k w_l w_p}\,]. \end{aligned}$$

II Theory of Homogeneous Isotropic Turbulence and the Condition of Pseudo-Similarity

In a wind-tunnel the mean velocity of the air-stream behind the grid is constant. For an observer moving with this mean velocity, the equations of motion for the turbulent fluctuation (1.4) are the same as the Navier-Stokes equations of motion. The theory of homogeneous isotropic turbulence is based upon the vortex solution of the Navier-Stokes equations which forms the element of turbulence. This vortex is axially symmetrical and is randomly distributed as regards to its position and orientation of its axis of symmetry in the turbulent fluid. Then the turbulent velocity correlation of any order at one point or between two distinct points can be calculated by averaging the products of the velocity components at two points within the vortex over its position and the orientation of its axis in the fluid.

But the Navier-Stokes equations are non-linear partial differential equations of the second order and have different kinds of solutions. The theory was first applied to the case in the final period of decay in which the non-linear terms in the Navier-Stokes equaitons can be neglected[7], and a similarity condition was assumed to select the right kind of vortex to represent the turbulence element. The already well-known double velocity correlation could be thus explained. Furthermore, the theory predicted the triple velocity correlation which was published in 1965[8] and verified experimentally by Bennett and Corrsin ten years later[9].

For very high Reynolds number flows in the initial period of decay near the grid in the wind-tunnel, there are solutions of the Navier-Stokes equations. By using another condition of similarity a different vortex solution was obtained, yielding the law of decay of turbulence and the spread of Taylor's scale of micro-turbulence λ in agreement with experiments. The double and triple velocity correlations thus calculated have also qualitative agreements with measuments[10].

The condition of pseudo-similarity was proposed to correlate the two kinds of similarity in the initial and final periods of decay for isotropic homogeneous turbulence[11,12]. A large number of theoretical calculations of the decay of turbulent energy, the spread of Taylor's scale of micro-turbulence λ, the double and triple velocity correlations from the initial to the final period of decay, together with the three-dimensional and one-dimensional energy spectrum functions and the energy transfer spectrum function etc. all agree with experimental measuments very well.

This condition of pseudo-similarity for homogeneous isotropic turbulence can be written in terms of λ and R_λ, the turbulence Reynolds number, as follows[5]:

$$\frac{\lambda}{\upsilon}\frac{d\lambda}{dt} = \frac{1}{R_0} R_\lambda^2 + 2 \tag{2.1}$$

with

$$R_\lambda = \frac{q\lambda}{\upsilon} \tag{2.2}$$

in which R_0, a Reynolds number, is a constant. In this relation when R_λ is large, namely, in the initial period of decay, the number 2 in (2.1) can be ignored. Then we have λ^2 and q^2 satisfying,

$$\lambda^2 = 10\,\upsilon\, t, \qquad q^2 \sim 1/t. \tag{2.3}$$

In the final period of decay, R_λ in (2.1) can be neglected and we have

$$\lambda^2 = 4\,\upsilon\, t, \qquad q^2 \sim \frac{1}{t^{5/2}}\,. \tag{2.4}$$

The experimental proof of the condition (2.1) from the initial to the final period of decay has been carried out in the turbulence wind-tunnel of Peking University (to be published).

For general shear turbulent flows the condition of pseudo-similarity (2.1) can be put in the following form,

$$\frac{\lambda}{\upsilon}\left(\frac{\partial\lambda}{\partial t} + U^j\lambda_{,j}\right) = \frac{1}{R_0}\{R_\lambda^2 - [R_1 + \frac{\kappa_1}{\upsilon}\lambda^2(g^{kl}U^j_{,k}U_{j,l})^{1/2} + \frac{\kappa_2}{\upsilon}\lambda^3(g^{ij}g^{kl}g^{mn}\Omega_{ik,m}\Omega_{jl,n})^{1/2}]^2 + 2R_0\}\,, \tag{2.5}$$

where $\Omega_{ik} = U_{i,k} - U_{k,i}$, and κ_1, κ_2, R_0, and R_1 are constants, and q and R_λ are defined by (1.8) and (2.2) respectively.

In the first order approximation for the channel flow, the wake and jets, the equations of mean motion (1.3), the double velocity correlatioon (1.5) and the condition of pseudo-similarity (2.5) have been used, while terms in the triple correlation have been ignored[5].

III The Plane Turbulent Wake

Now we consider the problem of the plane turbulent wake in which there are already measurements of the triple velocity correlation. For the second order approximation, besides the equations of mean motion and of the double velocity correlation, and the condition of pseudo-similarity, we have to use the dynamical equations of the triple and quadruple orders (1.6) and (1.7) in addition. The rigorous method of solution is to solve simultaneously for the mean velocity U_i, the double, triple and quadruple velocity correlations and the turbulence scale λ from all the equations, (1.3), (1.5), (1.6), (1.7), and (2.5) with (1.11) - (1.19), while neglecting the terms in the quintuple correlation. But this method of solution is complicated.

A simpler method of approach is to find the solution by successive substitution. This method was used in the first order approximation of the solution for the plane wake[13]. To start with in the first step we find the approximate expression of the Reynolds shear τ_{xy} to be proportional to the gradient of the mean velocity $\partial U/\partial y$, a result which was obtained before[14]. This gives us $\overline{w_1 w_2}^{(0)}$. Then the mean velocity denoted by $U^{(0)}{}_1$ is solved and the other components of the double velocity correlation, $\overline{w_i w_k}^{(0)}$, and the scale of turbulence $\lambda^{(0)}$

can be obtained from the equations of the double velocity correlation and the condition of pseudo-similarity, the terms in the triple correlation being ignored. The theoretical values of the above functions $U^{(0)}_i$, and $\overline{w_i w_k}^{(0)}$ thus calculated agree with experimental measurements very well.

The next step in this first order approximation is to use the above values of $\overline{w_i w_k}^{(0)}$ and $\lambda^{(0)}$ to solve for $U^{(1)}_i$, $\overline{w_i w_k}^{(1)}$ and $\lambda^{(1)}$. The new functions $U^{(1)}_i$, $w_i w_k^{(1)}$ and $\lambda^{(1)}$ obtained by this substitution differ very little from those obtained in the first step and this process can be repeated.

In this first order approximation for the solution of the plane wake, Mr. Chen Shi-yi who carried out this investigation, has also solved simultaneously the equations of mean motion (1.3) and of the double correlation (1.5), neglecting the terms in triple correlation, under the condition of pseudo-similarity (2.5), by using a computer. The results thus obtained also agree quite well with those obtained in the above method of successive substitution.

The rigorous method of finding the solution of the plane wake in the second order approximation, as mentioned before, is to solve simultaneously the equations of mean motion, of the double, triple and quadruple correlations together with the condition of psudo-similarity. To avoid this, we use again the method of successive substitution. Since in the first order approximation we have already obtained values of the mean velocity and double correlation which agree with experiment quite well, we can use them as known values and put them into the equations of $\overline{w_i w_k w_l}$ and $\overline{w_i w_k w_l w_p}$ and look for their solutions.

But to solve the equations for the triple and quadruple correlations together is still complicated. Hence we can separate their solutions by assuming an often used relation between the quadruple correlation with the double correlation in homogeneous isotropic turbulence based upon the quasi-normal distribution condition, which is given by

$$\overline{w_i w_k w_l w_p} = \overline{w_i w_k}\;\overline{w_l w_p} + \overline{w_i w_l}\;\overline{w_p w_k} + \overline{w_i w_p}\;\overline{w_k w_l} \qquad (3.1)$$

and solve the equations of triple correlation first. All of the six non-vanishing components of the triple correlation for the plane wake thus computed agree with experiment very well (cp. Fig.'s 1-6)[15].

The next step is to put the known triple correlation obtained above into the equation of quadruple correlation, neglect the quintuple correlation and seek its solution which can then be compared with the relation (3.1). Their agreement is comparatively satisfactory (cf. Fig.'s 7,8,9, agreeing well or not so well with (3.1), e.g.,out of its total nine non-vanishing components). Since the solution of $\overline{w_i w_k w_l w_p}$ from its dynamical equations agrees fairly well with (3.1), there is no need to put this solution of $\overline{w_i w_k w_l w_p}$ into the equations of triple correlation and find its solution again.

On the other hand, if we want to have more accurate values of U_i, $\overline{w_i w_k}$ and λ, we can put the third order velocity correlation obtained into the equations for $\overline{w_i w_k}$ and solve for U_i, $\overline{w_i w_k}$ and λ from (1.3), (1.5) and (2.5). This method of substitution can obviously be extended to get higher order approximations.

Discussion

In the above methods of successive approximation in treating the general shear turbulence problem there are some points which need further discussion. The solution of the turbulent velocity fluctuation given in the form (1.8) is a special form of solution which has the property of pseudo-similarity. This is analogous to the kind of similarity discovered by Prandtl in the steady viscous incompressible laminar flows along a semi-infinite plane, through a channel and a pipe, and in wakes and jets. But it is well known that there are also other different kinds of solutions for laminar viscous flows. We have the similar situation in incompressible turbulent flows. For example, for homogeneous shear flows, there are types of flow which satisfy the pseudo-similarity condition while there are other flows which have different properties[16].

In the first order approximation in solving the plane wake problem, we considered before the mean velocity and the double velocity correlation together, neglected the terms in the triple correlation and found their solutions from their dynamical equations under the condition of pseudo-similarity. Likewise, in the second order approximation we now consider the correlations of the triple and quadruple orders together, add their dynamical equations to the first set and find their solutions, while neglecting correlation terms of the fifth order. This method of solving correlation equations of an odd order and of the following even order together gives better approximation of the former and can be extended to find approximate solutions of still higher orders.

In the Appendix we have obtained the three viscous dissipation terms in (1.5), (1.6), (1.7) based upon the hypotheses introduced. Their justification also depends upon experiments.

Appendix

In the computations of the integrals in the turbulent pressure gradient and velocity fluctuation correlations we used the first term in the series expansion of $U^m_{,n}$ as an approximation in the 1945 paper[2]. This expansion has been improved to consider the second term[5] as follows:

$$U'^m_{,n} = U^m_{,n} + \xi^r U^m_{,nr} \tag{1}$$

Then in (1.5) we have

$$\begin{aligned}[U'^m_{,n} (\overline{w'^n w_i})'_{,m}]'_{,k} &= U^m_{,n} (\overline{w'^n w_i})'_{,mk} \\ &\quad + U^m_{,nr} [(\overline{w'^n w_i})'_{,m} \delta^r_k + \xi^r (\overline{w'^n w_i})'_{,mk}].\end{aligned} \tag{2}$$

Similarly in (1.6) and (1.7), we have respectively

$$\begin{aligned}[U'^m_{,n} (\overline{w'^n w_k w_l})'_{,m}]_{,i} &= U^m_{,n} (\overline{w'^n w_k w_l})'_{,mi} \\ &\quad + U^m_{,nr} [(\overline{w'^n w_k w_l})_{,m} \delta^r_i + \xi^r (\overline{w'^n w_k w_l})_{,mi}],\end{aligned} \tag{3}$$

$$[U'^m_{,n} (\overline{w'^n w_k w_l w_p})'_{,m}]_{,i} = U^m_{,n} (\overline{w'^n w_k w_l w_p})'_{,mi} \tag{4}$$

$$+ U^{m}_{,nr}\,[\overline{(v'^{n}v_k v_l v_p)}'_{,m}\,\delta^{r}_{i} + \xi^{r}\,\overline{(v'^{n}v_k v_l v_p)}'_{,mi}\,].$$

Putting the above relations into the integrals of (1.5), (1.6) and (1.7) respectively and considering the functions defined in (1.9), we then obtain (1.11), (1.12); (1.13), (1.14) and (1.15) and (1.16). Here we note that the constant d'_{iklp} in (1.16) involves differentiations leading to macro-lengths in the denominator and quintuple correlations under its integral. Therefore d'_{iklp} can be set equal to zero.

We found before[2] the viscous dissipation terms in the equations of double correlation (1.5) by assuming that the double velocity correlation between two points P and P', $\overline{w_i w_k'}$, could be expanded in powers of ξ^j and the coefficients of $\xi_l\xi_m$ in the expansion should be linear functions of $\overline{w_n w_p}$, δ_{rs} and their products. The double correlation $\overline{w_i w_k'}$ should furthermore satisfy the equation of continuity

$$(\overline{w_i w'^{j}})_{,j} = 0 \qquad (5)$$

From $\overline{w_i w_k'}$ the dissipation term in (1.17) was obtained by taking partial differentions of $\overline{w_i w_k'}$ with respect to x^m and x'^n, setting x'^n equal to x^n and then contracting by g^{mn}.

The viscous dissipation terms in (1.18) and (1.19) can be obtained in the same way. For the term in (1.18) we assume that in the expansion of $\overline{w_i w_k w_l'}$ in powers of ξ^j, the coefficients of $\xi_m\xi_n$ should be linear combinations of $\overline{w_p w_q w_r}$, δ_{st} and their products. Likewise $\overline{w_i w_k w_l'}$ should satisfy the equation of continuity,

$$(\overline{w_i w_k w^{,l}})_{,l} = 0 \tag{6}$$

By the method used to obtain (1.17), we can get (1.18).

Formula (1.19) can be obtained by the same method and similar assumption explained for getting the dissipation terms in the equations of the double and triple correlations in (1.17) and (1.18) respectively. This method can be extended to find similar viscous dissipation terms for still higher approximations.

REFERENCES

[1] Chou, P. Y., *Chin. Journ. of Phys.*, **4** (1940), 1-33.

[2] Chou, P. Y., *Quart. of Appl. Math.*, **3** (1945), 38-54.

[3] Rotta, J. C., *Zeit. für Phys.*, **129** (1951), 547-572; **131** (1951), 51-77.

[4] Launder, B. E., Stress Transport Closure-into the Third Generation, *Turbulent Shear Flows.* (1979), 259-266. Springer-Verlag (Berlin) for the references there in.

[5] Zhou (Chou), Pei-yuan, *Scientia Sinica* (A), **28** (1985), 405-421.

[6] Wu, Zhong, M. S. Thesis, Department of Mechanics, Peking University, Beijing, China, 1986.

[7] Chou, Pei-yuan and Tsai, Shu-tang, *Acta Mechanica Sinica* (*Chin. Jour. of Mech.*), **1** (1957), 3-14.

[8] Huang, Yong-nian, *Acta Mechanica Sinica,* **8** (1965) 122-132.

[9] Bennett, J. C. and Corrsin, S., *Phys. of Fluids,* **21** (12) (1978), 2129-2140.

[10] Chou, Pei-yuan, Shih, Hsun-kang and Li, Sung-nian, *Acta Scientiarum Naturalium Universitatis Pekinensis,* **10** (1965),39-52.

[11] Chou, Pei-yuan and Huang, Yong-nian, *Scientia Sinica,* **18** (1975), 199-222.

[12] Huang, Yong-nian and Zhou (Chou), Pei-yuan, *Scientia Sinica,* **24** (1981), 1207-1230; also *Proc. Indian Acad. of Sci.* (Eng. Sci.), **4** (1981), 177-197.

[13] Chen, Shi-yi, M. S. Thesis, Department of Mechanics, Peking University, Beijing, China, 1984.

[14] Chou, Pei-yuan, *Scientia Sinica,* **8** (1959), 1095-1119.

[15] Fabris, G., *Phys. of Fluids,* **26** (1983), 422-427.

[16] Huang, Yong-nian, *Papers on Theoretical Physics and Mechanics in Commemoration of Professor Zhou (Chou) Pei-yuan's Eightieth Birthday,* (1982), 137-159; *Turbulence and Chaotic Phenomena in Fluids* (edited by T. Tatsumi) (1984), 357-364.

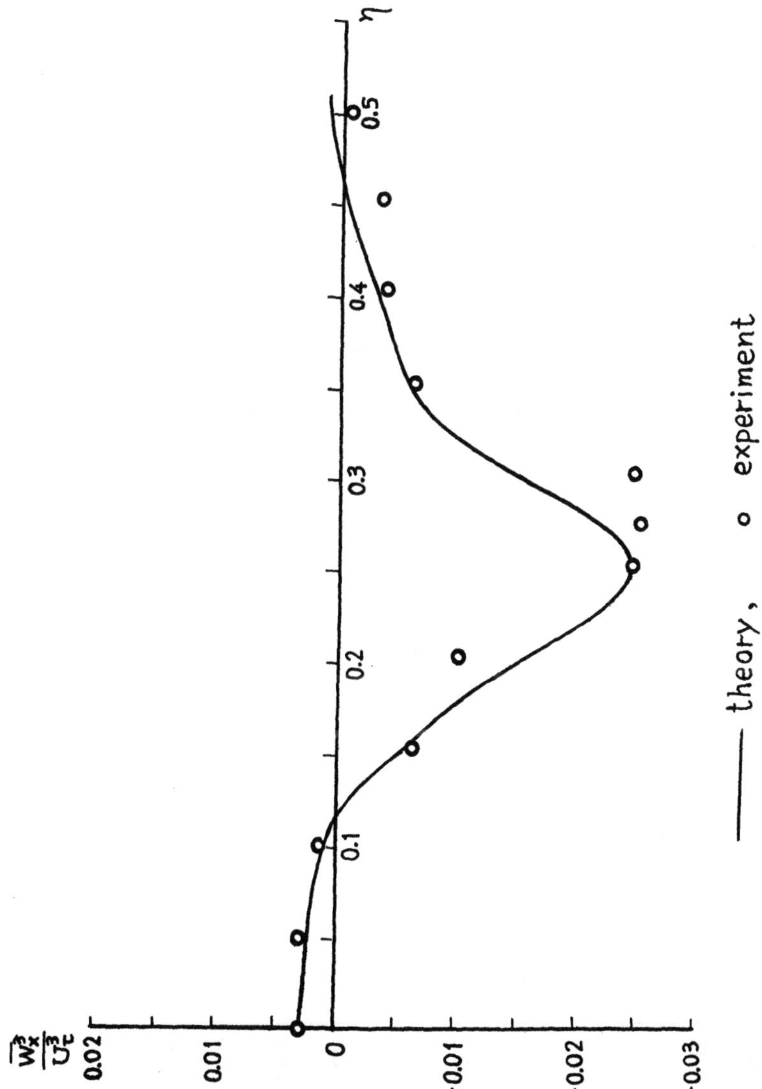
$\overline{W_x^3}/U_\tau^3$
0.02
0.01
0
-0.01
-0.02
-0.03
0.1
0.2
0.3
0.4
0.5
η
— theory, o experiment

Fig. 1

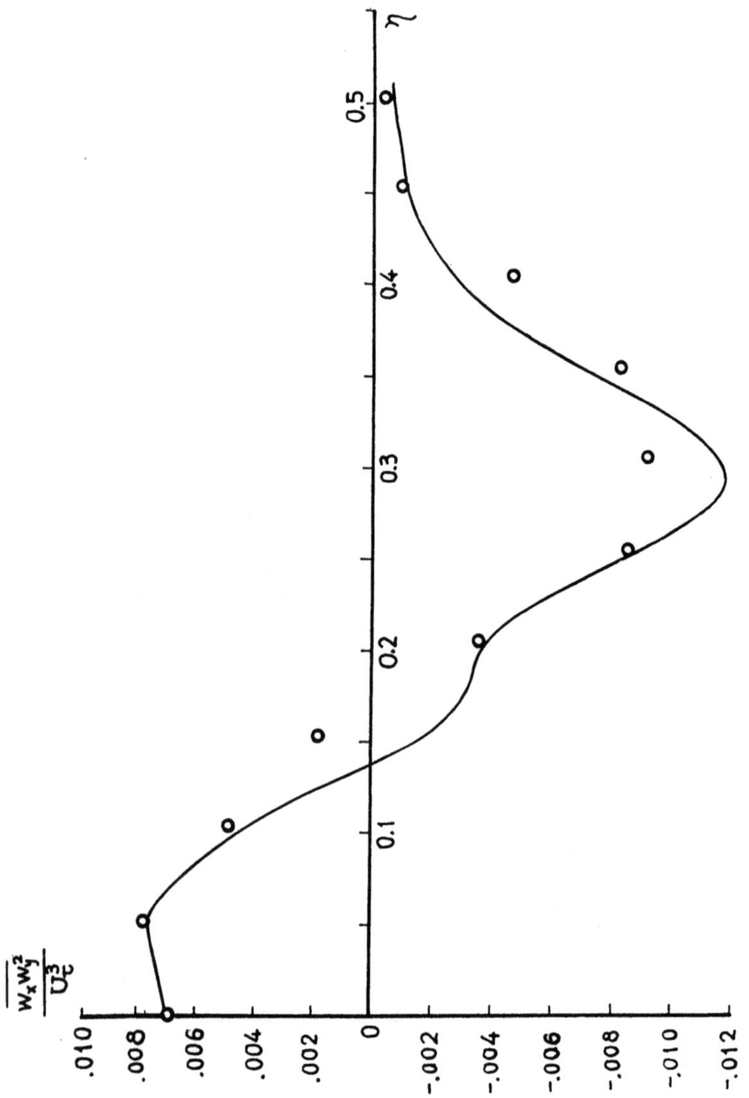

— theory, o experiment

Fig. 2

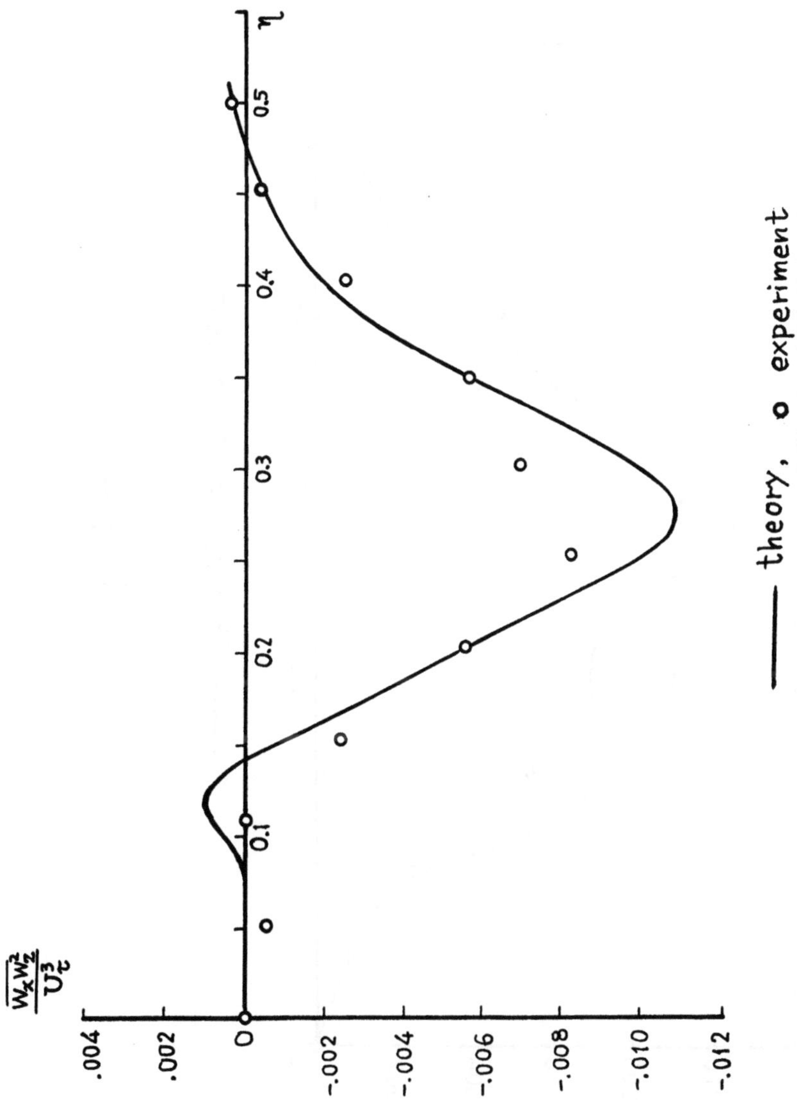

$\frac{\overline{w_x w_z^2}}{U_\tau^3}$
.004
.002
0
-.002
-.004
-.006
-.008
-.010
-.012
0.1
0.2
0.3
0.4
0.5
η
— theory, o experiment

Fig. 3

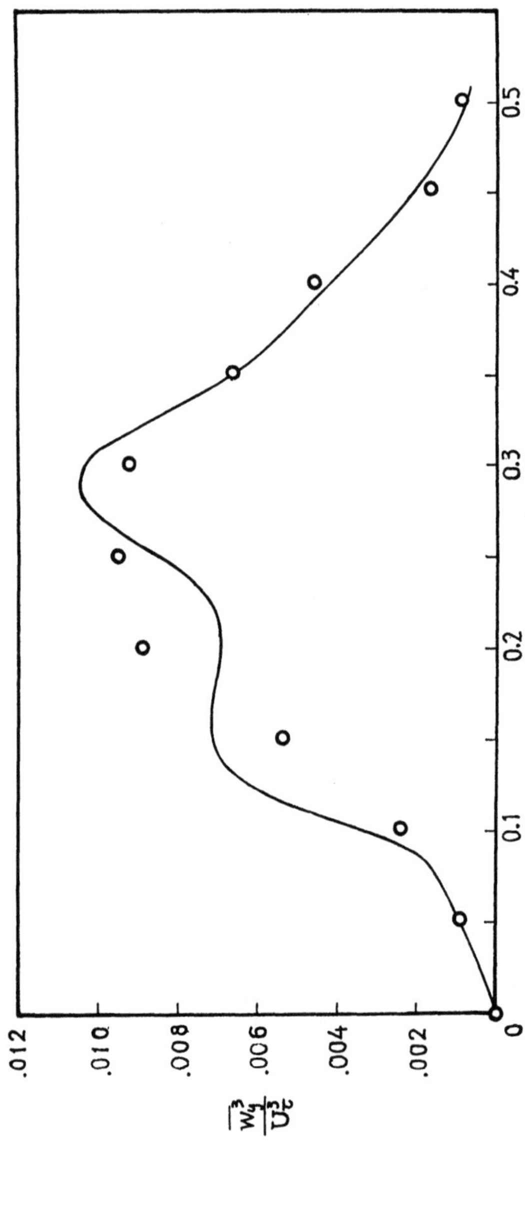

— theory, o experiment

Fig. 4

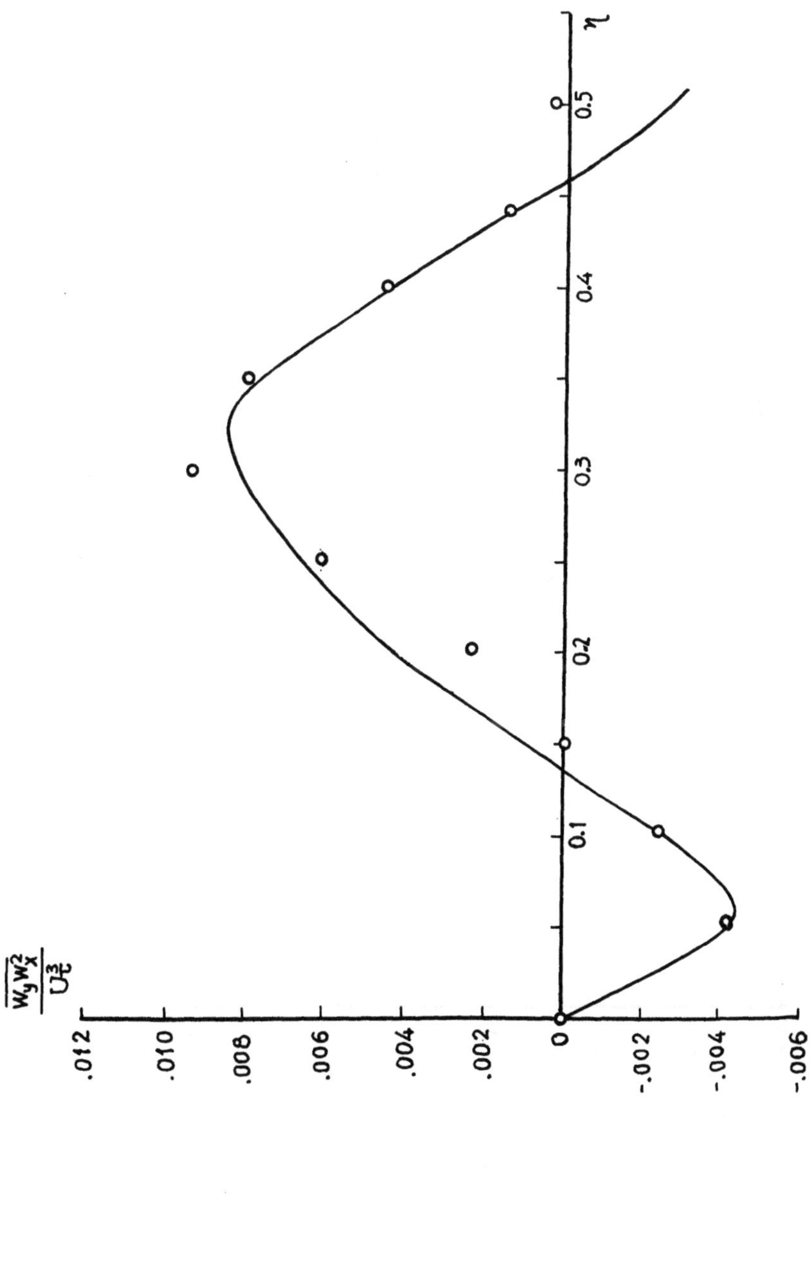

— theory, o experiment

Fig. 5

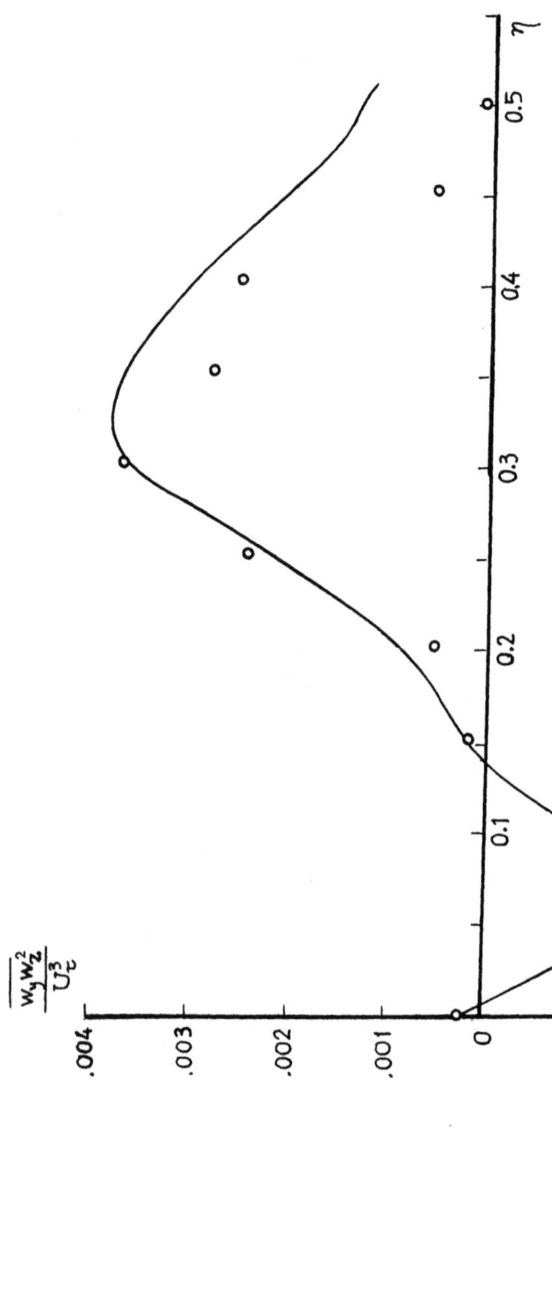

—— theory, o experiment

Fig. 6

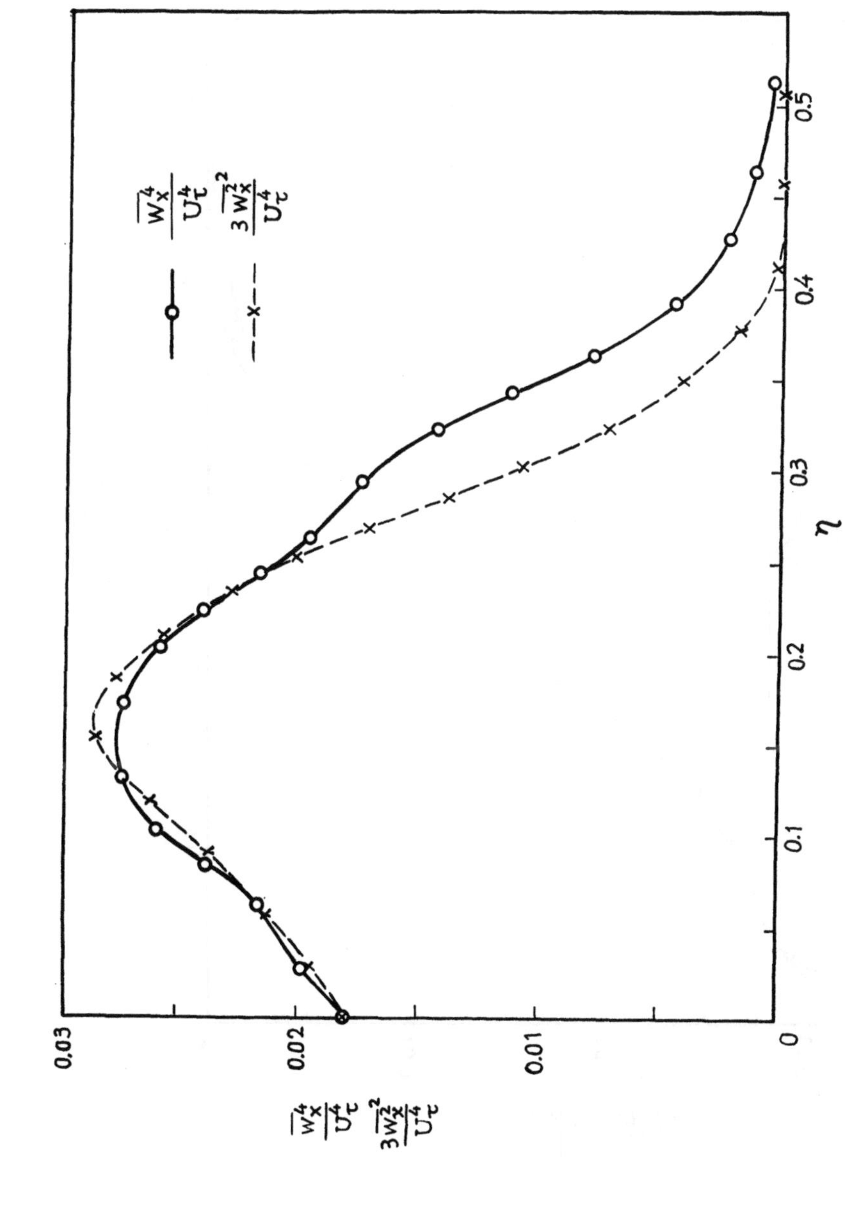

$\frac{\overline{w_x^4}}{U_\tau^4}$
$\frac{3\overline{w_x^2}^2}{U_\tau^4}$
0.03
0.02
0.01
0
0.1
0.2
0.3
0.4
0.5
η

Fig. 7

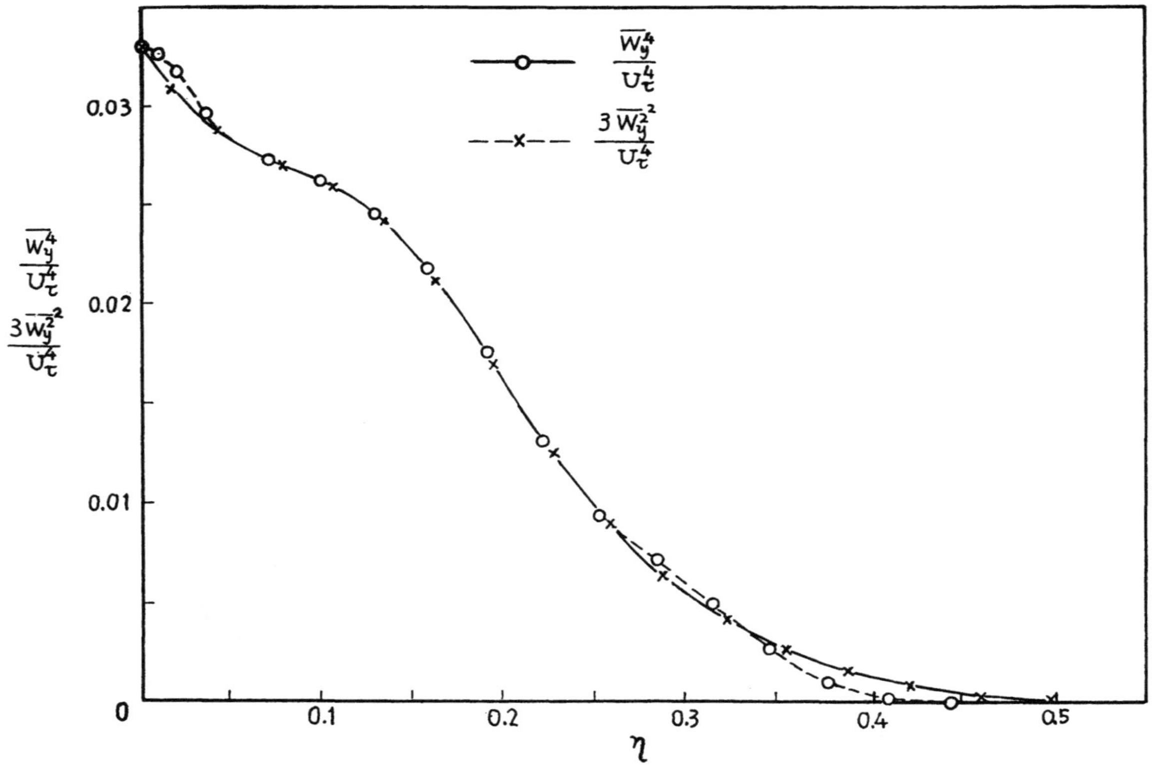
0.03
0.02
0.01
0
0.1
0.2
0.3
0.4
0.5
η
$\frac{\overline{W_y^4}}{U_\tau^4}$
$\frac{3\overline{W_y^2}^2}{U_\tau^4}$

Fig. 8

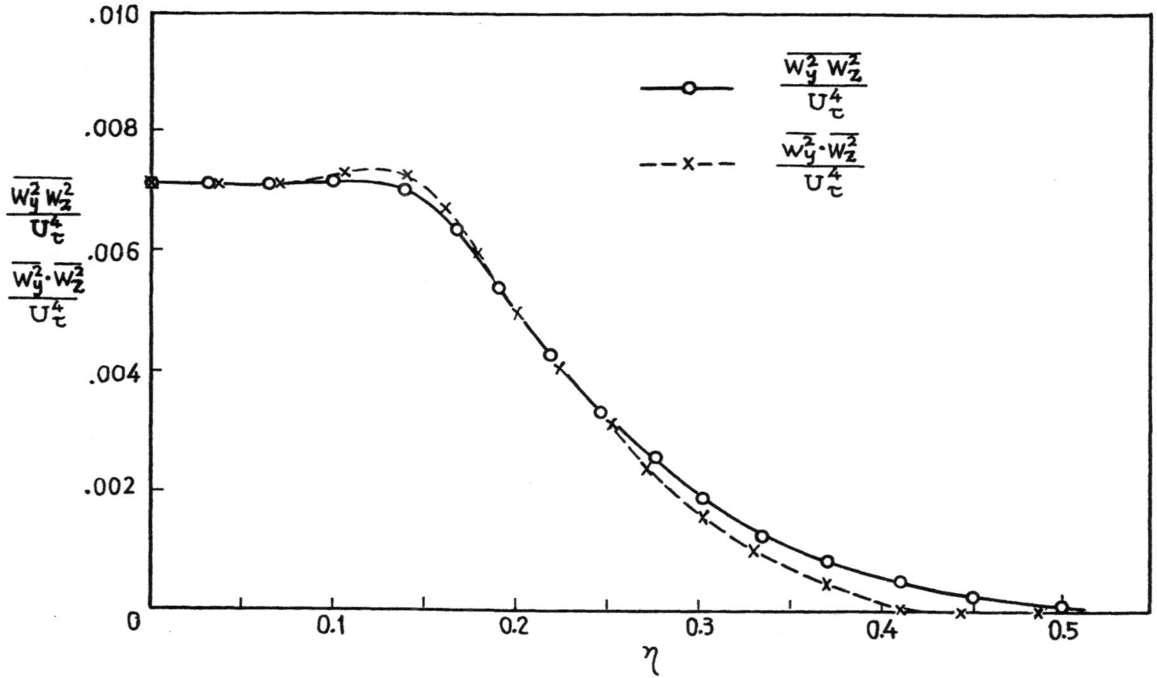
.010
.008
.006
.004
.002
0
0.1
0.2
0.3
0.4
0.5
η
$\frac{\overline{W_y^2 W_z^2}}{U_\tau^4}$
$\frac{\overline{W_y^2}\cdot\overline{W_z^2}}{U_\tau^4}$

Fig. 9

Three-dimensional Flows with Imbedded Longitudinal Vortices

P. BRADSHAW and A. D. CUTLER

Department of Aeronautics
Imperial College, London

Summary

Results for several projects on vortex flows are presented. The theme is the interaction between turbulent boundary layers and longitudinal vortices generated by lateral skewing of the flow further upstream. An obvious example is the passage of foreplane vortices over the main wing of a canard aircraft: however almost any flow round a surface-mounted obstacle will skew the flow in the boundary layer which has already developed on the surface, and thus produce longitudinal vorticity. The work to be described includes measurements downstream of various delta-wing vortex generators, as well as studies of vortex flows in wing-body junctions, wind-tunnel contractions, and impinging jets. Microcomputer-controlled traverse gear and data-logging procedures are essential for this kind of experiment.

Introduction

Dr. Rotta has made two crucial contributions to the calculation of three-dimensional ("3D") turbulent flows. First, in 1951 [1] he introduced to the Western literature the concept of calculation methods for turbulent flows based on the exact Reynolds-stress transport equations. Secondly, in 1977 [2] he pointed out that three- dimensionality of the mean flow has a direct effect on turbulence structure, and suggested how turbulence models could be modified to account for this effect.

Any method of predicting Reynolds stress is in some sense a model of the Reynolds-stress transport equations - for instance, the well-known mixing-length formula is obtained if all transport terms are neglected, and the Algebraic Stress Model (also known as "Algebraic Second Moment" model, ASM) arises if transport terms in the different Reynolds-stress equations are assumed to be proportional to the corresponding terms in the turbulent energy equation. We fear that nothing less than explicit term-by-term modelling of the Reynolds stress transport equations will suffice for engineering prediction methods in general turbulent flows - and in our pessimistic moments we fear that nothing short of large-eddy simulation [3] will suffice. If the future lies with large-eddy simulations, or full turbulent simulations including even the smallest eddies, then experimental fluid dynamics is a dying subject. At present, however, the computing costs and storage requirements of time-dependent simulations are much too high for routine engineering calculations, and contributions to Reynolds-stress transport equation models are still needed. These contributions can take two forms; theoretical/computational work on model development, and experiments intended to contribute understanding and/or test data for model development.

The large effect of three dimensionality on turbulence structure - even in the simple case of a 3D wing boundary layer - implies that methods developed for 2D flows cannot be reliably extended to 3D merely by taking model equations that proved satisfactory for the x direction in 2D flows, and applying them to the spanwise (z-wise) direction. Mathematically this would be a simple rotation of axes (or tensor indices) to produce - for example - a transport equation for the crossflow-plane shear stress $\overline{vw}$ with the same constants as the modelled transport equation for $\overline{uv}$ in 2D flow.

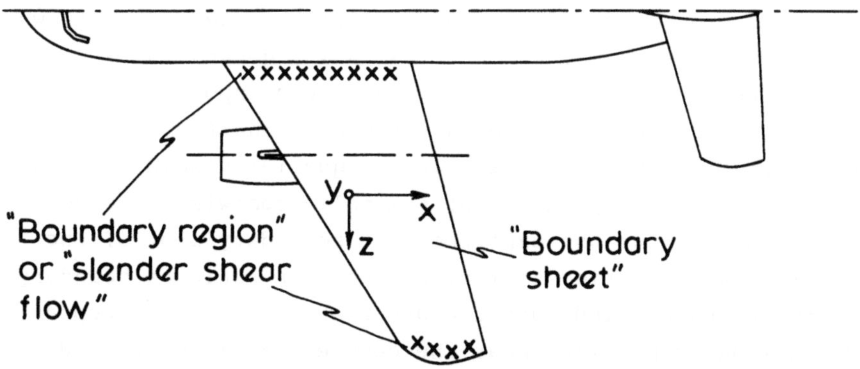

Fig.1. Two types of three-dimensional shear layer

The essential feature of a three dimensional flow is the appearance of longitudinal mean vorticity. The definition of "longitudinal" may be non-unique, but cannot be chosen to make the longitudinal vorticity zero everywhere, as in a 2D flow. Figure 1 shows the two forms of longitudinal vorticity (a) "cross flow", as in the boundary layer on a swept wing far from root or tip, in which the V component velocity (normal to the surface) is small and the longitudinal vorticity is approximately $\partial W/\partial y$, and (b) flows in which $\partial V/\partial z$ is of the same order as $\partial W/\partial y$, so that velocity vectors in the yz plane show circulation around a vortex core. Important though 3D wing boundary layers are, there are many fields of engineering in which three dimensionality is even stronger, with the appearance of longitudinal vortices and not merely crossflow.

The problems of measuring and modelling 3D wing boundary layers have by no means been entirely solved, but in this paper we propose to concentrate on flows with identifiable longitudinal vortices. A review of both kinds of 3D flow by one of us can be found in [4], so here we would like to concentrate on the work being done by members of our own group; supporting "review" material can be found in [4]. Also, Cousteix, in the present volume and in [5], has discussed calculation methods in detail: we will not repeat this material but merely comment on the implications of our experimental results for the future development of calculation methods.

Our research programme reflects our belief that secondary flows of Prandtl's second kind, which are actually generated by Reynolds-stress gradients, are very much less important in engineering practice than secondary flows of the first kind, in which longitudinal vorticity appears by quasi-inviscid "skewing" - rotation in the x,z plane - of nominally-spanwise (z-wise) vortex lines. In our one experiment on secondary flow of the second kind [6], we tried to find the maximum ratio of boundary-layer thickness to wind-tunnel width that would give acceptably two-dimensional conditions near the centre plane; but we had a good deal of difficulty with secondary flows of the first kind, which were generated in the wind tunnel contraction and tended to overwhelm the secondary flows of the second kind (generated in the working-section corners) which we were trying to study.

Most of our work has related to imbedded vortices in the boundary layer on a solid surface, since longitudinal vortices in free shear layers - e.g. wing wakes - exhibit the quasi-inviscid effects of vortex rollup, which generally confuse investigations of turbulence structure. However, two of our experiments have been strongly influenced by the rollup procedure: in [7], a cross-stream jet interacted with the boundary layer on the solid surface on which it impinged, and the longitudinal vorticity was due partly to the "horseshoe" vortex, which acquires its vorticity from the impingement-plane boundary layer, and partly to the streamwise vortices in the bent-over jet: see [8] for a useful comparison between the bent-over jet vortices and the trailing vortices of a lifting wing - both being a result of cross-stream flux of momentum. In our recent experiments [9] on a vortex pair imbedded in a turbulent boundary layer, the vortices were actually the trailing vortices of a delta wing mounted ahead of the test plate, and were not completely rolled up before they merged with the test-plate boundary layer.

As with most fluid dynamic phenomena, vortices are sometimes welcome, sometimes unwelcome. The trailing vortices behind a wing have been described as "the price we pay for lift" and there is a definite lower limit to the induced drag, which depends on the rate at which kinetic energy is fed into the vortices. A trailing vortex approximately conserves its circulation until it merges with the vortex from the opposite wing, so it is easy to see that an aircraft of small span can experience unwelcome rolling moments if it flies into one of the trailing vortices from an aircraft of large span (the practical result of this fluid-dynamic phenomenon being the mandatory minimum distance between a heavy aircraft and a following light aircraft in an airport approach pattern). Wing vortices have a welcome effect in the case of delta wings, including double-delta wings or wings with strakes, in which vortices generated at the leading edge of the wing pass over the top of the wing and generate extra lift by purely inviscid mechanisms (perhaps, at the same time, helping to produce a spanwise crossflow in the wing boundary layer and thus reduce the likelihood of boundary layer thickening and separation near the wing root). On swept-forward wings, the natural crossflow in the boundary layer leads to an accumulation of boundary layer fluid in the wing root, so that some form of special device, such as a strake, is needed near the root to clean up the boundary layer, whether or not the device produces any significant nonlinear vortex lift.

Vortices behind surface-mounted obstacles, such as tower buildings, submarine "sails" or conning towers, and aircraft wing roots (regarded as obstacles in the fuselage boundary layer) may or may not be welcome. In the "horseshoe" vortex pair behind a surface-mounted obstacle, the flow in the region between the vortices (hereafter called the "common" flow: see Fig. 2) is directed downwards towards the surface. In a wing-body junction, this simply convects low-momentum fluid from the wing wake into the fuselage boundary layer, increasing the likelihood of separation towards the rear of the fuselage. If the height of the surface-mounted obstacle is small, the "common flow" convects free-stream fluid down towards the surface, so the streamwise velocity near the surface may be larger than if the obstacle were not there (the so-called "negative wake" effect): there is of course an overall momentum loss corresponding to the drag of the obstacle.

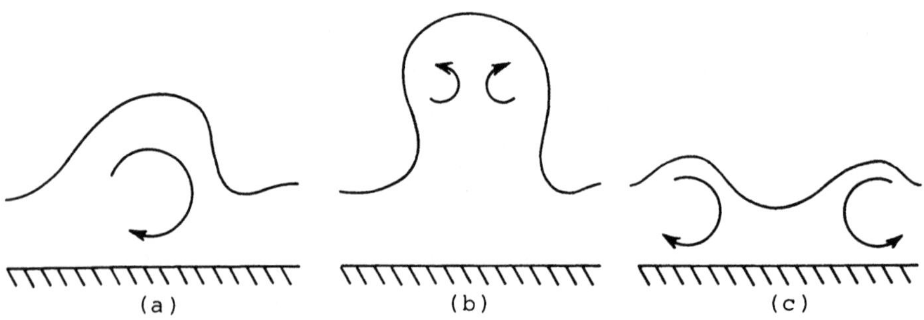

Fig.2. Three types of imbedded vortex (axial view):
(a) Single vortex; (b) Vortex pair with "common flow" upwards; (c) Vortex pair with "common flow" downwards

The horseshoe vortex around a bridge pier has the same sign as the wing-body junction vortex, and is unwelcome because of the erosion it causes around the front and side of the pier - but this is an effectively inviscid effect, occurring before the horseshoe vortex has been significantly affected by Reynolds stresses.

The vortices produced by "vortex generators" are by definition intentional and welcome, and indeed play an essential part in the performance of some aircraft. Usually, these vortex generators are added as curative measures at a late stage in design or testing of the aircraft. A sad but salutary example is the installation of vortex generators on the nacelles of the Boeing 757 and 767 aircraft [10]. These aircraft have large-diameter turbofan engines mounted under the wings, and an important design goal was to minimise the distance between the engine nacelle and the wing, to reduce the length and weight of the landing-gear legs required to give adequate ground clearance. The optimisation of the nacelle/wing design for cruise conditions was a triumph of computational fluid dynamics - the pylons between the engines and the wings are much shorter than in the previous generation of aircraft with under-wing engines. Unfortunately, at the higher angles of attack typical of takeoff and landing conditions, the flow over the nacelle interfered strongly with that over the wing, and vortex generators on the nacelle were needed to suppress flow separation. Similar vortex generators are fitted on the underwing nacelles of the DC 10 aircraft [11], and in both cases it has been possible to mount the vortex generators so that they are aligned with the flow under cruise conditions, thereby producing negligible induced drag, while still producing the required vortices when the aircraft is at a higher angle of attack.

Vortices form on the lee side of a body of revolution at high incidence (such as a manoeuvring missile or the forebody of an aircraft at high angle of attack [12]). There is an approximate analogy between the flow in the cross-sectional plane in these cases and the generation of a vortex street behind an impulsively-started circular cylinder in cross flow, which indicates that the vortex pattern in the high angle of attack cases may be asymmetrical, introducing a highly unwelcome side force. The "bilge vortices" underneath a ship appear because the bow behaves rather like a body at high angle of attack, but instead of travelling round to the lee side the vortices remain near the corners of the ship´s cross section, producing non-axisymmetric flow into the propeller - with resulting noise, vibration and structural fatigue.

This review of the mechanisms of formation of longitudinal vortices which then interact with turbulent boundary layers is simply intended to demonstrate their importance in engineering situations, and to justify presenting a fairly detailed overview of our work on this subject in the rest of the paper. The key to understanding secondary flows is that the generation of almost any strong vortex is an effectively-inviscid process (secondary flow of Prandtl´s first kind) but that subsequent diffusion of the vortex, and its interaction with other parts of the flow field, depends on turbulent stresses.

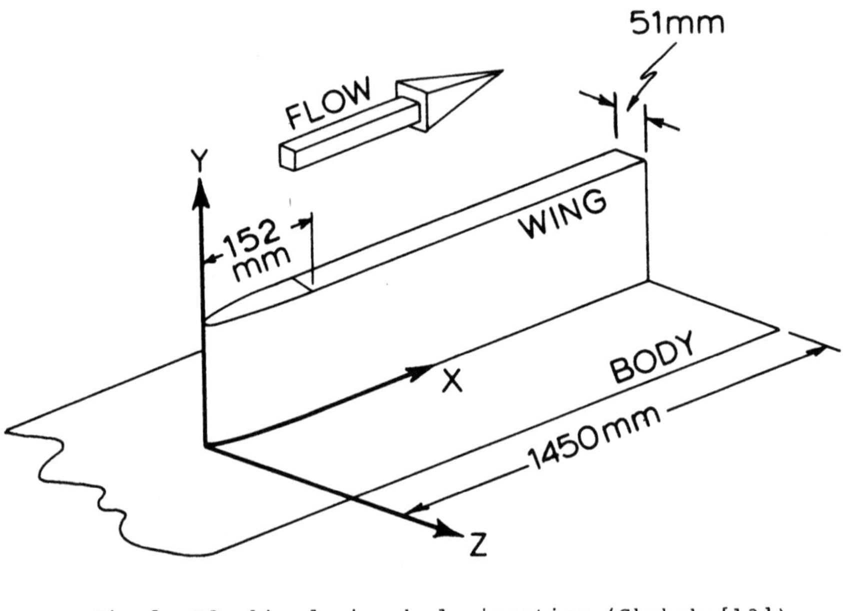

Fig.3. Idealized wing-body junction (Shabaka[13])

Imperial College Work

(a) Wing/body junction.

Work on imbedded vortices in our group in the Department of Aeronautics began with the measurements of Shabaka [13] on the flow in an idealized wing/body junction (Fig. 3): this was formed by a quasi-two-dimensional aerofoil, protruding from a flat surface and having a semi-elliptical nose followed by a long parallel section. Thus, although the horseshoe vortex was necessarily formed in the region of pressure gradient near the nose of the aerofoil, the subsequent diffusion of the vortex under the action of Reynolds stresses took place in nominally zero pressure gradient. Any failures of prediction methods to match the experimental results could not be blamed on pressure gradient (it seems to need saying frequently that the mean pressure does not appear in the Reynolds-stress equations in incompressible flow, so that mean pressure gradient cannot affect Reynolds stress directly). Briefly, Shabaka's results showed that the vortex decayed only slowly with distance downstream, and its diameter remained roughly equal to the thickness of the boundary layer on the body. See Fig. 4, in which we can see that W is larger near the surface than at the edge of the vortex core - this seems to be a common feature for vortices near solid surfaces, but is only partly explained by the notional presence of an image vortex below the surface.

Clearly, if a lump of fluid bearing a certain Reynolds-stress pattern is rotated, its contribution to the Reynolds-stress tensor will also be rotated. This effect appears explicitly in the Reynolds-stress transport equations, where some of the so-called "production" or "generation" terms actually represent rotation of the stress tensor by the mean-velocity gradients. A simple example is the term $\overline{uw}\partial V/\partial z$ in the $\overline{uv}$ transport equation, representing the rotation of contributions to $\overline{uw}$ by $\partial V/\partial z$ (it is interesting, but not very meaningful, to note that $\partial W/\partial y$ does not appear, although like $\partial V/\partial z$ it contributes to the x-component vorticity). Although the effects of the vortex on the Reynolds-stress tensor might be

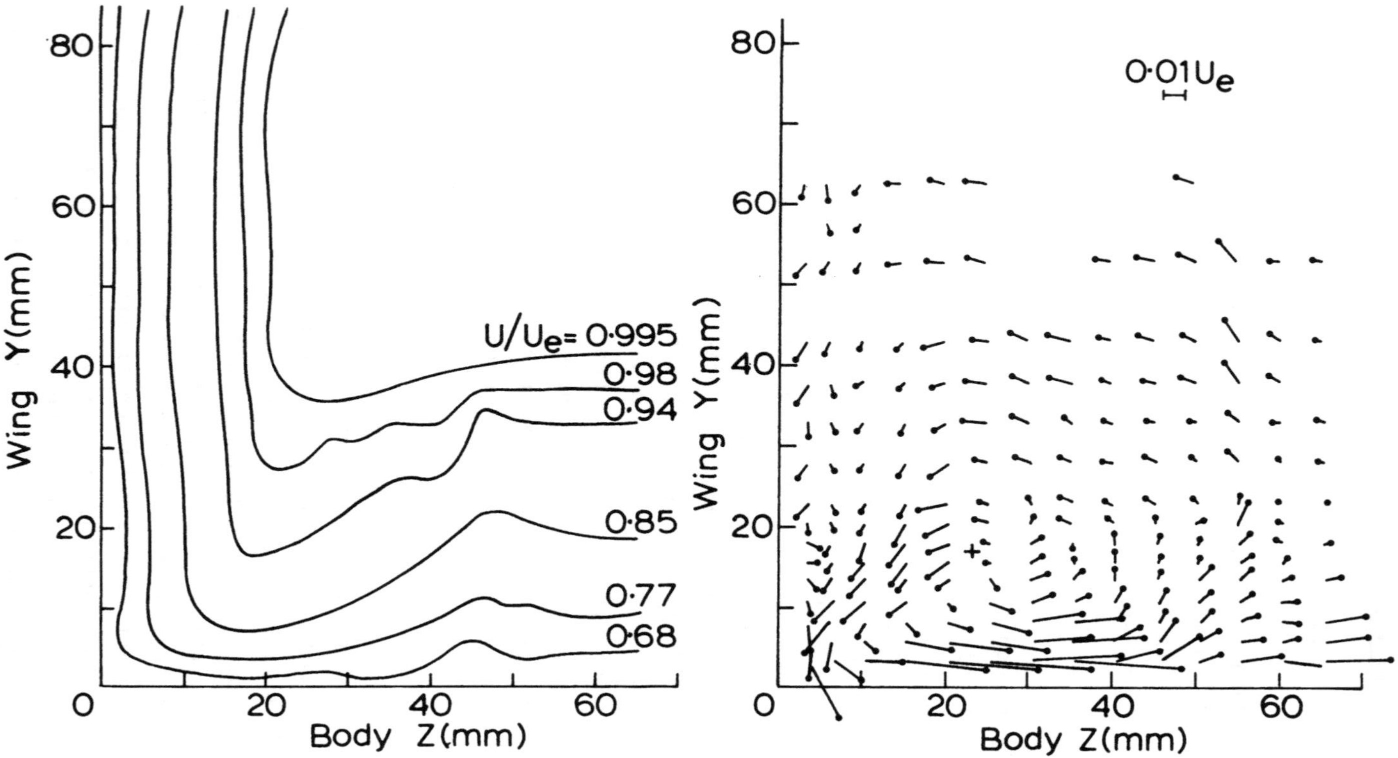

(a) Axial velocity contours (b) Secondary flow vectors

Fig.4. Flow in wing-body junction [13]: y axis along wing

qualitatively represented by these extra generation terms in the Reynolds-stress transport equations, quantitative behaviour is considerably more complicated. Detailed Reynolds-stress measurements will not be presented in the present review, but are to be found in our published papers.

(b) Single vortex imbedded in flat-plate boundary layer

Shabaka's experiment was originally conceived as an idealization of a common problem in axial turbomachinery, the flow in a blade/hub junction. As the flow was so complicated, (the vortex being constrained by two perpendicular solid surfaces) it seemed valuable to perform an experiment on the interaction of an isolated vortex with the boundary layer on a flat surface. Accordingly, we [14] installed a half-delta-wing vortex generator in the settling chamber of a wind tunnel (Fig. 5), so that the vortex which passed through the wind tunnel contraction

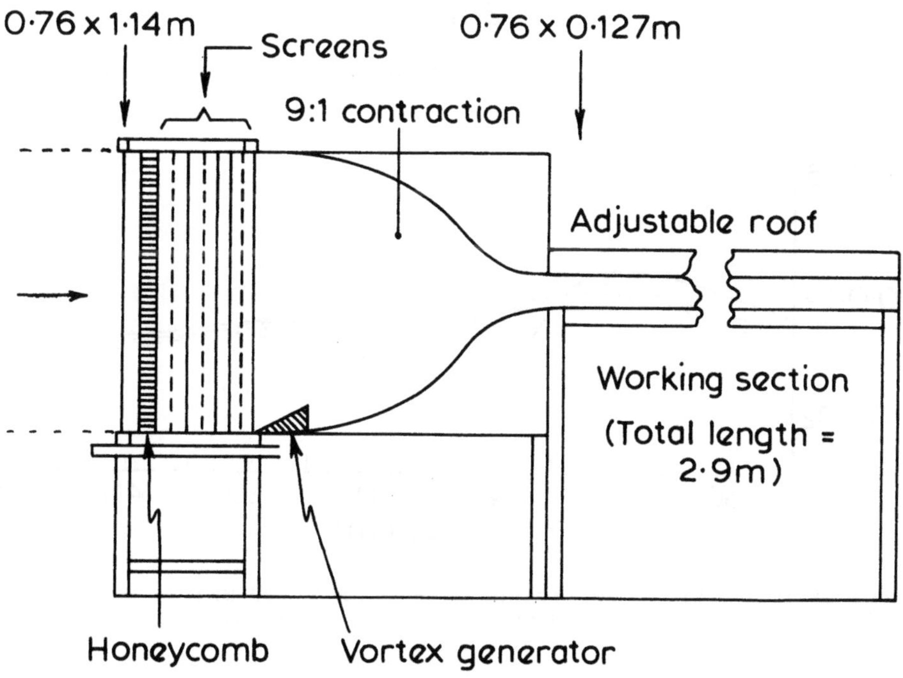

Fig.5. Vortex generator in blower tunnel settling chamber

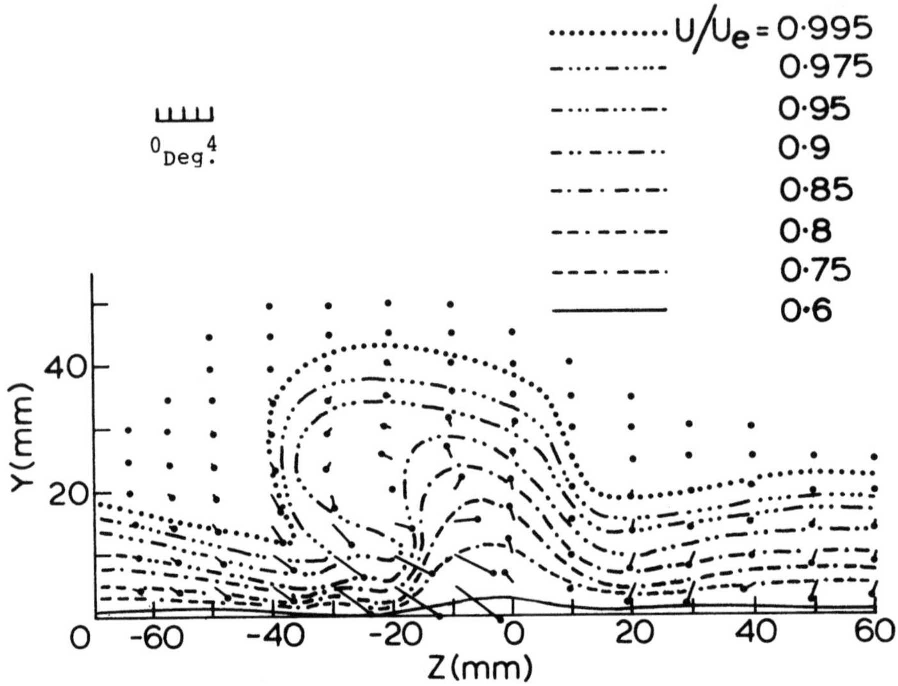

(a) Axial velocity contours and secondary-flow vectors, x = 1330 mm

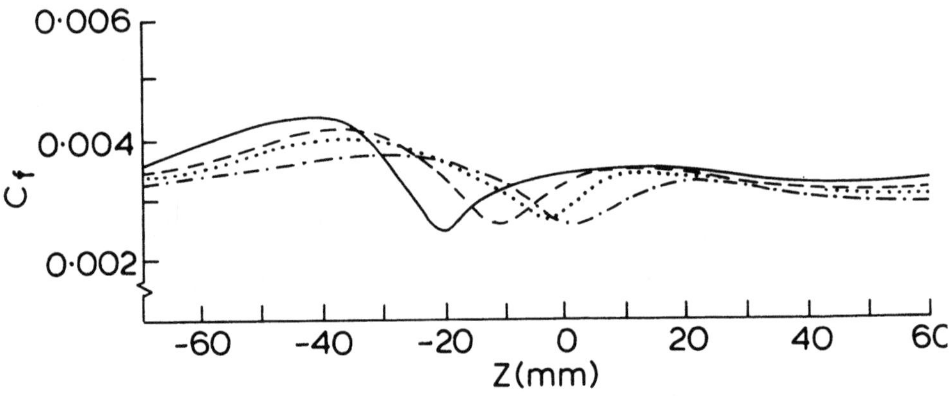

(b) Skin-friction coefficient. ———, x = 720 mm; — —, x = 1330 mm; · · · · ·, x = 1940 mm; —·—·, x = 2550 mm

Fig.6. Flow in single imbedded vortex [14]

and entered the working section had the same circulation as that leaving the vortex generator, while the total-pressure deficit in the wake of the vortex generator was very small compared with the dynamic pressure in the working section. Thus, the vortex entering the working section was as pure as possible, the only shortcoming being that the wind tunnel contraction was two-dimensional (the width of the wind tunnel normal to the paper in Fig. 5 being constant) so that in principle the vortex was flattened from its initial circular shape while passing through the contraction. In fact, no trace of flattening was found in measurements in the working section, and the simplicity of the setup contrasts with that used by Westphal, Eaton and Pauley [15] in which half-delta-wing vortex generators were mounted in the working section and produced a highly disturbed flow. Our measurements in the single imbedded vortex have already been published [14] so in the present paper we show only sample results, including the mean velocity components and the spanwise distribution of skin-friction coefficient (Fig. 6).

(c) Imbedded vortex pairs.

Next, the arrangement of half-delta wing vortex generators mounted in the settling chamber was extended to a pair of generators, producing a vortex pair with the "common flow" between the vortices directed upwards, away from the floor of the working section. In this case the vortices tended to drift away from the surface by self-induction and by interaction with the pair of image vortices below the solid surface (Fig. 7). However, the ratio of the vortex height to the boundary layer thickness remained roughly equal to two, and boundary layer fluid was entrained into the vortices. Although the temperature-tagging technique used in some of our other experiments was not employed systematically in this case, it was clear that, even within the limited length of the test section, boundary-layer fluid first entered the common flow and was then lifted away from the surface and circulated around the outside of each vortex. Detailed results [16] show that Reynolds-stress correlation coefficients, although still quite large, are significantly different from those expected in 2D boundary

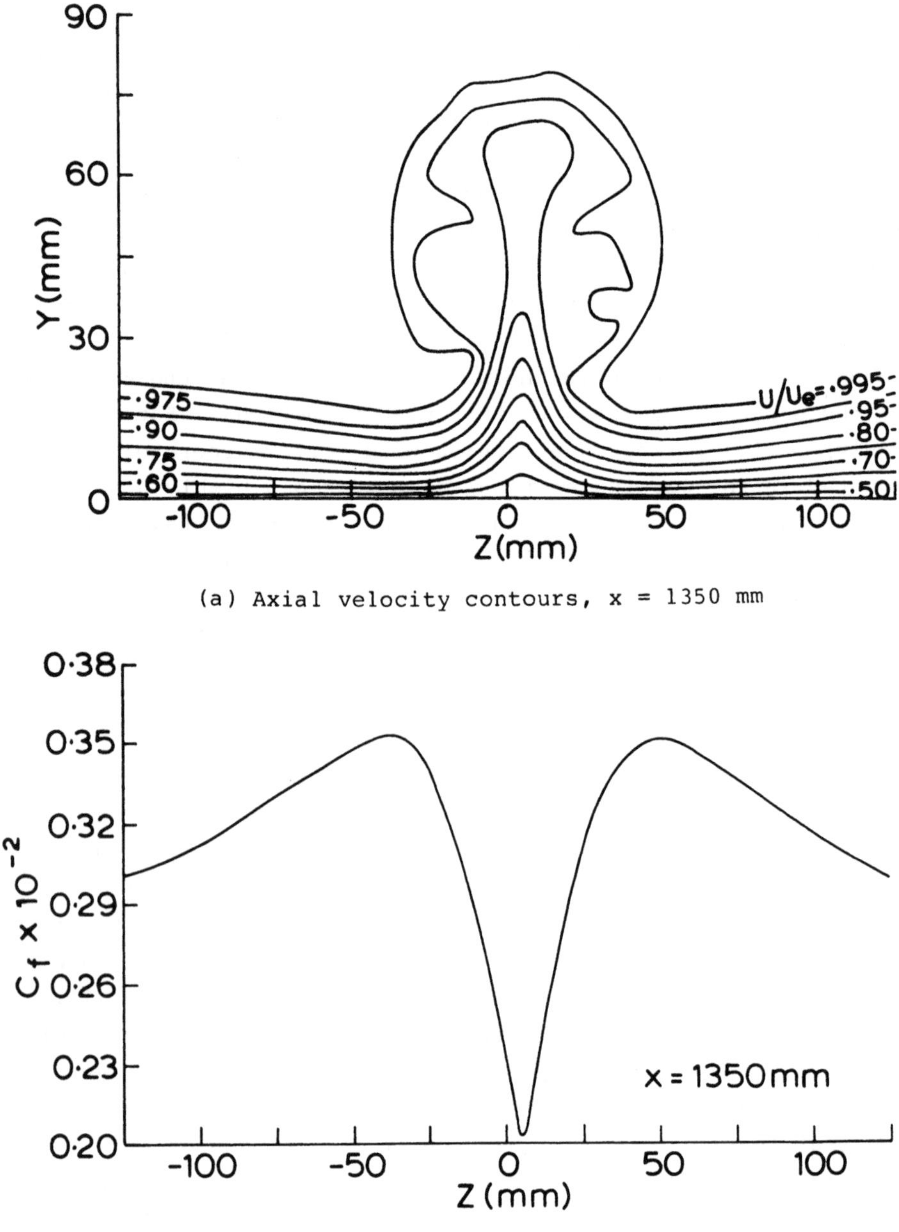

(a) Axial velocity contours, x = 1350 mm

(b) Skin-friction coefficient, x = 1350 mm

Fig.7. Flow in imbedded vortex pair, common flow upwards [16] layers because of the rotation of the Reynolds-stress tensor; also, the "transport velocities" of the different Reynolds stress components, defined as the rate of transport of Reynolds stress divided by the Reynolds stress itself, show extremely strange behaviour. If turbulent transport of Reynolds stress were due to identifiable large eddies, one would expect transport velocities to be roughly equal for the different Reynolds stresses. In the case of the turbulent energy equation, the transport velocity near the outer edge of the shear layer is formally equal to the entrainment velocity because "transport" terms dominate, but in the individual Reynolds-stress equations the pressure-strain terms near the edge of the shear layer are of the same order as the transport terms, so the individual equations do not reduce to "turbulent transport" equals "mean transport". The concept of turbulent transport by the large eddies is more plausible physically than the alternative gradient diffusion model, but even the large-eddy transport hypothesis appears to be inadequate in turbulent flows that are grossly distorted by longitudinal vortices. Transport velocities for one of our more recent experiments [9] are presented below.

In a companion experiment to the work on a vortex pair with the "common flow" upwards, we [17] studied a vortex pair with approximately the same strength but with the "common flow" downwards. When the "common flow" is upwards, the vortices themselves move upwards (qualitatively as in inviscid flow) but when the common flow is downwards the vortices move downwards: when they encounter the surface, the vortices necessarily move apart in the spanwise direction, rapidly becoming so far apart that they are essentially independent. The flow near each vortex is sufficiently close to that in a single imbedded vortex that a calculation method capable of predicting the one would almost certainly be able to predict the other. However, the boundary layer between the vortices is interesting in itself, as an example of the effects of strong lateral divergence - although the setup is not the best for a "clean" experiment on the laterally-diverging flow. A remarkable feature of the boundary layer between the vortices in this experiment was that the level of shear stress in the outer part of the boundary layer near the centre plane was abnormally low, although the distribution of turbulent energy appeared normal. This seemed to be a genuine effect of the lateral divergence, and not entrainment of vortex fluid with significant turbulent energy but negligible shear stress: in general [18], lateral divergence tends to increase turbulent shear stress more than turbulent energy.

The above experiments all involved generating vortices by inserting obstacles (wing roots, half-delta wings) into the flow. We [19] have also studied a naturally-occurring vortex pair, with "common flow" upwards, on the flat floor of a two-dimensional wind-tunnel contraction. Here, longitudinal vorticity is generated by lateral skewing of the contraction boundary layer (secondary flow of the first kind): the vorticity is amplified by longitudinal acceleration, and then concentrated into a vortex pair, with "common flow" upwards, by lateral convergence. The flow is of course symmetrical about the floor centre line: the tunnel is wide enough that the vortices do not contain fluid from the sidewall boundary layers. The process of formation of the vortex pair is quite rapid, with the interesting result that the boundary layer at the contraction exit shows little evidence of distortion by the vortices. In particular, there is no trace of a decrease in surface shear stress on the centre line, which would be expected as the result of lateral convergence of the flow close to the surface (induced by the vortex pair). In fact, the surface shear stress at the contraction exit (Fig. 8, x = 0) shows a maximum near the centre line: the explanation is that the longitudinal vorticity near the surface is of the opposite sign to that in the vortex above it, as a consequence of the no-slip condition on W. The expected minimum in surface shear stress is slow to develop: it is noticeable at x = 750 mm in Fig. 8, being about 8 times the boundary layer thickness at exit - a warning that even the mean flow´s response to perturbations can be quite slow.

In the experiments described above, secondary-flow velocities were no more than ten percent of the axial velocity - which is an operational definition of a "weak" vortex: as far as the experimenter is concerned, a weak vortex is one in which

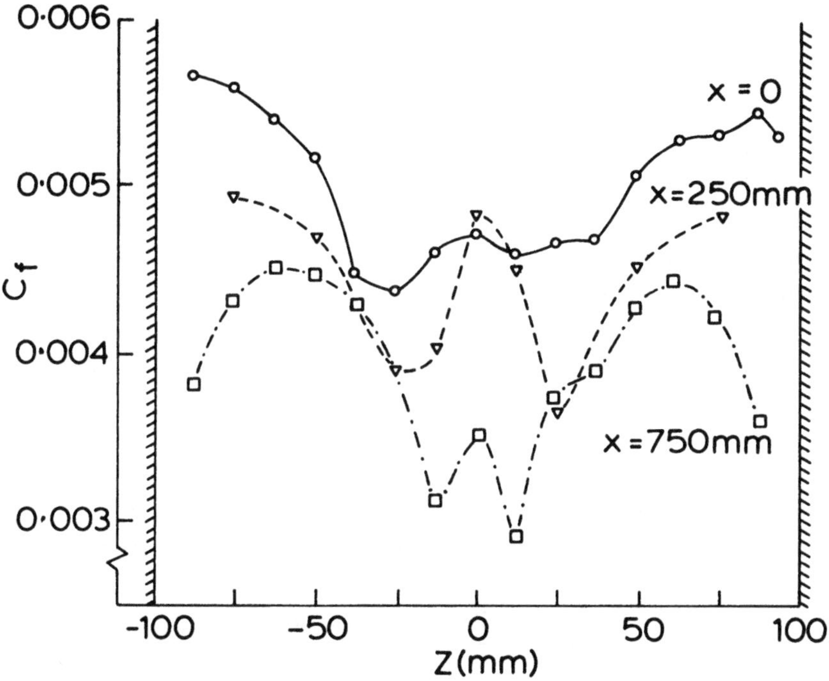

Fig.8. Naturally-occurring vortex pair (common flow upwards) emerging from wind-tunnel contraction [19]: skin friction coefficient at various distances from exit, boundary layer thickness 100 mm approx.

experiments can be made without needing to yaw the hot-wire probe into the local flow direction at each point. It is difficult to produce strong vortices with vortex generators in the settling chamber, because the generators become physically too large. Also, a single half-delta vortex generator mounted on the tunnel floor inevitably produces a large side force, which leads to large crossflow in the working section, even well outside the region of direct viscous interaction with the vortex. Therefore, although it is possible to produce weak vortices of a fairly pure generic type with vortex generators in the settling chamber, interactions between strong vortices and boundary layers are likely to be considerably influenced by the method used to generate the vortices, which should therefore be chosen to be relevant to a particular engineering configuration.

The experiment currently being carried out by one of us (A.C.), as briefly described in [9], is nominally a strong-vortex sequel to [14] on a single weak vortex: in fact we chose to generate a vortex pair with the common flow downwards (so that the vortices rapidly moved apart in the spanwise direction) as in [17]. The configuration (Fig. 9) is an idealization of a strake-wing or double-delta aircraft, with a gap between the highly-swept front portion and the main plane (in this case completely unswept). Thus the foreplane vortices passed over the top of the flat plate representing the main plane, while the non-rolled-up part of the foreplane wake passed underneath the main plane rather than contaminating the boundary layer on the latter. In order to ensure that the boundary layer on the test plate near the centre plane remained turbulent, despite being thinned by strong lateral divergence, it was desirable to make the model as large as possible while ensuring that the laterally-diverging vortices

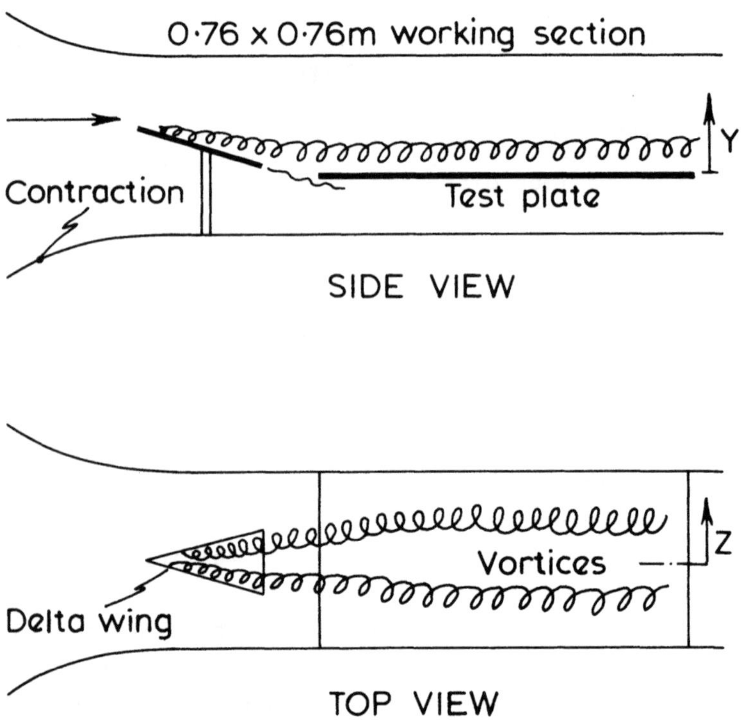

Fig.9. Schematic of delta-wing vortex generator and test plate [9]: strong vortex pair with common flow downwards.

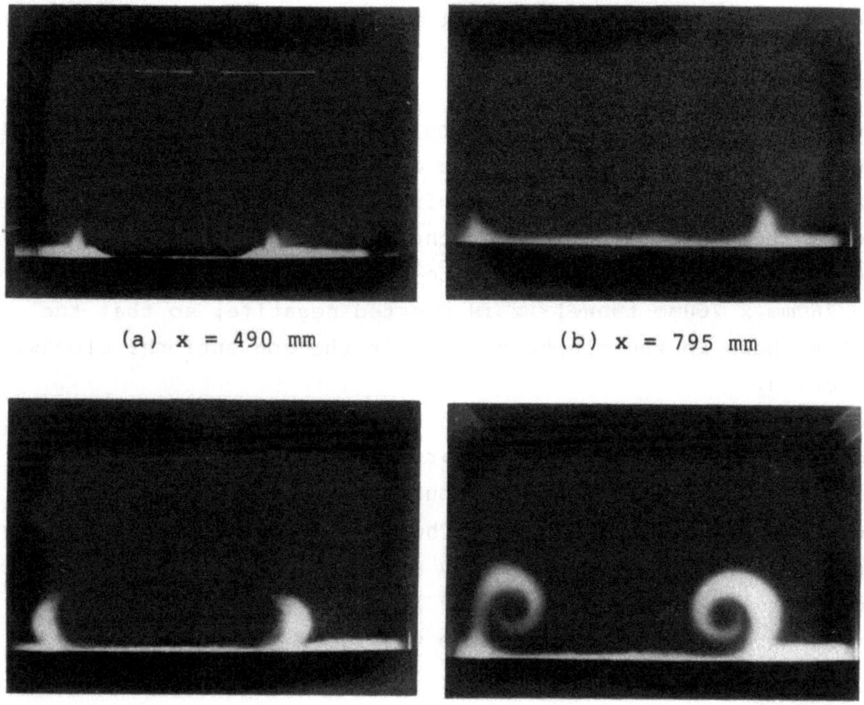

(a) x = 490 mm

(b) x = 795 mm

(c) x = 1100 mm

(d) x = 1400 mm

Fig.10. Illuminated cross-section of strong vortex pair: smoke-wire along leading edge of test plate (smoke fills boundary layer)

did not reach the tunnel walls before the end of the test section. To ensure this, and to simulate a canard aircraft configuration, we mounted the delta wing fairly close to the test plate, far too close for the vortices to be completely rolled up before encountering the test-plate boundary layer. Specifically, although the part of the delta-wing wake near the centre line does pass underneath the test plate as intended, some of the fluid leaving the trailing edge of the delta wing spirals around the leading-edge vortex without being completely rolled up. Now because the leading-edge vortices induce a strong lateral divergence of the flow on the top surface of the delta wing, the longitudinal vorticity shed from the outer part of the delta-wing trailing edge is of the opposite sign to the main vortex: the flow a short distance downstream of the trailing edge contains a primary vortex from each leading edge, with a secondary vortex of opposite sign spiralling around it.

Figures 10 to 14 show some of the results of the experiment. In the graphs, S is the span of the delta wing, 267 mm, and x is measured from the leading edge of the main plate. The spanwise coordinate z has its origin on the geometrical centre line of the delta wing, which is 75 mm to one side of the centre line of the 760mm x 760mm tunnel: z is plotted negative, so that the vortex shown in the graphs rotates in the conventional clockwise direction.
The long-exposure photographs in Fig. 10 show illuminated cross-sections of the flow at various streamwise distances: smoke was introduced into the boundary layer near the leading edge, effectively filling it. The extraction of boundary-layer fluid by the vortices is clearly seen: note also the thinness of the laterally-diverging boundary layer between the vortices, compared to that outboard of the vortices.

Fig. 11 shows the vortices, with smoke introduced by heating a kerosene-coated wire spanning the tunnel very close to the trailing edge of the delta wing: smoke was emitted only over the span of the delta wing, not across the full width of the tunnel. The main vortices pass above the smoke wire. The smoke from near the wing tips enters the secondary upper-surface vortex and forms the thick "eyebrows" seen in Figs. 11(a) and (b): their predominant vorticity is that of the secondary vortex, opposite to that of the main vortex, and the fluid is therefore slow to roll up into the main vortex. The almost smokeless "eyes" are the vortex cores: smoke-free or dye-free cores are common features of vortex pictures, and are often said to be due to outward centrifuging of heavy particles, but our results show quite conclusively that this is not a necessary mechanism. As many flow visualizations and other studies have shown, vortex cores are usually non-turbulent, being stabilized by the large positive radial gradient of angular momentum: therefore, fluid can be diffused in or out of the core only by molecular transport, which is much weaker than turbulent transport. It follows that a delta-wing vortex core can be filled with smoke

(a) x = 0 (plate l.e.)

(b) x = 185 mm

(c) x = 490 mm

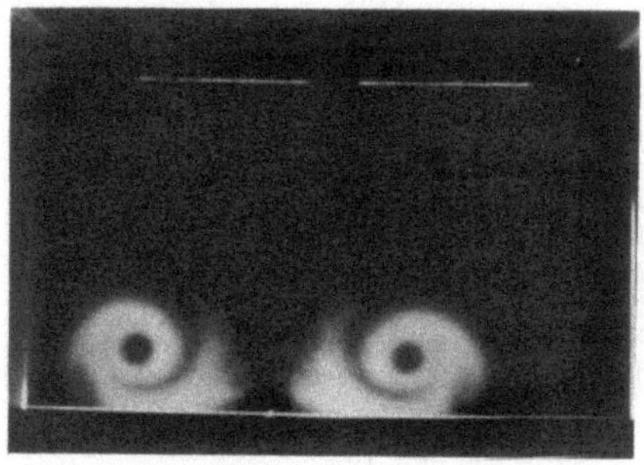

(d) x = 795 mm

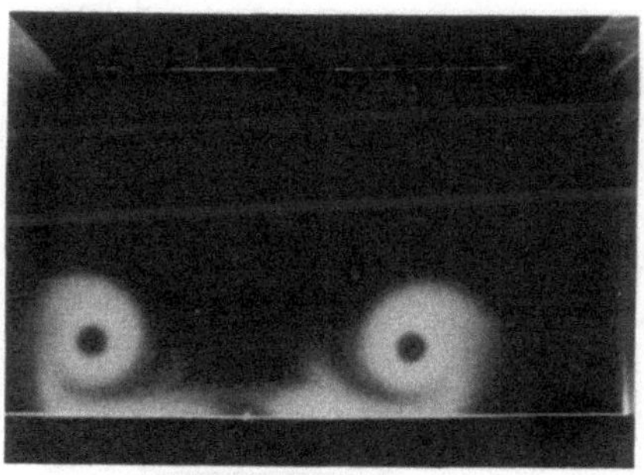

(e) x = 1100 mm

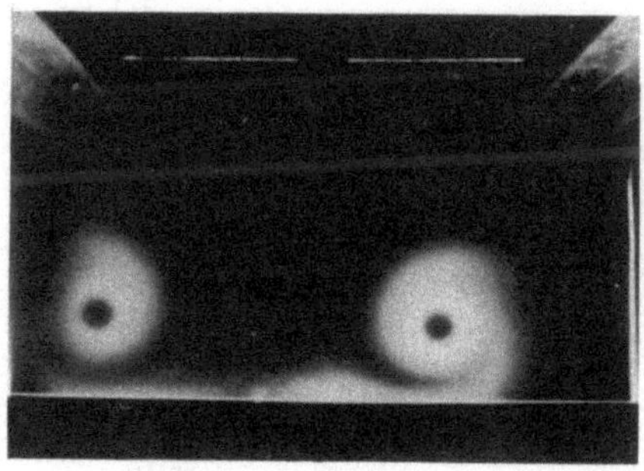

(f) x = 1400 mm

Fig.11. Illuminated cross-section of strong vortex pair: smoke-wire along trailing edge of delta wing. Bright line shows test plate position

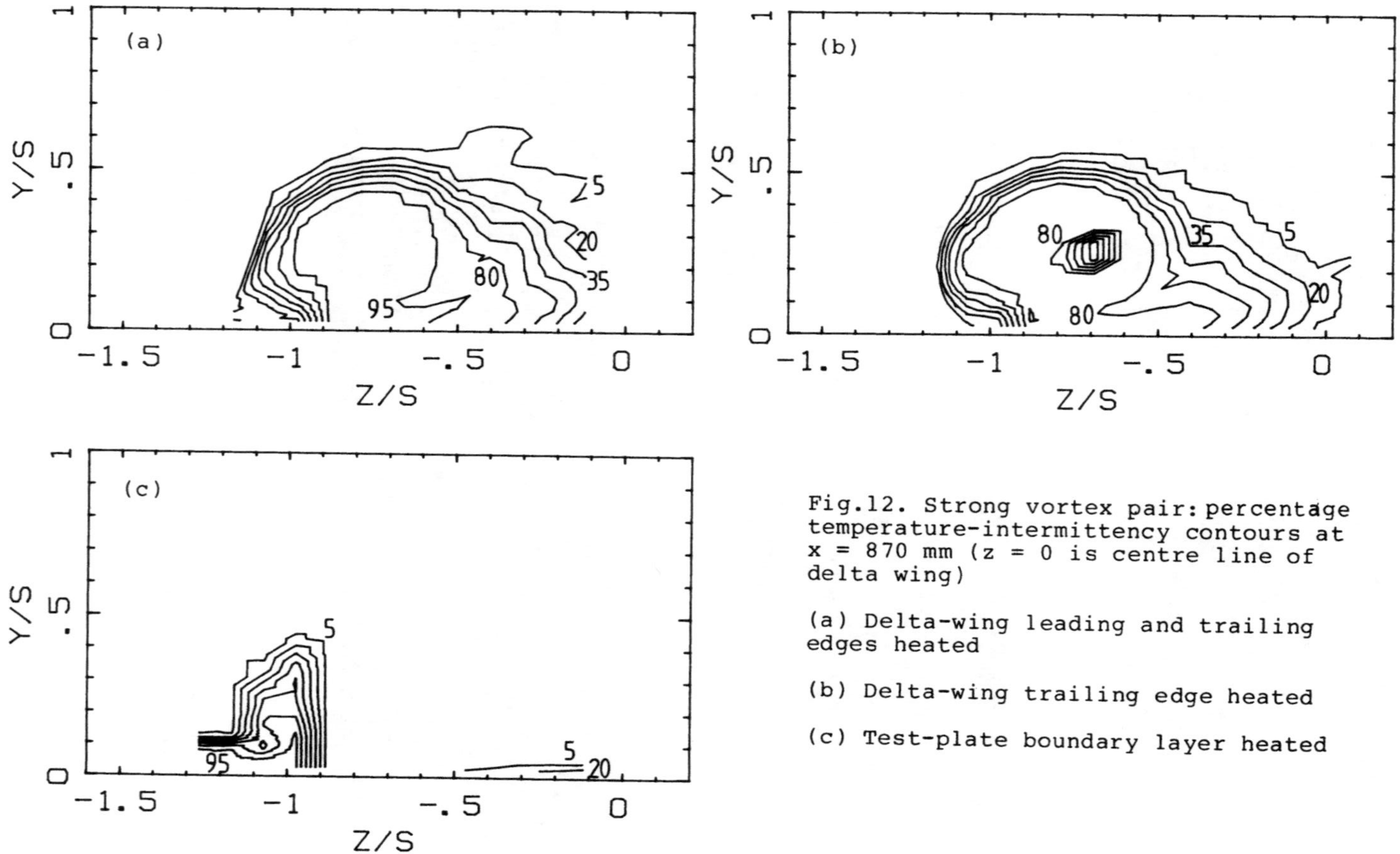

Fig.12. Strong vortex pair: percentage temperature-intermittency contours at x = 870 mm (z = 0 is centre line of delta wing)

(a) Delta-wing leading and trailing edges heated

(b) Delta-wing trailing edge heated

(c) Test-plate boundary layer heated

or other contaminant only if the contaminant is introduced over most of the length of the leading edge, because the vortex sheet that forms the core originally separates from the leading edge.

To demonstrate that centrifugal effects are not the cause of "empty" vortex cores, Fig. 12 shows the intermittency of the temperature fluctuation, measured with weak heat sources in various places in the flow - another example of our use of heat for a kind of quantitative flow visualization. (The temperature signal is roughly an "on-off" one, so the intermittency is roughly proportional to the mean temperature rise at a given point.) Fig. 12(a) shows that with the whole delta wing (including the leading edge) heated, the intermittency factor rises almost to unity in the core, which is centered at about $y/s = 0.25$, $z/s = -0.7$. Fig. 12(b), with heat introduced only at the delta-wing trailing edge, shows an intermittency factor of less than 0.2 in the core - that is, heated air reaches the core for less than 20 percent of the time. This is entirely consistent with the photographs of smoke-free cores in Fig. 11. Now smoke may be heavier than the surrounding fluid, but heated air is undoubtedly lighter, and any centrifugal effect on heated air would drive it INTO the vortex core: the consistency of Fig. 11 ("heavy" smoke) with Fig. 12 ("light" heated air) suggests that centrifugal effects are not a necessary mechanism for the absence of contaminant from vortex cores if the contaminant is introduced outside the core. (Undoubtedly, sufficiently heavy particles will centrifuge out, and this may explain the observation of empty cores in atmospheric tornadoes - though the usual pictures found in textbooks simply show a black outer region contaminated by ground erosion, and a white inner core of water vapour condensed under low pressure.)

Figures 13 and 14 show a selection of hot-wire results. The contour plots were generated by a simple algorithm which does not contain any smoothing: the vector plots show the fineness of the measurement grid, and give a better impression of the smoothness of the data. A fairly sophisticated microcomputer system was used in this experiment, both to control a four-degree-of-freedom traverse gear and to process the hot-wire

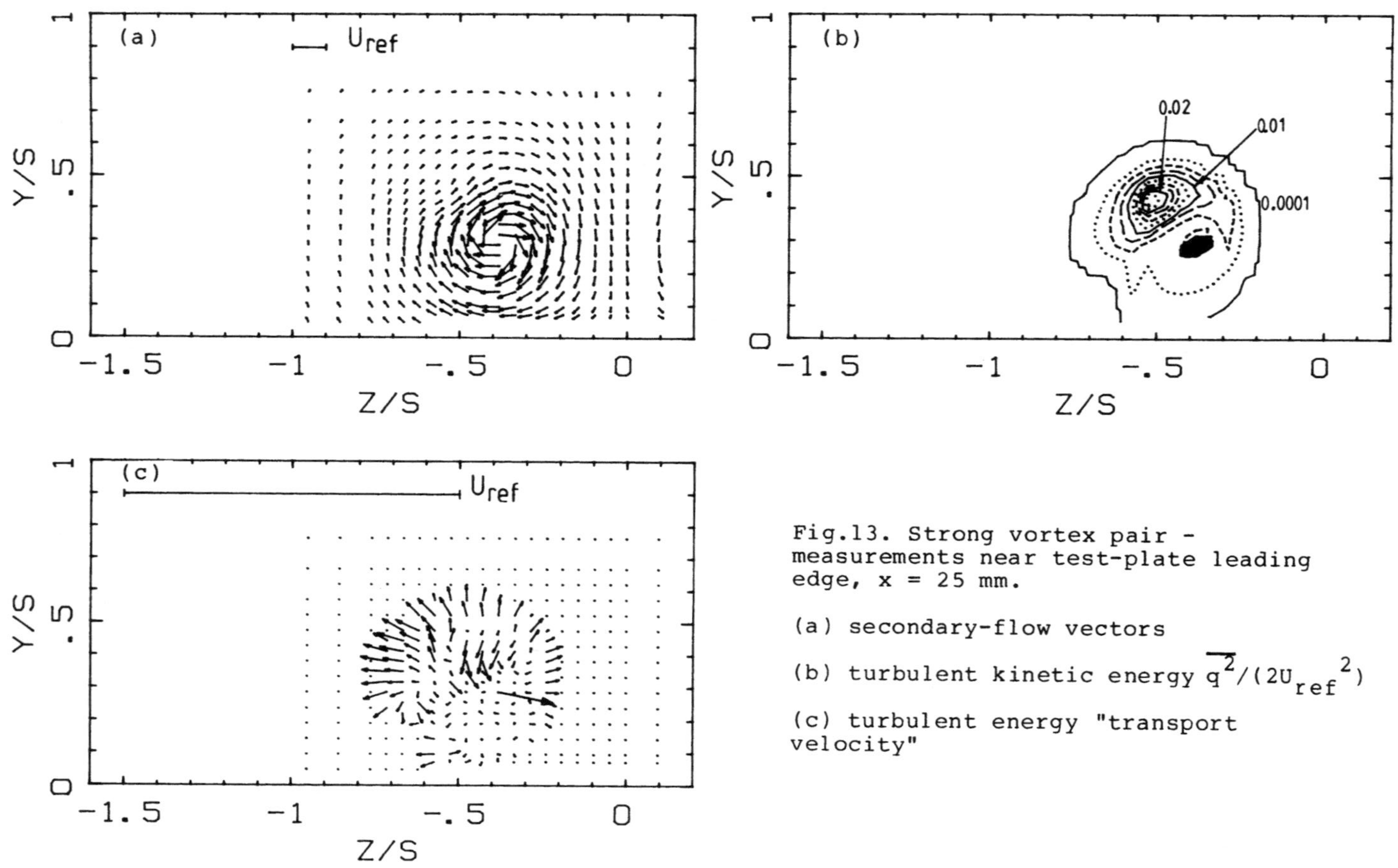

Fig.13. Strong vortex pair - measurements near test-plate leading edge, x = 25 mm.

(a) secondary-flow vectors

(b) turbulent kinetic energy $\overline{q^2}/(2U_{ref}^2)$

(c) turbulent energy "transport velocity"

signals: the system is a second edition of that originally developed [20] for the impinging-jet experiment to be described below. This improvement in technique, coupled with the fact that the crossflow velocities and turbulent stresses are an order of magnitude larger than in the weak-vortex experiments, has greatly reduced the scatter of the hot-wire data (especially for secondary-flow velocity) compared to our earlier work.

Figure 13(a) shows the secondary-flow velocity vectors close to the leading edge of the main plate (the boundary layer being too thin to be resolved). Note that the circumferential velocity in the core is almost as large as the reference velocity (tunnel speed). Figure 13(b) shows the turbulent kinetic energy at the same station. The main vortex core is the black area on the contour plot, where the turbulent energy rises rapidly to a peak of 0.09 in the units shown, and the pen plotter merges the contours: the smaller peak near $z/s = -0.5$ is the "eyebrow" of non-rolled-up fluid seen at about the same streamwise position in Fig. 11(a). There is no inconsistency in arguing that turbulent diffusion of mass in the core is small although fluctuation levels are large: the fluctuations appear to be low-frequency pulsations, associated either with the very large axial velocity in the core - itself a proof that mixing is small - or with wandering of the vortex, although the vortex seemed very steady.

Fig. 13(c) shows one of the most complicated quantities measured, the transport velocity of turbulent energy: defining $q^2 = u^2+v^2+w^2$ instantaneously, the y-component of transport velocity is $\overline{q^2v}/\overline{q^2}$, and similarly for the z-component. This definition neglects transport due to pressure fluctuations: this seems to be a good approximation in simple boundary layers, but pressure transport may be significant in more complex flows. Note that each vector in Fig. 13(c) is a combination of six hot-wire measurements; the smoothness of the results in Figs. 13(c) and 14(c) is good evidence for the general consistency of the results.

Figure 14 shows the same quantities as Fig. 13, but measured at

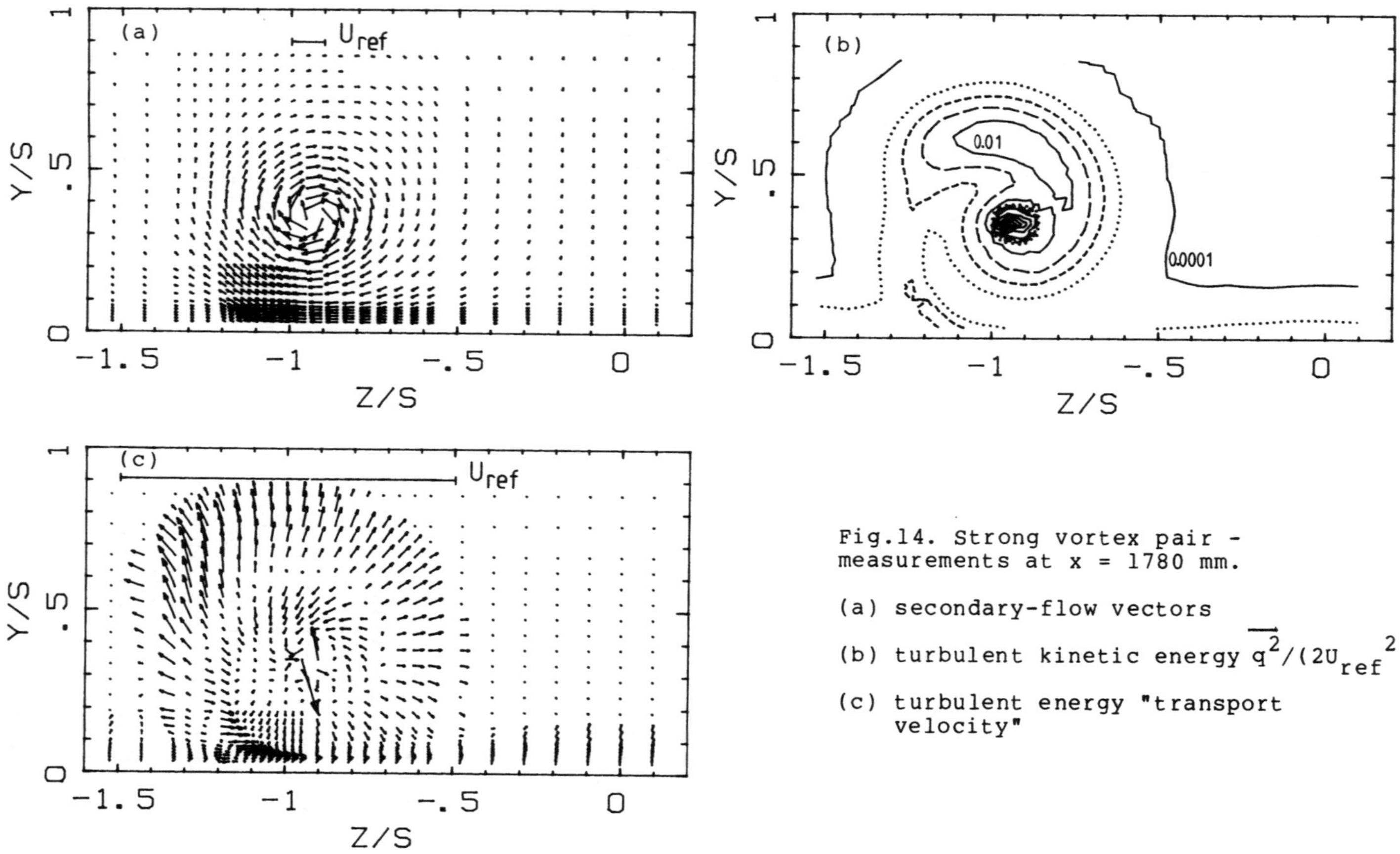

Fig.14. Strong vortex pair - measurements at x = 1780 mm.

(a) secondary-flow vectors

(b) turbulent kinetic energy $\overline{q^2}/(2U_{ref}^2)$

(c) turbulent energy "transport velocity"

a distance downstream of the leading edge equal to about 7 times the delta-wing span. The vortex has moved outwards from the centre line but is slightly further above the surface. The circumferential velocity and turbulent energy in the core have decreased only slightly (peak turbulent energy is about 0.07 on the scale of Fig. 14(b)). Figure 14(b) shows, at $z/s = -1.2$, the "tongue" of highly turbulent fluid which the vortex continually extracts from the inner layer of the boundary layer, and Fig. 14(c) shows the transport of turbulent energy out of the tongue in the positive and negative z directions. Note that the turbulent transport velocities near the perimeter of the vortex are very much larger than those in a boundary layer (for example at the extreme left of the figure), as well as having a rather complicated behaviour as in our weak-vortex experiments discussed above.

(d) Impinging jets.

In [7] we reported preliminary results of an experiment in which a cross-stream jet impinged on a solid surface (a configuration which corresponds to a vertical take off aircraft hovering in a cross-wind). As shown in [8], any cross-stream jet generates a longitudinal vortex pair with "common flow" downwards: the mechanism is qualitatively the same as the generation of a vortex pair by a lifting wing - the lift corresponding to the cross-stream momentum flux of the jet - but there are significant quantitative differences. If the bent-over jet with its vortex pair impinges on a solid surface, more longitudinal vorticity is generated, because the initially two-dimensional boundary layer on the solid surface is deflected sideways by the impinging jet (as in the formation of the "horseshoe" vortex around a solid obstacle). The lift-generated vortices and the obstacle-generated vortices are of the same sign, and they rapidly merge to form a single vortex pair with the "common flow" downwards. Figure 15 shows the surface streamline patterns, visualized by an oil/dye mixture on the surface: the dark area in the centre is a stagnation region, where high transfer rates have totally ablated the oil/dye mixture.

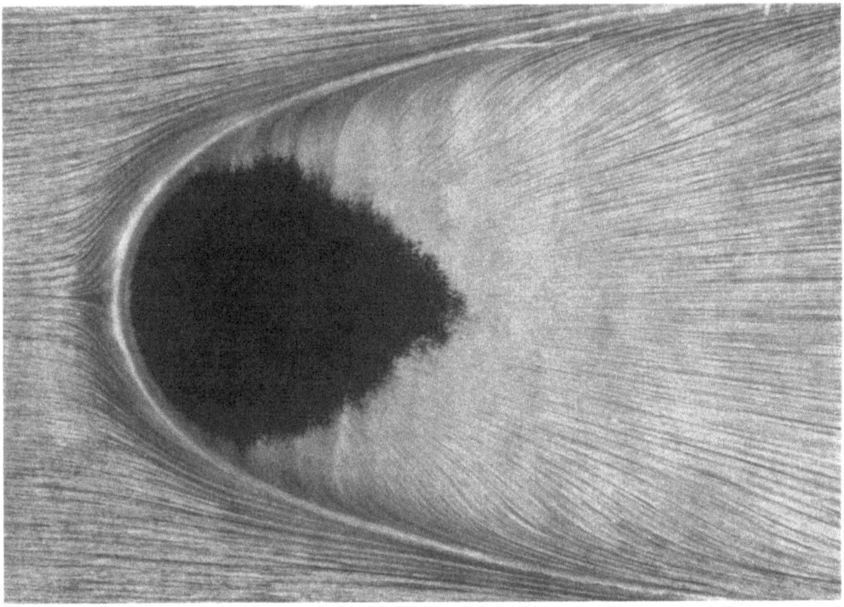

Fig.15. Surface oil flow under impinging jet in cross-stream [7]

Away from the central region, surface streamline traces are fairly clear: the strongly marked white parabolic contour is a region of maximum streamline convergence, just outside the horseshoe vortex region.

Some of the quantitative results of the impinging-jet experiment, including mean-velocity and turbulence contours in a cross-flow plane, are given in [7]. The main conclusion of this, as of all the other imbedded-vortex experiments, is that gradient- transport approximations for diffusion of momentum and turbulence stress in the cross flow plane are not even qualitatively reliable. The alternative simple hypothesis about the turbulence stress tensor is that it is rotated, without change or diffusion, by the mean vorticity: this, like the eddy-viscosity hypothesis, is a qualitatively plausible but quantitatively inaccurate representation of turbulence processes.

A genuine obstacle flow was investigated by Handford [21] who made measurements with a pulsed-wire anemometer behind a surface- mounted body of semi-circular cross section. Most of the measurements were made, deliberately, with a sharp nose on the body so as to minimise the generation of streamwise vorticity. In the sharp-nosed case, where streamwise vorticity was small, the shear-layer edge diffused outwards in a simple fashion, but where significant streamwise vortices were generated in the corner between the body and the solid surface (the "common flow" between the two corner vortices being downwards) significant distortions of the shear-layer edge occurred. In this experiment, a pulsed-wire anemometer was used for detailed measurements in the recirculating flow downstream of the blunt base of the surface-mounted body, providing a very much cheaper and more convenient method of measurement than the laser-doppler anemometer.

Detailed measurements are given in [21]. The most noticeable feature of the measurements downstream of the body is that the distance to reattachment is little more than twice the body radius, compared to about six step heights in the flow over a 2D backward facing step. Let us, however, equate the entrainment flow rate to the "leakage" flow out of the separated-flow region, i.e. equate the product of the entrainment velocity and the surface area of the recirculation zone to (constant) x (external flow velocity) x (cross-sectional area of body). Then, if the ratio of entrainment velocity to external-stream velocity is the same in 2D and in 3D, the distance to reattachment in flow over a surface-mounted body of semi-circular cross-section is simply one half that of the distance to reattachment behind a 2D backward-facing step. This accounts for most, but not all, of the difference. The rapid reattachment leads to strong streamwise and spanwise pressure gradients in the reattachment region, and thus to the creation of large longitudinal vorticity close to the surface. The diffusion of these vortex sheets tends to overwhelm the diffusion of the comparatively weak, larger-scale vorticity from the separated flow region. In general, however, the half-body of revolution flow is about the simplest test case that could be devised for 3D recirculating-flow calculation methods, and we hope to carry out measurements in the wholly-axisymmetric flow to establish, by difference, the effect of the no-slip condition at the solid surface.

Conclusions

Two-dimensional aerodynamics has survived in the laboratory and the academic environment for many years longer than in real-life aerospace engineering. Swept wings have been in use for forty years, and "slender" aircraft, which may well be defined as those whose wing/body junction occupies a large fraction of the body length, have been with us for twenty or thirty years. Those other slender bodies, ships, have been with us for three thousand years or more. The recent laboratory- and academic interest in turbulent flows with strong longitudinal mean vorticity is not before time. Turbulence models for "two-dimensional" flows concentrate on the prediction of the uv shear stress in the xy plane, and have had some reasonable success. However, even in non-slender swept wing flows, it is necessary to predict the cross-stream plane shear stress, $\overline{vw}$, as well as $\overline{uv}$, and experiments [22,23] have demonstrated physical effects which are not represented by conventional turbulence models. When a crossflow is applied to an initially two dimensional turbulent boundary layer, the direction of the shear-stress vector differs from that of the velocity gradient, and there is also a significant reduction in the magnitude of turbulence shear stress . Now "crossflow" merely implies a longitudinal vorticity $\partial W/\partial y$, whereas an operational definition of a "slender" flow is one in which the longitudinal vorticity receives significant contributions from $\partial V/\partial z$ as well as $\partial W/\partial y$. If the contributions from these two velocity gradients are roughly equal, identifiable longitudinal vortices ("vortex lines") occur, as distinct from, or in addition to, "vortex sheets" with large $\partial W/\partial y$. Present-day turbulence models are not capable of qualitatively representing the effects of imbedded longitudinal vortices. Qualitatively, a longitudinal vortex will rotate the Reynolds-stress tensor, so that a $\overline{uv}$ correlation rotates towards the $\overline{uw}$: it is notorious that large values of

$\overline{uw}$ occur in 3D flows, to the extent that streamwise derivatives of $\overline{uw}$ can be significant compared to the cross-stream gradients $\partial\overline{uv}/\partial y$ and $\partial\overline{vw}/\partial y$. The generation of vw is not explained by this rotation mechanism: in a "vortex sheet" flow, the main contribution to the generation of $\overline{vw}$ is $\overline{v^2}\partial W/\partial y$, while if identifiable longitudinal vortices occur then, by definition, the generation term $\overline{w^2}\partial V/\partial z$ is also significant; we note that since $\overline{v^2}$ and $\overline{w^2}$ are non-zero even in statistically two-dimensional turbulent flows, these "generation" terms in the $\overline{vw}$ equation become significant as soon as longitudinal vorticity is created by quasi-inviscid processes.

In practical engineering flows, Reynolds stresses tend to oppose the generation of longitudinal vorticity, rather than actually contributing to longitudinal vorticity as in secondary flows of Prandtl's second kind. However, although secondary flows of Prandtl's first kind are generated by quasi-inviscid mechanisms and are therefore independent of turbulence processes, the subsequent diffusion and decay of these "skew induced" secondary flows depends entirely on Reynolds stress tensor.

To calculate such flows in detail, there seems no alternative but a complete set of Reynolds-stress transport equations, modelled term by term. Since the diffusion process is so complicated, it seems very unlikely that the "algebraic stress model" or "algebraic second moment" scheme, which assumes that the ratio of the rate of turbulent diffusion of a Reynolds stress to the Reynolds stress itself is the same for all Reynolds stresses, will be adequate. Furthermore, we may expect that the anomalous behaviour of the pressure-strain term, identified and modelled by Rotta for the case of boundary layers with comparatively mild streamwise vorticity, will be even worse in flows with identifiable streamwise vortices.

References

1. Rotta, J.C.: Statistische Theorie nichthomogener Turbulenz. Z. Phys. 129 (1951) 547-572.

2. Rotta, J.C.: A family of turbulence models for three-dimensional boundary layers. In Durst, F.; Launder, B.E.; Schmidt, F.W.; Whitelaw, J.H. (eds.) Turbulent Shear Flows I, 267-278. Berlin: Springer-Verlag 1979.

3. Rogallo, R.S.; Moin, P.: Numerical simulation of turbulent flows. Ann. Rev. Fluid Mech. 16 (1984) 99-137.

4. Bradshaw, P.: Turbulent secondary flows. Ann. Rev. Fluid Mech. 19 (1987); 53-74.

5. Cousteix, J.C.: Three-dimensional and unsteady boundary-layer computations. Ann. Rev. Fluid Mech. 18 (1986) 173-196.

6. Brederode, V.A.S.L.; Bradshaw, P.: Influence of the side walls on the turbulent centre-plane boundary layer in a square duct. J. Fluids Engg 100 (1978) 91-96.

7. Shayesteh, M.V.; Shabaka, I.M.M.A.; Bradshaw, P.: Turbulence structure of a three-dimensional impinging jet in a cross stream. AIAA-85-0044. New York: AIAA 1985.

8. Broadwell, J.E.; Breidenthal, R.E.: Structure and mixing of a transverse jet in incompressible flow. J. Fluid Mech. 148 (1984) 405-412.

9. Cutler, A.D.; Bradshaw, P.: The interaction between a strong longitudinal vortex and a turbulent boundary layer. AIAA-86-1071. New York: AIAA 1986.

10. Stein, D.E.: Vortex generators. Boeing Airliner Oct.-Dec. (1985) 5-10.

11. Shevell, R.S.: Aerodynamic anomalies - can CFD prevent or correct them? J. Aircraft 23 (1986) 641-649.

12. Peake, D.J.; Tobak, M.: Three-dimensional interactions and vortical flows with emphasis on high speeds. AGARDograph 252. Paris: AGARD 1980.

13. Shabaka, I.M.M.A.: Turbulent flow measurements in an idealized wing-body junction. Ph.D. thesis, Imperial College (1979) and AIAA J. 19 (1981) 131-132.

14. Shabaka, I.M.M.A.; Mehta, R.D.; Bradshaw, P.: Longitudinal vortices imbedded in turbulent boundary layers. Part 1. Single vortex. J. Fluid Mech. 155 (1985) 37-57.

15. Westphal, R.V.; Eaton, J.K.; Pauley, W.R.: Interaction between a vortex and a boundary layer in streamwise pressure gradient. In Launder, B.E.; Schmidt, F.W.; Whitelaw, J.H. (eds.) Turbulent Shear Flow 5. Berlin: Springer-Verlag 1986.

16. Mehta, R.D.; Bradshaw, P.: Longitudinal vortices imbedded in turbulent boundary layers. Part 2. Vortex pair with "common flow" upwards. Submitted to J. Fluid Mech. 1985.

17. Shibl, A.; Bradshaw, P.: Longitudinal vortices imbedded in turbulent boundary layers. Part 3. Vortex pair with "common flow" downwards. Paper in draft 1986.

18. Bradshaw, P.: Effects of streamline curvature on turbulent flow. AGARDograph 169. Paris: AGARD 1973.

19. Mokhtari, S.; Bradshaw, P.: Longitudinal vortices in wind tunnel wall boundary layers. Aero. J. 87 (1983) 233-236.

20. Shayesteh, M.V.; Bradshaw, P.: Microcomputer controlled traverse gear for three-dimensional flow explorations. To appear in J. Physics E 1987.

21. Handford, P.M.: Measurements and calculations in three dimensional separated flow. Ph.D. thesis, Imperial College 1986.

22. Van den Berg, B.; Elsenaar, A.; Lindhout, J.P.F.; Wesseling, P.: Measurements in an incompressible three-dimensional turbulent boundary layer, under infinite swept wing conditions, and comparison with theory. J. Fluid Mech. 70 (1975) 127-148.

23. Bradshaw, P.; Pontikos, N.S.: Measurements in the turbulent boundary layer on an "infinite" swept wing. J. Fluid Mech. 159 (1985) 105-130.

Recent Progress in Three-dimensional Turbulent Boundary Layer Research at BIAA

S.J. Lu, C.H. Lee, Q.X. Lian, Z.Q. Zhu, H.Y. Teng

Fluid Mechanics Institute
Beijing Institute of Aeronautics and Astronautics
Beijing, China

1. Introduction

The studies of three-dimensional turbulent boundary layers at Beijing Institute of Aeronautics and Astronautics were initiated in the late seventies. The earlier works had been less emphasized on the fundamental aspects of the problem, and mostly engineering oriented. In the early eighties, major efforts were turned to the investigation of turbulence structures in incompressible boundary layers, and the research activities began to diversify thereafter to include the study of coherent structure, 3-D turbulent boundary layer modeling, research on drag reduction, investigation of shock/turbulent boundary layer interaction, and the development of computational techniques for solving 3-D turbulent boundary layer problems. Some of the recent development will be summarized in the present paper.

Flow visualization study of coherent structures in turbulent boundary layers using hydrogen bubbles technique in water tunnel has been carried out since the early eighties. The main results will be presented in the following section. Some development in 3-D turbulent boundary layer research conducted in low speed wind tunnel using hot wires will be discussed in section 3.1. The modified mixing length formulation correlated from the experimental data had been tested by numerical experiments. The results will be discussed together with some other computational works in sec.3.2 for comparison purpose. In the last section, a brief summary of other research activities at BIAA will be presented.

2. Investigation of coherent structures in turbulent boundary layer

2.1 Experimental facilities and test conditions

A water channel was designed for the investigation of coherent structures of turbulent boundary layers in 1980. The test section of the water channel is .4x.4 m^2 and it has a long test section of 6.8m. A long flat plate of 6m can be set in the water channel and the boundary layer is tripped at the leading edge of that plate. Studies of the coherent structures in both flows with and without adverse pressure gradient were made. The flow with adverse pressure gradient was provided by a two-dimensional convergent-divergent passage in which a curved side wall was set opposite to the flat plate. The throat of the passage was set at about 4.8m (x=0) from leading edge of the flat plate. This test configuration and free stream conditions are shown in Fig.1. The details of the test facilities and the test results were described in [1-4].

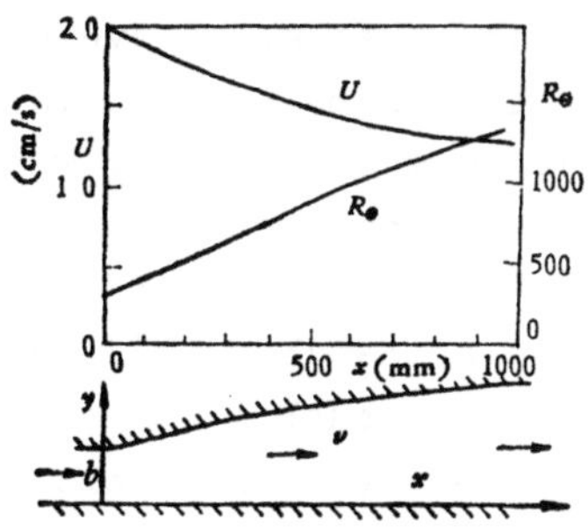

Fig.1 The free stream velocity U and the Reynolds number based on momentum thickness along the flat plate

The divergent section of the two-dimensional passage was designed such that the pressure gradient was steepest at the beginning of divergent section where the thickness of the boundary layer was thinnest, so that it could provide a strong adverse pressure gradient without separation of the boundary layer, and the gradient of pressure decreased gradually along the divergent section to avoid flow separation in the downstream part where the boundary layer was much thicker. Thus there was no flow separation occurred at the exit even if separation bubbles can be observed at the middle part of this divergent section, so the main free stream was stable.

Hydrogen bubble technique was used for the visual investigation of the coherent structures. The platinum wire used for generating hydrogen bubbles was .025 mm in diameter and the applied pulsating current had an adjustable frequency and adjustable voltage, the frequency used in most of these tests was about 9

Hz, and the voltage ranged from 50 to 200 volts.

Laser anemometer was used for measuring the time-mean velocity profile, the turbulence intensity and other quantities of the turbulent boundary layer.

The measured time-mean velocity profile at x=200mm, x=300mm and x=500mm is shown in Fig.2.

At x=200mm, $d\bar{u}/d\bar{y}$ is smaller on the wall, and there is a point of inflection on the profile. From the visual studies, separation bubble occurred on the wall between x=200mm and x=300mm. Due to the special design of the shape of the divergent passage, the velocity profile became flat again downstream of x=500mm, where the shape parameter H was far below the separation criterion of Sandborn[5], and no separation bubble was observed in these investigations. Most of tests were made in the region x=200mm to x=300mm, where the turbulent boundary layer was at the initial stage of separation but the downstream boundary layer is not separated. Therefore the free stream of the divergent passage could be more stable than the fully separated divergent flow, yet the experimental results were compared to the experiment of Simpson[6,7] made in a fully separated divergent flow.

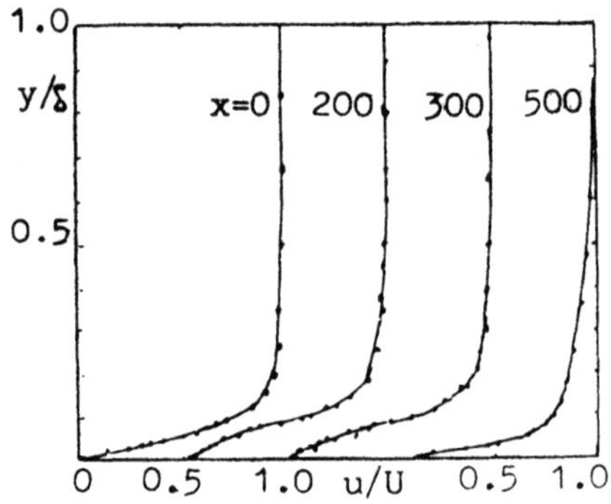

Fig.2 Time-mean velocity profile of the turbulent boundary layer along the flat side of the divergent passage

For the investigations of turbulent boundary layers in the flow without pressure gradient, velocity profiles and momentum thicknesses were measured by both LDA and hydrogen bubble techniques.

2.2 The investigations of the coherent structures of TBL with strong adverse pressure gradient and in the vicinity of separation

A. Plane views of near wall structures [1,2]

Plane views of the structures were obtained by the platinum wire placed parallel to wall and normal to the free stream. The hydrogen bubble time-lines shedded by the wire were initially parallel to the wall. In early 60's Kline[8] and recently Smith [9] used hydrogen bubble technique to investigate the plane views of the coherent structures of turbulent boundary layers in flows without pressure gradient. The main results of our studies to the TBL in flows with adverse pressure gradient are:

1) The width of the low speed streaks was found much narrower than that of the high speed streaks in Kline's experiment. Whereas the width of low speed streaks is found much wider in our studies. It is about the same order of the high speed streaks or even wider as shown in Plate 1.

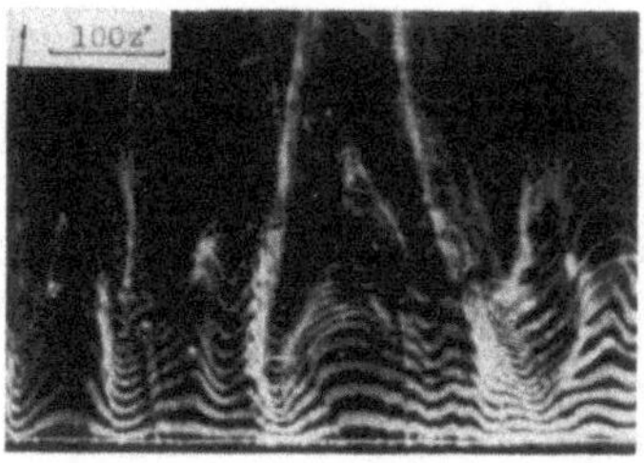

a)

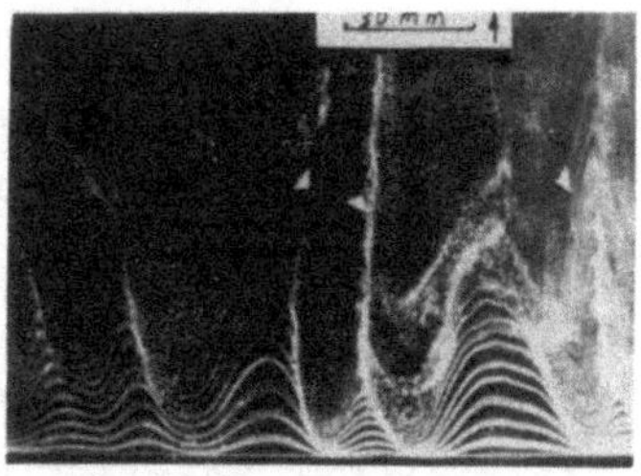

b)

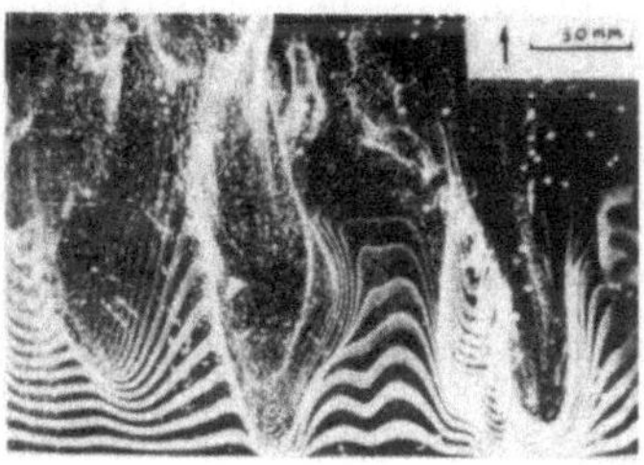

c)

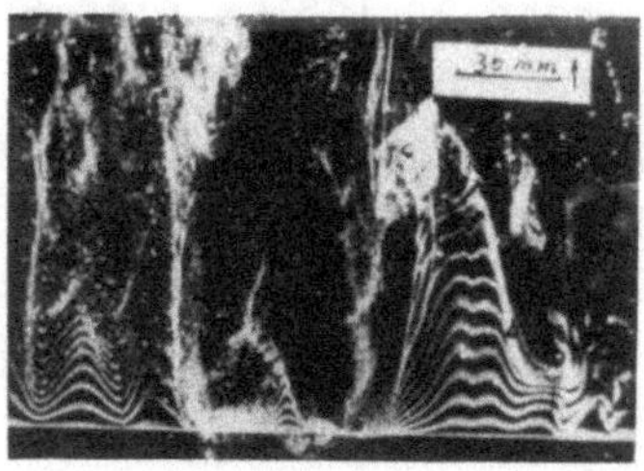

d)

Plate 1. Plane views of structures of turbulent boundary layers: a) dp/dx=0, R =1200, $y^+=8.4$ b) in divergent passage x=100mm, y=1mm c) in divergent passage x=200mm, y=4mm d) in divergent passage, x=300mm, y=4mm

2. The low speed streaks is very narrow and,perhaps, due to the technique used for the generation of hydrogen bubble streak time-lines, it is hard to observe the velocity distribution inside a low speed streak on the photographs of Kline[8] ; and it is also difficult to discriminate whether the long streaks are originated inside or on the border of the low speed streaks. In the present experiment with low speed streaks being so wide that there is no difficulty to observe or even to measure the velocity distribution inside them. The photographs show that the spanwise velocity distribution either inside a low speed streak or inside a high speed streak is relatively uniform, and abrupt changes of velocity occur only in a narrow region between low speed streak and high speed streak. It is also possible to observe that the chaotic flow and the long streak are originated in this region and extended to the far downstream. Finally, it seems to be rolled up vortex. This fact might provide an explanation to the question[10] about the lives of the hydrogen bubbles for the long streaks being much longer than those for the low speed or high speed streaks. The hydrogen bubbles might be concentrated to the core of a vortex due to the rotation of fluid and due to the fact that the hydrogen bubble is lighter than the surrounding fluid, the concentrated hydrogen bubbles should have a longer existing duration.

The hydrogen bubble time-lines are smooth either inside low speed streaks or inside high speed streaks. That implies the local flow is not chaotic, the chaotic flow occurs only along the border between high speed streak and low speed streak as shown by the mark 'Δ' in Plate 1(b) and 1(c) where the hydrogen bubble lines are not smooth, and sometime even twisted as shown by mark 'c' in Plate 1(d).

The low speed streak is not symmetrical in general. On one side of the low speed streak the transition to the high speed streak takes place gradually and du/dz is not very large, while on the other side du/dz becomes very large, and the long streak is originated along this side. The formation of vortex, generation of chaotic flow and bursting also take place along this side.

B. Side views of the coherent structures[3,4]

The side views were obtained by hydrogen bubbles generated by the platinum wire normal to the wall at various x stations. Side views of a TBL had been investigated by Kim, et al[11], streamwise and transverse vortices had been observed in his experiment. But the relation between the side view structures and the plane view structures had not been investigated. In our experiment the structures of TBL with adverse pressure gradient are different from those observed by Kim, and some of relations between side views and plane views had been studied.

1. The typical photographs of the low speed streak, high speed streak and separation bubble in side view are shown in Plate 2. In the high speed streak the instantaneous velocity profile is almost linear. The hydrogen bubble time-lines are almost straight lines with uniform space, which implies that the local velocity along x direction and the vorticity are uniform normal to the wall. The velocity is also uniform along z direction as mentioned from the plane view.

A typical side view of a low speed streak is shown in Plate 2b. It is interesting to note that the hydrogen bubble time-lines

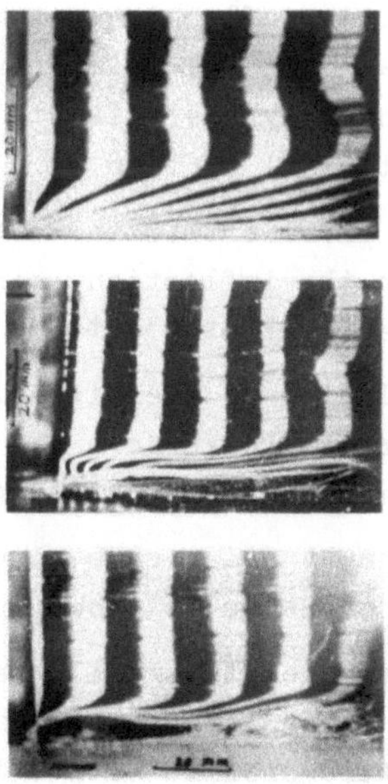

a) high speed streak

b) low speed streak

c) separation bubble

Plate 2 Typical side views of TBL in the vicinity of separation (x=210mm) are almost perpendicular to the wall inside the low speed streak. From Plate 2b and Plate 1, it is seen that the velocity inside a low speed streak is approximately uniform not only in x and z directions but also in y direction, and the transverse vorticity and the shear stress are almost zero inside it, yet over the low speed streak there is an inflection point and the local du/dy (the vorticity and the shear stress) increases abruptly to a value much larger than those in the high speed streak. The inflection point and the concentration of vorticity might be the sources of instability and the generation of chaotic flow. These results give us a three dimensional picture of a typical high speed streak, and low speed streaks as well.

2. More than 10000 frames of motion picture films were taken in this experiment, cyclic changes of the side view were observed and the typical cycle is summarized as follows:

a) The side view is a high speed streak with the thin instantaneous velocity profile and large du/dy, and u(y) is almost linear as shown in Plate 2a and 3a.

b) The instantaneous velocity profile gradually thickens and meanwhile du/dy is decreasing. An inflection point begins to appear above the inner layer, and the side view becomes a profile of a low speed streak. Generally there is still no chaotic flow where Plate 3c is a typical picture.

c) The instantaneous velocity profile thickens even further and the hydrogen bubble lines near the inflection point begin to oscillate or twist into vortices. (Plate 3d to 3e)

d) The amplitude of the oscillation or the size of the vortices increases rapidly or the vortex breaks down and a highly chaotic and thick profile are formed. (Plate 3f to 3g)

e) The chaotic fluid is blown away and a side view of high speed streak similar to Plate 2a reappears as shown in Plate 3h.

The instantaneous thickness of the velocity profile varies substantially during this cycle. The thinest profile is about 14mm thick and thickest one is the chaotic profile with a thickness about 42mm as shown in Plate 3f.

This cycle lasted about 6.1 sec., there is no fixed period of

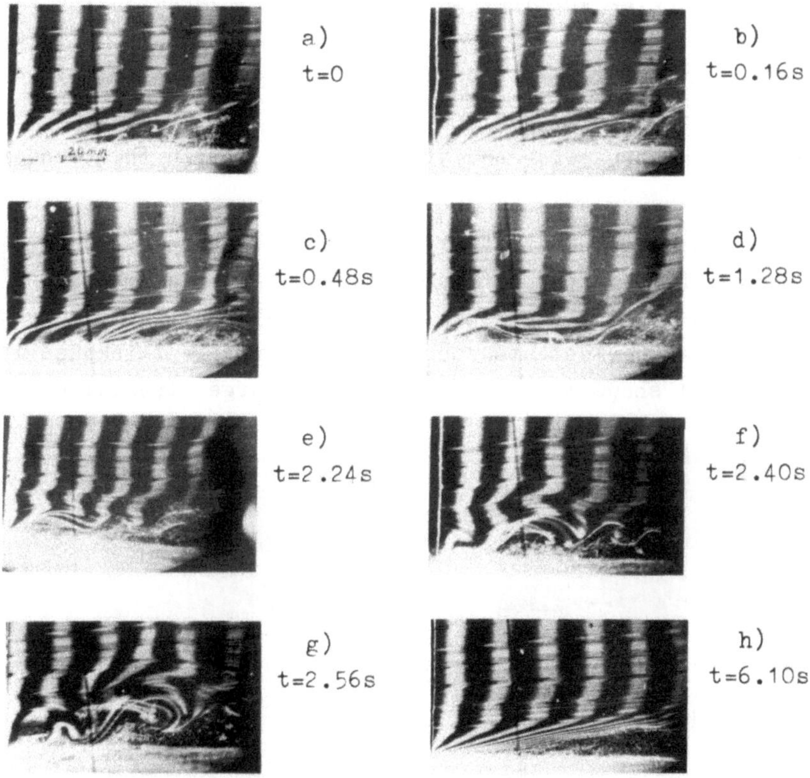

Plate 3 The side view of a typical cycle at x=200mm

the cyclic changes. During a cycle, the low speed streak usually has a much longer duration than that of a high speed streak; and a violently chaotic motion of the fluid is generated abruptly that, just as a burst, has only a short duration. The average period t_B of the observed cycles at x=200mm station is 2.4 sec., and $t_B U/\delta=12.1$ (U=17.7cm/s, δ=3.5cm). This value is close to the result of Simpson's experiment where $t_B U/\delta$ is 8.35 to 11.7 [6],and it is about the double of the normalized bursting period $t_B U/\delta=5$ in the flow without pressure gradient (cf. Hinze [12]). The observation of present experiment may provide an explanation to the increase of the normalized period of bursting; as we have seen previously the size of low speed streaks, and will be seen later the sizes of transverse vortices and streamwise vortices that become very large in the flow with adverse pressure gradient, especially in the vicinity of separation, the large size of the structures may result in a longer period of bursting.

2.3 The large streamwise and transverse vortex in the boundary layer

In the vicinity of separation the size of vortex may become very large, for example, at x=200mm, the time-mean profile thickness is 35mm, the maximum vortex size might be as large as 30mm while the instantaneous velocity profile thickness is 50mm. Plate 4 shows the development of a large streamwise vor-

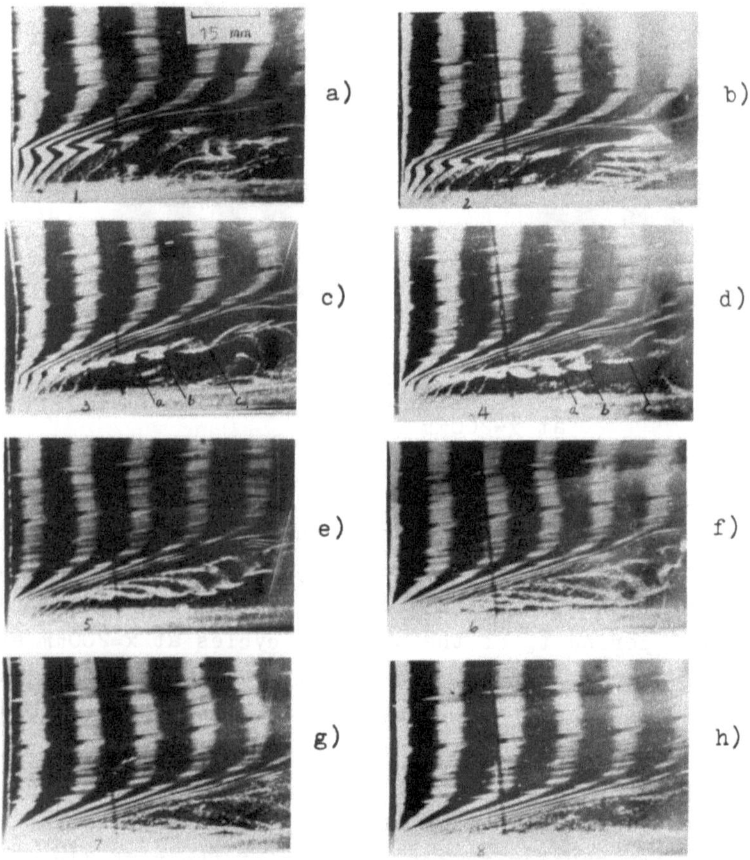

Plate 4 Motion pictures of a large streamwise vortex at x=200mm, 6.25 frames/s.

tex. The film was taking at the rate of 6.25 frames per second. Before the time of Plate 4a the motion picture shows the side view is a low speed streak similar to Plate 2b. On Plate 4a the velocity profile is still a low speed streak, but the hydrogen bubble lines below the inflection point begin to twist. On Plate 4b and 4c the streamwise vortex is formed below the inflection point. This vortex has a large streamwise velocity, streamwise and normal coordinates (x,y) of three traceable points marked 'a,b,c' in this Plate were measured and the results are shown in Plate 4, dx/dt calculated from this figure is 8.5 to 10cm/s. But after Plate 4e, dx_a/dt decreases gradually to 3.1cm/s, while dy_a/dt increases to 3.0cm/s. On Plate 4f the vortex axis at point 'a' is inclined about 45° to the wall, this is just a feature of the leg of a hair pin vortex, as predicted theoretically by Theodorsen [13] and verified by visualization in a paper of Head [14]. Plate shows clearly the effect of a large streamwise vortex: The high speed fluid above the low speed streak is brought to wall region by the revolving of the fluid around the vortex. The flow velocity is increased from about 3cm/s in Plate 4a to 8cm/s in Plate 4e at y=5mm, and the whole wall region becomes a high speed streak on Plate 4h.

3. Studies in three-dimensional turbulent boundary layer theory

Three-dimensional turbulent boundary layer, especially with flow separation, is of considerable interest since the most of boundary layer is of this type in practice. In BIAA, the theoretical studies of the subject have been carried out systematically through experimental and computational works since the late seventies. The main progress will be presented in the followings.

3.1 The behavior of 3-D turbulent boundary layer in a junction flow

One of the typical three dimensional flows known as junction flow will exist in a corner formed by an airfoil erected on a flat plate. The strong adverse pressure gradient imposed by the airfoil causes the fully developed turbulent boundary layer on the flat plate to be skewed and separated afterwards. An understanding of such a flow physically is important for improving aerodynamic performances. On this basis, we try to find the effects of pressure gradient and the streamline curvature on the 3-D turbulent boundary layer behavior and to provide some information about the modification of the eddy viscosity or mixing length expressions such as Michel, Cebeci-Smith or Mellor-Herring's models [15,16]. Usually these models, which are essentially 2-D turbulent models, are used in 3-D flow computations.

A. The 3-D turbulent boundary layer behavior in attached flow region of junction flow [17].

Experimental set up and measuring stations selected in the attached flow region are shown in Fig.3. The pressure distributions at the edge of boundary layer along the x-direction were shown in Figs.4 and 5 respectively. It can be seen that the pressure gradient and the streamline curvature are increased as the flow approaching the airfoil model.

The mean resultant velocity distributions with parameters u^+ and y^+ at the measuring stations are typically shown in Fig.6, in which a flat plate velocity distribution without the model was also included. If the Coles' velocity law [18]

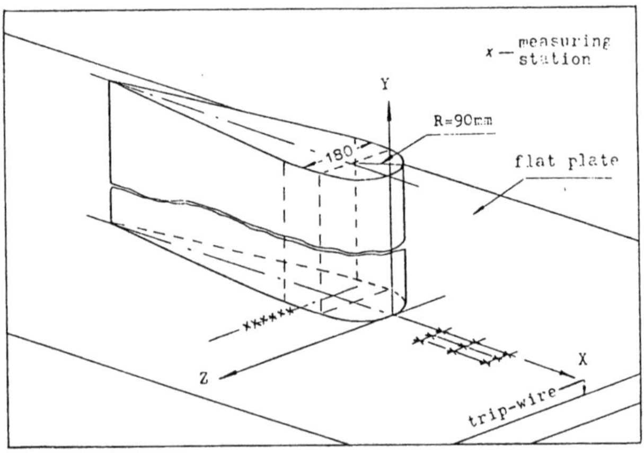

Fig.3 Sketch of model-installation and measuring stations

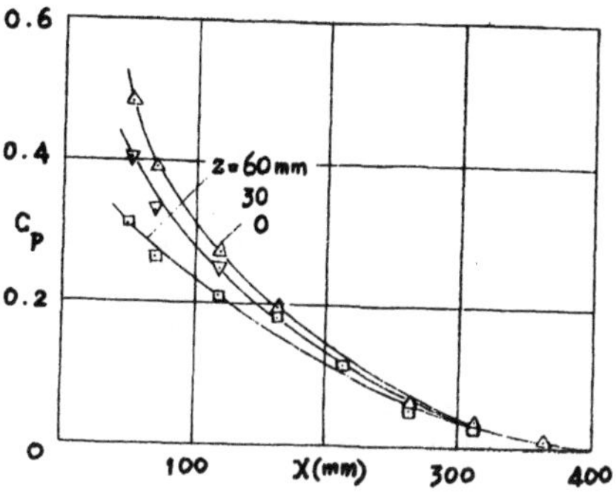

Fig.4 Pressure distribution along x-direction

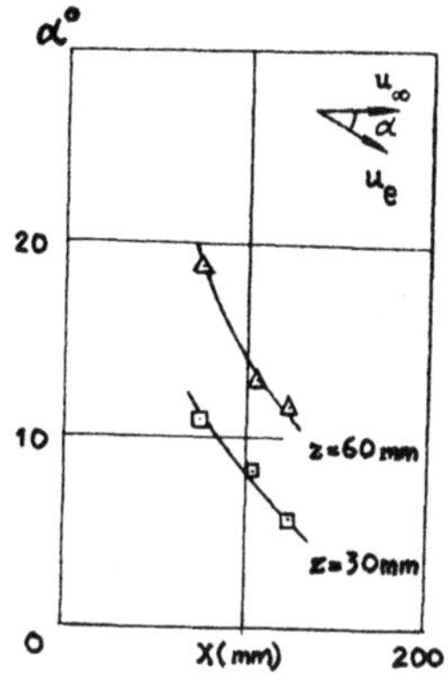

Fig.5 Flow angle at the edge of boundary layer along x-direction

$$u^+ = A\ln y^+ + B + \frac{\pi}{A} W\left(\frac{y}{\delta}\right) \qquad (1)$$

is used to fit the experiment results, it can be found that constants A,B and π vary at different measuring stations, i.e. with the variations on pressure gradients and 3-D curvatures of streamlines as shown in Fig.7. It is more interesting to note that the coefficient 1/A is decreased with the increase of the adverse pressure gradient and the decrease of the curvature of streamline. The physical explanation of the dependency of coefficient A upon the pressure gradient is described as the following. In the boundary layer near the wall

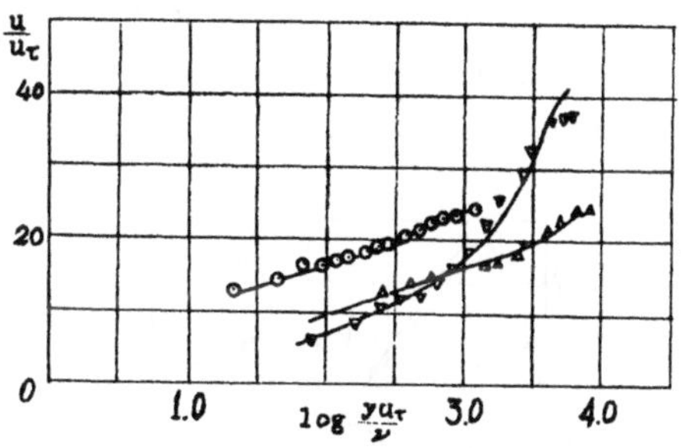

Fig.6 Mean velocity distribution at stations
▽ 75-0 △ 75-60
⊙ flat plate

$$\partial p/\partial s \approx \partial \tau/\partial y \qquad (2)$$

and the boundary layer approximation $\partial p/\partial y=0$ stays valid at least in the attached flow region. Therefore $\partial\tau/\partial y$=const. If

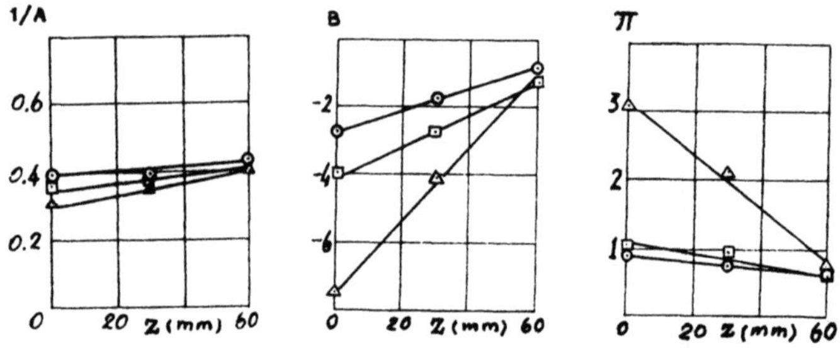

Fig.7 Variation of coefficients A,B and π at different stations ▲ x=75mm ◻ x=105mm ⊙ x=125mm

we consider $\tau = \tau_t = -\rho\overline{u'_x u'_y}$ except in the viscous sublayer region, $\partial\tau/\partial y$=const. means that the turbulent shear stress should be linearly varied with y as verified by measurements shown in Fig.8. If we use the mixing length concept, i.e. $\tau_t = l^2\left(\frac{\partial u}{\partial y}\right)^2$ and $l = ky$, the velocity distribution can be integrated as [19]

$$u^+ = \frac{1}{\kappa}\left[\ln y^+ - B - \frac{1}{2}\frac{\nu}{\rho u_\tau^3}\frac{\partial p}{\partial s}y^+\right]. \qquad (3)$$

If it is compared with eq.(1) for inner layer, then, whenever $\partial p/\partial s > 0$, there is $1/A < k$, which implies that the stronger the adverse pressure gradient the smaller the coefficint 1/A is. This conclusion can also be seen from mixing length distribution shown in Fig.9. It is clear that the slope of mixing

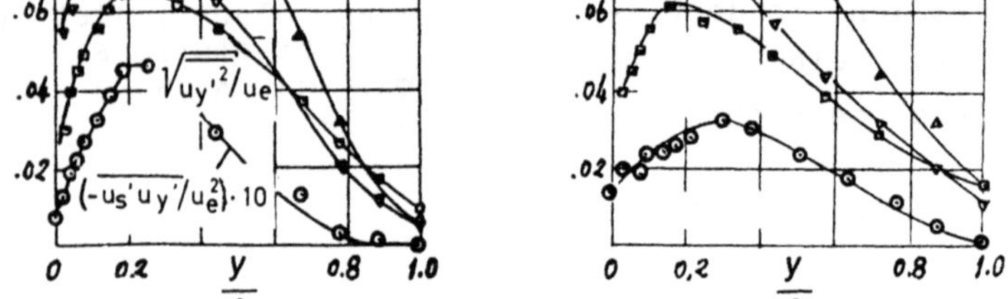

Fig.8 variation of the turbulent stress at different stations at low speed

length distributions near the wall region decreases with increasing the pressure gradient and with decreasing the curvature of streamline.

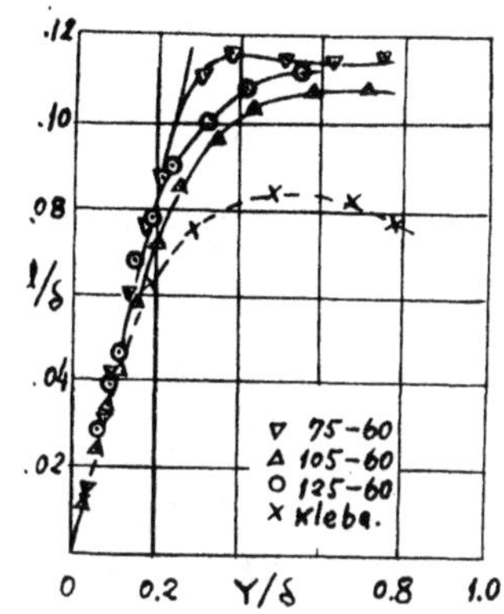

Fig.9 mixing length distributions at different stations

The typical turbulent stress distributions are shown in Fig.8. The values of the turbulent stresses are increased with increasing the pressure gradient from the comparison of the plots at stations 75-0 and 125-0. It appears that the peaks of the turbulent stresses are further away from the wall as the separation region is approached. The comparison of the plots at stations 75-0 and 75-60 shows that the turbulent stress distributions are more flattened and the peak values are decreased as the curvature of the streamline is increased due to the fact that, as the pressure gradient being practically unchanged, the difference of the streamline curvatures are rather significant at those stations.

B. An improved algebraic model for 3-D flow with adverse pressure gradient[20]

From above discussions it can be seen that the coefficient k in the mixing length expression should be a function of the streamwise pressure gradient and streamline curvature. The influence of the pressure gradient toward the mixing length can be estimated using expression

$$l = c\delta \tanh\left(\frac{ky}{c\delta}\right)\left[1-\exp\left(-\frac{y}{A}\right)\right] \tag{4}$$

where coefficients k and c are determined by experiments with adverse pressure gradients. By varying 2D-wedge angle, the pressure gradient or $\beta=(dp/dx)/(\delta^*/\tau_w)$ at various points can be changed. We found that at points further away from the apex of the wedge, the velocity profiles are independent of β, where the boundary layer flow can be treated approximately as in equilibrium state. Thus, a rule for the variations of k and c with increasing β in the region of higher adverse pressure

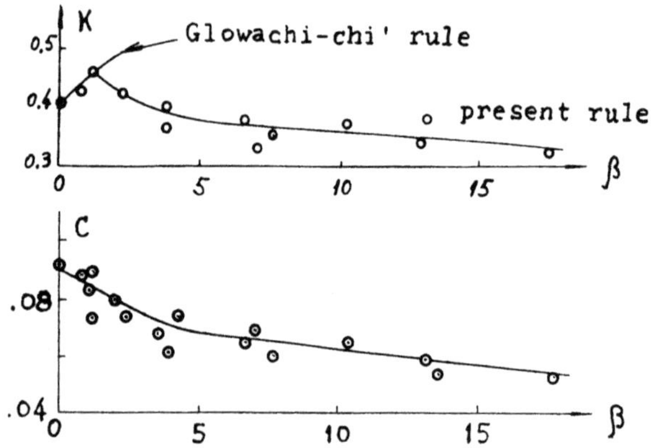

Fig.10 Variation of parameters k and c with coefficient $\beta=(\delta^*/\tau_w)(\partial p/\partial x)$

gradient can be obtained by measuring the boundary layer profile on a flat plate under the influence of the wedge angle (see,e.g., Fig.10).

In order to evaluate the effectiveness of this modified mixing length expression, 3-D turbulent boundary layer behavior of the flow along a flat plate with a circular cylinder erected upon it has been calculated. The computational results will be given in sec. 3.2. However, it should be pointed out that the computational values for the transverse stress τ_{zt} differ from the experimental data to some extent. It appears that the effect of streamline curvature have to be considered in the algebraic models.

C. The 3D turbulent boundary layer behavior near and within the separation region at low speed

The measuring stations at x=-135mm where in the separation region were selected as shown in Fig.3. These stations are in the junction flow region where the pressure gradient and 3D curvature are very small. However a separation-induced horse shoe vortex can be seen in this region as shown in Fig.11. There are a main vortex whose core is at z=50mm, y=13mm and a secondary vortex near the corner.

The induced horse shoe vortex has large effects on the turbulence behavior as well as on the mean flow. The turbulent stress distributions are shown in Fig.12 in which the experiment results without the model are also included for comparison. If the measuring stations are closer to z-position of the vortex core, the profiles of turbulent stresses would have peaks at the y-position of the vortex core, while the turbulent stresses near the airfoil decrease tremendously compared to the values on the flat plate. The peak values of the turbulent stresses are larger in this experiment compared with McMahon's results.

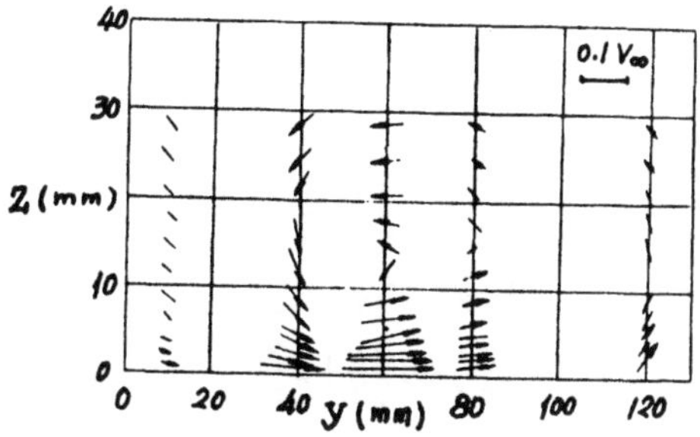

Fig.11 Velocity vectors in crosssection at x=-135mm

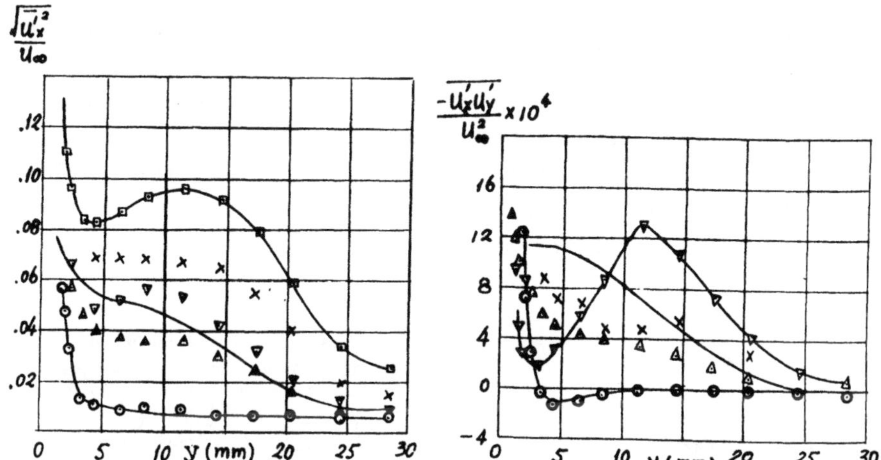

Fig.12 Distributions of turbulent stresses at various stations on x=-135mm ● z=10mm ▽ z=40mm ▫ z=60mm X z=80mm Δ z=120mm

While the turbulent stresses near the airfoil are smaller than those of McMahon's. This again reflects that the separation-induced vortex dominates the junction flow and the vortex in our case is stronger than that in McMahon's due to the rather thick airfoil model in the present experiment. The influence of the vortex on the mean and fluctuation values seems to be contributed from the vortex induced entrainment and the turbulent energy redistributed by the vortex action.

3.2 Computations of incompressible 3D turbulent boundary layer

On the boundary layer assumption the governing equations for incompressible 3D turbulent boundary layer are

$$\begin{aligned} &\frac{\partial u}{\partial x}+\frac{\partial v}{\partial y}+\frac{\partial w}{\partial z}=0 \\ &u\frac{\partial u}{\partial x}+v\frac{\partial u}{\partial y}+w\frac{\partial u}{\partial z}=-\frac{1}{\rho}\frac{\partial p}{\partial x}+\frac{\partial}{\partial y}\left(\nu\frac{\partial u}{\partial y}-\overline{u'v'}\right) \\ &u\frac{\partial w}{\partial x}+v\frac{\partial w}{\partial y}+w\frac{\partial w}{\partial z}=-\frac{1}{\rho}\frac{\partial p}{\partial z}+\frac{\partial}{\partial y}\left(\nu\frac{\partial w}{\partial y}-\overline{w'v'}\right) \end{aligned} \tag{5}$$

and boundary conditions can be formulated as

$$\begin{aligned} &y=0: \quad && u=v=w=0 \\ &y=\delta: \quad && u=u(x,z), \quad w=w(x,z). \end{aligned} \tag{6}$$

The incompressible 3D boundary layer flow induced by a circular cylinder erected on a flat plate is one of the typical test cases for calculation(Fig.13). R.Dechow and K.C.Felsch[21] have carried out extensive measurements for such a flow along a streamline AB of the outer flow(Fig.14) giving the profiles of the mean velocity vector, the turbulent shear stress,etc. Comparison is made between their measured data [21] and the results of our calculation[22] using different external pressure distributions at the edge of the boundary layer. These pressure distributions are obtained from either the potential flow theory or the measured values on the wall as well as at the edge of the boundary layer. From this comparison it is possible to investigate the effect of the external pressure distribution on the boundary layer behavior more or less in detail.

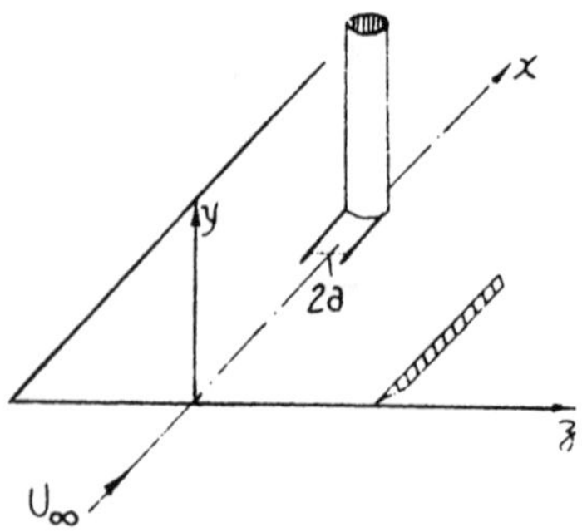

Fig.13 Geometry of the flow problem

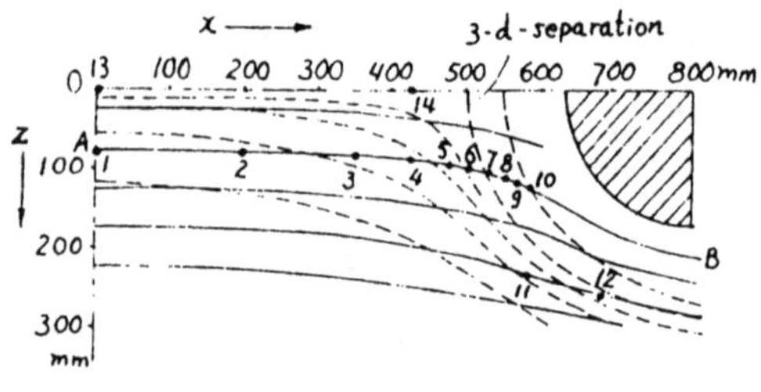

Fig.14 Streamline pattern and the position of measured station

In the governing equations the expressions $-\frac{1}{\rho}\frac{\partial p}{\partial x}$ and $-\frac{1}{\rho}\frac{\partial p}{\partial z}$ are obtained from the values of the inviscid flow at edge of the boundary layer. In the first case, the inviscid flow outside boundary layer may be approximated by the potential flow around an infinite circular cylinder, an exact classical solution[23] provides the velocity components along x and z directions:

$$U = U_\infty \left\{ 1 + \frac{z^2 - (x - x_1)^2}{(1/a^2)\,[(x - x_1)^2 + z^2]^2} \right\} \qquad (7)$$

$$W = -\frac{2U_\infty z\,(x - x_1)}{(1/a^2)\,[(x - x_1)^2 + z^2]^2} \qquad (8)$$

where x_1 denotes the x coordinate of the cylinder axis, a is the cylinder radius. And the expressions $-\frac{1}{\rho}\frac{\partial p}{\partial x}$ and $-\frac{1}{\rho}\frac{\partial p}{\partial z}$ can be readily derived from:

$$u\frac{\partial u}{\partial x} + w\frac{\partial u}{\partial z} = -\frac{1}{\rho}\frac{\partial p}{\partial x}$$

$$u\frac{\partial w}{\partial x} + w\frac{\partial w}{\partial z} = -\frac{1}{\rho}\frac{\partial p}{\partial z}. \qquad (9)$$

In the second case, the velocity components are obtained from a measured pressure distribution which is specified at a finite number of locations. In analogy to [24] we choose to solve the differential equation expressing the irrotational condition, i.e.,

$$\frac{\partial u}{\partial z} - \frac{\partial w}{\partial x} = 0 \qquad (10)$$

together with the relationship given by

$$c_p = 1 - u^2 - w^2. \qquad (11)$$

The box scheme has been used to solve these equations. And we use bi-B-spline to smooth the measured values for numerical procedure. A finit difference method, viz. the three dimensional box method has been used in the present study for numerical calculation of transformed governing equations of the boundary layer.

For comparison, calculations using different external pressure distribution have been performed with the same flow configuration up to the region immediately ahead of the separation line. The comparison of some integral boundary layer parameters along the streamline AB were depicted in[22]. As an example, it is given here the comparison of the displacement thickness δ^*, in which c_{pE} measured pressure value at the edge of boundary layer;

c_{pw} - measured pressure on the wall; c_{pt} - the pressure from the potential flow theory. It can be seen that in the flow field upstream of the separation line, especially before station 4, fairly good agreement with experimental data is obtained. From this good agreement it seems that the conventional boundary layer concept may still be applied. But in the flow region near the separation line, i.e. near the stations 5,6,7 the agreement with experimental data becomes worse. For example, in Fig.15 it is clear that the experimental data of displacement thickness increases sharply as it approaching to the separation line. Whereas the variation of the calculated results using the measured pressure values is rather slow to some extent. It is apparent that this discrepancy is not mainly caused by the difference of the pressure distribution. It can probably be reasoned that the boundary layer assumption, namely, $\frac{\partial p}{\partial y}=0$, starts to be invalid. From the measurements[21] it can also be found that in this region $\frac{\partial p}{\partial y}$ becomes larger. A higher order approximation is needed for analysing the flow in this local region.

The algebraic eddy viscosity has been employed for turbulence model due to its simplicity. In order to consider anisotropic feature of turbulence an additional parameter T, which was proposed by J.C.Rotta[25] and G.R.Schneider[26], is introduced. Based on this concept, the integral boundary layer parameters for different values of the anisotropy parameter T were computed in[22] for the 3D turbulent boundary layer. As an example, the computed displacement thickness is depicted in Fig.16.

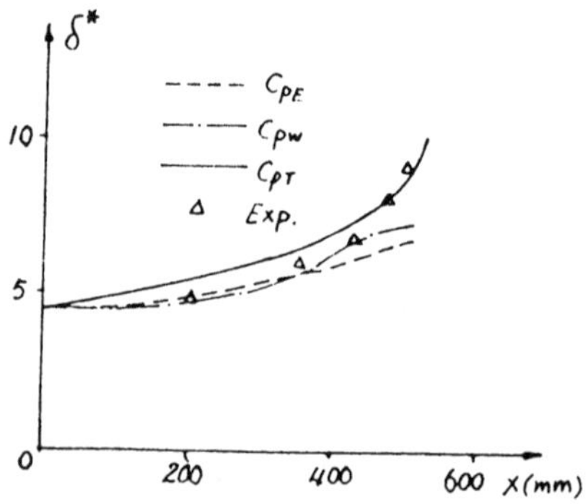

Fig.15 Displacement thickness along AB

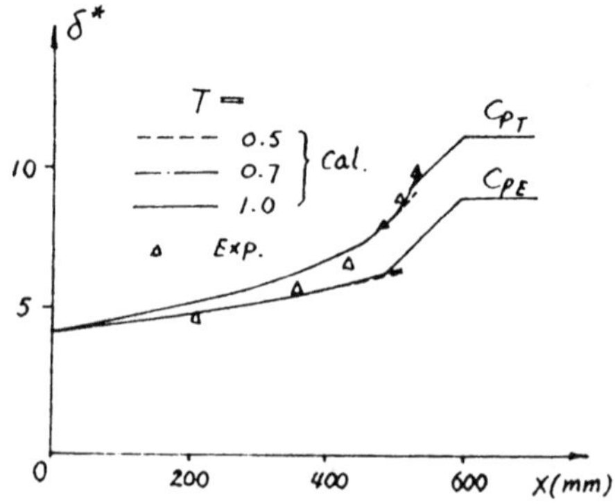

Fig.16 Displacement thickness along AB

It can be seen that the difference in results due to different anisotropical parameters (i.e. T=0.5,0.7,1.0 respectively) is very small for both the potential pressure distribution c_{pt} and the measured pressure distribution c_{pE}. This implies that the effect of anisotropy on the integral boundary layer parameters is insignificant. It may be reasoned that 3-dimensionality produced by the cylinder ahead of the separation line may not be strong and T=const. is probably valid. This can also be found in the measurements. Fig.17 shows the measured variations of T throughout the boundary layer at several stations [21]. From these comparisons one may conclude that a simple anisotropy eddy viscosity turbulence formulation and the assumption T= const. is valid for the calculation of 3D turbulent boundary layer in the region upstream of the separation line.

In [27] the calculations similar to above mentioned configuration were made for various eddy viscosity formulations due to Cebeci-Smith, Michel, Mellor-Herring. The agreement among computed results of the integral boundary layer parameters from those three turbulent models and the experimental data is fairly good in the 3D field aside from the region near the separation line. Whereas the agreement of shear stresses is not so satisfactory.

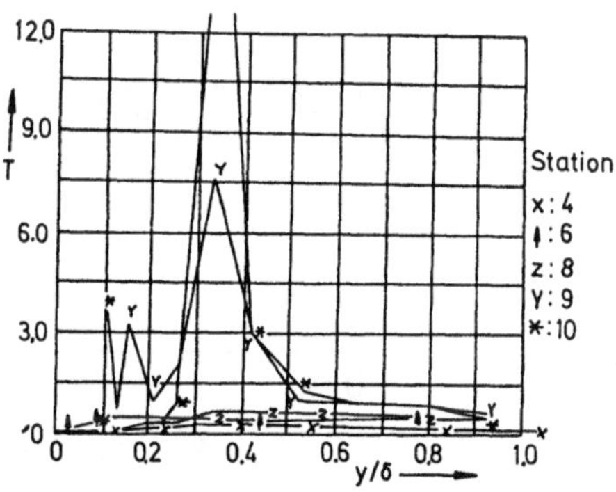

Fig.17 Variation of anisotropy parameter T across the boundary layer

Therefore a improved algebraic eddy viscosity formulation (4) based on the experimental work presented in sec.3.1, was employed for calculating the 3D turbulent boundary layer behavior of Dechow's flow configuration [21]. Besides employing this improved algebraic model, the effects of nonequilibrium of boundary layer on the parameters k and c are also considered in the calculation. The results for skin friction and turbulent shear stresses are shown in Fig.18

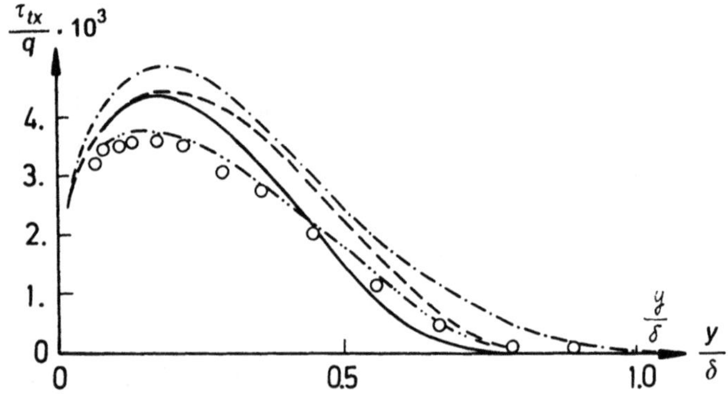

a) at station No.14

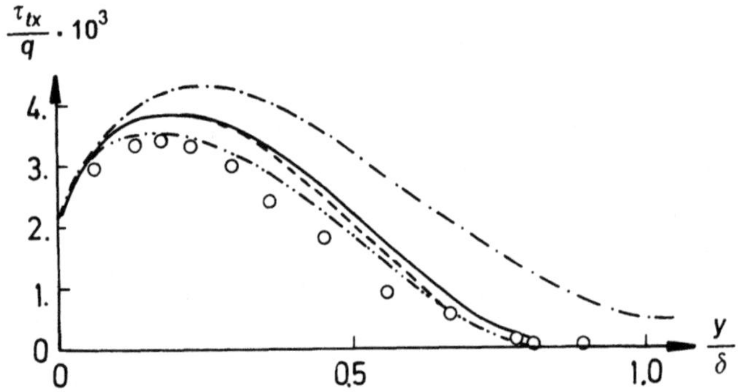

b) at station No.4

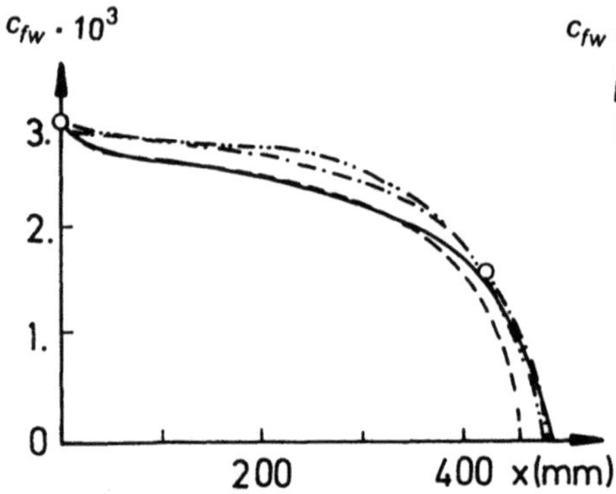

c) along z=o

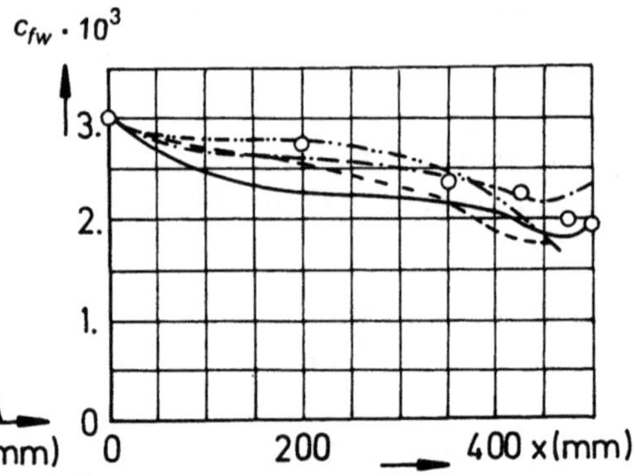

d) along streamline AB

Fig.18 Comparisons of calculated results τ_{tx}/q and c_f with experimental data. ⊙ experimental data[21]
— Cebeci-Smith ---Michel —·-Mellor-Herring
—··-present

It is seen that there are some rather good improvements in the distributions of skin friction and turbulent shear stresses τ_{xt}.

Another effort was made for improving calculation of 3D boundary layer on an infinit swept wing[28]. Using the anisotropic turbulence model by introducing an additional constant parameter T and the relations

$$k=0.41-c_1\Delta R_{ex}$$
$$k_o=0.085-c_2\Delta R_{ex} \qquad (12)$$

where $\Delta R_{ex}=R_e(U_\infty,x)-R_e(u_e,x)$

$R_e(U_\infty,x)=U_\infty x/\nu \qquad R_e(u_e,x)=u_e x/\nu$

$c_1=1.134\times10^{-7} \qquad c_2=5.041\times10^{-8}$

improved results can be obtained. Fig.19 depicts the comparison of c_f, θ_{11}, H, β_w. The agreement for the calculated results against experimental data is better than those for Schneider's results.

4. Some brief remarks

Currently, flow visualization studies of turbulence structures are still the major interests at BIAA. Research activities that

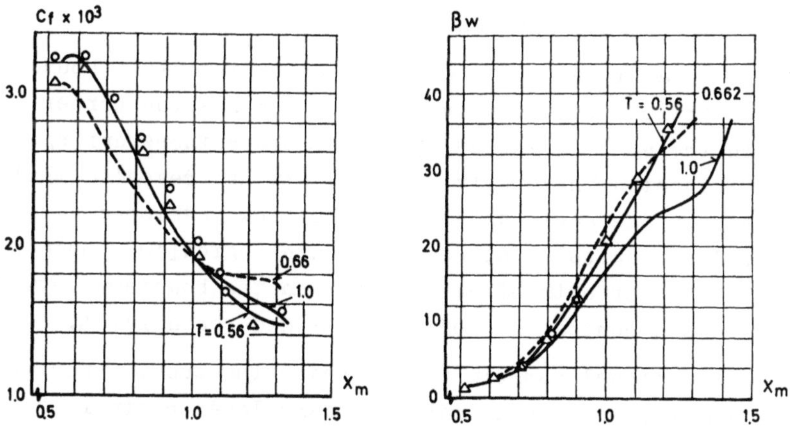

a) $c_f \sim x_m$ b) $\beta_w \sim x_m$

Fig.19 Comparison of calculated results with experimental data ---- Schneider's calculated result with T=0.7 — present results with various T • ▲ experimental data

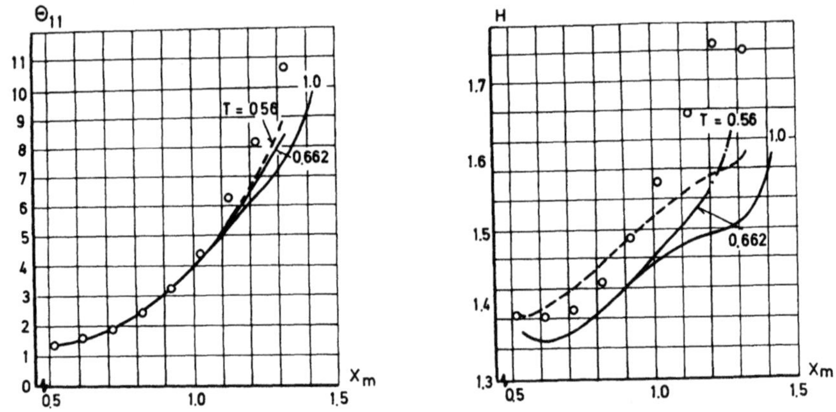

c) $\Theta_{11} \sim x_m$ d) $H \sim x_m$

Fig.19 continued

are under way include the unsteady, separated flows, 3-D turbulent mixing, structures of starting vortex, shock/turbulent boundary layer interaction, etc. The problem of starting vortex is essentially a part of the coherent structure research program, and some preliminary results were given in [29]. The efforts for investigating the 3-D mixing of flows with different turbulence structures will be directed toward the development of a turbulent mixing model.

The experimental investigation on 3-D shock/turbulent boundary layer interactions generated by a swept compression corner was started in the early eighties by one of the authors during his visit at Princeton University. Experiments had revealed that, depending on wedge angle and swept, the flow developed either cylindrical or conical upstream influence region[30]. Some results about the effects of Mach number on the boundary between cylindrical and conical region are obtained recently[31].

On the computational aspect, a drag reduction problem for transonic flows over wings utilizing boundary layer control is currently being tackled. An inverse problem using finite element-difference technique in conjunction with inviscid/boundary layer iterative scheme has been developed to solve the problem. A preliminary results for 2-D flows over airfoils have been obtained[32], and currently we are attempting to extend the numerical procedure to the 3-D case.

Flow field measurements on 3-D turbulent boundary layer are continued for the purposes of modeling as well as drag reduction. The k-ε model is now being investigated, and numerical algorithms utilizing the model are also under developing for solving the parabolized Navier-Stokes equations [33].

5. References

1. Lian, Q.X., Journal of Mechanics (ACTA Mechanica SINICA), 15 (1983), 414-418. (in chinese)
2. Lian, Q.X., Proc. of 2nd Asian Congress of Fluid Mechanics Oct. 1983, pp. 70-78.
3. Lian, Q.X., Journal of Mechanics (ACTA Mechanica SINICA), 17 (1985), 78-85. (in chinese)
4. Lian, Q.X., ACTA Mechanica SINICA, vol. 1, no. 1, Mar., 1985, pp.71-80
5. Sandborn, V.A., et al., J. Basic Eng., 83(1961), 317.
6. Simpson, R.L., et al., JFM 79 (1977), 535.
7. Shiloh, K., et al., JFM 113 (1981), 75.
8. Kline, S.J., et al., JFM 30 (1967), 741.
9. Smith, C.R., JFM 129 (1983), 27-54.
10. Kline, S.J., et al., AFOSR-TR-80-029 (1980).
11. Kim, H.T., et al., JFM 50 (1971), 133.
12. Hinze, J.C., Turbulence, 2nd edition, (1975), 667-668.
13. Theodorsen, T., The structure of turbulence, 50 Jahre Grenzschichforsung, ed. H.Gortler, W.Tollmin (1955), 55.
14. Head, M.R., et al., JFM 107 (1981), 297-338.
15. Michel, R., et al., AGARD CP-93, (1972)
16. Cebeci, T., and Smith, A.M.O., Analysis of Turbulent boundary layers, Academic Press, (1974)
17. Hsing, T.D., and Teng, H.Y., AIAA paper-84-1529, (1984)
18. Coles, D.E., Proc. of the Stanford Conference on Turbulent Boundary Layer Prediction, vol.2, (1968)
19. Berg, B.Van den, JFM 70 (1975)

20. Hsing, D.T., Teng, H.Y., and Kuo, W.H., Proc. of 3rd Asian Congress of Fluid Mechanics Sept. 1986, pp.422-425.

21. Dechow, R., et al., Mitteilungen des Instituts fur Stromungslehre und Stromungsmaschinen, Universitat (TH) Karlsruhe, Heft 21 Marz, (1977).

22. Zhu, Z.Q., AIAA paper-84-1673, (1984).

23. Milne-Thomson, L.M., Theoretical Hydrodynamics, 5 edition MacMillan & Co. Ltd., (1968).

24. Cebeci, T., and Meier, H.U., Z.F.W. 6 (1982), Heft 6.

25. Rotta, J.C., Proc. 1st Symposium on Turbulent Shear Flow, April 1977, vol.1 pp. 10.27-10.34.

26. Schneider, G.R., DLR-FB 77-73 (1977).

27. Wang, L., M.S. Thesis, BIAA (1983).

28. Zhu, Z.Q., et al., ACTA Aerodynamica SINICA no.1 (1983) (in chinese)

29. Lian, X.Q., Proc. of the Conf. on the Advancements in Aerodynamics, Fluid Mech. and Hydraulics June, 1986, pp. 314-321.

30. Settles, G.S., and Teng, H.Y., AIAA Journal, vol.22, no.2, 1984, pp. 194-200.

31. Teng, H.Y., A preprint, (1986).

32. Lee, C.H., and Jin, S., A preprint, (1986).

33. Zhao, H.S., Proc. of 3rd Asian Congress of Fluid Mech. Sept. 1986, pp.298-301.

Three-dimensional Turbulent Boundary Layer Calculations

J. COUSTEIX - R. MICHEL

ONERA/CERT
Department of Aerothermodynamics
TOULOUSE (FRANCE)

SUMMARY

In this paper, problems arising in three-dimensional boundary layer calculations are reviewed and discussed. The emphasis is placed on unsolved problems such as : turbulence modelling, laminar-turbulent transition, occurrence of singularities in boundary layer computations.

Introduction

The development of a three-dimensional boundary layer calculation method involves two kinds of problems. The first problems are associated with the use of a curvilinear non-orthogonal axis system which is needed to avoid any restriction in the application of the calculation method. The theory of general coordinates is well known but the practical implementation is not at all straightforward and requires familiarity with tensorial algebra. Particular attention must be paid to the preprocessing of the geometrical data and of the boundary conditions. The postprocessing of the computed data is also an essential part of the calculation method as it should help in the analysis of the results, as discussed by HIRSCHEL /1/.

The second kind of problems are more fundamental and their solution is not known or is incomplete. An example is the numerical technique used to integrate the equations. This question should not be considered as solved. As stressed by CEBECI /2/, improvements in the accuracy and economy of calculation methods become increasingly important as the methods are increasingly used for design purposes. In addition, the development of new types of computers (vectorial and parallel computers) will lead to adapted algorithms.

In this paper, some other unsolved problems are reviewed and discussed. Three main topics will be covered : 1) *calculation methods and turbulence modelling ;* 2) *calculation of the laminar-turbulent transition ;* 3) *occurrence of singularities* in boundary layer computations. The detailed presentation of these questions would need a great deal of space. Here, the discussion is deliberately oriented towards a few aspects which are important from a practical point of view.

1. Definitions - Boundary layer equations

1.1. Boundary layer equations

Sometimes, boundary layer calculation methods are developed in a particular axis system, especially as regards the axes in the plane of the surface. For example, a streamline coordinate system has often been used but this is not flexible. It is better to not specify the axis system in advance. It is also recommended to use *a curvilinear non-orthogonal axis system.*

The most efficient way to handle the mathematical operations is to work with the tensorial formalism /3/. However, the use of the tensorial notation is rather rare in the literature on boundary layer theory. Here we will conform to these habits but it should be pointed out that tensorial analysis underlies our treatment.

The X- and Z-axes form two families of curves drawn on the surface and the y-axis is normal to the wall. λ is the angle between the X- and Z-axes (fig. 1). The metric elements along X and Z are h_1 and h_2.

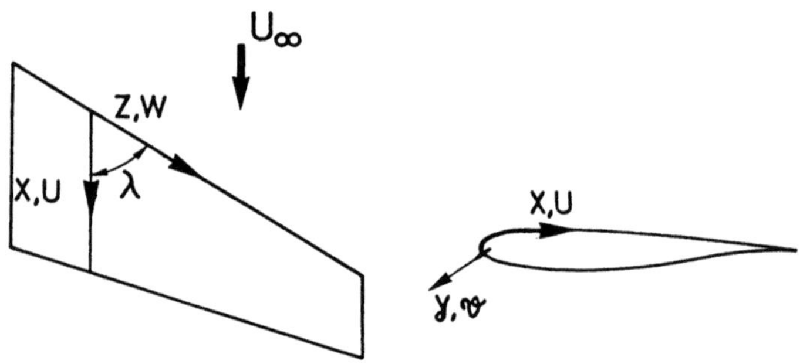

FIG. 1 - BOUNDARY LAYER AXIS SYSTEM

The physical velocity components along X, Z, y are respectively U, W, v.

For an incompressible flow, the boundary layer equations are :

$$\frac{1}{q}\frac{\partial}{\partial X}(\rho U \frac{q}{h_1}) + \frac{1}{q}\frac{\partial}{\partial Z}(\rho W \frac{q}{h_2}) + \frac{\partial}{\partial y}\rho v = 0 \qquad (1a)$$

$$\rho\frac{U}{h_1}\frac{\partial U}{\partial X} + \rho\frac{W}{h_2}\frac{\partial U}{\partial Z} + \rho v\frac{\partial U}{\partial y} - \frac{\cos\lambda}{\sin\lambda}K_1\rho U^2 + \frac{K_2}{\sin\lambda}\rho W^2 + K_{12}\rho UW$$

$$= a_1\frac{\partial P}{\partial X} + a_2\frac{\partial P}{\partial Z} + \frac{\partial}{\partial y}(\mu\frac{\partial U}{\partial y} - \rho\langle u'v'\rangle) \qquad (1b)$$

$$\rho\frac{U}{h_1}\frac{\partial W}{\partial X} + \rho\frac{W}{h_2}\frac{\partial W}{\partial Z} + \rho v\frac{\partial W}{\partial y} + \frac{K_1}{\sin\lambda}\rho U^2 - K_2\frac{\cos\lambda}{\sin\lambda}\rho W^2 + K_{21}\rho UW$$

$$= b_1\frac{\partial P}{\partial X} + b_2\frac{\partial P}{\partial Z} + \frac{\partial}{\partial y}(\mu\frac{\partial W}{\partial y} - \rho\langle w'v'\rangle) \qquad (1c)$$

In addition, the normal pressure derivative $\partial P/\partial y$ is assumed zero within the boundary layer.

K_1 and K_2 are the geodesic curvatures of the X- and Z-axes defined as :

$$K_1 = \frac{1}{h_1h_2\sin\lambda}\left(\frac{\partial}{\partial X}(h_2\cos\lambda) - \frac{\partial h_1}{\partial Z}\right) ; \; K_2 = \frac{1}{h_1h_2\sin\lambda}\left(\frac{\partial}{\partial Z}(h_1\cos\lambda) - \frac{\partial h_2}{\partial X}\right)$$

The other coefficients are :

$$q = h_1h_2\sin\lambda$$

$$K_{12} = \frac{1}{\sin\lambda}\left[-(K_1 + \frac{1}{h_1}\frac{\partial\lambda}{\partial X}) + \cos\lambda\,(K_2 + \frac{1}{h_2}\frac{\partial\lambda}{\partial Z})\right]$$

$$K_{21} = \frac{1}{\sin\lambda}\left[-(K_2 + \frac{1}{h_2}\frac{\partial\lambda}{\partial Z}) + \cos\lambda\,(K_1 + \frac{1}{h_1}\frac{\partial\lambda}{\partial X})\right]$$

$$a_1 = -\frac{1}{h_1\sin^2\lambda} \; ; \; a_2 = \frac{\cos\lambda}{h_2\sin^2\lambda} \; ; \; b_1 = \frac{\cos\lambda}{h_1\sin^2\lambda} \; ; \; b_2 = -\frac{1}{h_2\sin^2\lambda}$$

According to the boundary layer approximations, it is assumed that the metric coefficients h_1, h_2 and λ are independent of the coordinate y. The coefficients h_1, h_2 and λ are functions of X and Z.

The wall boundary conditions are the no-slip conditions U = W = v = 0 (in the case of an impermeable wall).

The outer edge conditions $U = U_e$, $W = W_e$ are such as :

$$\rho\frac{U_e}{h_1}\frac{\partial U_e}{\partial X} + \rho\frac{W_e}{h_2}\frac{\partial U_e}{\partial Z} - \frac{\cos\lambda}{\sin\lambda}K_1\rho U_e^2 + \frac{K_2}{\sin\lambda}\rho W_e^2 + K_{12}\rho U_eW_e = a_1\frac{\partial P}{\partial X} + a_2\frac{\partial P}{\partial Z} \qquad (2a)$$

$$\rho \frac{U_e}{h_1}\frac{\partial W_e}{\partial X} + \rho \frac{W_e}{h_2}\frac{\partial W_e}{\partial Z} + \frac{K_1}{\sin\lambda}\rho U_e^2 - K_2 \frac{\cos\lambda}{\sin\lambda}\rho W_e^2 + K_{21}\rho U_e W_e = b_1\frac{\partial P}{\partial X} + b_2\frac{\partial P}{\partial Z} \quad (2b)$$

The outer edge conditions can be determined either from an inviscid flow calculation or from an experiment. In the first case, the two components of the external velocity are given. In the second case, it is often more difficult to deduce these data because, in general, only the wall pressure distribution is measured. Equations (2) are integrated to deduce the two components of the external velocity from the wall pressure. This is not a straightforward task because initial conditions are difficult to generate. Most of the available procedures are devoted to fuselage-like bodies /4,5,6/.

The technique we developed /6/ proceeds in two steps. Equations (2) are first integrated upstream by starting not too far downstream of the stagnation point. The starting conditions (direction of external streamline) are guessed and the right conditions are obtained when the calculated streamline passes through the stagnation point. The second step consists of integrating equations (2) following the stream by using the starting conditions determined in the first step.

Calculation methods are based on the solution of equations (1) but this set of equations is not closed : the turbulent shear stress $-\rho\langle u'v'\rangle$ and $-\rho\langle w'v'\rangle$ need to be represented by some turbulence model.

Practical calculation methods are also developed from an integrated form of equations (1). The integration is performed between the wall and the boundary layer edge. Such a set of equations is the basis of *integral methods*. In fact, an infinity of global equations can be imagined : the most evident equations are the integrated equations (1) but, sometimes, other equations are used, for example the global kinetic energy equation.

The most often used integral equations are the integrated forms of the continuity and momentum equations. Without suction or blowing at the wall, these equations are :

$$C_E = \frac{1}{u_e q}\frac{\partial}{\partial X}\left[\frac{u_e q}{h_1}\left(\delta\frac{U_e}{u_e} - \Delta_1\right)\right] + \frac{1}{u_e q}\frac{\partial}{\partial Z}\left[\frac{u_e q}{h_2}\left(\delta\frac{W_e}{u_e} - \Delta_2\right)\right] \quad (3a)$$

$$\frac{C_{fX}}{2} = \frac{1}{u_e^2 q}\left[\frac{\partial}{\partial X}\left(\frac{u_e^2 q}{h_1}\Theta_{11}\right) + \frac{\partial}{\partial Z}\left(\frac{u_e^2 q}{h_2}\Theta_{12}\right)\right] + \frac{\Delta_1}{u_e h_1}\frac{\partial U_e}{\partial X} + \frac{\Delta_2}{u_e h_2}\frac{\partial U_e}{\partial Z}$$
$$- K_1 \frac{\cos\lambda}{\sin\lambda}\left(\frac{U_e}{u_e}\Delta_1 + \Theta_{11}\right) + \frac{K_2}{\sin\lambda}\left(\frac{W_e}{u_e}\Delta_2 + \Theta_{22}\right) + K_{12}\left(\frac{U_e}{u_e}\Delta_2 + \Theta_{12}\right) \tag{3b}$$

$$\frac{C_{fZ}}{2} = \frac{1}{u_e^2 q}\left[\frac{\partial}{\partial X}\left(\frac{u_e^2 q}{h_1}\Theta_{21}\right) + \frac{\partial}{\partial Z}\left(\frac{u_e^2 q}{h_2}\Theta_{22}\right)\right] + \frac{\Delta_1}{u_e h_1}\frac{\partial W_e}{\partial X} + \frac{\Delta_2}{u_e h_2}\frac{\partial W_e}{\partial Z}$$
$$- K_2 \frac{\cos\lambda}{\sin\lambda}\left(\frac{W_e}{u_e}\Delta_2 + \Theta_{22}\right) + \frac{K_1}{\sin\lambda}\left(\frac{U_e}{u_e}\Delta_1 + \Theta_{11}\right) + K_{21}\left(\frac{W_e}{u_e}\Delta_1 + \Theta_{21}\right) \tag{3c}$$

In these equations, u_e is the resultant external velocity :

$$u_e^2 = U_e^2 + W_e^2 + 2\,U_e W_e \cos\lambda$$

The definition of the various global quantities is :

δ : boundary layer thickness.

C_E is the entrainment coefficient : $C_E = \frac{U_e}{u_e}\frac{1}{h_1}\frac{\partial\delta}{\partial X} + \frac{W_e}{u_e}\frac{1}{h_2}\frac{\partial\delta}{\partial Z} - \frac{V_e}{u_e}$

$$\frac{C_{fX}}{2} = \frac{\tau_{WX}}{\rho u_e^2} \;;\quad \frac{C_{fZ}}{2} = \frac{\tau_{WZ}}{\rho u_e^2} \;;\quad \Theta_{11} = \int_0^\delta \frac{U(U_e - U)}{u_e^2}\,dy \;;\quad \Theta_{22} = \int_0^\delta \frac{W(W_e - W)}{u_e^2}\,dy$$

$$\Theta_{12} = \int_0^\delta \frac{W(U_e - U)}{u_e^2}\,dy \;;\quad \Theta_{21} = \int_0^\delta \frac{U(W_e - W)}{u_e^2}\,dy$$

$$\Delta_1 = \int_0^\delta \frac{U_e - U}{u_e}\,dy \;;\quad \Delta_2 = \int_0^\delta \frac{W_e - W}{u_e}\,dy$$

In these definitions, τ_{WX} and τ_{WZ} are the wall shear stress components along X and Z.

Obviously, the set of global equations (3) contains too many unknowns. The solution is obtained with the help of closure assumptions which give additional relationships between the unknowns.

The boundary layer equations are often written in a *streamline coordinate system*. The x-axis is formed by the external streamlines and the z-axis by the orthogonal lines. The basic equations are slightly simplified since :

$$\lambda = \pi/2 \quad W_e = 0 \quad U_e = u_e$$

In the streamline coordinate system, the velocity components are u and w and the boundary layer characteristics are denoted by small letters ($\delta, \theta \ldots$) :

$$\frac{C_{f_X}}{2} = \frac{\tau_{W_X}}{u_e^2} \; ; \; \frac{C_{f_Z}}{2} = \frac{\tau_{W_Z}}{u_e^2} \; ; \; \delta_1 = \int_o^\delta (1 - \frac{u}{u_e}) \, dy \; ; \; \delta_2 = \int_o^\delta - \frac{w}{u_e} \, dy$$

$$\theta_{11} = \int_o^\delta \frac{u}{u_e} (1 - \frac{u}{u_e}) \, dy \quad ; \quad \theta_{22} = \int_o^\delta - \frac{w^2}{u_e^2} \, dy$$

$$\theta_{12} = \int_o^\delta \frac{w}{u_e} (1 - \frac{u}{u_e}) \, dy \quad ; \quad \theta_{21} = \int_o^\delta - \frac{uw}{u_e^2} \, dy$$

The shape parameter of the streamwise velocity profile is : $H = \frac{\delta_1}{\theta_{11}}$.

The ratio of the wall shear stress components defines the angle β_o (or β_w) between the limiting wall streamline and the external streamline (fig. 2).

$$\tan \beta_o = \frac{\tau_{W_Z}}{\tau_{W_X}} = \left(\frac{\frac{\partial w}{\partial y}}{\frac{\partial u}{\partial y}} \right)_{y=0} = \lim_{y \to 0} \frac{w}{u} \tag{4}$$

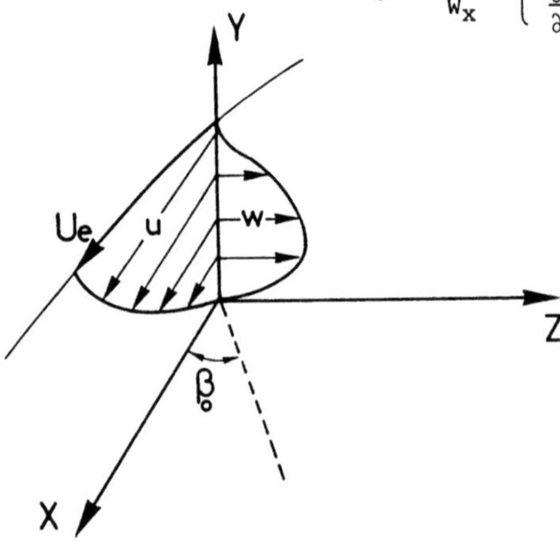

FIG. 2 - VELOCITY PROFILES IN THE STREAMLINE COORDINATE SYSTEM

Let us notice that *the true displacement thickness* δ^* *results from the solution of a partial differential equation.* Without suction or blowing at the wall, this equation is :

$$\frac{\partial}{\partial X} \left[\frac{u_e q}{h_1} (\delta^* \frac{U_e}{u_e} - \Delta_1) \right] + \frac{\partial}{\partial Z} \left[\frac{u_e q}{h_2} (\delta^* \frac{W_e}{u_e} - \Delta_2) \right] = 0 \tag{5}$$

The boundary layer effects can also be represented by *an equivalent wall transpiration velocity* $v^*(0)$:

$$\frac{v^*(0)}{u_e} = \frac{1}{u_e} \left[\frac{1}{q} \frac{\partial}{\partial X} (\frac{u_e q}{h_1} \Delta_1) + \frac{1}{q} \frac{\partial}{\partial Z} (\frac{u_e q}{h_2} \Delta_2) \right] \tag{6}$$

1.2. Nature of the system of equations

If the pressure distribution is prescribed, it has been shown that *the system of equations (1), with the second derivatives neglected, is hyperbolic.* All the streamlines including the skin friction lines are characteristics. Associated with the diffusive nature of equations along normals to the wall, these characteristics determine *zones of influence and dependence.* A perturbation at a point P is carried along the y-axis by the diffusion process and along the boundary layer streamlines by convection in the downstream direction. The domain of influence of a point P is the downstream part of a volume delimited by surfaces normal to the wall containing the outermost streamlines which cross the normal to the wall passing through P. The domain of dependence of P is the upstream part of the volume defined in the same way as above (fig. 3).

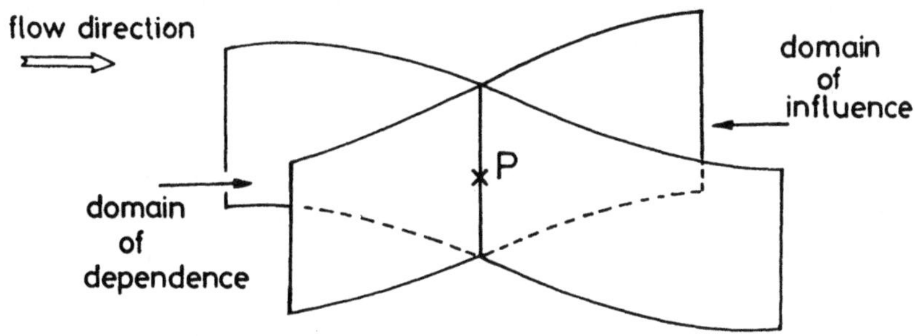

FIG. 3 - DOMAINS OF INFLUENCE AND OF DEPENDENCE

The practical consequences of these features are important. A first consequence is the COURANT-FRIEDRICHS-LEWY condition : one of the rules to construct a correct numerical approximation is that the numerical domain of dependence of a difference scheme should include the analytical domain of dependence of the differential equations. This means that the integration path of equations must be oriented to follow this rule.

A second consequence is that boundary conditions are needed along the sides of the computation domain through which the flow enters, but not along the parts of the boundary through which the flow leaves.
Particular solutions can be used to generate initial or boundary conditions. For example, at the stagnation point or along the attachment line of an

infinite swept wing, self similar solutions exist. Sometimes, particular solutions along planes of symmetry are used to produce boundary conditions.

2. INTEGRAL METHODS

For calculating classical boundary layers, it has been shown that integral methods remain a valuable engineering tool (see for example the 1968 and 1980-81 STANFORD Conferences /7, 8/).

Their principle is well known. The global equations (3) are the basis of the integral methods and they are completed with appropriate closure relationships /9/.

2.1. Closure relationships

The closure relationships can be classified into three categories : *1) representation of integral thicknesses ; 2) skin friction law ; 3) entrainment coefficient.*

Relationships for integral thicknesses are generally obtained from *a family of velocity profiles.* The generation of these velocity profiles is often based on the decomposition of the flow into a streamwise and a crosswise components. It is assumed that the streamwise velocity profile behaves like in two-dimensional flow. Therefore, the procedures developed in two-dimensional flow apply : let us cite the use (in turbulent flow) of power law velocity profiles (MYRING /10/, SMITH /11/, OKUNO /12/), self similar solutions (COUSTEIX /9/), COLES profiles (STOCK /13/).

The streamwise skin friction coefficient and the entrainment coefficient are also generally calculated from *two-dimensional relationships.*

The modelling of crosswise velocity profiles is often very empirical. An example is the MAGER representation /14/ :

$$\frac{w}{u} = \tan\beta_o \left(1 - \frac{y}{\delta}\right)^2 \tag{7}$$

Another example is the GRUSCHWITZ-JOHNSTON triangular polar plot /15, 16/ :

$$\begin{aligned} w &= u \tan\beta_o & \quad y < y_a \\ w &= C\,(U_e - U) & \quad y > y_a \end{aligned} \tag{8}$$

If the apex of the triangle is assumed fixed at a location y_a (for example where $U_a^+ = U_a/U_\tau = 14$), this gives a relationship between C and β_o.

The above representations are used to obtain relationships for the crossflow integral thicknesses and for the limiting angle β_o.

In the method we developed, the crossflow velocity profiles have been generated from an extension, to three-dimensional boundary layers, of self-similar solutions /9/.

Another method for constructing velocity profiles is *a vectorial representation* as proposed by COLES /17/. This is an extension of the two-dimensional law of the wall-law of the wake. The boundary layer velocity vector is given by :

$$\vec{Q} = \vec{Q}_\tau \left(\frac{1}{\chi} \ln \frac{y|\vec{Q}_\tau|}{\nu} + C + \frac{\overline{\overline{\pi}}}{\lambda} \omega\left(\frac{y}{\delta}\right)\right) \tag{9}$$

where $\vec{Q}_\tau$ is parallel to the skin friction vector $\vec{\tau}_W$ and its modulus is $\sqrt{\tau_W/\rho}$. $\overline{\overline{\pi}}$ is a tensor such that $\overline{\overline{\pi}}\ \vec{Q}_\tau$ is parallel to the external velocity. In this formulation, the velocity vector turns from the direction of the limiting wall streamline to the direction of the external velocity. A similar model has been used by CROSS /18/ and LE BALLEUR /19/.

2.2. Discussion

From the brief presentation of the various hypotheses used to construct the closure relationships in turbulent flow, three main difficulties can be identified : *1) modelling of the entrainment coefficient ; 2) representation of crossflow velocity profiles ; 3) the three-dimensional law of the wall.*

The entrainment coefficient is very important because it describes the rate at which the outer fluid is entrained into the boundary layer. This process controls the growth of the boundary layer to a large extent.

The entrainment is intimately associated with the turbulence structure in the outer part of the boundary layer. Now the experiments of ELSENAAR-BOELSMA /20/ and BRADSHAW-PONTIKOS /21/ have shown that the turbulence structure, compared to the case of a two-dimensional boundary layer, is modified by the three-dimensionality of the flow. Both sets of experiments consist of generating a three-dimensional boundary layer on a swept wing by applying a crosswise pressure gradient to a developed two-dimensional turbulent boundary layer. It has been shown that the transport velocity for the turbulent kinetic energy decreases and the magnitude of the shear stress decreases compared with an equivalent two-dimensional boundary layer. Another result is that the main axes of the stress tensor are not aligned with the main axes of the velocity gradient tensor. It is also probable that the decrease in the turbulence activity is associated with *a decrease in the entrainment*. This is not taken into account when it is assumed that the entrainment coefficient follows the same rules as in two-dimensional flow. These effects are the results of the action of the crossflow development on large eddies which are distorted when the mean velocity profile starts to be skewed.

When a developed two-dimensional boundary layer is suddenly submitted to a crosswise pressure gradient, the crossflow develops immediately in the outer part of the boundary layer (but, of course, the no-slip condition at the wall enforces $w = 0$ there). In such a case, it seems that the process of crossflow generation is purely inviscid. A proof of this is tentatively provided by the application of the SQUIRE-WINTER formula. This formula is based on inviscid considerations. Assuming a linear polar plot :

$$\frac{w}{u_e} = C\left(1 - \frac{u}{u_e}\right) \tag{10}$$

the evolution of the slope is related to the turning angle α of the external streamlines. For small turning angles, we have :

$$C = 2(\alpha - \alpha_o) \tag{11}$$

In addition, if eq. 10 is valid in the whole boundary layer, we have :

$$C = -\delta_2/\delta_1 \tag{12}$$

The comparison between eq. 11 and eq. 12 for the experiments of Van den BERG et al. /22/ is given in figure 4. By adjusting the value of α, a very good agreement is obtained.

This supports the idea that *the turbulence in the outer part of the boundary layer is strongly out of equilibrium* since the mean flow changes rapidly /23/. This could explain the effects of three-dimensionality on the structure of turbulence and on the entrainment coefficient. It can be argued that the effects will be stronger if the crossflow is generated in a fully developed two-dimensional boundary layer as it is the case of Van den BERG et al experiments and BRADSHAW-PONTIKOS experiments. In practical situations, this happens if the crosswise pressure gradient becomes significant in a region where a two-dimensional boundary layer is already well developed : this is the case of a modern supercritical swept wing, and of an obstruction placed in a thick two-dimensional boundary layer.

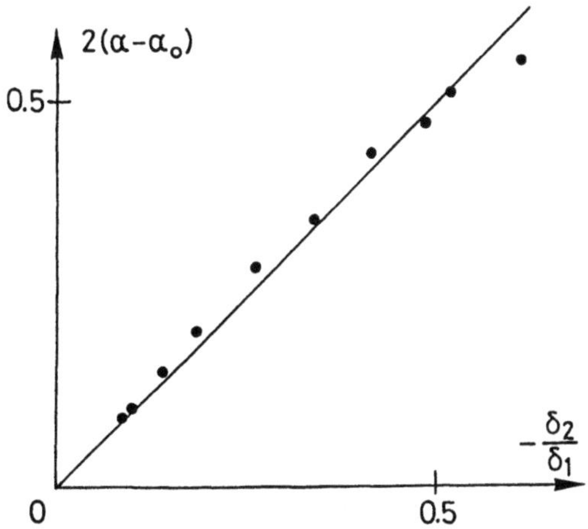

FIG. 4 - CALCULATION OF THE SLOPE OF THE HODOGRAPH IN THE OUTER PART OF THE BOUNDARY LAYER FROM THE SQUIRE-WINTER FORMULA. EXPERIMENTS : VAN DEN BERG ET AL /22/

The second difficulty to construct the closure reltationships is *the modelling of crossflow velocity profiles*. Indeed this is a challenge because these profiles can exhibit very different shapes. For example, S-shaped profiles can appear downstream of an inflexion point of the external streamlines. In most integral methods, the representation of such profiles is either very poor or even ignored, so that the angle β_o is incorrectly predicted.

In fact, this problem is more or less associated with the third problem which is the absence of any counterpart of the two-dimensional *law of the wall*. This is very unfortunate because, in two-dimensional flow, this is a key for the success of the representation of velocity profiles and the skin friction law ; it is also a key in turbulence modelling because all models use it implicitly or explicitly. Several attempts have been made to extend the law of the wall to three-dimensional boundary layers /60/. Generally the behaviour of the velocity profile is obtained by solving the basic equations with a similarity hypothesis and by using a turbulence model, for example a mixing length model (Van den BERG, /24/). In two-dimensional flow, the reasoning is not at all the same : the logarithmic law of the wall comes from the analysis of the double-layer structure of the boundary layer and turbulence models are devised to reproduce the properties of the velocity profile. In addition, measurements of EAST-SAWYER /25/ in two-dimensional equilibrium turbulent boundary layers with various pressure gradients have shown that the logarithmic law of the wall seems to be the most universally valid and that the mixing length varies from flow to flow. In particular, the slope of the mixing length in the inner region of the boundary layer varies over a rather wide range.

Recently, GOLDBERG-RESHOTKO /26/ have performed an asymptotic analysis of the three-dimensional boundary layer. They found that the direction of the velocity is constant in the inner layer. At very large REYNOLDS number, this result is certainly true but, in practice, it is insufficient because the experimental results show that the velocity direction can vary rapidly near the wall. In JOHNSTON's model, the apex of the polar plot is around $U^+ = 14$; this means that it cannot be assumed that the velocity direction in the logarithmic region is equal to the wall limiting angle β_o. Indeed, substantial variations of the flow angle can exist in the inner region.

3. Turbulence modelling

The solution of the averaged NAVIER-STOKES equations needs a turbulence model which has to reproduce the behaviour of REYNOLDS stresses as well as possible. Many turbulence models have been developed and, for practical applications, other criteria than the good representation of REYNOLDS stresses have to be taken into account in the choice of a turbulence model. In most applications, the quantities of interest are related to the mean flow : these quantities are the displacement thickness, the skin friction coefficient, the shape of limiting streamlines. In addition, the cost of the results is a non-negligible parameter. This means that the choice of the appropriate model is often a matter of compromise and that the accuracy of the calculation of REYNOLDS stresses is not always the determinant criterion.

In two-dimensional flow, the application of *mixing length* or *eddy viscosity* models to classical boundary layers has proved to be a very valuable tool and their extension to the three-dimensional case is very tempting. In the most straightforward extension, we write :

$$\tau_{t_X} = -\rho \langle u'v' \rangle = \rho\, \nu_t \frac{\partial U}{\partial y} \qquad \tau_{t_Z} = -\rho \langle w'v' \rangle = \rho\, \nu_t \frac{\partial W}{\partial y} \qquad (13)$$

The eddy viscosity ν_t can be expressed by using an extension of the CEBECI-SMITH model (CEBECI /2/, /27/) or by using a mixing length formulation (COUSTEIX et al /28/). This latter scheme is :

$$\nu_t = F^2 l^2 \left[\left(\frac{\partial U}{\partial y}\right)^2 + \left(\frac{\partial W}{\partial y}\right)^2 + 2 \frac{\partial U}{\partial y} \frac{\partial W}{\partial y} \cos\lambda \right]^{1/2} \tag{14a}$$

The mixing length is expressed by the same formula as in two-dimensional flow :

$$\frac{l}{\delta} = 0.085 \tanh \left(\frac{\chi}{0.085} \frac{y}{\delta}\right) \qquad \chi = 0.41 \tag{14b}$$

and the viscous damping function is given by :

$$F = 1 - \exp\left[- \frac{1}{26\chi\mu} \left(\left(\tau_{t_X}^{\ 2} + \tau_{t_Z}^{\ 2} + 2\tau_{t_X}\tau_{t_Z}\cos\lambda\right)^{1/2}\rho\right)^{1/2} \right] \tag{14c}$$

The above mixing length scheme reduces to the classical formulation in the two-dimensional case and the expression of the REYNOLDS stresses is independent of the axis system. Obviously these conditions are not sufficient to insure the success of the model.

The eddy viscosity can also be calculated from a k-ε model :

$$\nu_t = C_\mu \frac{k^2}{\varepsilon} \tag{15}$$

where the turbulent kinetic energy and its dissipation rate ε are calculated from transport equations :

$$\frac{Dk}{Dt} = P - \varepsilon + \frac{\partial}{\partial y} \left(\frac{C_\mu}{\sigma_k} \frac{k^2}{\varepsilon} \frac{\partial k}{\partial y}\right) \tag{16a}$$

$$\frac{D\varepsilon}{Dt} = C_{\varepsilon_1} P \frac{\varepsilon}{k} - C_{\varepsilon_2} \frac{\varepsilon^2}{k} + \frac{\partial}{\partial y} \left(\frac{C_\mu}{\sigma_\varepsilon} \frac{k^2}{\varepsilon} \frac{\partial \varepsilon}{\partial y}\right) \tag{16b}$$

with

$$P = - \langle u'v' \rangle \frac{\partial U}{\partial y} - \langle w'v' \rangle \frac{\partial W}{\partial y} - \left[\langle u'v' \rangle \frac{\partial W}{\partial y} + \langle w'v' \rangle \frac{\partial U}{\partial y} \right] \cos\lambda$$

and

$$C_\mu = 0.09 \qquad C_{\varepsilon_1} = 1.44 \qquad C_{\varepsilon_2} = 1.92 \qquad \sigma_k = 1 \qquad \sigma_\varepsilon = 1.3$$

It is easy to show that, near the wall, the k-ε model reduces to the mixing length model. Indeed, near the wall, the classical approximations give :

$$P = \varepsilon$$

$$C_{\varepsilon_1} P \frac{\varepsilon}{k} - C_{\varepsilon_2} \frac{\varepsilon^2}{k} + \frac{\partial}{\partial y} \left(\frac{C_\mu}{\sigma_\varepsilon} \frac{k^2}{\varepsilon} \frac{\partial \varepsilon}{\partial y}\right) = 0$$

From these equations and eq. 15, we deduce :

$$\varepsilon = C \frac{k^{3/2}}{y} \qquad C^2 = \frac{C_\mu}{\sigma_\varepsilon (C_{\varepsilon_2} - C_{\varepsilon_1})}$$

$$- \langle u'v' \rangle = \frac{C_\mu}{C} k^{1/2} y \frac{\partial U}{\partial y} \quad ; \quad - \langle w'v' \rangle = \frac{C_\mu}{C} k^{1/2} y \frac{\partial W}{\partial y}$$

$$\frac{\tau}{k} = C_\mu{}^{1/2}$$

$$\tau = \left(\frac{C_\mu{}^{3/4}}{C}\right)^2 y^2 \left[\left(\frac{\partial U}{\partial y}\right)^2 + \left(\frac{\partial W}{\partial y}\right)^2 + 2 \frac{\partial W}{\partial y}\frac{\partial U}{\partial y} \cos\lambda\right]$$

with

$$\tau = (\langle u'v' \rangle^2 + \langle w'v' \rangle^2 + 2 \langle u'v' \rangle \langle w'v' \rangle \cos\lambda)^{1/2}$$

A number of applications of the above models, including comparisons with available experiments, have been performed by various authors /2/, /9/, /27/. Concerning the mean flow properties, the general conclusion could be that the predictions are not so wrong as is sometimes stated. However, as in two-dimensional flow, these models become inappropriate as separation is approached. In two-dimensional flow, it has been stressed by CEBECI /2/ that the normal pressure gradient could have a very important role but, in three-dimensional boundary layers, these effects have not been investigated. It is probable that some of the inaccuracies attributed to turbulence modelling are in fact the result of neglecting the normal pressure gradient. Flows involving separation are mainly pressure driven and the turbulence model is less important.

On the other hand, the application of the above models to *shear driven flows* leads to fairly good results. This is the case for the experiments of BRADSHAW-TERRELL /29/ (see EAST /30/) who studied the relaxation of an initially three-dimensional boundary layer. Figure 5 shows that the mixing length and the k-ε models give the correct decay of the crossflow angle β_o. Figure 5 also shows results obtained with the integral method. Clearly the rate of decay of β_o is underpredicted at the beginning of the relaxation. This illustrates the greater flexibility of the solution of local equations as compared to integral methods.
However there is no doubt that the above turbulence models are not satisfactory. Experimental data (EAST /31/, JOHNSTON /32/, ELSENAAR-BOELSMA /20/, BRADSHAW-PONTIKOS /21/) have shown that several of the hypotheses included in these models are wrong. Firstly the experimental results show that the "shear stress vector" $\vec{\tau}$ (τ_{t_x}, τ_{t_z}) is not aligned with the "velocity gradient $\frac{\partial \vec{V}}{\partial y}$ $(\frac{\partial U}{\partial y}, \frac{\partial W}{\partial y})$; secondly, they show that the magnitude of shear stress is reduced compared with an equivalent two-dimensional boundary layer. These effects are not reproduced in the models. *More sophisticated models are needed:* below we discuss suggestions made by Rotta in 1979.

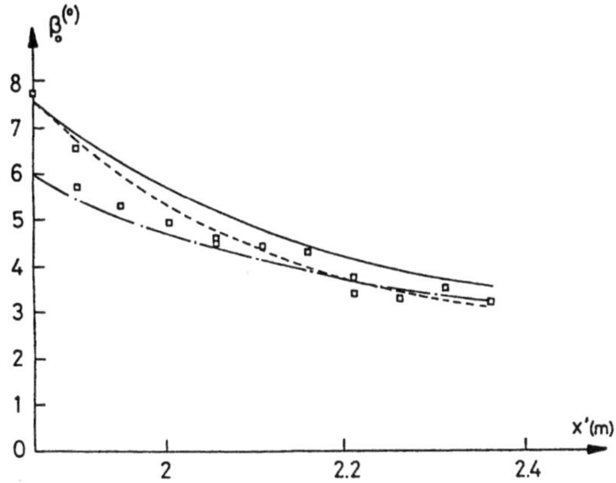

FIG. 5 - CALCULATION OF BRADSHAW-TERRELL EXPERIMENTS /29/ (45° INFINITE SWEPT WING) : RELAXATION OF AN INITIALLY THREE-DIMENSIONAL BOUNDARY LAYER TOWARDS A TWO-DIMENSIONAL BOUNDARY LAYER. x' : DISTANCE FROM LEADING EDGE ALONG TUNNEL CENTERLINE ——— MIXING LENGTH --- k-ε ——·—— INTEGRAL METHOD

A possible solution is to use *a full transport equation model* for the REYNOLDS stresses. In a cartesian axis system, these equations are :

$$\frac{D}{Dt}\langle u_i'u_j'\rangle = P_{ij} + \phi_{ij} - D_{ij} + T_{ij} \tag{17}$$

where P_{ij} is the so-called production term :

$$P_{ij} = -\langle u_i'u_k'\rangle \frac{\partial U_j}{\partial x_k} - \langle u_j'u_k'\rangle \frac{\partial U_i}{\partial x_k}$$

and ϕ_{ij}, D_{ij}, T_{ij} are respectively the velocity-pressure correlation, the destruction term and the transport term.

Several versions of the modelling of these equations have been proposed in the literature. The model developed by LAUNDER et al./in ref. 8/ which includes the ROTTA formulation of the "return to isotropy" term $\phi_{ij,1}$, is (see for example, the 1980-81 STANFORD Conference /8/) :

$$T_{ij} = C_S \frac{\partial}{\partial x_k}\left(\langle u_k'u_l'\rangle \frac{k}{\varepsilon}\frac{\partial}{\partial x_l}\langle u_i'u_j'\rangle\right) \tag{18}$$

$$D_{ij} = \frac{2}{3}\varepsilon\,\delta_{ij}$$

$$\phi_{ij} = \phi_{ij,1} + \phi_{ij,2} + \phi_{ij,w}$$

$$\phi_{ij,1} = -C_1\frac{\varepsilon}{k}(\langle u_i'u_j'\rangle - \frac{2}{3}\delta_{ijk})$$

$$\phi_{ij,2} = -C_2\,(P_{ij} - \frac{1}{3}\delta_{ij}P_{kk})$$

$$\phi_{ij,w} = \left(C_1' \frac{\varepsilon}{k} (\langle u_k' u_m' \rangle n_k n_m \delta_{ij} - \frac{2}{3} \langle u_k' u_i' \rangle n_k n_j - \frac{3}{2} \langle u_k' u_j' \rangle n_k n_i) \right.$$

$$\left. + C_2' (\phi_{km,2} n_k n_m \delta_{ij} - \frac{3}{2} \phi_{ik,2} n_k n_j - \frac{3}{2} \phi_{jk,2} n_k n_i) \right) f$$

$$f = \frac{k^{3/2}}{2{,}5\, \varepsilon\, x_n}$$

where n_i are the components of a unit vector normal to the surface and x_m is the normal distance from the wall.

The transport equations are often simplified to give the so-called *algebraic stress models*. One version of these models is based on the following approximations :

$$\frac{D}{Dt} \langle u_i' u_j' \rangle - T_{ij} = \frac{\langle u_i' u_j' \rangle}{k} (\frac{Dk}{Dt} - T_{kk}) = \frac{\langle u_i' u_j' \rangle}{k} (P - \varepsilon) = P_{ij} + \phi_{ij} - \frac{2}{3} \delta_{ij} \varepsilon \quad (19)$$

where $P = P_{kk}/2$.

From these equations, we get :

$$\frac{\langle u_i' u_j' \rangle}{k} = \frac{2}{3} \delta_{ij} + \frac{(1 - C_2) \left(\frac{P_{ij}}{\varepsilon} - \frac{2}{3} \delta_{ij} \frac{P}{\varepsilon} \right) + \frac{\phi_{ij,w}}{\varepsilon}}{C_1 + \frac{P}{\varepsilon} - 1} \quad (20)$$

If the boundary layer approximations are strictly applied, the velocity gradient components other than $\partial U/\partial y$ and $\partial W/\partial y$ are neglected. The expressions forthe REYNOLDS stress become :

$$\langle u'v' \rangle = \frac{C_2 (1 - \frac{3}{2} C_2' f) - 1}{C_1 + \frac{P}{\varepsilon} - 1 + \frac{3}{2} C_1' f} \frac{k^2}{\varepsilon} \frac{\langle v'^2 \rangle}{k} \frac{\partial U}{\partial y} \quad (21)$$

$$\langle w'v' \rangle = \frac{C_2 (1 - \frac{3}{2} C_2' f) - 1}{C_1 + \frac{P}{\varepsilon} - 1 + \frac{3}{2} C_1' f} \frac{k^2}{\varepsilon} \frac{\langle v'^2 \rangle}{k} \frac{\partial W}{\partial y}$$

$$\frac{\langle v'^2 \rangle}{k} = \frac{\frac{2}{3} (C_1 - 1 + \frac{P}{\varepsilon} C_2) - \frac{4}{3} C_2' C_2 \frac{P}{\varepsilon} f}{C_1 + \frac{P}{\varepsilon} - 1 + 2 C_1' f}$$

It can be noted that the above expressions lead to an isotropic eddy viscosity : the shear stress vector is aligned with the velocity gradient $\frac{\partial \vec{V}}{\partial y}$. In addition, it seems that no three-dimensional effect is included in this model. ABID-SCHMITT /33/ have applied this model to the experiments of Van den BERG et al./22/. The agreement with the experimental results is slightly better than with the standard k-ϵ model especially for the shear stress profiles. The improvement has been attributed to a reduction in the eddy viscosity due to a reduction of $\langle v'^2\rangle/k$. In two-dimensional flows, with adverse pressure gradient, the same behaviour has been noticed, so the improvement is not solely a consequence of three-dimensional effects. Apparently, there is a contradiction with experimental results.

In addition, velocity gradient components other than $\partial U/\partial y$ and $\partial W/\partial y$ have been neglected in the transport equations. It is not clear if this hypothesis is justified, because it is known that small extra rate of strain can strongly affect the development of REYNOLDS stresses. It is known that the boundary layer hypotheses are not necessarily applicable with the same degree of approximation in the REYNOLDS stress equations as in the mean flow equations. This is the case, for example, of a two-dimensional boundary layer developing on a longitudinally curved wall. If the curvature is not too large, the curvature terms in the mean flow equations are small whereas it is essential to retain them in the REYNOLDS stress equations. Indeed, even if the extra rate of strain $\partial V/\partial x$ introduced by longitudinal curvature is small with respect to $\partial U/\partial y$, its effect on the REYNOLDS stress is comparatively large. Now three-dimensionality is characterized by a lateral curvature of the streamlines which implies the existence of extra rates of strain, e.g. $\partial W/\partial x$, which are not necessarily very small compared to $\partial U/\partial y$ and $\partial W/\partial y$. The effect of these terms in the transport equations has not been investigated.

Other attempts have been proposed to improve the classical turbulence models. Let us mention the approach developed by ROTTA /34/ who suggested modifying the expression of *the pressure-strain* dependence upon the mean rate of strain. Modelling of the pressure-strain terms is certainly the key to the problem, because the mean-flow transport terms and the diffusion terms of the REYNOLDS stress equations are smaller and are not likely to explain three-dimensional effects on turbulence. The ROTTA proposal can be used in an eddy viscosity model. In a cartesian axis system, the result is :

$$- \langle u'v'\rangle = \nu_t \left(a_{XX} \frac{\partial U}{\partial y} + a_{XZ} \frac{\partial W}{\partial y}\right)$$

$$- \langle w'v'\rangle = \nu_t \left(a_{ZX} \frac{\partial U}{\partial y} + a_{ZZ} \frac{\partial W}{\partial y}\right)$$

$$a_{XX} = \frac{U^2 + TW^2}{U^2 + W^2}$$

$$a_{XZ} = a_{ZX} = (1 - T)\,\frac{UW}{U^2 + W^2}$$

$$a_{ZZ} = \frac{TU^2 + W^2}{U^2 + W^2}$$

where T is an anisotropy factor which represents the ratio of transverse and longitudinal eddy viscosities expressed in an axis system aligned with local boundary layer velocity.

Associated with a k-ε model, the eddy viscosity is :

$$\nu_t = C_\mu \frac{k^2}{\varepsilon}$$

The mixing length version of this model is :

$$\nu_t = l^2 \left\{ \left(\frac{\partial U}{\partial y}\right)^2 + \left(\frac{\partial W}{\partial y}\right)^2 + \frac{(T - 1)\left(W\frac{\partial U}{\partial y} - U\frac{\partial W}{\partial y}\right)^2}{U^2 + W^2} \right\}^{1/2} \tag{23}$$

From this expression, it appears that the eddy viscosity is reduced if $T < 1$. This is in qualitative agreement with experimental results of ELSENAAR-BOELSMA /20/. Unfortunately this model suffers from several drawbacks. As ROTTA has himself noted, the definition of the parameter T is not Galilean invariant, but it can be argued that the model is formulated for wall boundary layers and therefore the formulation is valid in an axis system tied to the wall. A second restriction is that the value of T is not universal. The few experiments in which values of T have been determined have shown a wide spread of T. If calculations are performed by using an average experimental value of T, the results are slightly improved but noticeable differences with experiments remain.

Figure 5 also shows that the BRADSHAW-TERRELL experiments /29/ are well calculated by classical models which correspond to $T = 1$. Another illustration is provided by the calculation of the wake developing behind a swept wing (COUSTEIX et al, /35/). This flow is also a shear-driven flow, and it is clear that the best agreement with experiments is obtained with $T = 1$ (figure 6) : a smaller value of T leads to a slower decay of the crossflow.

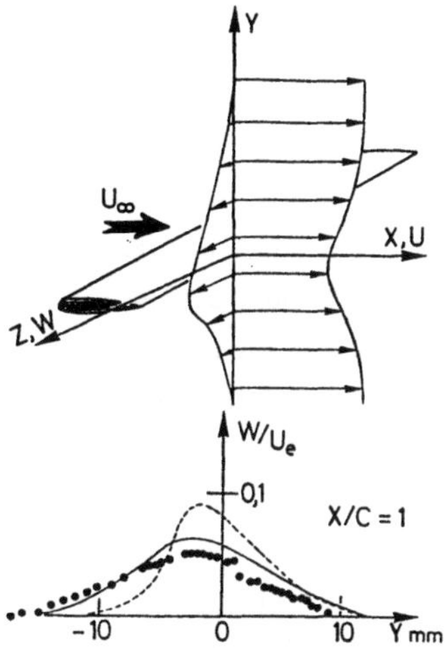

FIG. 6 - CROSSFLOW VELOCITY PROFILE IN A WAKE BEHIND A SWEPT WING /35/ (ONE CHORD LENGTH DOWNSTREAM OF THE TRAILING EDGE) - EXP. : • CALCULATION : k-ε - <u'v'> - <w'v'> MODEL —— T = 1 ----- T = 0.5

4. Calculation of the transition region

Recent reviews of problems arising from the laminar-turbulent transition are available in references /36/, /37/.

The practical calculation of a transition region involves two problems : *1) determination of the onset of transition ; 2) calculation of the transition region itself.*

The calculation of the transition region is often performed by using the so-called *intermittency method.* An intermittency function γ is applied to the turbulent shear stress :

$$\tau = \tau_\ell + \gamma\tau_t$$

where τ is the total shear stress, τ_ℓ is the viscous stress and τ_t is calculated by using a classical turbulence model. The function γ is equal to 0 in laminar flow and γ is equal to 1 in turbulent flow. It is interesting to notice that the best agreement with experiment is obtained if the function γ has an overshoot ($\gamma > 1$) near the end of the transition region.

Although this treatment is crude, the influence on the downstream boundary layer is not very large. It is more important to predict the location of *transition onset.*

In two-dimensional flow, so-called natural transition begins with the amplification of TOLLMIEN-SCHLICHTING waves. The *linear stability theory* enables us to calculate the characteristics of these waves as eigensolutions of the linearized NAVIER-STOKES equations.

One of the most successful transition criteria is based on the calculation of the amplification of TOLLMIEN-SCHLICHTING waves from stability theory. *The e^n method* (SMITH-GAMBERONI /38/, Van INGEN /39/) assumes that the onset of transition occurs when the total amplification of the most amplified waves reaches a critical value e^n. The exponent n has been determined experimentally as a function of the free stream turbulence level.

In three-dimensional flow, the problem is more complicated because *the crossflow can generate instabilities and transition, in regions where the streamwise boundary layer flow is stable*. This kind of instability was demonstrated by flight tests on swept wing aircraft at the RAE between 1951 and 1952 (GRAY /40/). Since then, several experimental studies on swept wings have confirmed the phenomenon and have contributed to its analysis /40/, /45/. The same phenomenon also occurs on other bodies than wings : rotating disk /41/, ellipsoid /46/, /47/. A peculiar feature of this instability is to generate a system of stationary vortices which are almost aligned with the external streamlines.

This kind of instability can be analyzed with the same linear stability theory as in two-dimensional flow. Basically this type of instability has the same nature as the TOLLMIEN-SCHLICHTING waves. The theory predicts the stationary waves very well but the theory also tells the existence of travelling waves which are more amplified.

The crossflow instability is of fundamental importance in the design of laminar swept wings. In particular, it is not useful to optimize the wing at zero sweep angle (in two-dimensional flow) because the crossflow transition can lead to a very different optimization when the wing is swept.

A few authors have tried to extend the e^n method to predict the transition in three-dimensional flow. The calculations are time-consuming because, compared to the two-dimensional case, the number of parameters determining the most-amplified waves increases. Thus, the number of successful applications of the e^n method in three-dimensional flow is very small. However, existing comparisons with experiments are encouraging as the critical value of n at the transition onset does not differ considerably from the two-dimensional values /48/, /49/.

For practical applications, other criteria have been developed. It is assumed that transition occurs either from streamwise instability or from crossflow instability, and it is also assumed that the two processes are uncoupled. Separate criteria are available for both mechanisms and the transition region starts when one of the two criteria is first satisfied.

As concerns the streamwise instability, the same criteria as in two-dimensional flow are applied along the external streamlines.

For the crossflow transition, a criterion was first proposed by OWEN-RANDALL /50/. Later, on the basis of a few experimental data, BEASLEY /51/ proposed another criterion formed with the crossflow REYNOLDS number $R_{\delta 2} = u_e \delta_2/\nu$. According to this criterion, crossflow transition occurs when $R_{\delta 2}$ reaches a critical value $(R_{\delta 2})_T = 150$. However, the experimental data which are now available clearly show that a single value of $R_{\delta 2}$ is not sufficient to correlate all the data. An additional parameter has been introduced to improve the correlation by ARNAL et al /52/ who proposed the following crossflow criterion (fig. 7) :

$$(R_{\delta 2})_T = \frac{300}{\pi} \operatorname{atan}\left(\frac{0.106}{(H-2.3)^{2.052}}\right); \quad 2.3 < H < 2.7 \qquad (24)$$

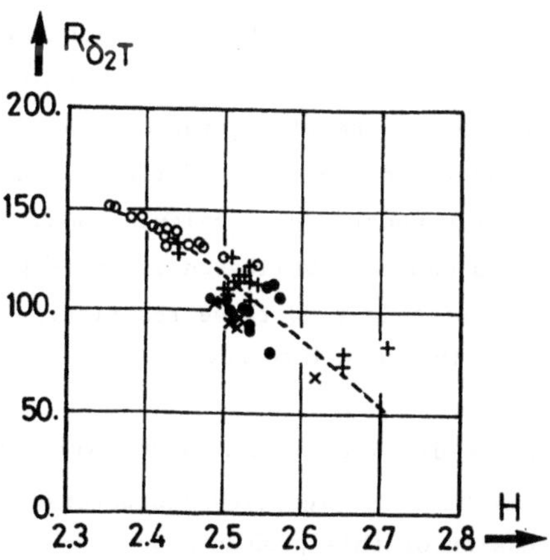

FIG. 7 - TRANSITION CRITERION FOR THE CROSSFLOW (FROM ARNAL ET AL. /52/)
o POLL x SCHMITT-MANIE + BOLTZ ET AL. • ARNAL ET AL.

On swept back wings, another mechanism of transition is very often met. This is *the leading edge contamination.* The turbulence coming from the fuselage is transported along the leading edge and leads to the transition of the boundary layer on the wing. The mechanism is not explained by the linear stability theory but a very simple criterion shows that transition does not occur if the REYNOLDS number based on the momentum thickness of the leading edge boundary layer is less than 100. Generally, when the flow is turbulent along the leading edge, it is also turbulent on the wing. Sometimes, a relaminarization occurs if the acceleration is strong enough downstream of the leading edge.

5. Singularities in boundary layer calculations

The boundary layer equations (either local or global) are completed by closure relationships and are associated with appropriate initial and boundary conditions. It is then assumed that the problem is mathematically well posed and numerical means are employed to solve it. However, it is not known a priori that a solution exists in any preassigned domain.

In two-dimensional (laminar) flow, the analysis of this problem has been performed by GOLDSTEIN /53/. A singularity has been identified at the point where the skin friction vanishes. The boundary layer calculations cannot proceed downstream and the vertical velocity becomes unbounded.

Obviously, this behaviour is not physical but it has been argued that it is an exaggeration of the actual behaviour of the flow. Then, the singularity exhibited in the boundary layer calculations is often taken as *an indicator of separation.* This interpretation has to be made with great care because the interaction between the boundary layer and the inviscid flow can have strong effects and modify the results of boundary layer calculations. In addition, the boundary layer assumptions can be invalidated if the separation is too much developed. However, keeping in mind these restrictions, the study of possible singularities remains very interesting.

The nature of the singularities of three-dimensional boundary layer equations is not completely known. Moreover, numerical difficulties encountered in calculations have obscured the discussion of this problem.

A first answer has been provided by COUSTEIX-HOUDEVILLE /54/, /55/ who analyzed the properties of an integral method in turbulent flow. This method is based on three global equations (continuity and momentum equations) together with closure relationships deduced from self-similarity solutions.

The equations form a system of three first order partial differential equations. The system has three characteristic directions which are always real and distinct. *The system is hyperbolic* (MYRING /10/, COUSTEIX-HOUDEVILLE /54/).

The calculation of characteristic roots of the set of global equations has been performed with approximate closure relationships. The angle of the characteristic directions with respect to the external streamline is $\tan^{-1}\gamma$. The three characteristic roots are :

$$\gamma_1 = \frac{C(H - 1)}{1 - \beta H} \tag{25}$$

$$\gamma_2 = C(H - 1)$$

$$\gamma_3 = \frac{C(H - 1)}{(2\alpha + \beta)\ H + 1}$$

$$\alpha = 0{,}631 \qquad \beta = -\alpha + \sqrt{\alpha^2 + \alpha} = 0{,}382$$

The coefficient C is a parameter defining simplified crossflow velocity profiles used for the calculation of γ. These simplified profiles are a linear representation of the polar plot $\frac{w}{u_e} = C(1 - \frac{u}{u_e})$ which is not too bad for evaluating the crosswise integral thicknesses.

From the closure relationships used in the integral method, it has been shown that the γ_1-characteristic line is very close to the limiting wall streamline. The other two characteristic lines lie between the limiting wall streamline and the external streamline.

The value of the characteristic roots depend on the value of the shape parameter H (fig. 8). In the range $1 \leq H \leq \infty$, the angle between the γ_2- (and γ_3-characteristic line and the external streamline is between $-\pi/2$ and $\pi/2$, but the angle between the γ_1-characteristic line and the external streamline can be larger :

$$-\pi \leq \tan^{-1} \gamma_1 \leq \pi$$

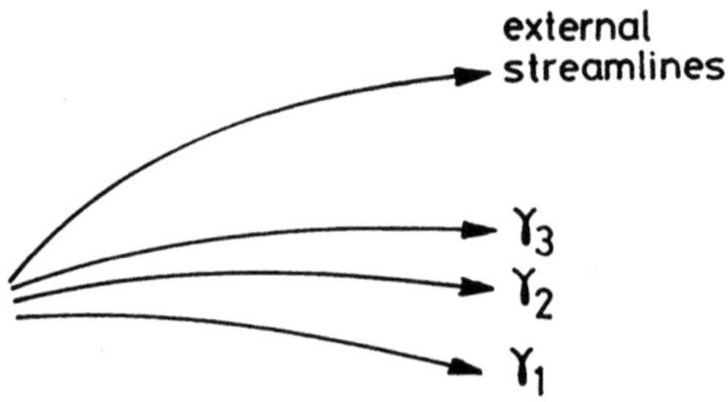

FIG. 8 - CHARACTERISTIC LINES OF THE SET OF GLOBAL EQUATIONS

The γ_1-characteristic line and the external streamline are at right angle when $H = H_c = 2,6$ (fig. 9). This value of H corresponds to a zero-streamwise skin friction. However, this point is not singular except if the flow has a locally two-dimensional behaviour.

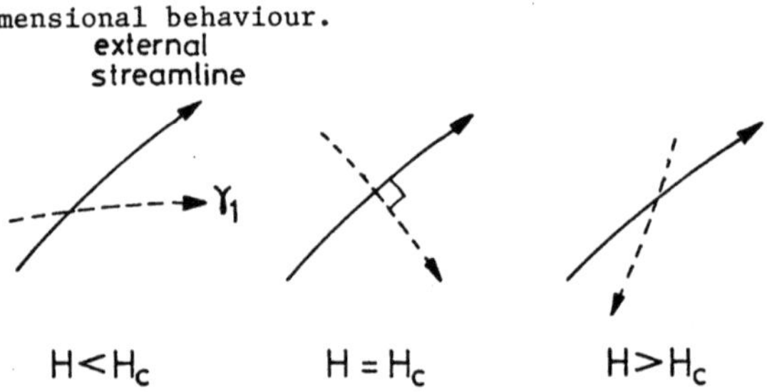

FIG. 9 - VARIATION OF THE DIRECTION OF THE γ_1 CHARACTERISTIC LINE WITH THE VALUE OF THE SHAPE PARAMETER

Generally, in the three-dimensional case, *the singularities are not local.* They occur in configurations leading to a focusing (convergence) of the characteristic lines belonging to the same family. In practice, the γ_1-characteristic lines (which are close to the wall streamlines) are more often likely to converge and to form an envelope. In the sense of the theory of characteristics, the γ_1-characteristic lines can form a shock. Indeed it has been shown analytically that the formation of a shock is possible. This shock constitutes a discontinuity in the evolution of certain boundary layer thicknesses. Obviously, as in the two-dimensional case, this behaviour is not physical and should be considered as an exaggeration of the actual flow.

An example of wall streamlines convergence has been provided by the calculations of the boundary layer on a wing root section (fig. 10) (AMSTERDAM Workshop, see LINDHOUT et al /56/). Similar results have been obtained by various participants in the AMSTERDAM Workshop. The calculated pattern of wall streamlines is in qualitative agreement with experimental flow visualizations, which show a strong convergence of the wall streamlines.

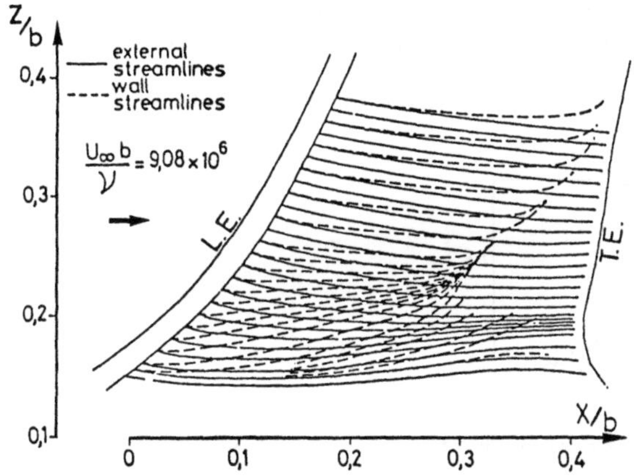

FIG. 10 - CALCULATION OF THE THREE-DIMENSIONAL BOUNDARY LAYER ON A WING ROOT SECTION (CALCULATED BY COUSTEIX-AUPOIX, IN LINDHOUT ET AL /56/)

In the experiment, the separation leads to a strong vortex flow. It is doubtful that a boundary layer calculation can reproduce the flow in the separated vortex flow, but it is interesting to notice that a rather simple method is able to give very useful information on separation. In particular, it should be pointed out that the location of the separated region is rather well predicted although no interaction between the boundary layer and the inviscid flow is accounted for.

The problem of occurrence of singularities has been studied numerically by several authors. In particular, CEBECI et al /57/ analyzed the flow around an ellipsoid of revolution. Figure 11 gives results in laminar flow ; the pattern of wall streamlines shows some similarities with figure 10. From their results, these authors define *a line of accessibility* for calculations that start from the forward stagnation point. This line terminates the calculations. Two points of the line of accessibility need to be considered. Between the "ok" (the most forward point of the line of accessibility - ok = "arrow" in Turkish) and the windward line of symmetry, calculations indicate that the line of accessibility is singular. On the leeward side, the main reason for the termination of calculations is that the singularity occurring at the ok is convected with the local velocity along any streamline passing through the normal to the surface containing the ok.

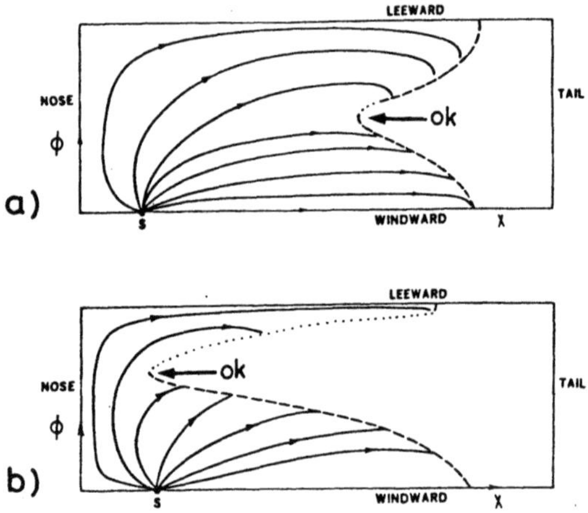

FIG. 11 - CALCULATION OF WALL STREAMLINES ON AN ELLIPSOID OF REVOLUTION IN LAMINAR FLOW /2/
a) $\alpha = 6°$ b) $\alpha = 30°$
ϕ : AZIMUTHAL ANGLE - x : DISTANCE ALONG THE AXIS OF SYMMETRY

Another piece of information is provided by the study carried out by J.C. WILLIAMS /58/. This author clearly demonstrated the possibility of formation of an envelope of wall streamlines. Particular solutions called *semi-similar solutions* have been analyzed. The equations are formulated in a cartesian axis system and a new system of coordinates is introduced :

$$\eta = \frac{y(U_o/\nu\ell)^{1/2}}{g^*(x^*,\ z^*)} \qquad \xi = \xi\,(x^*,\ z^*)$$

where U_o and ℓ are reference quantities. x^* and z^* are dimensionless variables :

$$x^* = \frac{X}{\ell} \qquad z^* = \frac{Z}{\ell}$$

The function g^* is a scaling function for the y-coordinate and represents a new X-variable.

New stream functions F and G are defined such that :

$$\frac{U}{U_e} = \frac{\partial F}{\partial \eta} \qquad \frac{W}{W_e} = \frac{\partial G}{\partial \eta} \tag{27}$$

The semi-similarity hypothesis is that F and G are functions of η and ξ only :

$$F = F(\eta,\ \xi) \qquad G = G(\eta,\ \xi)$$

The X- and Z-momentum equations become :

$$\frac{\partial^2 W_1}{\partial \eta^2} + \alpha_{11}\frac{\partial W_1}{\partial \eta} + \alpha_{12}W_1 + \alpha_{13} = \alpha_{14}\frac{\partial W_1}{\partial \xi} \tag{28}$$

$$\frac{\partial^2 W_2}{\partial \eta^2} + \alpha_{21} \frac{\partial W_2}{\partial \eta} + \alpha_{22} W_2 + \alpha_{23} = \alpha_{24} \frac{\partial W_2}{\partial \xi}$$

with

$$W_1 = \frac{U}{U_e} = \frac{\partial F}{\partial \eta} \quad ; \quad W_2 = \frac{W}{W_e} = \frac{\partial G}{\partial \eta}$$

The definition of the α_{ij} comes from the change of variables. In particular, we have :

$$\alpha_{14} = \alpha_{24} = H \frac{\partial F}{\partial \eta} + I \frac{\partial G}{\partial \eta} \tag{29}$$

with

$$H = g^{*2} \, U_e^* \frac{\partial \xi}{\partial x^*} \qquad I = g^{*2} \, W_e^* \frac{\partial \xi}{\partial z^*} \tag{30}$$

From definition of the semi-similar solutions, the number of independent variables in eq. (28) reduces to two. The integration of these equations starts at an initial station, let us say $\xi = 0$, and proceeds in the downstream direction ($\xi > 0$).

A singularity may occur in an equation which has the form of (28) when α_{14} changes sign in the interval of integration ; the problem is referred to as *singular parabolic*. At the wall, α_{14} is always zero but a singularity may occur if $(\frac{\partial \alpha_{14}}{\partial \eta})_{\eta=0}$ changes sign.

The condition $\alpha_{14} = 0$ is met at a point ξ_o if :

$$H(\xi_o) \frac{U}{U_e} = - I(\xi_o) \frac{W}{W_e} \tag{31}$$

Now, in the plane (x^*, z^*), the slope dz^*/dx^* of a line $\xi = \xi_o$ is given by :

$$\frac{\partial \xi}{\partial x^*} dx^* + \frac{\partial \xi}{\partial z^*} dz^* = 0$$

Combining this relation with eq. (30), we get :

$$\frac{H(\xi_o)}{U_e} dx^* = - \frac{I(\xi_o)}{W_e} dz^*$$

The comparison with eq.(31) shows that *the locus of singular points is a streamline* (possibly *a wall streamline*).

Basically, two kinds of singularities have been identified by WILLIAMS /58/.

In the first case, the wall streamlines become tangent to a singular line ; *this singular line forms an envelope of wall streamlines* and is itself a wall streamline. This situation can be identified with separation.

An example of this type of singularity is provided by the calculation of the boundary layer which develops under the following conditions :

$$\frac{U_e}{U_o} = 1 \quad ; \quad \frac{W_e}{U_o} = 1 - \xi \quad ; \quad \xi = \frac{X}{l} + \frac{Z}{l}$$

which fulfill the semi-similarity hypotheses.

A singularity is recorded at the point $\xi = 0{,}239$ and the wall streamlines drawn in the (x^*, z^*) plane have an envelope the equation of which is $x^* + z^* = 0{,}239$ (fig. 12). As in the case of the two-dimensional GOLDSTEIN singularity, the calculations cannot proceed downstream. It is clear that a close resemblance exists with figure 10.

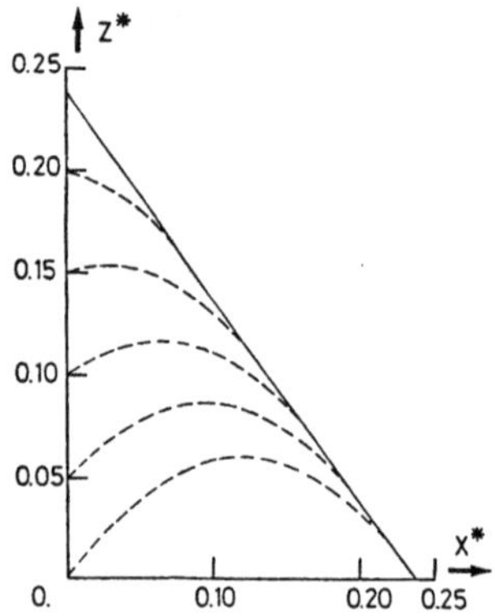

FIG. 12 - CALCULATION OF A THREE-DIMENSIONAL BOUNDARY LAYER WITH SINGULARITY /58/
---- LIMITING WALL STREAMLINES
—— LINE OF CONSTANT ξ CORRESPONDING TO SINGULARITY

WILLIAMS studied also the following form of external velocity distribution /58/ :

$$\frac{U_e}{U_o} = 1 \qquad \frac{W_e}{W_o} = \left(\frac{z}{l}\right)^{\beta}$$

Positive and negative values of β have been considered in the range $-2 \leq \beta \leq 0.8$. A second type of singularity has been identified when an external streamline ($\beta < 0$) or a wall streamline ($\beta > 0$) becomes tangent to a line ξ = cst. The significance of the singularity which occurs along this line is that the corresponding streamline marks *the boundary of the region where the flow is influenced by the given set of initial conditions* (at $\xi = 0$). The domain beyond this line cannot be calculated because it is influenced by downstream conditions ($\xi \to \infty$). This second kind of singularity is very similar to the STEWARTSON singularity in unsteady two-dimensional flow /59/. The STEWARTSON singularity occurs, for example, in the problem of a semi-infinite flat plate impulsively set into motion at time $t = 0$ with a uniform velocity U_w : for $U_w\, t/x < 1$, the flow develops as in the case of an infinite plate (RAYLEIGH solution) ; for $U_w t/x > 1$, the presence of the leading edge is felt and the solution tends to the BLASIUS solution at large times. WILLIAMS also noticed that a numerical scheme which takes into account the rules of influence might remove this type of singularity just as in the case of the STEWARTSON singularity.

Conclusion

Simple methods have been developed to calculate three-dimensional boundary layers. They are able to determine the flow characteristics of practical interest with an accuracy which enables us to perform a useful analysis of the flow behaviour. This does not mean that improvements are not highly desirable.

The numerical techniques have not been discussed here but this is certainly a field where progress can be made. More accurate and economic methods are needed especially if we consider the use of the full REYNOLDS stress transport equations. With the development of new types of computers, it can be expected that other numerical techniques will be more efficient.

The numerical techniques should not be unrelated to the mathematical behaviour of the equations. A better knowledge has been acquired in the last years but the complete determination of possible singularities is not yet available. Progress in this field is important to avoid misinterpretations of computed results and to develop viscid-inviscid interactive techniques.

In the foreseeable future, the REYNOLDS averaged methods will remain the basis of calculations and therefore the turbulence modelling will remain a major problem. Existing turbulence models including the REYNOLDS stress transport equation models have been developed on the basis of experimental data concerning two-dimensional flows. It has been assumed that the modelling is not influenced by the three-dimensionality of the mean flow because turbulence is always three-dimensional. In fact, it is frequent that two-dimensional flows possess particular properties. To a certain extent, it can be said that ad hoc models have been devised to reproduce these features. The result is that three-dimensional effects of the mean flow are not well taken into account. In addition, the physical understanding of these effects is poor. Fundamental experiments are needed and full simulations of basic three-dimensional turbulent flows could help to improve turbulence modelling.

Laminar-turbulent transition is not the easiest outstanding problem . Stability calculations and full simulations are very valuable to gain a better knowledge of transition processes but do not give a direct evaluation of transition location. Experimental data are the basis of practical tools. In particular, in flight, tests are necessary. Indeed the transposition of wind tunnel data to flight should be made with care because the transition location is very sensitive to the flow environment especially if a laminar flow is maintained at large REYNOLDS numbers.

Acknowledgements

The authors are grateful to P. BRADSHAW who improved this article significantly.

References

1. Hirschel, E.H. : Evaluation of results of boundary layer calculations with regard to design aerodynamics. AGARD FDP VKI Special Course "Computation of three-dimensional boundary layers including separation" (1986)

2. Cebeci, T. : Problems and opportunities with three-dimensional boundary layers. AGARD Report N° 719 "Three-dimensional boundary layers" (1984)

3. Hirschel, E.H. ; Kordulla, W. : Shear flow in surface oriented coordinates. Notes Num. Fluid Mech. Vol. 4 - BRAUNSCHWEIG/WIESSBADEN Vieweg (1981)

4. Vollmers, H. : Integration of streamlines from measured static pressure fields on a surface. AIAA Journal, Vol. 20, N° 10 (1982)

5. Cebeci, T. ; Meier, H.U. : Problems associated with the calculation of the flow around bodies of revolution at incidence. DFVLR IB 222 81 A 09 (1981)

6. Gleyzes, C. ; Cousteix, J. : Calcul des lignes de courant à partir des pressions pariétales sur un corps fuselé. La Rech. Aérosp. N° 1984-3 (1984)

7. Kline, S.J. ; Morkovin, M.V. ; Sovran, G. ; Cockrell, D.J. : Computation of turbulent boundary layers. 1968 AFOSR-IFP STANFORD Conference

8. Kline, S.J. ; Cantwell, B.J. ; LILLEY, G.M. : 1980-81 AFOSR-HTTM STANFORD Conference on Complex Turbulent Flows.

9. Cousteix, J. : Three-dimensional boundary layers. Introduction to calculation methods. AGARD FDP VKI Special Course on "Computation of three-dimensional boundary layers including separation" (April 1986)

10. Myring, D.F. : An integral prediction method for three-dimensional turbulent boundary layers. RAE TR 70147 (1970)

11. Smith, P.D. : An integral prediction method for three-dimensional compressible turbulent boundary layers. ARC R&M 3739 (1972)

12. Okuno, T. : Distribution of wall shear stress and crossflow in three-dimensional turbulent boundary layer on ship hull. Journ. Soc. Nav. Arch. JAPAN, Vol. 139 (1976)

13. Stock, H.W. : Calculation of three-dimensional boundary layers on wings and bodies of revolution. Proc. DEA Meeting "Viscous and interacting flow field effects" MEERSBURG (1979)

14. Mager, A. : Generalization of boundary layer momentum integral equation to three-dimensional flows including those of rotating systems. NACA Report 1067 (1952)

15. Gruschwitz, E. : Turbulente Reibungsschichten mit Sekundärströmungen. Ing. Arch. Bd 6, pp. 355 (1935)

16. Johnston, J.P. : On the three-dimensional turbulent boundary layer generated by secondary flow. Journal of Basic Engineering, Series D, Trans. ASME, Vol. 82, pp. 233-248 (1960)

17. Coles, D. : The law of the wake in the turbulent boundary layer. J. Fluid Mech., Vol. 1, pp. 191-226 (1956)

18. Cross, A.G.T. : Calculation of compressible three-dimensional turbulent boundary layers with particular reference to wings and bodies. British Aerospace Brough YAD 3379 (1979)

19. Le Balleur, J.C. : Numerical viscid-inviscid interaction in steady and unsteady flows. 2nd Symp. on Numerical and Physical Aspects of Aerodynamic Flows - T. Cebeci, Ed. (1983)

20. Elsenaar, A. ; Boelsma, S.M. : Measurements of the REYNOLDS stress tensor in a three-dimensional turbulent boundary layer under infinite swept wing conditions. NLR TR 74095 U (1974)

21. Bradshaw, P. ; Pontikos, N.S. : Measurements in the turbulent boundary layer on an "infinite" swept wing. J. Fluid Mech., Vol. 159, pp. 105-130 (1985)

22. Van den Berg, B. ; Elsenaar, A. ; Lindhout, J.P.F. ; Wessling, P. : Measurements in an incompressible three-dimensional turbulent boundary layer under infinite swept wing conditions and comparison with theory. J. Fluid Mech., Vol. 70, pp. 127-149 (1975)

23. Bradshaw, P. : Physics and modelling of three-dimensional boundary layers. AGARD FDP VKI Special Course on "Computation of three-dimensional boundary layers including separation" (1986)

24. Van den Berg B. : A three-dimensional law of the wall for turbulent shear flows. J. Fluid Mech., Vol. 70, p. 149 (1975)

25. East, L.F. ; Sawyer, W.G. : An investigation of the structure of equilibrium turbulent boundary layers. AGARD CP N° 271 (1979)

26. Goldberg, U.C. ; Reshotko, E. : Scaling and modelling of three-dimensional end wall turbulent boundary layers. NASA CR 3792 (1984)

27. Cebeci, T. : An approach to practical aerodynamic configurations. AGARD FDP VKI Special Course on "Computation of three-dimensional boundary layers including separation" (April 1986)

28. Cousteix, J. ; Quémard, C. ; Michel, R. : Application d'un schéma amélioré de longueur de mélange à l'étude des couches limites turbulentes tridimensionnelles. AGARD CP N° 93 on Turbulent Shear Flows (1977)

29. Bradshaw, P. ; Terrell, M.G. : The response of a turbulent boundary layer on an infinite swept wing to the sudden removal of pressure gradient. NPL Aero Rep. 1305 (1969)

30. East L.F. : Computation of three-dimensional turbulent boundary layers. EUROMECH 60 - TRONDHEIM FFA TN AE 1211 (1975)

31. East, L.F. : Measurements of the three-dimensional incompressible turbulent boundary layer induced on the surface of a slender delta wing by the leading edge vortex. ARC R&M 3768 (1973)

32. Johnston, J.P. : Measurements in a three-dimensional turbulent boundary layer induced by a swept forward-facing step. J. Fluid Mech., Vol. 42, Part 4 (1970)

33. Abid, R. ; Schmitt, R. : Etude critique de modèles de turbulence appliqués à une couche limite tridimensionnelle décollée. La Rech. Aérosp. N° 1981-6 (1981)

34. Rotta, J.C. : A family of turbulence models for three-dimensional boundary layers. Turbulent Shear Flows I, BERLIN, Springer Verlag, pp. 267-278 (1979)

35. Cousteix, J. ; Aupoix, B. ; Pailhas, G. : Synthèse de résultats théoriques et expérimentaux sur les couches limites et sillages turbulents tridimensionnels. ONERA N.T. 1980-4 (1980)

36. Special Course on Stability and Transition of Laminar Flow. AGARD Report N° 709 (1984)

37. Arnal, D. : Three-dimensional boundary layers : laminar-turbulent transition. AGARD FDP VKI Special Course on Computation of Three-dimensional Boundary Layers including separation (April 1986)

38. Smith, A.M.O. ; Gamberoni, N. : Transition, pressure gradient and stability theory. Douglas Aircraft Co. Rept. ES 26388, EL SEGUNDO, CA (1956)

39. Van Ingen, J.L. : A suggested semi-empirical method for the calculation of the boundary layer transition region. Univ. of Tech., Dpt. of Aero. Eng., Rept. UTH 74, DELFT (1956)

40. Gray, W.E. : The effect of wing sweep on laminar flow. RAE Technical Memorandum 255 (1952)

41. Gregory, N. ; Stuart, J.T. ; Walker, W.S. : On the stability of three-dimensional boundary layer with application to the flow due to a rotating disk. Philosophical Transactions of the Royal Society of LONDON, Series A, Vol. 248 (1955)

42. Boltz, F.W. ; Kenyon, G.C. ; Allen, C.Q. : Effects of sweep angle on the boundary layer stability characteristics of an untapered wing at low speeds. NASA TN D-338 (1960)

43. Schmitt, V. ; Manie, F. : Ecoulements subsoniques et transsoniques sur une aile à flèche variable. La Rech. Aérosp. N° 1979-4 (1979)

44. Poll, D.I.A. : Some observations of the transition process on the windward face of a long yawed cylinder. J. Fluid Mech., Vol. 150 (1985)

45. Michel, R. ; Arnal, D. ; Coustols, E. ; Juillen, J.C. : Experimental and theoretical studies of boundary layer transition on a swept infinite wing. 2nd Symp. IUTAM on Laminar-Turbulent Transition - Springer Verlag (1984)

46. Eichelbrenner, E. ; Michel, R. : Observations sur la transition laminaire-turbulent en trois dimensions. La Rech. Aérosp. N° 65 (1958)

47. Meier, H.U. ; Kreplin, H.P. ; Vollmers, H. : Development of boundary layer and separation patterns on a body of revolution at incidence. 2nd Symp. on Numerical and Physical Aspects of Aerodynamic Flows, LONG BEACH (1983)

48. Hefner, J.N. ; Bushnell, D.M. : Status of linear boundary layer stability theory and the e^n method, with emphasis on swept wing applications. NASA Technical Paper 1645 (1980)

49. Malik, M.R. ; Poll, D.I.A. : Effect of curvature on three-dimensional boundary layer stability. AIAA Journal, Vol. 23, N° 9 (1985)

50. Owen, P.R. ; Randall, D.G. : Boundary layer transition on a swept back wing : a further investigation. RAE TM 330 (1953)

51. Beasley, A. : Calculation of the laminar boundary layer and the prediction of transition on a sheared wing. ARC R&M 3787 (1973)

52. Arnal, D. ; Habiballah, M. ; Coustols, E. : Théorie de l'instabilité laminaire et critères de transition en écoulements bi et tridimensionnels. La Rech. Aérosp. N° 1984-2 (1984)

53. Goldstein, S. : On laminar boundary layer flow near a position of separation. Q. J. Mech. Appl. Math., 1, pp. 43-69 (1948)

54. Cousteix, J. ; Houdeville, R. : Singularities in three-dimensional turbulent boundary layers calculations and separation phenomena. AIAA Journal, Vol. 19, N° 8, pp. 976-985 (1981)

55. Cousteix, J. : Three-dimensional and unsteady boundary layer computations. Ann. Rev. Fluid Mech., Vol. 18, pp. 173-196 (1986)

56. Lindhout, J.P.F. ; Van den Berg, B. ; Elsenaar, A. : Comparison of boundary layer calculations for the root section of a wing. The September 1979 AMSTERDAM Workshop Test Case - NLR MP 80 028 U (1981)

57. Cebeci, T. ; Khattab, A.K. ; Stewartson, K. : Three-dimensional laminar boundary layers and the ok of accessibility. J. Fluid Mech., Vol. 107, pp. 57-87 (1981)

58. Williams, J.C. : Singularities in solutions of the three-dimensional laminar boundary layer equations. J. Fluid Mech., Vol. 160, pp. 257-279 (1985)

59. Stewartson, K. : On the impulsive motion of a flat plate in a viscous fluid. Q.J. Mech. Appl. Math., Vol. 4, pp. 182-198 (1951)

60. Pierce, F.J. ; McAllister, J.E. ; Tennant, M.H. : A review of near-wall similarity models in three-dimensional turbulence. J. Fluids Engg. 105 (1983)

Numerical Simulation of Turbulent Flows Using Navier Stokes Equations

WOLFGANG SCHMIDT
Aerodynamics, Dornier GmbH
D-7990 Friedrichshafen, FRG

Summary

The numerical simulation of turbulent flow fields by solving the Navier Stokes equations is no longer limited to basic research applications. New high speed vector computers along with fast numerical algorithms and better physical models allow pioneering application even in industry. The emphasis in the following article will be on the discussion of the requirements and so far limitations in numerical simulation from an industrial point of view. It will be pointed out clearly that our physical understanding of flow topology and turbulence is still limited. Much more detailed experimental investigations are mandatory to validate numerical simulations.

Introduction

Numerical simulation can be used as a tool to generate scientific understanding of the mechanisms involved in, and the behaviour of flows of interest as well as a tool for designing the hardware of engineering components and configurations. In industrial application such a simulation must be seen as a computer aided engineering job with its specific requirements. Geometry preparation, mesh discretization, postprocessing can be more expensive than just the number crunching using efficient numerical algorithms and vector computers.

There are three major motivations for extensive numerical flow simulations. One is to provide important new technological capabilities that cannot easily be provided by experimental facilities. Because of basic limitations, experiments suffer e.g. from wall interference, flow angularity, Reynolds number limitations, dynamic problems and insufficient techniques for local measurements. Numerical flow simulations, on the other hand, have none of these fundamental limitations and/or error sources, but have their own: no method can produce results beyond the validity of the physical model on which the mathematical modelling is based, the numerical method may require excessive computer time and memory, and last but not least the complexity in geometrical details of complete configurations can easily exceed available mesh generation strategies.

A second compelling motivation concerns total configuration analysis time and cost. It is evident that the time to design, build, and test a wind tunnel model or even a test engine is limiting the configurational space in industrial analysis. The time of response to questions of the advanced design team for experimental data in general is too long. Numerical simulations require no model construction time and parametric studies can easily be verified by computations.

The third major motivation for using computational methods in industry is the supplementary function of Computational Fluid Dynamics (CFD) with respect to experimental research, and the accompanying quality growth of CFD-results which become competitive to experiments in this sense. While the economy of achieving many global data in experiments, e.g. series of forces and moments in wind tunnel tests, never can be exceeded by computations, the calculations allow a much easier insight into details of flowfield. This can become extremely important during a design cycle since otherwise essential delays can occur.

For industry computational fluid dynamics (CFD) has to be integrated as a computer aided engineering (CAE) application. The task within limited time and cost is to generate a more or less complex input, process a flow solver, and produce an easy to understand output in engineering form, e.g. an analog output in form of plots. The task can only be fulfilled efficiently if soft- and hardware for pre- and postprocessing, mesh generation software and numerical algorithms are of high standards along with fast number crunching computer hardware. The weakest point in this loop is limiting the capabilities. Faster computers and algorithms automatically lead to an imbalance if the man power cost and time of numerical simulations is not reduced consequently by using expert systems. By the same time it should be kept in mind that the quality of the input (e.g. geometry) to a large extend is dominating the quality of the results accuracy.

Reviewing the last decades it can be found that panel methods and small disturbance and/or full potential methods combined with boundary layer methods have been the main simulation tools for inviscid and viscous flows. Since the beginning of the present decade Euler and Navier Stokes methods are getting more and more attention. This is in accordance with the needs for simulations of very complex flowfields of present day and future aerospace vehicle designs at its outer limits of the flight envelope. Solving the Navier Stokes equations is the highest level of laminar and turbulent time averaged flow modelling in continuous flows. Shocks as well as separation can be treated properly and even dynamic effects such as stall and buffet or other self-induced oscillation caused by separation can be simulated. For hypersonic flows thermal and flow field analysis can be made in flow regimes for which no wind tunnel facilities are available. Such simulations are only limited by our physical understanding and the deduced physical models for the gas, turbulence and the boundary conditions (heat transfer). A second class of limitations is coming from the numerical modelling. The numerical descretization of a model must be as accurate as a wind tunnel model and the mesh scaling characteristics of the generated grid must allow for the resolution of the correct physics. This implies not only fine meshes in viscous layers to resolve the diffusion terms, but also fine meshes to resolve convective terms where large gradients occur.

The most difficult task, however, is the validation of numerical results since for local flow quantities experimental data are very rare at mostly unknown accuracy bandwidth. Despite such concerns, there is a strong demand for Navier Stokes simulations in order to face the challenge of the next generation aerospace-vehicles with improvements in :

- wing performance (next generation of transonic laminar wing)
- wing fuselage fairing design
- fuselage afterbody analysis for low drag
- airframe-engine integration
- vortex-wake interactions
- analysis and management of dynamic flow physics
- hypersonic flows with heat management

A second challenge for numerical simulation is based on its capabilities to predict wind tunnel corrections and extrapolate wind tunnel results to free flight conditions.

The next chapters will show some of the design principles we are following to establish large scale numerical simulation in aerodynamics.

Mesh Generation

We are generally interested in calculating flows over and through configurations of extreme geometric complexity with flow regions of large gradients. Any non-appropriate mesh distribution will lead to large discretization errors even for the best numerical flow solver. The objective for mesh generation has to be along the following lines:

- for configuration flexibility mesh generation should be an user-oriented interactive tool
- manpower time and cost has to meet design cycle requirements
- geometry data have to be obtained via access to a geometry data base in advanced design
- discretization with appropriate mesh-scaling laws requires user-skill.

It has to be kept in mind that the mesh-capability is limiting the configuration complexity, not the flow solver.

Triangular (2D) and tetrahedral (3D) mesh generation - as discussed in Ref.[1] - seems to give the largest flexibility even for complete aircraft configurations, however, with penalties on computer time and storage requirements. Secondly, the resolution of thin viscous layers with regular quadrilateral meshes seems to allow more easily for the correct physical scaling. So far, we concentrate on multi-block-structured meshes with quadrilateral (2D) and hexahedral (3D) elements as described in Ref.[2], [3] and [4]. The general strategy of generating such meshes consists of seven successive steps:

- numerical definition of configuration (CAD-data)
- generate body surface grid
- locate far field and block boundaries
- generate surface grid on block faces including far field
- generate volume grids by specific scheme
- control grid and smooth if necessary
- mesh adaption depending on solution

The strong influence on non-adequate mesh resolution can be seen from FIG. 1 taken from Ref.[5]. Separation point as well as wake size and structure depend heavily on the resolution.

Flow Solvers

Since more than twenty years numerous numerical algorithms have been developed for solving the compressible Navier Stokes equations. A comprehensive survey is given in Ref.[6] and [7]. There exists no preference of any algorithm in principle, but all good schemes should have some design principles in common.

- Conversion of partial differential equation and boundary conditions in algebraic equations without loss of physical sense.

- Different physical problems might require different optimal algorithms, search for problem optimal methods.

- Algorithms and programming should not depend on the type of configuration that is to be analyzed.

- Solver and computer architecture should both allow for high speed processing.

Therefore, we solve the nonlinear time-dependent equations by a finite volume scheme using multi-stage time stepping for time-wise integration. We prefer a finite volume approach since it guarantees an optimal representation of physics based on the integral form of the conservation laws for mass, momentum and energy including correct treatment of discontinuities. Its application is convenient since it permits contour conformal meshes with simple Cartesian representation of coordinates and momentum components, it avoids numerical differentiation as far as possible, and it is very forgiving with respect to mesh in-homogenities.

One of the principal objectives which motivated the design of the METS (Multigrid-Explicit Time-Stepping) Schemes [8],[9], [10]had in fact been the need to find an algorithm which would perform effectively with vector, pipeline, or parallel computer architectures. It seems worthwhile to accept some limitations on the rate of convergence to a steady state in return for factors of 20 or more in processing speed which might be realized through the use of long vectors in a pipeline machine, or by the introduction of multiple parallel processors. Accordingly, we had concentrated on extracting the maximum possible efficiency from explicit time stepping schemes, which in principle would allow simultaneous processing of every point in the entire flow field. In the event, through the introduction of measures such as variable time steps based on the local stability limit, enthalpy damping, residual averaging, and multigrid time stepping, it has proved possible to obtain convergence sufficient for engineering predictions in about 20 cycles for the Euler equations. Mesh aspect ratios in Navier Stokes solutions cause more cycles so far. The logical structure of the multiblock scheme which is required to support the previously mentioned block structured mesh for complex shapes is to some extend discussed in Ref. [3] and [9]. This block structure allows large scale computing with high resolution, like some million mesh points for three-dimensional problems, since only one block with some 10.000 points at a time has to be solved in core per processor used.

Turbulence Models

For over 100 years intelligent people have worked long hours and have written thousands of technical papers on the last unsolved problems of classical physics, namely turbulence. However, as Peter Bradshaw said:

"... A numerical procedure without a turbulence model stands in the same relation to a complete calculation scheme as an ox does to a bull".

Therefore, we have to live with a variety of models, none of which is perfect. Details on the models mainly in use can be found in Ref.[11] to [14].

Exact numerical simulation of turbulence for complex realistic configurations seems to be unattainable. Direct simulations of the full time-dependent Navier Stokes equations without any modelling, however, have been obtained already for some simple shapes at relatively low Reynolds numbers for incompressible flows. Ref.[15] is presenting such results for a channel flow. FIG. 2 shows the results for the rms fluctuations and the Reynolds-stresses from this reference compared with experimental data. The agreement is very promising and such approaches might lead to a better understanding of turbulence models.

So far we have to use the traditional modelling approach which consists of a time averaging of turbulent flow quantities. It is very important to keep in mind that the averaging time must be long compared with that of predominant turbulent frequencies, but by the same time short compared with the main flow field unsteadiness. This can also be a critical task in turbulence measurements.

Most commonly used are zero equation eddy viscosity models, such as the Cebeci-Smith model or derivatives. In such models the turbulent shear stresses are replaced by the product of an effective viscosity and a mean rate of strain. Velocity and length scales are specified in terms of equilibrium between mean flow and turbulence. Although these models are suffering from classical boundary layer assumptions and difficulties in length scale estimates, they are the most commonly used in Navier Stokes simulations.

In general, there is no evidence that higher order models perform better in practical applications. Appropriate estimates of length scales seem to be a key source of good modelling as shown in Ref.[16].

Validation of numerical solutions and engineering applications

The good quality of two-dimensional results obtained with the previously discussed METS-Scheme solving the Euler equations has been documented in detail in Ref.[17]. The various comparisons in that report clearly demonstrate the problems in validating numerical solutions to non-linear flow problems for which no exact solutions exist in general. Our approach for the Euler-solvers has been to compare subsonic Euler solutions with potential flow results which should agree exactly, analyze total pressure losses or entropy production in numerical solutions, analyze the mesh-dependency of a numerical solution, and to compare Euler solutions for shock free airfoil designs. For viscous flows, which are governed by the Navier Stokes equations, no validation against exact solutions is possible, even in two-dimensional flows. We can only compare with experimental results which often suffer from tunnel wall effects and other possible error sources. Therefore, thorough investigations are necessary in order to evaluate the consistency and quality of results in comparison with sufficiently detailed experiments. We are still lacking high quality experimental results on local quantities.

In the course of applications we would like to differ between three classes of Navier Stokes simulations, namely thin layer equations, full Reynolds-averaged and parabolized Navier Stokes. Thin layer Navier Stokes methods assume that the viscous terms have only to be resolved normal to the surface and normal to wakes, streamwise and crosswise diffusion is neglected. This assumption is very similar to boundary layer theory, but allows for normal pressure gradients. Limitations occur in regions of massive separation and for low Reynolds numbers and rarefied gases. However, most published Navier Stokes solutions are essentially of thin layer type, since the meshes used do not allow for resolution of the viscous terms in streamwise and crossflow directions. By definition thin layer methods require meshes that are essentially aligned with the surfaces and wakes. G. S. Deiwert has demonstrated the capabilities of such simulation during his 3-D afterbody studies [18]. The detailed simulation of the very complex flowfield pattern are quite striking as can be seen in FIG. 3. It is quite clear that such experiments will help understanding the flow details and designing improved shapes.

A second quite impressive example has been taken from Ref.[19]. This turbulent transonic flowfield analysis over a wingbody combination is pioneering in its information about separation analysis in the wing body juncture and the interaction with the shock induced wing separation. FIG. 4 taken from Ref.[19] is giving only some impression of the vast amount of information contained in such a simulation. It is also quite clear that such results cannot be obtained by any simpler approach not solving the Navier Stokes equations. Comparisons with solutions of the full time averaged equations still have to prove the accuracy. Also the importance of the form of turbulence model (Baldwin-Lomax) being used is not clear yet.

Finally, the analysis of wing tip vortices as discussed in Ref.[20] is shown as a typical Navier Stokes Simulation of large importance not only to transport wing, but also to blade design for rotors. FIG. 5 summarizes some typical results and gives an impressive view of the flow phenomena occuring at wing tips and its dependency on wing tip shape. Such simulations provide an extremely valuable insight in the physics for the aerodynamic engineer designing such wings. Obtaining the same amount of information from experimental analysis would be impossible on a routine basis.

Going to the class of full Reynolds-averaged Navier Stokes simulations we have to face the problems that the mesh used and the number of meshpoints have to allow for resolution of the corresponding length scales in all directions. We have to ensure that the numerical scheme used also is producing much less numerical viscosity than physics. Finally, we have to worry about flow unsteadiness and turbulence. These are the main reasons why for this class of simulations basically only two-dimensional and axisymmetric results are available so far. 3-D results will require some million mesh points which still demands larger and faster computers.

The first example has been taken from Ref.[21]. Jet mixing at missile afterbodies is a common problem in missile aerodynamics. Base pressure, boattail separation and the estimate of Mach-discs are typical design questions. FIG. 6 shows such results and also indicates good agreement with experimental data. However, further improvements can be expected by adjusting the mesh to the jet and wake structure.

A classical task in aircraft design is the analysis of airfoils. Ref.[22] is giving a comparison of different tools including a very careful Navier Stokes simulation. FIG. 7 presents the results for the RAE 2822 section. The main reasons for the high quality of the results are the very fine mesh being adapted to the viscous layer and the use of a turbulence model with consistent length scale estimates. Such results raise questions to what is more accurate - the numerical simulation or the wind tunnel result.

While all applications so far have dealt with external flows, FIG. 8 taken from Ref.[23], is comparing results for a separated turbine-cascade flow. The multigrid-method by N_i has been used to solve the Euler as well as the Navier Stokes equations. Separation on the pressure side is evident, causing clear differences between Euler and Navier Stokes. Using a boundary layer method along with the predicted pressure distribution is giving quite good agreements even in velocity profiles compared with Navier Stokes profiles for the suction side; the pressure side, however, is showing distinct differences.

Another extremely complex and difficult to resolve test case for numerical simulations is the analysis of the Porsche 956 center section in ground effect with its mix of external flow and channel flow with nozzles and diffusers that cause large separation. FIG. 9 taken from Ref.[24] is presenting some of the very interesting results using adaptive block structured meshes and multistage time-stepping with grid refinement as a zonal approach. Although the method is using a simple zero-order turbulence model, the large separation zone on the tunnel floor is well predicted as well as the pressure distribution. We must admit that correct length scale estimates for this case caused quite some difficulties, but the final results simulate the correct wind tunnel floor effects on the section flow field rather nicely. Separation from smooth surfaces is very demanding in mesh resolution.

The class of parabolized Navier Stokes (PNS) methods is extensively being used for steady supersonic external flows. If we neglect the viscous terms in marching (streamwise) direction and the streamwise convective flux derivative has positive time-like behaviour with respect to the remaining spatial derivatives (supersonic flow), we can integrate in marching direction directly, reducing the time-like solution of a three-dimensional problem to a set of two-dimensional cross plane solutions. The crucial point in such a method, however, is the treatment of pressure gradients in locally embedded subsonic flow regions. Such methods have been used extensively for supersonic and hypersonic space plane analysis. A well proven and extremely accurate result of a PNS simulation for a relatively simple body but still very complex flowfield has been taken from Ref.[25]. FIG. 10 portrays shock position and viscous boundary in comparison with experimental results by Tracy. The agreement even for the thick separated viscous layer on the leeside is very promising. Such methods are mandatory for analyzing complex hypersonic flow fields along with future supersonic transport as well as hypersonic space vehicles.

To conclude steady flow analysis we would like to show another very complex three-dimensional simulation of an internal fluid flow problem. In Ref.[26] the space shuttle main engine power head flow field has been simulated to understand potential sources of total pressure losses. The incompressible approximation was solved by using artificial compressibility. The results shown in FIG. 11 give some impression of the flowfield details obtained via such a simulation. It is quite obvious that the very large value of such numerical results stems from the detailed local information about all flowfield variables. Effects can be studied quite in detail that by experiments are very hard to analyse. Although our understanding of turbulence is quite limited, such simulations are quite valuable. Detailed validation, however, is still not possible since we are missing corresponding experimental results.

Since most Navier Stokes methods basically solve the time-dependent equations in order to construct a stable iterative solution procedure by time-stepping, the simulation of time-dependent processes in a time-accurate sense is easily possible. FIG.12 taken from Ref.[27] presents results of an axisymmetric turbulent Navier Stokes simulation of a compression stroke in a Diesel engine cylinder having a considerable swirl prescribed initially. In this solution to our knowledge for the first time a trial has been made to resolve the actual viscous layers developing at the cylinder walls while previous solutions used wall laws for the near wall regions. Some quite interesting flow details were revealed during earlier crank angles.

Conclusions

The last ten years have been very exciting for researchers in computational fluid dynamics. Advances in solution algorithms, complex mesh generation, physical understanding supported by CFD, computer power and graphics packages have led not only to research applications of Navier Stokes simulations, but also to pioneering applications in industry. Direct full time dependent simulations even seem to help understand the last open question : turbulence. However, it is getting very difficult to find or generate accurate experimental data to validate numerical simulations in the range of practical interest.

Figures

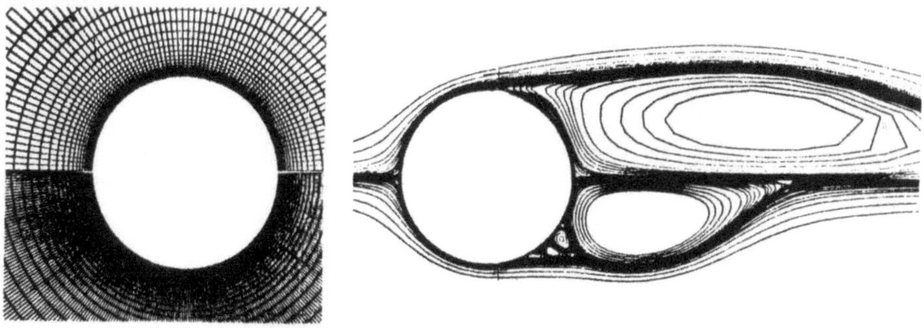

Fig. 1: Mesh Dependency of Separated Flow for Turbulent Flows Around a Circular Half-Cylinder [5]

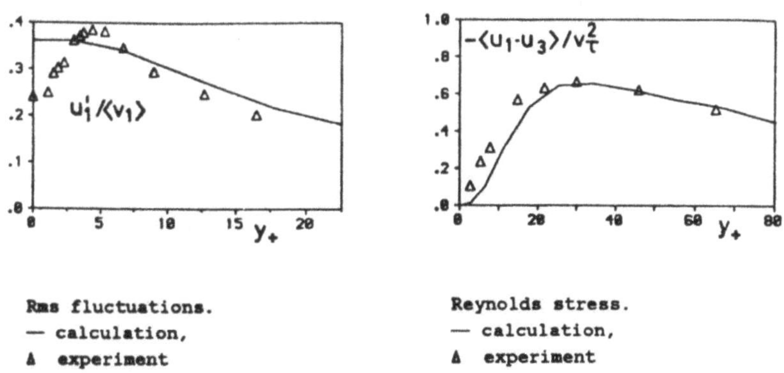

Fig. 2: Direct Numerical Simulation of Turbulence for Channel Flow [15]

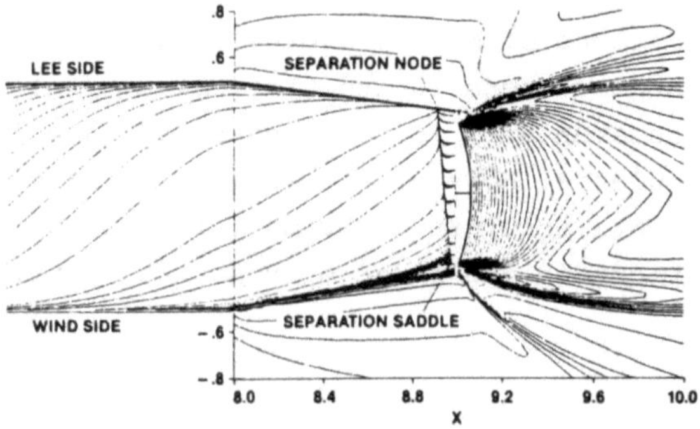

Afterbody flow detail: surface streamlines and density contours on bilateral plane of symmetry.

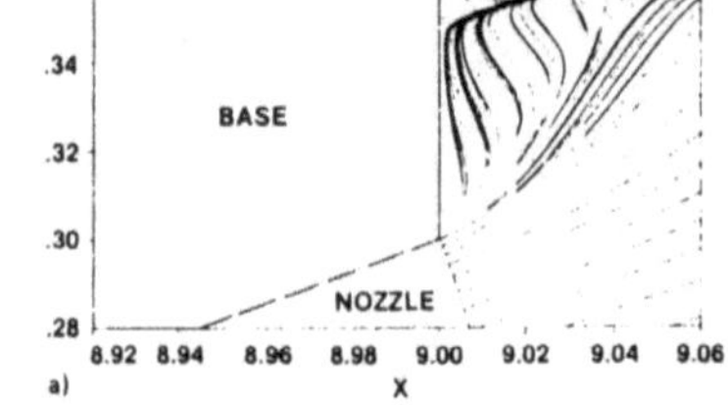

Fig. 3: Afterbody-Jet Interference Analysis at Incidence [18]

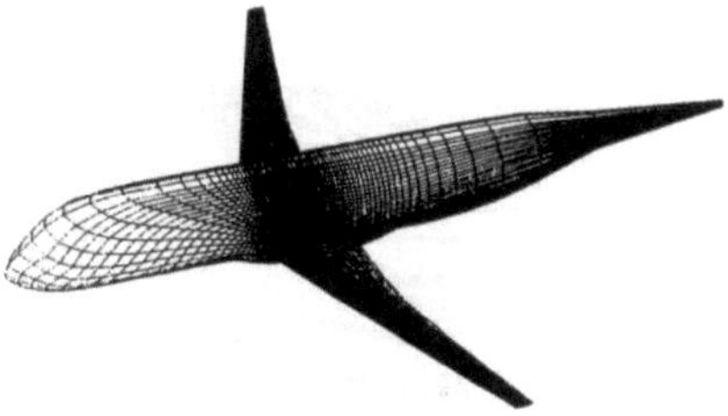

Computational Grid (Surface)

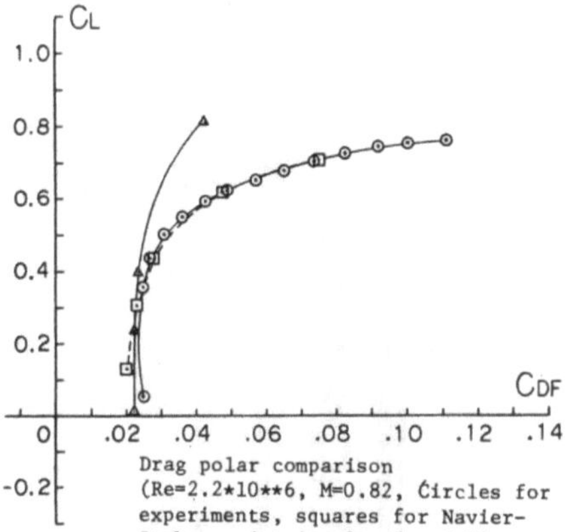

Drag polar comparison (Re=2.2*10**6, M=0.82, Circles for experiments, squares for Navier-Stokes and triangles for full potential)

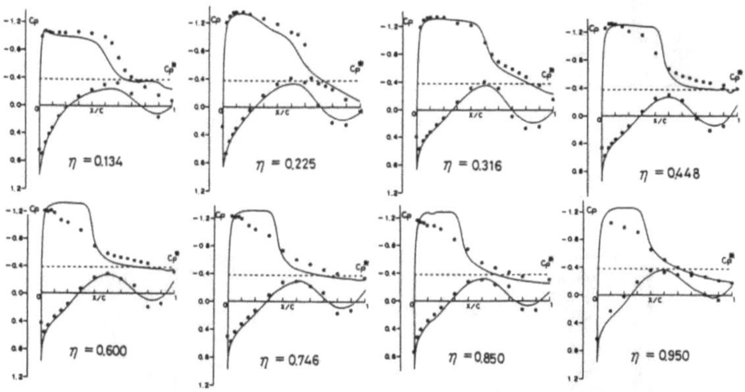

Wing pressure distribution (Solid lines for Navier-Stokes and circles for experiments, Re=2.2*106, M=0.82, α=6.00° in NS and 5.76° in exp.)**

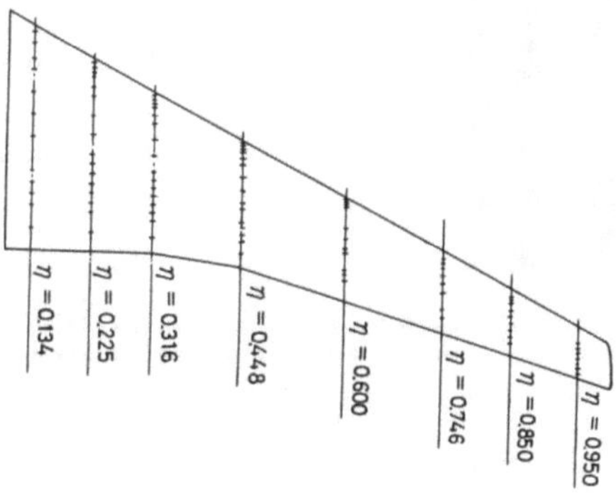

Pressure tap locations

<u>Fig. 4:</u> Transonic Wing-Body Analysis [19]

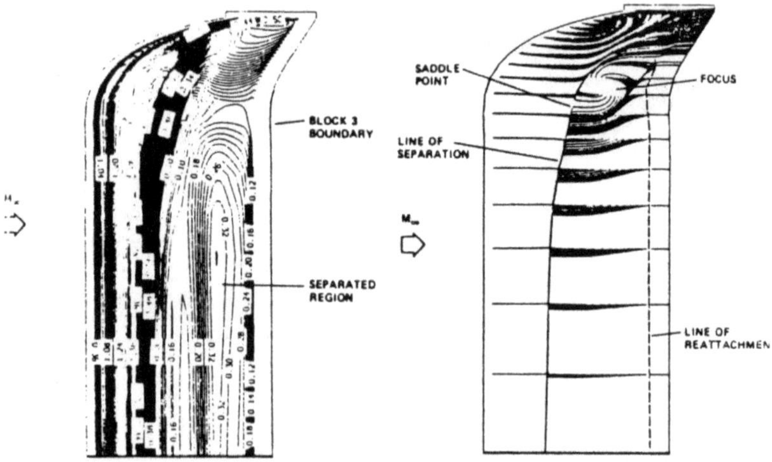

Planform view of the upper surface Mach number contours for the ONERA wing.

Surface oil flow pattern for ONERA wing showin separation and reattachment lines.

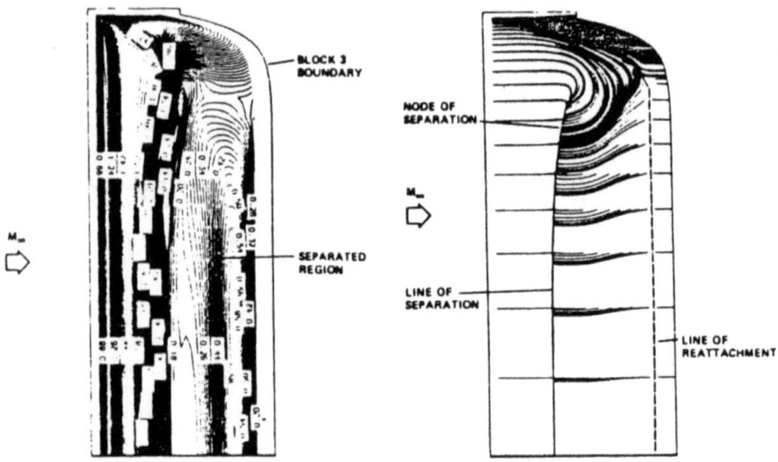

Planform view of the upper surface Mach number contours for the modified ONERA wing. $M_\infty = 0.85$, $\alpha =$ 5 degrees, and Re = 8.5 million.

Surface oil flow pattern for the modified ONERA wing showing the separation and reattachment lines.

Fig. 5: Numerical Simulation of Tip Vortices of Wings in Subsonic and Transonic Flows [20]

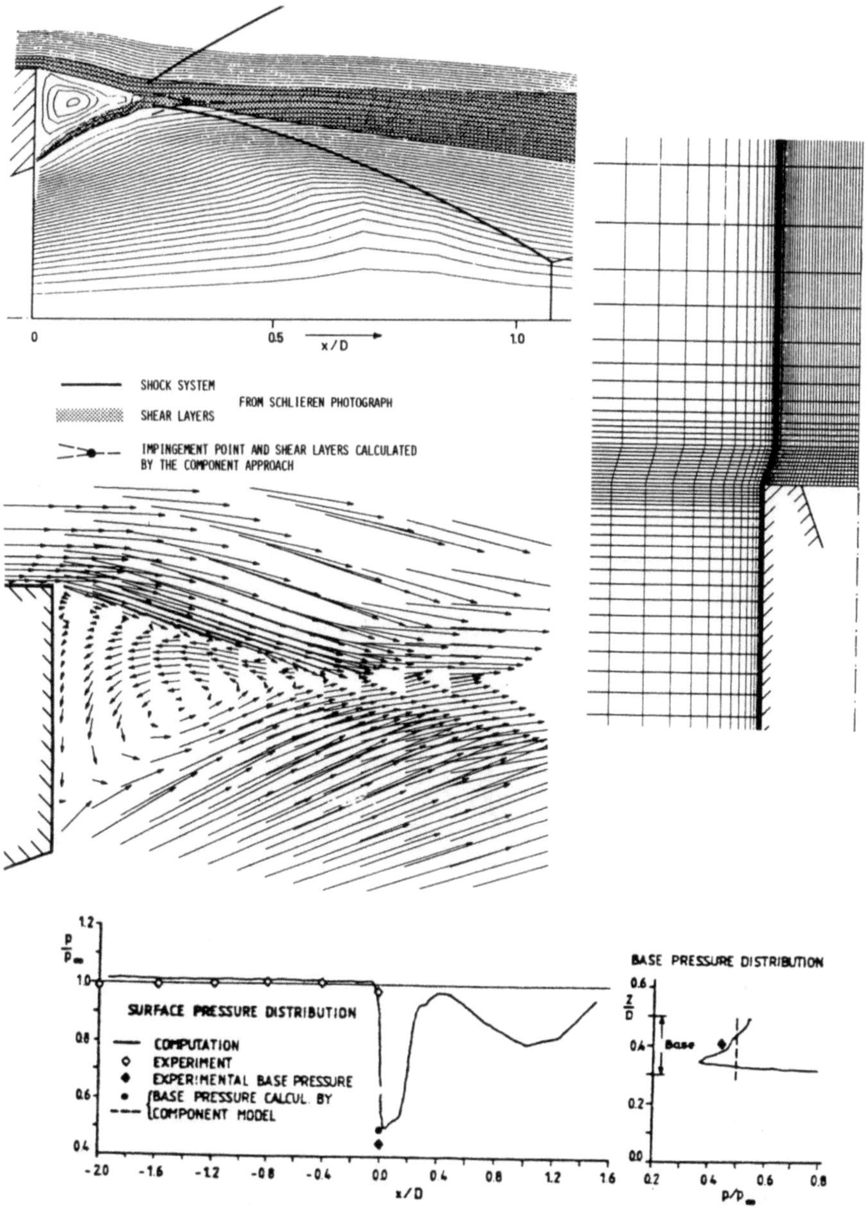

Pressure distributions on body surfaces and mesh dividing line for Test Case 9
$M_\infty = 2.01$; $\beta_r = 0°$; $M_j = 2.5$; $\beta_j = 20°$; $R_j/R = 0.6$; $p_j/p_\infty = 1.0$

Fig. 6: Axisymmetric Turbulent Base Flow Analysis at Subsonic Speed [21]

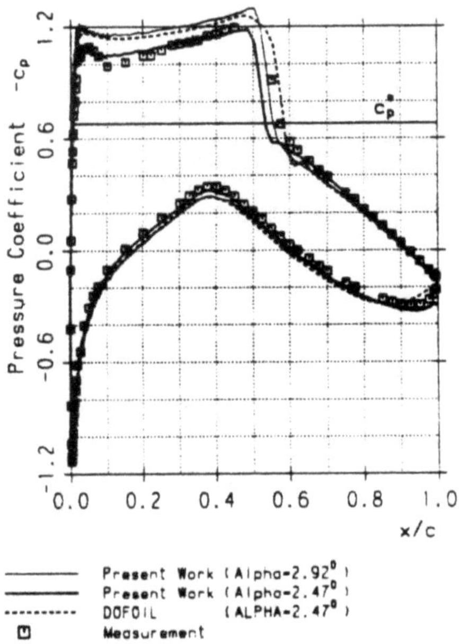

Pressure Coefficient for RAE 2822 Airfoil (Case 6), Ma = 0.725, Re = 6.5x10⁶

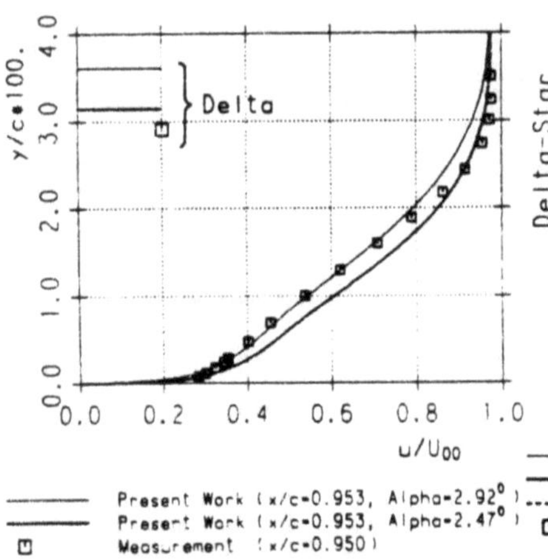

Boundary Layer Velocity Profile for RAE 2822 Airfoil (Case 6) at x/c = 0.95, Ma = 0.725, Re = 6.5x10⁶

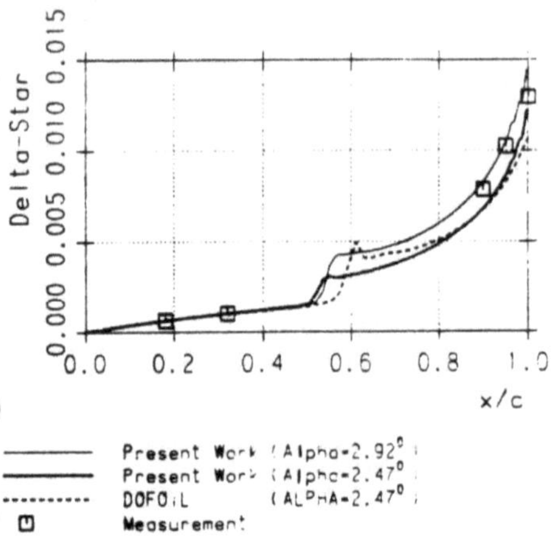

Boundary Layer Displacement Thickness on Upper Surface for RAE 2822 Airfoil (Case 6), Ma = 0.725, Re = 6.5x10⁶

Fig. 7: Adaptive Mesh High Resolution of Transonic Aerofoil Flow [22]

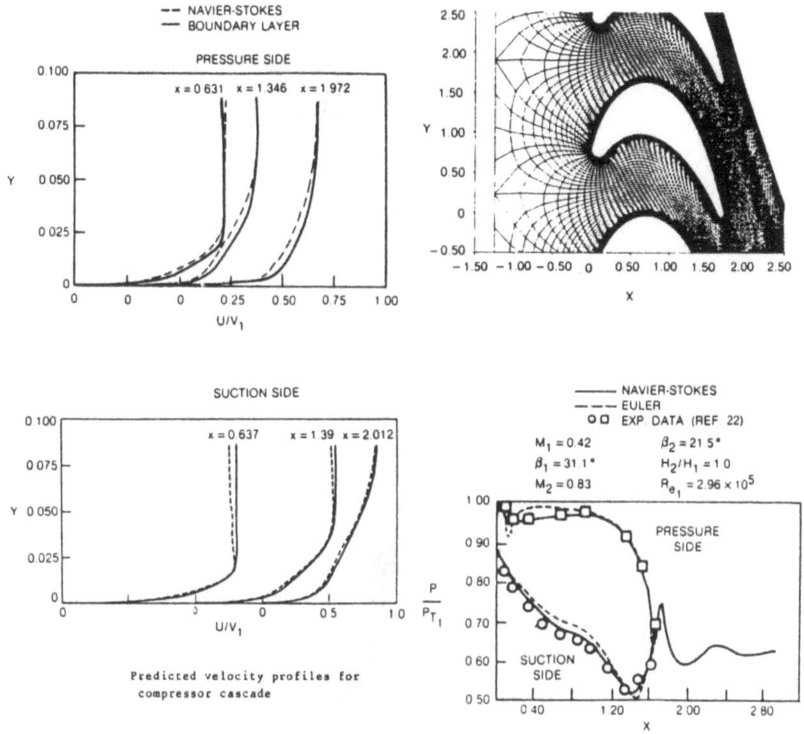

Fig. 8: Separated Turbulent Turbine-Cascade Flow Simulation [23]

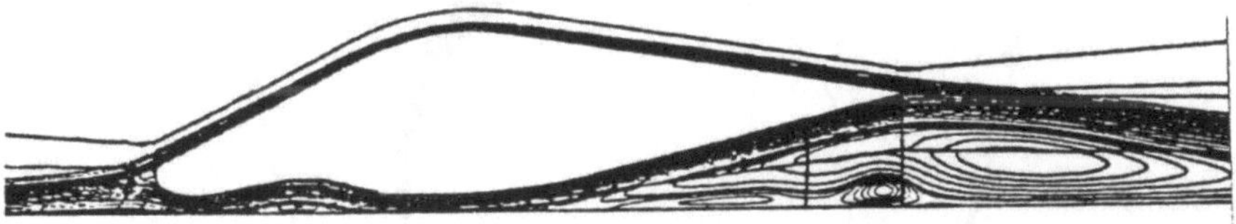

Calculated Streamlines, Fixed Ground (Wind Tunnel)

Fig. 9: Adaptive Mesh High Resolution Analysis of Racing Car Center Section Turbulent Flow with Wind Tunnel Floor Boundary Layer [24]

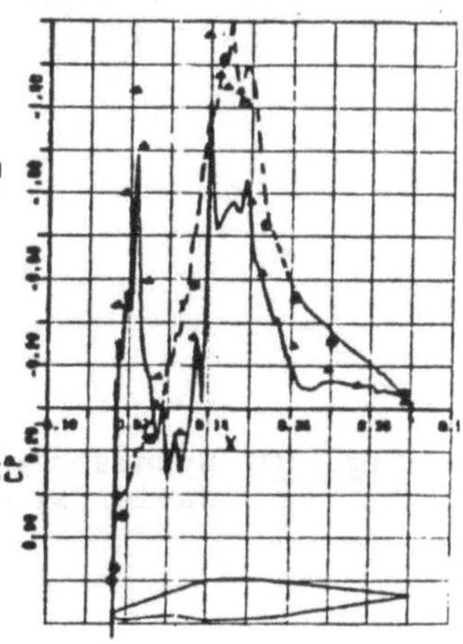

CALCULATION LOWER SURFACE

CALCULATION UPPER SURFACE

2D-EXPERIMENT UPPER SURFACE

2D-EXPERIMENT LOWER SURFACE

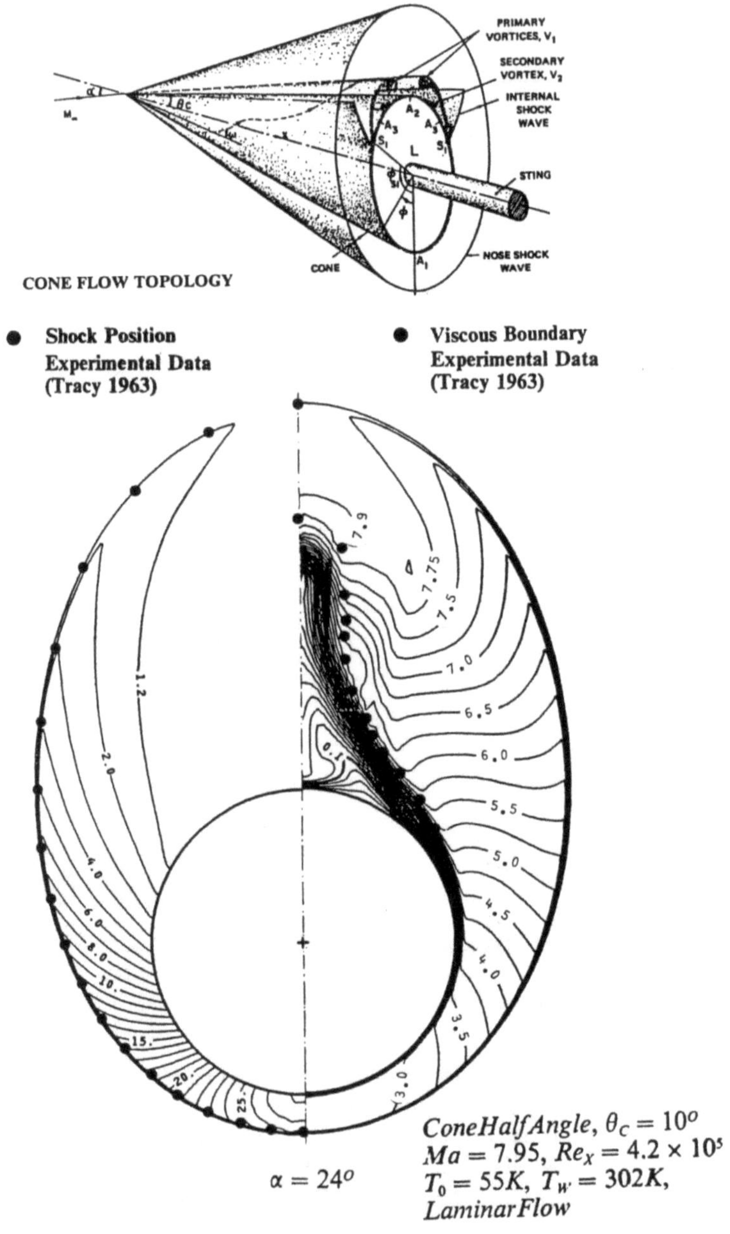

Fig. 10: Hypersonic Cone Flow Analysis Using Parabolized Navier Stokes Simulation [25]

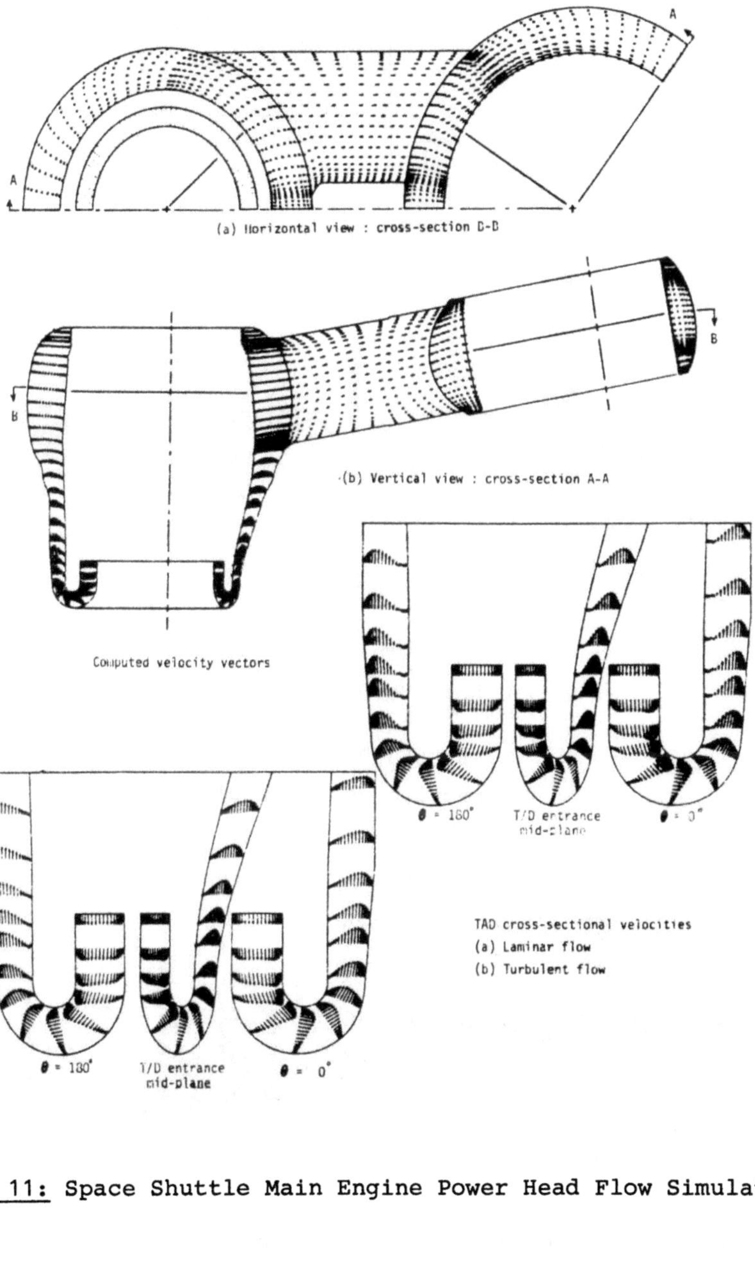

<u>Fig. 11:</u> Space Shuttle Main Engine Power Head Flow Simulation [26]

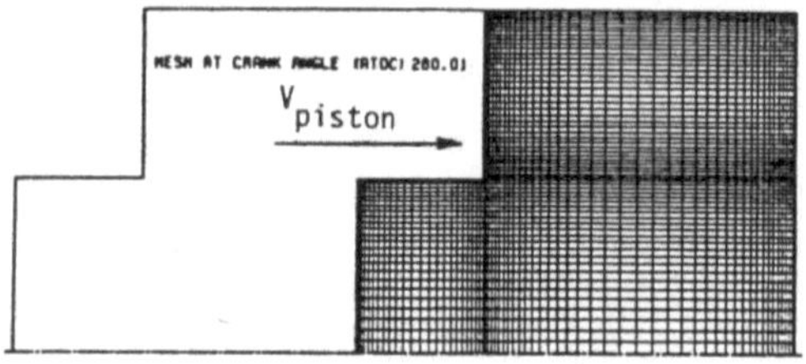

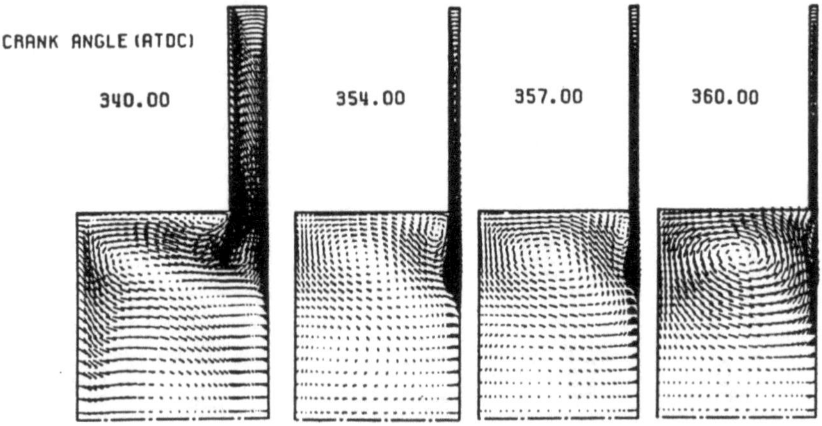

Flow Field after Compression Stroke in an Axisymmetric Diesel Engine Cylinder with Initial Swirl (60 m/s Maximum Piston Speed)

VELOCITY FIELD

Direction Vector Field

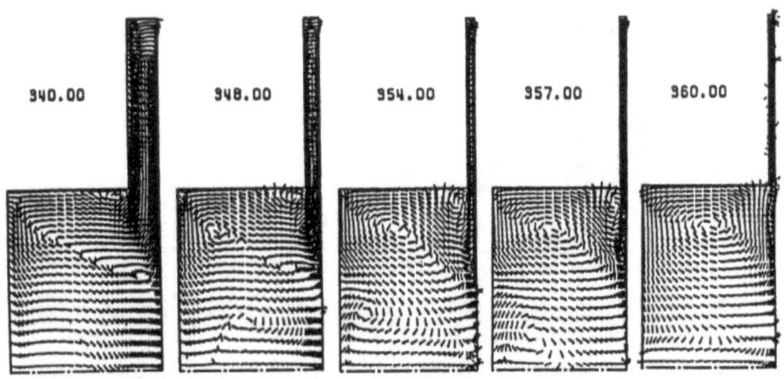

Fig. 12: Time-Accurate In-Cylinder Turbulent Simulation with Swirl [27]

References

1. Jameson, A.; Baker, T.J.; and Weatherhill, N.P.
Calculation of Inviscid Transonic Flow Over a Complete Aircraft, AIAA-86-0103 (Jan. 1986)

2. Fritz, W.
Two Dimensional Euler and Navier-Stokes Solutions of Flow Over the Mid Section of a Car, 2nd IAVD Congress on Vehicle Design and Components, Geneva, 1985

3. Leicher, S.
Analysis of Transonic and Supersonic Flows Around Wing-Body Combinations, ICAS 84-1.2.2 (1984)

4. Seibert, W.
An Approach to the Interactive Generation of Block Structured Volume Grids Using Computer Graphics Devices, First International Conference on Numerical Grid Generation in Computational Fluid Dynamics, Landshut, Germany (1986)

5. Fritz, W.
Two Dimensional Navier-Stokes Solutions for Section Shapes, to be published, 1987

6. Shang, J.S.
An Assessment of Numerical Solutions of the Compressible Navier-Stokes Equations, J. Aircraft, Vol. 22, No. 5, May 1985, pp. 353-370

7. Schmidt, W.; Jameson, A.; Pulliam, T.; Thompson, J.F.
Navier-Stokes Flow Simulations, AIAA Professional Study Seminar, June 1986, San Diego

8. Jameson, A.; Schmidt, W.; Turkel, E.
Numerical Solution of the Euler Equations by Finite Volume Methods Using Runge-Kutta Time-Stepping Schemes, AIAA Paper NO. 81-1259 (June 1981)

9. Jameson, A.; Leicher, S.; Dawson, J.
Remarks on the Development of a Multiblock Three-Dimensional Euler Code for Out of Core and Multiprocessor Calculations, in: Progress and Supercomputing in Computational Fluid Dynamics, Birkhäuser-Verlag (1985)

10. Jameson, A.; Schmidt, W.
Some Recent Developments in Numerical Methods for Transonic Flows, Comp. Meth. in Applied Mech. and Eng. 51 (1985) 467-493

11. Bradshaw, P. et al
Engineering Calculation Methods for Turbulent Flow, Academic Press, 1981, ISBN 0-12-12.4550-0

12. Rodi, Wolfgang
Turbulence Models and their Application in Hydraulics
IAHR Publication, 1980, Delft

13. Murphy, J.D.
Turbulence Modelling
NASA-TN-85889, 1984

14. Marvin, Joseph
Turbulence Modelling for Computational Aerodynamics
AIAA Journal Vol. 21, No. 7, July 1983, p.941

15. Gilbert, N.; Kleiser, L.
Low Resolution Simulations of Transitional and Turbulent Channel Flow.
To appear in Proc. of Int. Conf. on Fluid Mechanics, Beijing, July 1-4,1987, China

16. Stock, H.W.; Haase, W.
An analytical eddy viscosity model for attached and slightly detached flows in Navier Stokes computations.
AV BF30-46/86, 1986, Dornier GmbH

17. Yoshihara, H.; Sacher, P. (editors)
Test Cases for Inviscid Flow Field Methods,
AGARD-AR-2111 (1985)

18. Deiwert, G.S.
Numerical Simulation of Three-Dimensional Boattail Afterbody Flowfields. AIAA J., Vol.19, No. 2, 1981, p.582

19. Miyakawa, J.; Takanashi, S.; Fuji, K.,and Amano, K.
Searching the Horizon of Navier-Stokes Simulation of Transonic Flows. AIAA-Paper 87-0524

20. Srinivasan, G.R.; McCroskey, W.J. et al
Numerical Simulation of Tip Vortices of Wings in Subsonic and Transonic Flows. AIAA-Paper 86-1095

21. Wagner, B.
Calculation of Turbulent Flow about Missile Afterbodies Containing an Exhaust Jet.
Zeitschrift für Flugwissenschaften und Weltraumforschung, Vol. 9, pp. 333-338, 1985.
(Results also in "Aerodynamics of Aircraft Afterbody", AGARD Advisory Report, No. 226, 1986)

22. Haase, W.; Echtle, H.
Computational Results for Viscous Transonic Flows Around Airfoils. AIAA-Paper 87-0422

23. Davis, R.L. et al
Cascade Viscous Flow Analysis Using the Navier Stokes Equations. AIAA-Paper 86-0033, 1986

24. Fritz, W.
Porsche 956 2D-Section Navier-Stokes Calculations.
To be published, 1987

25. Rieger, H.
Solution of Some 3-D Viscous and Inviscid Supersonic Flow Problems by Finite-Volume Space Marching Schemes.
AGARD Fluid Dynamics Panel Symposium on Aerodynamics of Hypersonic Lifting Vehicles, April 6-9, 1987, Bristol U.K.

26. Yang, R.-J.; Chang, J.L.C.,and Kwak, D.
A Navier-Stokes Flow Simulation of the Space Shuttle Main Engine Hot Gas Manifold. AIAA-Paper 87-0368

27. Haase, W.; Misegades, K.; Wagner, B.
Numerische Strömungssimulation für den Kompressionshub in einem Dieselmotor (Numerical Flow Simulation during the Compression Stroke in a Diesel Engine).
Dornier Internal Report BF10/84 B, revised version 1986.

Subject Index

MIX
Papier aus verantwortungsvollen Quellen
Paper from responsible sources
FSC® C105338

If you have any concerns about our products,
you can contact us on
ProductSafety@springernature.com

In case Publisher is established outside the EU,
the EU authorized representative is:
Springer Nature Customer Service Center GmbH
Europaplatz 3, 69115 Heidelberg, Germany

Printed by Libri Plureos GmbH
in Hamburg, Germany